J. H. Edwards.

Best regards
Cedric Smith

BIOMATHEMATICS

OTHER BOOKS ON STATISTICS AND MATHEMATICS

Games, gods and gambling: the origins and history of probability and statistical ideas .. .. F. N. DAVID
Combinatorial chance .. .. F. N. DAVID and D. E. BARTON
A statistical primer .. .. F. N. DAVID
An introduction to the theory of statistics .. G. U. YULE and M. G. KENDALL
The advanced theory of statistics .. .. M. G. KENDALL and A. STUART
Rank correlation methods .. .. M. G. KENDALL
Exercises in theoretical statistics .. .. M. G. KENDALL
Rapid statistical calculations .. .. M. H. QUENOUILLE
The design and analysis of experiment .. .. M. H. QUENOUILLE
Random processes and the growth of firms .. .. .. J. STEINDL
Sampling methods for censuses and surveys .. .. F. YATES
Statistical method in biological assay .. .. .. D. J. FINNEY
The mathematical theory of epidemics .. .. .. N. T. J. BAILEY
Probability and the weighing of evidence .. .. .. I. J. GOOD

GRIFFIN'S STATISTICAL MONOGRAPHS AND COURSES:

No. 1: *The analysis of multiple time-series* .. .. M. H. QUENOUILLE
No. 2: *A course in multivariate analysis* .. .. .. M. G. KENDALL
No. 3: *The fundamentals of statistical reasoning* .. M. H. QUENOUILLE
No. 4: *Basic ideas of scientific sampling* .. .. .. A. STUART
No. 5: *Characteristic functions* .. .. E. LUKACS
No. 6: *An introduction to infinitely many variates* .. E. A. ROBINSON
No. 7: *Mathematical methods in the theory of queueing* A. Y. KHINTCHINE
No. 8: *A course in the geometry of* n *dimensions* .. .. M. G. KENDALL
No. 9: *Random wavelets and cybernetic systems* .. .. E. A. ROBINSON
No. 10: *Geometrical probability* .. M. G. KENDALL and P. A. P. MORAN
No. 11: *An introduction to symbolic programming* .. .. P. WEGNER
No. 12: *The method of paired comparisons* .. .. .. H. A. DAVID
No. 13: *Statistical assessment of the life characteristic: a bibliographic guide* .. .. W. R. BUCKLAND
No. 14: *Applications of characteristic functions* E. LUKACS and R. G. LAHA
No. 15: *Elements of linear programming: with economic applications* R. C. GEARY and M. D. MCCARTHY
No. 16: *Inequalities on distribution functions* .. .. H. J. GODWIN
No. 17: *Green's function methods in probability theory* .. J. KEILSON
No. 18: *The analysis of variance* .. .. A. HUITSON

Descriptive brochure from Charles Griffin & Co. Ltd.

BIOMATHEMATICS

the principles of mathematics for students of biological and general science

CEDRIC A. B. SMITH, M.A., PH.D.

Weldon Professor of Biometry in the University of London

FOURTH EDITION
IN TWO VOLUMES

Volume 1

ALGEBRA, GEOMETRY, CALCULUS

CHARLES GRIFFIN & COMPANY LIMITED
LONDON

First edition by W. M. Feldman, 1923
Second " " " " 1935
Third " " C. A. B. Smith, 1954
Fourth " " " " 1966

PRINTED IN GREAT BRITAIN BY BUTLER AND TANNER LTD,
FROME AND LONDON

" He who knows mathematics and does not make use of his knowledge, to him applies the verse in Isaiah (c. 5, v. 12), 'They regard not the work of the Lord, neither consider the operation of His hands.' "

THE TALMUD

"The laws by which God has thought good to govern the Universe are surely subjects of lofty contemplation; and the study of that symbolical language by which alone these laws can be fully deciphered is well deserving of his [man's] noblest efforts."

PROFESSOR SEDGWICK

" The living and the dead, things animate and inanimate, we dwellers in this world and this world wherein we dwell, are bound alike by physical and mathematical law."

D'ARCY W. THOMPSON

" The ultimate aim of embryology is the mathematical derivation of the adult from the distribution of growth in the germ."

WILHELM HIS

PREFACE TO THE FOURTH EDITION

Over forty years ago *Biomathematics* first appeared under the authorship of Dr W. M. Feldman. At that time the idea of applying mathematical and statistical methods to biological and medical problems must have seemed rather novel and unusual to many biologists. Today the situation is very different. Statistics is widely used and its power is appreciated. A mathematical theory of genetics has been developed; its ramifications are extending into fields such as the study of evolution, the growth of populations, "racial variation" in blood-groups and similar characters, and the genetical determination of various types of disease. In the future there must be a greatly extended mathematical treatment of biochemical and biophysical problems, and the linking of the physical, psychological and social properties of living organisms.

Because mathematics now has this greatly increased importance in the life sciences, this fourth edition has been enlarged and divided into two volumes. But I hope that the spirit of Feldman's original text remains, even if the letter is altered. The first volume deals with such problems as measurements of distances, volumes, areas, forces, velocities, changes, etc., and the graphical presentation of data, i.e. those subjects coming under the traditional names of geometry, algebra, trigonometry, and calculus. The second volume deals mainly with some more modern developments, such as numerical analysis, matrices, probability, and statistics. The two volumes can very largely be read independently of one another, although the second presupposes some knowledge of standard results in algebra and calculus which are explained in the first volume.

Although this book is intended in the first instance for biological and medical workers, much of the mathematics in it is just as important in such fields as the physical sciences, psychology, economics, and social science. Students of these subjects and of mathematics itself should therefore also find the book of value. Many of the illustrative examples have a biological slant, but this should not deter those who are interested in learning a little about the life sciences as well as their own speciality.

The first volume is mainly intended for the reader who has perhaps half-forgotten the mathematics he learnt at school, and who wishes to revise and extend it. The second, which is largely devoted to probability and statistics, is intended for the reader who wishes to have more understanding of the reasons for statistical methods than is provided by non-mathematical texts, without going into great technical detail and rigorous formulation. There may, however, be some points in both volumes which will also interest more advanced readers. Some particular problems have been dealt with in a more condensed form than the main part of the text, and may well be omitted, especially at first reading. I have not marked these, leaving the reader to exercise his own discretion. It has not always been easy to find naturally occurring examples of processes which showed the mathematical treatment in its clearest form, so, for one reason or another, I have occasionally substituted artificial examples. But an attempt has been made to keep these as near to reality as possible.

Some of the characters appearing in this book (including Herlock Soames, Professor Moronami, the Ngboglus, the Society Hostess and the Universities of Camburgh and Edinford) are evidently intended to be entirely fictional. Should they coincide in name or description with any real person or institution, this is entirely fortuitous, and apologies are herewith tendered. Apologies are also due should any acknowledgments have been inadvertently omitted or wrongly made.

It is a great pleasure to record the very considerable assistance I have received from friends and colleagues in the preparation of this edition, and who have contributed to many of the ideas incorporated in the book. But any faults I trust are my own, and not to be attributed to them. I am also grateful for a number of letters pointing out errors in the previous (third) edition. I would particularly wish to express my thanks for useful suggestions to Dr Vivian Feldman (son of the original author), Dr C. C. Spicer, Professor R. S. Krooth, Professor H. Grüneberg and Professor S. H. Boyer, who have read large parts of the manuscript; to Mr E. B. Butler and Dr A. S. Wiener among many others who have made particularly useful suggestions and comments; to my wife, my colleagues at University College, and the publishers, all of whom have been both encouraging and extremely patient during the difficult period of the revision of the text; and to Mr A. J. Lee, who has made a beautiful job of a few of the most complicated diagrams. I am also indebted to the following authors, editors and publishers for kind permission to use and adapt tables and diagrams: Sir Ronald Fisher, Dr F. Yates and Messrs Oliver and Boyd; Dr M. G. Kendall and Messrs Charles Griffin; Dr B. Woolf and Professor L. S. Penrose; and Professor E. S. Pearson.

Professor H. Grüneberg and Dr S. B. Holt have kindly allowed me to use some of their data, and Mr M. C. K. Tweedie has let me give an account of his method of estimation, not yet fully published. Finally, I have to thank the publishers and printers for the excellent appearance of the finished book.

CEDRIC A. B. SMITH

CONTENTS OF VOLUME 1

Chapter		
	Glossary of mathematical symbols used in Volume 1	xi
1	INTRODUCTORY	1
2	BRUSH UP YOUR ARITHMETIC	9
3	SOME POINTS IN ALGEBRA	33
4	COMPARISONS OF MAGNITUDES	73
5	SHAPES AND NUMBERS	79
6	LOGARITHMS	116
7	GRAPHICAL AIDS TO CALCULATION	152
8	RATES OF CHANGE	175
9	THE CALCULATION OF SMALL CHANGES	211
10	RELATIONS INVOLVING RATES OF CHANGE	238
11	LENGTHS, AREAS, AND VOLUMES	269
12	ACCELERATION: MAXIMUM AND MINIMUM VALUES	325
13	SERIES	356
14	DIRECTED MAGNITUDES	383
15	SOME USEFUL INTEGRALS	418
16	PHYSICAL AND CHEMICAL MAGNITUDES	447
	APPENDIX TABLES: Explanation of the tables	493
	Greek alphabet—Common logarithms—Sines and cosines—Tangents and cotangents	
	Answers to problems	505
	Index	515

GLOSSARY OF MATHEMATICAL SYMBOLS USED IN VOLUME 1

WITH COMMONLY RECOGNIZED VARIANTS, AND GUIDE TO PRONUNCIATION

(For further details, see main Index)

C, constant of integration, 241
d, D, ∂, see below
$e = \exp 1$, base of natural logarithms, 143
$i = \sqrt{-1}$, 394
$\boldsymbol{O}$, zero vector, 383
r, polar co-ordinate, 81, 391
x, y, cartesian co-ordinates, 51, 80
$z = x + yi$, complex number, 391, 399
δ or Δ (delta), small change or difference, 178, 211
θ (theta), polar angle, 81, 391
π (pi) $= 3{\cdot}141592\ldots$, circumference/diameter, 33, 137
Σ (sigma), sum of . . ., 281, 361, 380
ω (omega), complex cube root of 1, 394
Letters in ***bold type*** denote vectors, 383
$+$ ("plus"), addition; for vectors, 386; for complex numbers, 396
$-$ ("minus"), subtraction; for vectors, 388; for complex numbers, 396
$\times$ or . ("times" or "into"), multiplication; for vectors, 387; for complex numbers, 392
$\div$ or : or / ("by" or "over"), division, for complex numbers, 395
$\neq$, not equal to, 5
$\simeq$, approximately equal to, 138
$\rightarrow$, ("tends to"); lim (limit), 139, 189
$>$, greater than; $\geqslant$, greater than or equal to, 74
$<$, less than; $\leqslant$, less than or equal to, 74
∞, infinity, 53, 189, 318
$|x|$, ("mod x"), modulus or absolute value, 75; for vectors, 389
$\bar{z}$ ("z bar"), conjugate complex, 400

$\underline{|n} = n!$ ("n factorial" or "factorial n") $= \Gamma(n+1)$, 342

$(n-)^r$ ("n minus to the r") $= n^{(r)} = {}^nP_r$ ("n, P, r"), descending factorial, $(n+)^r = n^{[r]}$, ascending factorial, 369

$(n-)_r$, ("n minus, subscript r") $= {}^nC_r$ ("n, C, r") $= \binom{n}{r}$ ("binomial n, r", or "n choose r"), reduced descending factorial, 370

$[y]_a^b$, change in value of y, 269

$f(x)$, ("f of x" or "f, x"), function of x, 49; $f(x, y)$, 50

$f^{-1}(x)$, ("f to the minus one [of] x"), $\sin^{-1} x$, etc., inverse function, 150

$\sin x$ ("sine x"), $\cos x$, $\tan x$, $\operatorname{cosec} x$, $\sec x$, $\cot x$, trigonometric functions, 82, 410

$\sinh x$ ("shine x" or "sinsh x") $= \operatorname{sh} x$, $\cosh x = \operatorname{ch} x$, $\tanh x$ ("than x" or "tansh x") $= \operatorname{th} x$, $\operatorname{cosech} x$, $\operatorname{sech} x$, $\coth x$, hyperbolic functions, 144, 410

$\log x = \log_{10} x$, $\operatorname{antilog} x = 10^x$, common logarithm and antilogarithm, 116

$\ln x$ ("ellen x") $= \log_e x$ [or sometimes simply, $\log x$], natural, napierian, or hyperbolic logarithm, 142, 403

$\exp x = e^x$, exponential function (natural antilogarithm), 143, 405

$Dy = D_t y$ ("D [subscript] t, y") $= dy/dt$ ("$d\,y$ by $d\,t$") $= y_t = y'$ ("y dash" or "y prime") $= \dot{y}$ ("y dot"); $f'(t)$, derivative, derivate, or differential coefficient, 186

$D_{t|u} y$ (or simply $D_t y$) $= y_{t|u}$ (or y_t) $= \partial y/\partial t$ ("curly d y by d t"); $f_t(t, u)$; partial derivative, 214

$D_t^2 y$ ("D t squared [of] y") $= \left(\dfrac{d}{dt}\right)^2 y = \dfrac{d^2y}{dt^2}$ "d two y by d t squared") $=$ $y_{tt} = y''$ ("y double dash"), second derivative, 327

$\int v\,dt$ ("integral v d t") $= D_t^{-1} v$, indefinite integral, 241

$\int_a^b v\,dt$ ("integral from a to b of v, d t"), definite integral, 269

$\int_{-\infty}^{\infty} v\,dt$ ("integral from minus infinity to infinity of v, d t"), infinite integral, 318

CHAPTER ONE

INTRODUCTORY

1.1 Mathematics and biology

Pure Mathematics, as far as one can exactly define it, is the science of number and shape: or, as one may say, Arithmetic and Geometry in the most general sense of these words. Recently, too, pure mathematicians have taken a great interest in the more abstract ideas of classes, properties, relations, and so forth; or, as we may say, in Logic. Applied Mathematics consists simply in the application of the discoveries of Pure Mathematics to observed phenomena: and since we already have the sciences of "Biophysics" and "Biochemistry", we may perhaps speak of "Biomathematics", or "Mathematical Biology", meaning the application of mathematical methods to Biology. This will accordingly cover a wide range of subjects: in fact, whenever we count or measure we are using mathematical ideas of the simplest kind; when we draw a graph to represent a set of observations, calculate the rate of a biochemical reaction, or try to find the chances of passing on a hereditary disease, we are using more complicated ideas, but ones which are based in a straightforward way on the simpler ideas of measuring and counting. It is the purpose of this book to explain some of the more complicated mathematical ideas of interest to biologists, and to show how to use them, and how they arise fairly naturally from the simpler ones. It would be dishonest to claim that it is possible to make all ideas and methods entirely painless to the reader who is completely new to the subject, but an endeavour will be made to smooth away the difficulties, and as far as possible to give biological examples of the use of each process as it is introduced.

We may summarize the growth of scientific knowledge somewhat in this fashion. As a rule, an experimenter will have some hypothesis or hypotheses at the back of his mind before making any experiments. These hypotheses may be quite precise, or sometimes they may be extremely vague and tentative—but usually, only when his field of research is entirely new and uncharted will he explore in a random and undirected manner. Naturally he will be most interested to know how these hypotheses can be related to observable quantities, and to know how he can best plan his experiments to test them. If counting or measurement is involved, it will usually be convenient to express the hypotheses by mathematical formulas. Strictly speaking, these really amount to no more than a form of shorthand: thus the equation

$x + y = z$ states no more than the sentence "if we add the number 'x' to the number 'y' we obtain the number 'z' ". All such equations could be translated into ordinary language, though sometimes very clumsily and lengthily. But this form of shorthand is extremely convenient in the next stage of the process, in which we work out the consequences of our hypotheses. Here we can play about with the formulas according to standard rules, without having to bother ourselves too much with what the formulas mean in practice. This formal manipulation means that we can deduce various consequences without the labour and tedium of a considerable amount of heavy reasoning. When in this way we have discovered what results may be expected to follow from our hypotheses, we shall be able to see what are the most useful experiments we can perform, and what sort of calculations we must make when we have obtained our experimental results. Finally, if these turn out to agree with our hypotheses, we can provisionally accept them as proved, and from them predict further consequences which may not for the time being be verifiable by observation.

For example, when Mendel began his famous experiments on heredity, he must have had the idea that interesting results could be obtained by counting the numbers of different kinds of offspring obtained from two given parents. Further investigation showed that, for example, if we self-fertilize a pea plant which is obtained from a cross between a strain having yellow seeds and one having green seeds, then in the next generation we find approximately 3 times more yellow seeds than green ones. If we now assume that in the long run we get exactly 3 times as many yellows as greens, the theory of probability tells us that in an experiment in which 1000 offspring are obtained, it is most likely that the number of yellows will be between 714 and 786: in fact, that will be true in 99 out of every 100 such experiments. In this way we can verify that Mendel's laws do in fact hold to within the accuracy of our measurements. And once we know that Mendel's laws are true, we can predict the effect of many generations of inbreeding on our plants.

1.2 Plan of this book

We begin our subject by recalling some simple arithmetical facts: they occupy this chapter and the next. Various useful calculating devices and methods of computation are considered, both with paper and by the use of various mechanical devices, such as desk calculators, punched card equipment, and so on. Graphical methods of calculation such as slide-rules and nomograms are explained rather later, namely in Chapter 7. Chapter 3 sets out in a systematic way the fundamental rules of algebra, and will be useful for purposes of revision. Elementary algebra as explained in most books deals largely with equations, or equalities, but in practice we often require to know of two quantities which is the larger one and which the smaller; hence it is important to discuss "inequalities" or the comparisons of magnitudes. This is done rather briefly in Chapter 4. In Chapter 5 we see how algebra and geometry are bound up with one another, so that many problems in algebra can be represented graphically, and many geometric problems can be solved algebraically. This leads on naturally to the study of logarithms in Chapter 6; these are numbers which are not only useful in assisting with arithmetical calculations but also have many other applications to the study of rates of growth, chemical reactions, and other phenomena, and in addition have a geometric interpretation. In Chapter 7 these logarithms are applied to slide-rules, nomograms, and similar calculating devices. In Chapter 8 we begin the study of rates of change, or "differential calculus"; this can be applied to all kinds of problems in which changes are taking place, e.g. growth, chemical reactions, and motion, as well as to finding what are the maximum and minimum values of a variable quantity. When looked at in reverse this becomes "integral calculus", and can rather unexpectedly be used for such problems as finding lengths, areas, volumes, work, energy, centres of gravity and moments of inertia.

In Chapter 13 we go on to the study of series. These sometimes occur naturally in nature, as when an organism such as a bacterium reproduces itself with "geometric" rapidity, the whole population doubling itself in each generation. But for the most part series are mathematical devices which help us to solve otherwise intractable problems. We continue by introducing yet another useful device, the "square root of minus one": this turns out to be very much less mysterious than it sounds. Chapter 16 is a recapitulation of some of the more important facts of physics and chemistry, together with a discussion of the appropriate mathematical methods. In Chapter 17, which begins Volume 2, we go on to consider the various ways of solving equations. We find that any equation can be solved, at least by numerical methods. This leads on to the study of "matrices" which are nothing more than collections of numbers. They have been used increasingly often in biological calculations in the last thirty years, both in statistical work and in the theory of artificial selection.

In Chapter 19 we discuss the "laws of chance" which are so important for geneticists. These laws are applied in Chapter 20 to the study of natural variation, including the discussion of averages, range or spread, and the correlation between different characters. In Chapter 21 we discuss ways of drawing conclusions from statistical data, and these methods are applied in Chapter 22 to some common types of experimental design.

The book ends with an account of an arithmetical notation by means of which calculations can be greatly simplified in many branches of Mathematics. This notation is actually 200 years old, having been (apparently) first discovered by John Colson, F.R.S. But it has been strangely neglected.

At all points an endeavour has been made to keep the argument in as simple a form as possible. But the problems under discussion are of very different degrees of difficulty. And some fairly difficult sections have been added in places where the author has felt it appropriate to explain some particularly tricky point in greater detail and with greater rigour. These sections may be omitted at a first reading with considerable advantage. Other sections deal with special devices used in solving particular problems; they are again not needed for a general understanding of the subject, but may be useful for reference when required. Thus the reader may well wish to pass over the latter parts of Chapters 3 and 5 (say from Sections 3.8 and 5.10 onward) as these deal with more specialized topics, and are written rather concisely, whereas in Chapters 6, 7, and 8 there are discussions of logarithms, nomograms and the calculus, subjects of much more general applicability. However, these harder sections have not been specially marked in any way: it has been felt better to leave the reader quite free to make his own choice according to his individual preference.

1.3 Accuracy

Pure mathematics is often looked upon as being a perfectly exact science, the physical sciences as being next, and the biological sciences still less so. In a limited sense this is true: when we are dealing with numbers in their own right we have no experimental error to contend with, and our formulas will often be exactly true. On the other hand many biological measurements will be limited to 2- or 3-figure accuracy. Even when greater accuracy would be possible, there would often not be much point in it. For example, there would be no useful purpose served in measuring the weights of the inhabitants of a town to the nearest milligram, for all these weights would be substantially altered within a few hours. Thus the accuracy of our results is limited by the inaccuracy of our measurements.

But perhaps it is worth pointing out that accuracy of reasoning is just as desirable in biology as in mathematics, physics, or any other science. We may take two examples to illustrate this, one being biological, and one mathematical. It is known that in certain species of cave fish the eyes are degenerate. We might explain this as follows: since the fish live in practically complete darkness, they have no use for their eyes, and so do not exercise them; and since they are now atrophied, it seems plausible that in the course of thousands of years the eyes have degenerated through lack of use. Following this line of argument, we might even be able to construct a quantitative theory which would explain the observed facts to a high degree of accuracy. Nevertheless no such effect has been observed in animals whose heredity has been studied experimentally, and there seems every reason to believe that the explanation is wrong. A more plausible theory would be that mutations tend to occur which damage the eyes; since the eyes are no longer used, natural selection does not eliminate the fish with defective eyes, and they pass on their defects to future generations. Which theory is accepted may not matter very much when we are only considering the particular case of these cave fish, but clearly it will be very important when we come to examine the problem of evolution in general. In the same way, an apparently slight error in a mathematical argument may lead to completely wrong conclusions. A famous example is the "proof" that $1 = 2$. Let us suppose that $x = 1$, and $y = 1$, so that $x = y$. Multiplying both sides of this equation by x, we obtain $x^2 = xy$, and on subtraction of y^2, $x^2 - y^2 = xy - y^2$. This is the same as saying that $(x + y)(x - y) = y\,(x - y)$, and on dividing by the common factor $(x - y)$ we obtain $x + y = y$, or since $x = 1$, $y = 1$, this proves that $2 = 1$. What has gone wrong? If we put in the values $x = 1$, $y = 1$, we see that all our equations are correct as far as the point where we asserted that $(x + y)(x - y) = y\,(x - y)$; we have broken the rules by dividing through by the common factor $x - y$, which is zero. Division by zero never gives a definite answer. When we say that $10/2 = 5$, we mean that 5 is the number which when multiplied by 2 gives the product 10. So $0/0$ ought to mean the number which gives the product 0 on multiplication by 0. But that is true for any number whatever. In other words, 0 divides into 0 exactly either once, twice, or as many times as we like. On the other hand, $1/0$ would be the number which when multiplied by 0 gives the product 1. This is impossible: no multiple of zero is ever equal to 1. Thus division by zero is an unfruitful operation, and it is customary among mathematicians to ban it entirely. The punishment for a breach of this law may be an absurd answer to a calculation. The correct procedure in the above case would be to say, "We know that $(x + y) \times (x - y) = y\,(x - y)$. There are now two possibilities, either $x-y=0$, or else $x - y$ is not equal to 0 [in symbols, $x - y \neq 0$], and we are allowed to divide by $(x - y)$, obtaining $x + y = y$. In fact we know that in our case it is the first possibility, $x - y = 0$, which is the correct one." Since Biomathematics rests, as it were, on the twin foundation stones of Biology and Mathematics, it is necessary to make sure that its foundations are secure on both sides. It is of no use building an elaborate mathematical calculus on an inadequate theory; and equally well the most perfect biological insight will be largely nullified by faulty calculation. Unfortunately in biological matters only experience can be a safe guide: but mathematically the pitfalls are for the most part relatively few in number, and well-defined, such as the inadvisability of division by zero, and they will be pointed out when they occur.

1.4 Experimental error

There are several kinds of inaccuracy affecting calculations. Besides the possibility of inexactness of reasoning—always to be avoided if possible—we have already mentioned the inaccuracy of measurements. Now there are some measurements which we can make practically as accurately as we wish, with sufficient trouble. Thus if we are measuring the height or weight of an animal, ordinary scales and balances are frequently good enough to give us the answer to an accuracy well beyond anything we require (though of course for special investigations this may not be true). In such a case it is usually good enough to specify the measurement only to a certain limited accuracy, and also to state the maximum error. Thus if a man's height is stated to be 181 centimetres, or 1·81 metres, measured to the nearest centimetre, we mean that the error of measurement does not exceed $\frac{1}{2}$ centimetre: his height lies somewhere between 180·5 and 181·5 centimetres. We then say that the measurement is correct to 3 *significant figures*. ("Significant figures" are those whose value is definitely known, but excluding zeros at the left-hand end of the number. Thus 181, 1·81 and ·000181 all contain 3 significant figures.) This definition should not be taken too literally or pedantically. A length which appears to be $180\frac{1}{2}$ cm might turn out with more accurate measurement to be nearer 180 cm than 181 cm, or perhaps vice versa. So in the absence of further information it could reasonably be taken as either 180 or 181 to 3 figures, if the error is not greater than *about* 5 units in the next place. Now, generally speaking, if each of our measurements is correct to (say) 3 significant figures, then the result of any calculation we do on them will also be correct to about 3 figures. However, this is not quite true, as there will be a slight increase in the error at each step in the calculations. Thus, if we multiply 181 by 2·53, we obtain 457·9300. But if these numbers are correct to 3 figures only, "181" may stand for any number between about 180·5 and about 181·5, and "2·53" may stand for any number between 2·525 and 2·535. The product lies between $180{\cdot}5 \times 2{\cdot}525 = 455{\cdot}7625$ and $181{\cdot}5 \times 2{\cdot}535 = 460{\cdot}1025$; roughly speaking, between 456 and 460. So the value 457·93 will no longer necessarily be correct to 3 figures: the error may be as large as 2 in the third figure. Again if we add $181 + 253 + 102 + 376$, where all the numbers have three-figure accuracy, the real total may be as small as $180{\cdot}5 + 252{\cdot}5 + 101{\cdot}5 + 375{\cdot}5 = 910$, or as great as $181{\cdot}5 + 253{\cdot}5 + 102{\cdot}5 + 376{\cdot}5 = 914$. Errors always tend to accumulate in this manner; but although this tendency cannot be entirely avoided, it can be kept down to a minimum by using one or two more figures in the calculations than are needed in the final result (in exceptional cases several more figures may be needed). Fortunately also the errors introduced in the different steps of the calculation will frequently cancel one another to a large extent, so that the error in the final answer will rarely be very large.

The number of "significant figures" may be considered as a rough way of stating the proportional error. If a measurement comes to 100·1, correct to 4 figures, we know that its error is not more than about ·05, or 1 part in 2000: but if it is 999·9, then the same error, ·05, amounts to only 1 part in 20,000. Thus a number is correct to 4 figures when the error lies between 1 part in 2000 and 1 in 20,000, five-figure accuracy corresponds to an error between 1 in 20,000 and 1 in 200,000, and so on.

A point worth noting is an inconsistency which sometimes occurs in the usual way of writing very large numbers, as, for example, the population of a large country. According to the 1931 census, the population of the United Kingdom was estimated as 46,038,357, or 46 millions, to the nearest million. Now when this figure is considered to the nearest million, it is usually written 46,000,000; the zeros here do not actually stand for zero itself, but rather for digits whose exact value does not matter much, or is unknown. Such digits are of course *not* significant figures. If now we consider the population of England and Wales, 39,952,377, to the nearest million, this must be written as 40,000,000. Here, however, the first zero *is* significant—the population is 40 million, and not 41 or 39 million: the other zeros are just put in to show that we are counting in millions. But there is nothing in the way the number is written to show that distinction. There are two ways in which we can overcome this difficulty. Firstly, we can write 40,000,000 as 40×10^6, thus showing clearly that we are counting in millions. Alternatively, we can invent a symbol, such as ?, or *, or ɔ to stand for a digit whose value is unknown or unimportant: the number 46 million could be written 46,ɔɔɔ,ɔɔɔ and the number 40 million as 40,ɔɔɔ,ɔɔɔ, thus making a clear distinction between significant and non-significant figures. Counted to the nearest hundred-thousand, the population of England and Wales would be 40,0ɔɔ,ɔɔɔ.

So far we have been dealing with cases in which the error is so small, in comparison with the quantity to be measured, that it is enough simply to state the number of significant figures correct in the calculations. The opposite case is that in which the error is of the same order of magnitude as the effect to be measured. For example, if we are trying to compare the yields of two varieties of wheat, we can easily find that the differences of fertility, even between different parts of the same field, are greater than the difference we are trying to measure. When this is so, special statistical methods have to be applied in order both to reduce the error as much as possible and also to make a careful estimate of its magnitude. We shall discuss these points later.

Finally we may notice one other frequent form of inaccuracy. Often it is not possible to take into account all relevant factors, either because we do not know all of them exactly, or because they would make the calculations too complicated. Instead we use a deliberately simplified model of the situation, knowing that we are introducing some error, but hoping that our model will reproduce the essential features of the situation. There is of course little harm in this sort of idealization, so long as one is aware that it is a deliberate simplification and has its limitations.

1.5 Rounding off

When a number is given to more figures than are required, or are justified by its accuracy, we usually omit the superfluous figures at the end, but increase the last figure retained if the following (omitted) figure is 5 or more. This is called "rounding off". Thus $\pi = 3{\cdot}141593$ becomes 3·142 to 4 significant figures, instead of 3·141, since the fifth figure is 5. This is because 3·141593 is nearer to 3·142 than to 3·141. Similarly, $\pi = 3{\cdot}1416$ to 5 figures. However, we have seen that a number like 180·50 could be considered equally well as 180 or 181 to 3 significant figures, since it lies half-way between. In such a case the rule is often adopted of rounding off the last digit retained to the nearest even value: e.g. 180·50 would become 180, not 181. The rounding off is then equally likely to be upwards, to the next higher figure, or downwards, so there is no systematic bias.

When a multiplication comes in a series of calculations of given accuracy, it is best to work out the product completely, and then round off. For example, a rectangle 1·23 cm $\times$ 4·56 cm has an area 5·6088 cm^2, or 5·61 cm^2 to 3 figures. For it is much easier to check an exact multiplication like $1{\cdot}23 \times 4{\cdot}56 = 5{\cdot}6088$ than an approximate product like 5·61; see Section 2.7. (But remember to keep about two extra figures in the intermediate steps of a long calculation to ensure sufficient accuracy in the final answer.)

1.6 Decimal places

The reader should be clear about the distinction between the numbers of significant figures and of decimal places. A length of 180·5 cm has 1 decimal place, but 4 significant figures. The same length expressed as 1·805 m has 3 decimal places, but still 4 figures. The accuracy is best expressed as the number of significant figures, which does not depend on the units used. But it is only common sense to take similar measurements to the same number of decimal places: in a series of comparable plants, it would be absurd to measure some heights to the nearest centimetre, and others to the nearest millimetre, unless there were two quite distinct groups of giants and dwarfs.

CHAPTER TWO

BRUSH UP YOUR ARITHMETIC

2.1 Simple counting

Arithmetic began with the discovery that it was possible to count groups of objects. Nowadays, when we have machines which will do thousands of multiplications and additions per second, and carry out a whole complicated programme of computation, we usually take such matters as simple counting entirely for granted. All the same, if the universe was constructed in such a way that when we counted a set of objects we sometimes got the answer 2, sometimes 4, and sometimes 5, then not only mathematics but also a great part of science as we know it simply would not exist. No doubt the discovery of counting was made by degrees, and not suddenly: but we could imagine it happening in some such way as this:

Jacob: Father, you know how difficult it is to make sure you've got all your sheep in the pen.

Isaac: Yes, son. My memory's pretty bad these days—and there are so many names to remember: Deborah, and Mary, and Rachel, and Beth, and . . .

Jacob: Well, I've found out an easier way.

Isaac: Tell me, my son.

Jacob: When the sheep walk into the fold I call them in turn "one, two, three, four, five. . . ."

Isaac: Where did you get these outrageous words, Jacob?

Jacob: They came to me in a dream, father. When the last sheep has gone in, I should have got to "fifty-seven". If I haven't, I know there's something wrong.

Isaac: Well, you can say what you like, son, but I like to stick to old, well-tried methods, with none of this new-fangled gibberish like "one, two, three". And what will you do when the sheep have lambs?

Jacob: I haven't thought that one out yet, father. But my method has always worked perfectly so far.

After the discovery of counting itself, the processes of addition, multiplication, subtraction, and division must each have been in their time first-class discoveries—though now they are only too painfully familiar to every schoolboy.

By counting groups of objects in this way we obtain the numbers zero, 0 (when there is nothing there) and the "positive integers" 1, 2, 3, 4. . . . These have many fascinating properties, but the most important one from the practical standpoint is that of "unique factorization".

2.2 Factorization of numbers

A number greater than 1 which is exactly divisible only by itself and 1 is called a "prime number": for example, 2, 3, 5, 7, 11 and 13 are primes. Other numbers (greater than 1) are "composite" and can therefore by definition be split up into factors, which are either prime, or else can themselves be split into smaller factors, and so on, until finally we must end up by resolving the number into factors which are all prime. Thus $60 = 2 \times 30 = 2 \times 5 \times 6 = 2 \times 5 \times 2 \times 3$ or alternatively $60 = 3 \times 20 = 3 \times 2 \times 10 = 3 \times 2 \times 5 \times 2$ or $60 = 10 \times 6 = 5 \times 2 \times 3 \times 2$. Now it is a remarkable fact that in whatever way we split up any number, we always arrive at the same prime factors in the end. Thus 60 decomposes into two 2's, 3, and 5, and although we may obtain these in different orders according to the different ways of doing the decomposition, we could never obtain factors like $3 \times 3 \times 7$. This is not an obvious fact (at least not to the writer) and a proof is given later. (Just to show that there are more difficult problems we mention Goldbach's conjecture, that every even number except 2 can be expressed as the sum of two prime numbers, for example, $4 = 2 + 2$, $6 = 3 + 3$, $8 = 3 + 5$, $10 = 3 + 7$, $12 = 5 + 7$, and so on. Although this conjecture sounds extremely simple, and no even number is known for which it is not true, nevertheless to supply a formal proof has so far baffled the ingenuity of the best mathematicians in the world.) As it is often very useful to be able to split a number up into its factors, we give below the standard tests for the most frequent prime factors.

A number is divisible by 2 if the last figure is even (0, 2, 4, 6, 8). It is divisible by $2^2 = 4$ if the number formed by the last two figures is divisible by 4 (e.g. 312 is divisible by 4 because 12 is), and by $2^3 = 8$ if the last three figures are divisible by 8, that is, if the last two figures are divisible by 8 and the preceding figure is even (e.g. 1216) or if the last two figures are divisible by 4, but not by 8, and the preceding figure is odd (e.g. 1312).

A number is divisible by 3 if the sum of all its figures is divisible by 3 (e.g. 1362, because $1 + 3 + 6 + 2 = 12$ is divisible by 3): and by $3^2 = 9$ if the sum of all its figures is divisible by 9.

A number is divisible by 5 if it ends in 0 or 5, by $5^2 = 25$ if it ends in 00, 25, 50 or 75, and by $5^3 = 125$ if it ends in 000, 125, 250, 375, 500, 625, 750, or 875.

A number is divisible by 11 if the total obtained by alternately adding and subtracting consecutive figures is divisible by 11. For example, 8481 is divisible by 11, since $8 - 4 + 8 - 1 = 11$.

Divisibility by 7 is best tested by straightforward division. Although there are dodges which enable the work to be slightly shortened, they scarcely seem worth the trouble of remembering. It is also rarely worth while testing for prime factors above 11.

2.3 Fractions, negative and irrational numbers

These tests for divisibility are useful in dealing with fractions. It must have been discovered quite quickly that numbers could be applied also to measuring, as when we talk of a weight of 2 kilograms or a length of 3 metres. But a new kind of number, called a "fraction", must be invented to express measurements like $\frac{1}{2}$ kilogram or $\frac{17}{100}$ ($= 17/100 = \cdot 17$) metre, thus supplementing the whole numbers (integers) used in counting. We assume that the reader knows the following rules:

(*a*) no fraction has zero denominator;

(*b*) a fraction is unaltered in value by multiplying or dividing both numerator and denominator by the same number (other than 0), as

$$\frac{2}{5} = \frac{2 \times 3}{5 \times 3} = \frac{6}{15};$$

(*c*) to multiply fractions, multiply their numerators to obtain the new numerator, and their denominators to obtain the new denominator, as

$$\frac{2}{3} \times \frac{5}{7} = \frac{10}{21};$$

(*d*) to divide by a fraction, exchange numerator and denominator of the divisor and multiply, as $\frac{2}{3}\Big/\frac{7}{5} = \frac{2}{3} \times \frac{5}{7} = \frac{10}{21}$;

(*e*) fractions with the same denominator may be added or subtracted by adding or subtracting their numerators, as

$$\frac{2}{5} + \frac{6}{5} - \frac{4}{5} = \frac{2 + 6 - 4}{5} = \frac{4}{5};$$

(*f*) $x/1 = x$, $0/y = 0$ (provided y is not 0).

Other properties of fractions follow from these six rules.

The addition and subtraction of fractions with different denominators is done by reducing them to a common denominator, by using rule (*b*), and then applying rule (*e*). Suppose we wish to find the sum

$$S = \frac{7}{46} + \frac{1}{69} + \frac{1}{12}.$$

A suitable common denominator must be a multiple of each of the three separate denominators, 46, 69, and 12. Although the product $46 \times 69 \times 12$ would be such a common multiple, it is a large number, and the calculation will be simplified if we can use instead the least common multiple (L.C.M.) of 46, 69, and 12. Now 46 has prime factors 2×23, while $69 = 3 \times 23$, $12 = 2 \times 2 \times 3$. Any multiple $46n$ of 46 must contain 2 and 23 as factors; similarly $69m$ must have factors 3 and 23, while any multiple of 12 must contain the factor 2 twice over, as well as the factor 3. The L.C.M. must therefore be $2 \times 2 \times 3 \times 23 = 276$, which is just sufficient to include all the factors of 46, 69 and 12. It can be written as $(2 \times 23) \times (2 \times 3) = 46 \times 6$, and similarly as 69×4 and 12×23. Hence, by rule (*b*),

$$\frac{7}{46} + \frac{1}{69} + \frac{1}{12} = \frac{7 \times 6}{46 \times 6} + \frac{1 \times 4}{69 \times 4} + \frac{1 \times 23}{12 \times 23}$$

$$= \frac{42}{276} + \frac{4}{276} + \frac{23}{276} = \frac{69}{276}.$$

Since $69 = 3 \times 23$ and $276 = 2 \times 2 \times 3 \times 23$, it appears that we can cancel out the common factors 3×23 from the numerator and denominator of $69/276$, giving finally $S = 1/(2 \times 2) = 1/4$.

In this way we find it necessary to extend our system of numbers to include fractions as well as whole numbers. Another important extension is required to include opposites. We know that "hot" is opposite to "cold", "up" to "down", "forward" to "backward", "positive electricity" is opposite to "negative", and so on. We therefore invent a new kind of number or symbol to represent the results of measurements where two opposites are involved: we call it a "negative number". If a measurement upward, say, of 4 feet, is represented by the number 4, then a measurement downwards of 4 feet is usually denoted by the symbol -4, the minus sign showing that it is reversed in direction. (It is suggested in Chapter 22 that this could also conveniently be written as $\rotatebox{180}{4}$, by inverting the symbol 4, and similarly $\rotatebox{180}{2}$ would stand for -2. This apparently trivial device is there shown to be extremely useful.) We can also have negative fractions, as a measurement downwards of $4\frac{1}{2}$ feet might be represented as $-4\frac{1}{2}$ or $-\frac{9}{2}$.

Unfortunately even this very comprehensive system of integers and fractions still has its defects, as Pythagoras observed. We would like to be able to express the length of any line as a number. But suppose that ABC is a triangle with a right angle at B, and that AB and BC are both of unit length. What is the length of AC? By Pythagoras's theorem, $AB^2 + BC^2 = AC^2$, that is $AC^2 = 2$, or $AC = \sqrt{2}$, if there is such a number. But there is no fraction x/y whose square is exactly equal to 2, so that we cannot express the length of AC by any whole number or fraction. We can most easily see this by splitting both x and y into prime factors; say, $x = pqr \ldots u$ and $y = PQR \ldots W$. Then if the equation $(x/y)^2 = 2$ was true, this would imply that $x^2 = 2y^2$, or $ppqqrr \ldots uu = 2PPQQRR \ldots WW$. But this equation cannot be true, since the left-hand side contains an even number of prime factors, the right-hand side an odd number, and we know that a number can be factorized in only one way. But although we cannot find any fraction to represent the length of AC exactly, we can approximate to it as closely as we like. If we draw the figure and actually measure the length of AC to 3 figures, we obtain the value $1{\cdot}41$, whose square $1{\cdot}41^2 = 1{\cdot}9881$ is very close to 2. To 4 figures we find $AC = 1{\cdot}414$, $AC^2 = 1{\cdot}999396$, and to 5 figures $AC = 1{\cdot}4142$, so that $AC^2 = 1{\cdot}99996164 = 2{\cdot}0000$ correct to 5 figures. Carrying on in this way we can, by taking a sufficient number of figures, obtain a number whose square is as near 2 as we wish.

The usual trick by which mathematicians get round these difficulties is to invent an "unending decimal" $1{\cdot}41421356\ldots$ to represent $\sqrt{2}$. The figures in this decimal can be obtained by any standard process for calculating square roots: since such a process can never terminate, we can calculate as many figures of the decimal as we wish. The interpretation we give to such an "unending decimal" is this: although we cannot find a terminating decimal whose square is *exactly* equal to 2, nevertheless by taking enough places in the sequence $1{\cdot}41421356\ldots$ we can find one whose square is as near 2 as we please. Thus we have seen that $1{\cdot}414^2$ differs from 2 by less than $\cdot 001$, and $1{\cdot}4142^2$ differs from 2 by less than $\cdot 0001$: we can continue in this way as far as we like. A similar situation occurs with a fraction such as $\frac{1}{3}$: this is $\cdot 33$ correct to 2 figures, $\cdot 333$ correct to 3 figures, and we may say that it is represented by the infinite (or "recurring") decimal $\cdot 333333\cdot\ldots$ In short we can say that an unending decimal is a convenient way of summarizing a process whereby we can obtain an answer to any degree of accuracy. We shall later come across many other such processes, especially in the calculus, where it is impossible or inconvenient to give the answer to a problem in a finite form, but where it is possible to give a method by which we can get as near to the answer as we wish. Such a process is called "convergent".

In practice, of course, it is impossible to make a measurement to unlimited accuracy: only in exceptional cases can we expect to attain anything like 7-figure accuracy. For this reason it may seem rather like splitting hairs to bring in *unending* decimals. One can imagine the reader asking, "Why shouldn't we simply take all our measurements and calculations to 7 figures only? That would be quite ample for my purposes, and avoid a great deal of complication." The answer is that some such line of approach would indeed be possible, but not quite in such a crude form as suggested. In a calculation such as $\sqrt{2} - \frac{3}{2}$, although $\sqrt{2} = 1{\cdot}414214$ to 7 figures, and $\frac{3}{2} = 1{\cdot}500000$, the difference $\sqrt{2} - \frac{3}{2} = -\cdot 085786$ is correct only to 5 figures. Two figures have already been lost, and the margin of safety may not be as adequate as it seems at first sight. We cannot in fact evade the difficulties so easily.

We shall not go further into this subject. The reader who is interested will find many books in which the matter is discussed more fully, the best being perhaps G. H. Hardy's classic, *An Introduction to Pure Mathematics* (10th edn., 1952, Cambridge). We shall merely state that it is convenient to consider unending decimals as numbers on an equal status with terminating decimals and integers. This enables us to speak of square roots, cube roots, logarithms and many other quantities as numbers; if we restricted ourselves to exact fractions that would not be possible. In addition we can show that these unending decimals can be added, subtracted, multiplied, divided, and otherwise used exactly like other numbers: that is indeed almost obvious to common sense. For example the equation $(x + 1)(x - 1) = x^2 - 1$ gives $(\sqrt{2} + 1)(\sqrt{2} - 1) = 2 - 1 = 1$. To test if this is correct, let us take $\sqrt{2}$ to 5 figures, 1·4142. Then $(\sqrt{2} + 1)(\sqrt{2} - 1) = 2{\cdot}4142 \times {\cdot}4142 = {\cdot}99996164 = 1{\cdot}0000$ (correct to 5 figures). If we took $\sqrt{2}$ to 7 figures, the equation would be correct to 7 figures, and so on.

Mathematicians usually speak of all decimals, terminating or unending, positive or negative, as "real numbers". The word "real" here is perhaps confusing, as it no longer has any special connection with the ordinary meaning of the word "real". The use of the term in this sense seems to have arisen historically by a misunderstanding we shall come across later; for the present the reader must simply accept it as the usual (and international) phrase.

It is also very convenient to suppose that the measure of a length or a weight is a real number. After all, we can measure a length correct to 2, 3, 4, 5 or 6 places. There is no apparent limitation to the number of figures we can obtain, apart from the difficulty of making the apparatus sufficiently delicate. We can accordingly imagine a length as represented by an unending decimal, again in the sense that in theory we can determine as many figures of the decimal as we please. Strictly speaking this may be wrong, as atomic and quantum theory suggests that there is a limit to the accuracy of any measurement. Strictly speaking, too, relativity suggests that Euclid's geometry may be false: but it makes little difference in practice. Throughout this book we shall talk as if all measurements of quantities such as length, area, angle, volume, mass, etc., were ideally expressible as "real numbers", i.e. as unending or terminating decimals, and as if Euclid's geometry was exactly true. Any errors we may commit are quite negligible for all or almost all biological purposes: and it may be added that many of our results can be proved by other methods which do not involve these assumptions, but which are considerably more difficult to follow.

2.4 Calculating machines

There are many modern devices for cutting down the labour and strain of calculation. Not only do these save work, but also they increase accuracy, since they are less liable to error and do not get tired like the human mind. Some of these devices, such as logarithms, nomograms and slide-rules, will be discussed in later chapters. Here we shall consider mechanical devices.

The simplest form of calculating instrument is the adding and subtracting machine. A very simple, inexpensive and efficient machine is the "Exactus", which consists of a number of bars pushed up or down by a stylus: it is operated by hand. Although it is primarily intended for addition and subtraction, it can readily be adapted for division in the following way.

First we multiply the divisor by 2, 3 and 6. Multiplication by 2 can be done either mentally or manually in the usual way, or else mechanically by adding the divisor to itself. To find three times the divisor, add the divisor to its double. To find the six times multiple, double the three times multiple. Thus supposing we wish to divide 3017400 by 243. We calculate

$$2 \times 243 = 486$$
$$3 \times 243 = 486 + 243 = 729$$
$$6 \times 243 = 2 \times 729 = 1458$$

These multiples, together with $1 \times 243 = 243$, are set out on a separate piece of paper (see Fig. 2.1 overleaf).

We now put into the adding machine the number to be divided, 3017400. Looking at the first three figures of this we see that the largest of the four multiples which can be subtracted (and not give a negative answer) is $1 \times 243 = 243$. So 1 is written down as the first figure of the quotient, and 243 is subtracted, leaving 58. The next figure is 7, so that we now have to see how many times 243 will go into 587. Looking at our four multiples, we see that we can subtract $2 \times 243 = 486$, leaving 101, and giving 2 as the second figure of the quotient. Bringing down the next figure, we have remainder 1014. From this we can subtract $3 \times 243 = 729$, leaving 285, and then 1×243, leaving 42; i.e. the next figure of the quotient is $3 + 1 = 4$. Similarly we find the last two figures to be 1 and $6 + 1 = 7$, giving quotient 12417, and the remainder is 69.

This process can be applied equally well to manual division (without a machine): it avoids the trouble of guessing how many times the divisor will go at each step, and also the labour of doing a multiplication sum each time.

Alternatively, by writing the multiples of 243 on a separate slip of paper, the necessity of rewriting them where they are printed overleaf in italics can be avoided: the slip is simply moved along so that the appropriate multiple to be subtracted can be seen at once at each stage, and the subtraction performed mentally.

$$
\begin{array}{lr|l}
 & & \quad 12\overset{1}{3}1\overset{1}{6} = 12417 \text{ (quotient)} \\
\hline
(1\times) & 243 & 3017400 \\
(2\times) & 486 & \mathit{243} \\
(3\times) & 729 & \overline{587} \\
(6\times) & 1458 & \mathit{486} \\
 & & \overline{1014} \\
 & & \mathit{729} \\
 & & \overline{285} \\
 & & \mathit{243} \\
 & & \overline{420} \\
 & & \mathit{243} \\
 & & \overline{1770} \\
 & & \mathit{1458} \\
 & & \overline{312} \\
 & & \mathit{243} \\
 & & \overline{69} \text{ (remainder)}
\end{array}
$$

Fig. 2.1—A method of long division

The Exactus can also be adapted for multiplication in a similar way, though rather clumsily. Thus to multiply 243 by 12417 we shall simply do the above process in reverse, working out the 1, 2, 3, and 6 times multiples of 243 as before, and then setting out 10000, 2000, 100, 300, 10, 1, and 6 times 243 and adding: (100 + 300 gives 400, 1 + 6 gives 7).

$$
\begin{array}{rr}
243 \times 10000 = & 2430000 \\
2000 & 486000 \\
100 & 24300 \\
300 & 72900 \\
10 & 2430 \\
1 & 243 \\
6 & 1458 \\
\hline
243 \times 12417 = & 3017331
\end{array}
$$

The other type of machine which is most useful for computation is the type which will do all four operations, addition, subtraction, multiplication, and division. There are a number of different makes of such machines, some operated by hand and others electrically, and choice between them is largely a matter of taste, convenience, and cost. Fundamentally, however, they all work on very similar principles. They have a set of levers or keys to put any given number into the machine, (usually) a register indicating this number, to check that it has been correctly set up, and also a product register and a revolution (or multiplier) register. The fundamental operation is simply that of addition: by turning a lever, or pressing a key, the number set up is added on to the number already in the product register, which then shows the total. By turning the lever twice, or holding the key down while the machine does two revolutions, the number is added on twice, i.e. multiplied by 2. By a backwards turn, or pressing of a subtraction key, the number is subtracted: by repeating this the number is multiplied by −2, and so on. The important feature of these machines is, however, that the product register is movable. Thus if we put the number 123 into the machine, and add it with the product register in its normal position, we shall obtain the total 123.

Product register (initially)	0000000000000000
Number in machine	123
Product register (after addition)	0000000000000123

If, however, we move the product register, say, 2 places to the right then after addition it will show the sum 12300, i.e. we have multiplied 123 by 100.

Product register (initially)	0000000000000000
Number in machine	123
Product register (after addition)	0000000000012300

By repeating the operation we can multiply by 200, and so forth. Thus to multiply by 298, all we need to do is to perform 8 additions with the product register in its normal position, 9 moving it one place to the right, and another 2 additions after moving it yet another place. The number 298 will then show in the revolutions register, and 298 times whatever number has been set up will be added on to the product register.

We can, however, do better. To multiply by 298 in this way requires $2 + 9 + 8 = 19$ additions. But 298 is 2 less than 300, and so we can multiply by 300 and subtract 2 (or perform 2 backward revolutions) in the units place: this involves only 5 revolutions, instead of 19. If we denote 2 backward revolutions by an inverted figure ↊, then we can say that to multiply by 298, it is better to consider it as 30↊. Similarly if we are to multiply by 37, it is best to consider it as 4↋, i.e. 4 revolutions in the tens position, and 3 backward ones in the units position. This is known as "short cut" multiplication, and in that way any number can be multiplied without doing more than 5 revolutions in any one place. It is the standard practice with calculating machines, and as we shall see in Chapter 23, the idea (originally due to J. Colson, 1726) has a wide field of application.

Division is done by ordinary calculating machines by effectively the same procedure as is usually taught in schools for manual division—but most electrical calculating machines have automatic division, i.e. the machine, once started, does the whole process on its own. This greatly increases speed and accuracy.

The newly developed electronic computers can do complete calculations at remarkable speed, but to discuss them adequately would need another book.

2.5 Manual calculation

Even in a laboratory fitted with modern electrical calculators there will be a number of calculations which are most efficiently done with pencil and paper: so it is desirable to be able to calculate rapidly and efficiently. This is not in fact difficult: here are a few hints, which may be helpful.

The general principle is to say as little as possible. Thus when adding $2 + 3 + 5$ say only the totals, "2, 5, 10", and not "2 *and* 3 *are* 5 *and* 5 *are* 10", which is much more laborious. When multiplying 123 by 7 say "1 (2) 6 (1) 8; answer 861", and not "7 3's are 21, 1 and carry 2, etc.". This requires some practice. It is worth mentioning that many rapid calculators find it most convenient to read numbers from right to left, as that is the direction in which most calculations are performed. Thus they think of 28 as "8, 2" and 135 as "5, 3, 1". But that is a matter of taste.

The usual method of doing multiplications as taught in schools has a mixture of multiplication and addition at each step; e.g. in working out 123×7 we find $3 \times 7 = 21$ (carry 2), $2 \times 7 + 2 = 16$ (carry 1), and $1 \times 7 + 1 = 8$. It is easier to do one thing at a time, thus:

$$\begin{array}{rrcr}
 & 06 & & .6 \\
 & 1804 & & 18.4 \\
 & 301202 & & 3.12.2 \\
642 \times 7531 = & 422006 & \text{or} & 422..6 \\
 & 2810 & & 281. \\
 & 14 & & 14 \\
\hline
 & 4834902 & & 4834902 \\
\hline
\end{array}$$

The working is as follows. We take the first figure, 6, of 642 and multiply it in turn by the digits 7, 5, 3, 1 of the second number, writing the answers 42, 30, 18, 06 diagonally upwards, i.e. each one place to the right and one line up from the previous one. (1×6 must be written 06 to keep the correct place: a dot may be used instead of 0, as in the second version.) The second figure, 4, of 642 is now multiplied by 7, 5, 3, 1, giving the products 28, 20, 12, 04, written diagonally upwards, and starting one place to the right of the left-hand end. The third figure, 2, of 642 similarly gives the bottom right diagonals, and a simple addition gives the product $642 \times 7531 = 4834902$. Although this involves more writing than the usual method, it is easier and quicker in practice.

Those who are good at arithmetic may find it easy to calculate 642×7531 in a single line, by "cross-multiplication". The working goes as follows, beginning from the *right*-hand end:

$$
\begin{aligned}
2 \times 1 &= 2;\ \text{write } 2,\\
4 \times 1 + 2 \times 3 &= 10;\ \text{write } 0,\ \text{carry } 1,\\
1 + 6 \times 1 + 4 \times 3 + 2 \times 5 &= 29;\ \text{write } 9,\ \text{carry } 2,\\
2 + 6 \times 3 + 4 \times 5 + 2 \times 7 &= 54;\ \text{write } 4,\ \text{carry } 5,\\
5 + 6 \times 5 + 4 \times 7 &= 63;\ \text{write } 3,\ \text{carry } 6,\\
6 + 6 \times 7 &= 48;\ \text{write } 48.
\end{aligned}
$$

Reading these figures from right to left gives the product 4834902.

It may be worth mentioning here that many people find it difficult to reconcile themselves to the rule that the product of two negative numbers is positive, e.g. that $(-2) \times (-3) = +6$. They agree that this is the rule they have been taught, and the one which they use in practice; but they have a lingering feeling that it is not quite natural. It therefore seems worth while devoting a little space to showing how it arises.

We first recall that multiplication by a positive integer is defined to mean repeated addition: e.g. 3 times 2 means $2 + 2 + 2 = 6$. Now if we write out the 3 times table we get a regular series of numbers:

$$
\begin{aligned}
3 \times 1 &= 3\\
3 \times 2 &= 6\\
3 \times 3 &= 9\\
3 \times 4 &= 12 \ldots
\end{aligned}
$$

Each number is greater than the preceding one by exactly 3; thus $3 + 3 = 6$, $6 + 3 = 9$, $9 + 3 = 12$, and so on. (Such a regular series is usually called an "arithmetic progression with common difference 3".) Now suppose that instead of going forward in this series, we read it backwards

$$
\begin{aligned}
3 \times 4 &= 12\\
3 \times 3 &= 9\\
3 \times 2 &= 6\\
3 \times 1 &= 3
\end{aligned}
$$

The numbers now diminish regularly by 3 at each step. If we continue the series in the same way we get

$$
\begin{aligned}
3 \times 0 &= 0\\
3 \times (-1) &= -3\\
3 \times (-2) &= -6
\end{aligned}
$$

and so on. These values still agree with our definition of multiplication; 3 times (-2) can be interpreted as "take away 2 three times over", which is the same as "take away 6", or in symbols, $(-2) + (-2) + (-2) = -6$. Now suppose that instead of this we take the series

$$\begin{aligned} 4 \times 3 &= 12 \\ 3 \times 3 &= 9 \\ 2 \times 3 &= 6 \\ 1 \times 3 &= 3 \end{aligned}$$

Again the numbers diminish by 3 at each step. Now the continuation of this series by analogy would give the following results:

$$\begin{aligned} 0 \times 3 &= 0 \\ (-1) \times 3 &= -3 \\ (-2) \times 3 &= -6 \\ (-3) \times 3 &= -9 \end{aligned}$$

and so on. However, the usual definition of multiplication as a repeated addition will not cover an expression like $(-2) \times 3$; we cannot take (-2) threes and add them together. We must, therefore, define what we are to mean when we write such an expression as $(-2) \times 3$; and the argument given above suggests that it is most natural to take this to be -6, the opposite of $2 \times 3 = +6$. In general, the same argument would suggest that $(-A) \times B$ should be defined to be $-(AB)$. This has a further justification, as follows. When A and B are both positive, we know that $A \times B = B \times A$. Now if this rule is to apply for negative numbers too, it should imply that $(-2) \times 3 = 3 \times (-2)$. But we have already shown that $3 \times (-2) = -6$; hence we should define $(-2) \times 3$ to be -6.

In the light of this argument, what should $(-2) \times (-3)$ be? It should be the opposite of $2 \times (-3) = -6$; and the opposite of -6 is $+6$. Hence $(-2) \times (-3)$ should be defined as $+6$. This argument may seem almost too brief to be true; we can write it out at greater length. Consider the series

$$\begin{aligned} 4 \times (-3) &= -12 \\ 3 \times (-3) &= -9 \\ 2 \times (-3) &= -6 \\ 1 \times (-3) &= -3 \end{aligned}$$

which conforms to the original definition of multiplication as repeated addition. Again we get a regular series, in which the numbers increase by 3 at each step. Continue it in the obvious way: it gives

$$\begin{aligned} 0 \times (-3) &= 0 \\ (-1) \times (-3) &= 3 \\ (-2) \times (-3) &= 6 \end{aligned}$$

and so on. We can condense the whole argument by writing out the multiplication table in full as follows:

	−3	−2	−1	0	1	2	3
−3 ×	9	6	3	0	−3	−6	−9
−2 ×	6	4	2	0	−2	−4	−6
−1 ×	3	2	1	0	−1	−2	−3
0 ×	0	0	0	0	0	0	0
1 ×	−3	−2	−1	0	1	2	3
2 ×	−6	−4	−2	0	2	4	6
3 ×	−9	−6	−3	0	3	6	9

The bottom right-hand corner is the ordinary multiplication table for positive numbers; e.g. $2 \times 1 = 2$, $2 \times 2 = 4$, $2 \times 3 = 6$, etc. Now in every row and every column of this table we get a regular arithmetic progression, with the numbers increasing or decreasing by the same amount at each step. This is the only possible way of arranging the table for this to remain true.

The rule is also needed to make sure that the ordinary laws of algebra hold equally well for negative numbers as for positive ones. For example, we have shown above that

$$(x + y)^2 = x^2 + 2xy + y^2.$$

For example, if $x = 4$ and $y = 2$ this becomes

$$6^2 = (4 + 2)^2 = 4^2 + 2 \times 4 \times 2 + 2^2 = 16 + 16 + 4$$

which is easily verified to be true. Now take instead $x = 4$ and $y = -2$. We get

$$[4 + (-2)]^2 = 4^2 + 2 \times 4 \times (-2) + (-2)^2$$

i.e. $$2^2 = 4 = 16 - 16 + (-2)^2$$

and this can only be true if we define $(-2)^2$ to mean 4.

It is not easy to find a simple "natural" example of the multiplication of two negative numbers with an immediate and obvious interpretation in real life, but perhaps the following will serve. Imagine a tank acting as a reserve; we can either pump water into it, or draw water out. If at some period the tank is being filled at the rate of G gallons per minute, then M minutes later the *extra* amount of water in the tank (as compared with its present state) will be GM gallons. For example, if it is filled at $G = 4$ gallons per minute for $M = 3$ minutes, it will gain $4 \times 3 = 12$ gallons. If we take G to be negative, the obvious interpretation is that water is being pumped out, not in. Thus if it is filled at (-4) gallons per minute, i.e. in fact 4 gallons are removed each minute, then after 3 minutes the gain is $(-4) \times 3 = -12$ gallons. If we take M to be negative, the interpretation is that we refer to the past, not the future. Thus if we fill the tank at the rate of $G = 4$ gallons per minute, and put $M = -3$, we find $GM = 4 \times (-3) = -12$, meaning that 3 minutes ago the tank was emptier by 12 gallons. If we take G and M to be both negative, say $G = -4$ and $M = -3$, the problem is: if the tank is being emptied at the rate of 4 gallons per minute, what was its state 3 minutes ago as compared with the present? The answer is clearly 12 gallons more full; so we must take $GM = (-4) \times (-3)$ to be $+12$.

Arguments of this kind can be used to suggest other extended definitions. For example, we define

$$\begin{aligned} 4^1 &= 4 &&= 4 \\ 4^2 &= 4 \times 4 &&= 16 \\ 4^3 &= 4 \times 4 \times 4 &&= 64 \\ 4^4 &= 4 \times 4 \times 4 \times 4 &&= 256 \end{aligned}$$

and so on. What would be the natural way to define 4^0 or 4^{-2}? Let us try reading the above series backwards:

$$4^4 = 256;\ 4^3 = 64;\ 4^2 = 16;\ \ldots$$

We see that each number is obtained from the preceding one by dividing by 4. If we were to continue similarly we would get

$$\begin{aligned} 4^1 &= 4^2/4 = 16/4 = 4 \\ 4^0 &= 4^1/4 = 4/4 = 1 \\ 4^{-1} &= 4^0/4 = 1/4 = \tfrac{1}{4} \\ 4^{-2} &= 4^{-1}/4 = \tfrac{1}{4}/4 = \tfrac{1}{16} \end{aligned}$$

etc. Thus it seems that the most natural definition of 4^0 is 1 (and the same holds for x^0 for all non-zero x). This surprises most people when they first meet it. However, it must be so if the usual rule $A^m A^n = A^{m+n}$ is to be preserved, for if we put $A = 4$, $m = 2$, $n = 0$ we would get

$$4^2 \times 4^0 = 4^{2+0} = 4^2$$

and this can only be true if we take 4^0 to be 1. Similarly we must take 4^{-2} to mean $1/16 = 1/(4)^2$. This again is the only possible interpretation compatible with the rule $A^m A^n = A^{m+n}$. For if we take $A = 4$, $m = 3$, $n = -2$ we would get

$$4^3 \times 4^{-2} = 4^{3-2}$$

i.e.

$$64 \times 4^{-2} = 4^1 = 4$$

whence $4^{-2} = 1/16$.

Another series we shall often meet later on is the series of "factorials"

$$\begin{aligned} \underline{|2} &= 2! = 1 \times 2 &&= 2 \\ \underline{|3} &= 3! = 1 \times 2 \times 3 &&= 6 \\ \underline{|4} &= 4! = 1 \times 2 \times 3 \times 4 &&= 24 \\ \underline{|5} &= 5! = 1 \times 2 \times 3 \times 4 \times 5 &&= 120 \end{aligned}$$

and so on. (Read $\underline{|2}$ or $2!$ as "factorial 2".) If we read this backwards, we see that we get from $\underline{|5} = 120$ to $\underline{|4} = 24$ by dividing by 5, from

$\underline{|4} = 24$ to $\underline{|3} = 6$ by dividing by 4, from $\underline{|3} = 6$ to $\underline{|2} = 2$ by dividing by 3. Continuing in this way we have

$$\underline{|1} = \underline{|2}/2 = 2/2 = 1$$
$$\underline{|0} = \underline{|1}/1 = 1/1 = 1$$

Hence $\underline{|0}$ is to be taken as 1 (and not as zero, as one might too hastily think). The series cannot be continued backwards any further, as it would give $\underline{|-1} = \underline{|0}/0 = 1/0$, and division by 0 is not possible.

2.6 Placing the decimal point

A number of computers are often a little uncertain about placing the decimal point in a multiplication or division sum. The rule is quite simple. First do the multiplication or division in full in the ordinary way, ignoring the point. Then, in multiplication, if the point is a places from the right-hand end in one of the numbers being multiplied, and b places from the right in the other, it will be $(a + b)$ places from the right in the product. In division, if the point is a places from the right in the dividend, and b in the divisor, then it will be $(a - b)$ in the quotient. Thus

$$123 \times 456 = 56088$$

so that

$$12{\cdot}3 + 4{\cdot}56 = 56{\cdot}088$$
$$[a = 1,\ b = 2,\ a + b = 3]$$

and

$$20000/147 = 136 \text{ (to three figures)}$$

so that

$$2{\cdot}0000/14{\cdot}7 = {\cdot}136$$
$$[a = 4,\ b = 1,\ a - b = 3]$$

This rule is the one which will be used when working with a calculating machine. Most calculating machines have indicators which can be used to mark the decimal point.

2.7 Checking arithmetic

Not only is no human computer infallible, but unfortunately some types of calculating machines are liable to occasional error. Even where the machine is perfect, the operator is not; so it is very important that all calculations should be checked.

When a long calculation is being performed it is usual to insert checks at various points in the process of computation. It is not usual to check every individual operation, although if the calculation is a specially important one, it is desirable to repeat it entirely and preferably by a different method, or in a different order. (A mere repetition of a calculation involves a risk of repeating an error.)

For short calculations there is a simple method of checking ordinary additions, subtractions, multiplications, and divisions, known as "casting out the nines". This is very similar to the method numerologists use when finding a person's "lucky number". Each letter in the person's name is given a number: $A = 1$, $B = 2$, $C = 3$, and so on: and all these numbers are added together. Thus John Smith would add up to $10 + 15 + 8 + 14 + 19 + 13 + 9 + 20 + 8 = 116$. The digits of the resulting total are again added $(1 + 1 + 6 = 8)$, and if necessary the process is repeated until finally the name is reduced to a single figure (here 8) which is the "lucky number" in question. We can apply this process to any number; 183728 would be reduced to $1 + 8 + 3 + 7 + 2 + 8 = 29$, which again would reduce to $2 + 9 = 11$, and finally to $1 + 1 = 2$. This final result we call the "reduced value". Now any addition, subtraction, or multiplication sum, if correctly done, will remain true when all numbers are replaced by their reduced values. For example, take the addition sum:

137	reduced	value	2
824	„	„	5
73	„	„	1
76	„	„	4
1110	„	„	3

and we see that $2 + 5 + 1 + 4 = 12$, with reduced value $1 + 2 = 3$, which agrees with the reduced value of the total. The mixed addition and subtraction $2432 - 1234 + 256 + 824 - 56 = 2222$ becomes $2 - 1 + 4 + 5 - 2 = 8$, which checks: and the multiplication $123 \times 456 = 56088$ becomes $6 \times 6 = 9$, which checks, since $6 \times 6 = 36$ has reduced value 9. To check a division, we have to rewrite it as *dividend* = *divisor* × *quotient* + *remainder*. Thus the division of 3017400 by 243 gives quotient 12417 and remainder 69, or $3017400 = 243 \times 12417 + 69$. Using reduced values this becomes $6 = 9 \times 6 + 6$, which checks on reduction.

The reason why this check works is essentially that the "reduced value" of a number is simply the remainder obtained on dividing the number by 9 (except that numbers which divide exactly by 9 have reduced value 9 and not 0). This is because the numbers 9, 99, 999, etc., all divide exactly by 9, so that $10 = 9 + 1$, $100 = 99 + 1$, 1000, etc., must all give remainder 1 on division by 9. By a similar argument 2, 20, 200, 2000 . . . all give 2 as remainder; 3, 30, 300, 3000 . . . all give remainder 3, and a number like $213 = 200 + 10 + 3$ gives the same remainder on division by 9 as does $2 + 1 + 3$. If a number A has reduced value a, it is simply a plus a multiple of 9, so that we can write

$$A = a + 9u$$

where u is an integer. Similarly if B has reduced value b, then $B = b + 9v$. Therefore the total $(A + B) = a + 9u + b + 9v$

$$= (a + b) + 9(u + v)$$

differs from $(a + b)$ by a multiple of 9, so that $(A + B)$ and $(a + b)$ have the same reduced value. Considerations like this explain why the check of casting out the nines works with addition or subtraction. For multiplication we have

$$\begin{aligned} AB &= (a + 9u)(b + 9v) \\ &= ab + 9ub + 9av + 81uv \\ &= ab + 9(ub + av + 9uv) \end{aligned}$$

so that AB and ab differ by a multiple of 9, and must have the same reduced value.

In casting out the nines we can ignore multiples of 9. Thus a number like 3994 can be reduced to $3 + 4 = 7$. (The direct calculation would be $3 + 9 + 9 + 4 = 25$, $2 + 5 = 7$.) For 2361, we can ignore the 3 and 6 which add to 9, and take the reduced value to be $2 + 1 = 3$. Again if we try to check the subtraction $210 - 35 = 175$ by casting out the nines, we find that $210 - 35$ reduces to $3 - 8 = -5$, whereas 175 reduces to 4. This is a perfectly good check, because -5 and 4 differ by 9, and must be considered as equivalent in terms of reduced values.

"Casting out the nines" is, of course, not an infallible check, though a very useful one. It is possible to make an error, or series of errors, which may pass the test. In particular it is easy to interchange 2 figures, writing for example 2659 instead of 2569, and this rather common type of error will not affect the reduced value and so will not be detected. A rather better check is that of "casting out the elevens". In this check the reduced value is calculated by alternately adding and subtracting the figures, starting from the *right*-hand end. Thus 137 will reduce to $7 - 3 + 1 = 5$, and 824 to $4 - 2 + 8 = 10$, which further reduces to $0 - 1 = -1$. The addition sum $137 + 824 + 73 + 76 = 1110$ reduces to $5 - 1 - 4 - 1 = -1$, which is a perfect check. In this method of checking it is multiples of eleven which are ignored, rather than multiples of nine: in all other respects it is used exactly like the "casting out the nines" check, and can be used for addition, multiplication, subtraction, and division. It will detect an interchange of two adjacent figures, provided that there is not by bad luck a compensating error elsewhere. It follows that a calculation carefully performed and checked by both the nines and elevens methods is most unlikely to be wrong.

2.8 Punched cards

Large quantities of data are tiresome to record and sort by hand. If they are put on punched cards then the sorting and at least part of the computation can be done mechanically, with a resulting improvement in speed and accuracy.

There are two well-known types of punched cards: the "Cope-Chat" type, and the other type which is represented by the "Hollerith" and "Powers-Samas" systems.

The Cope-Chat system is simplicity itself: the only mechanical equipment required consists of a pack of cards, a punch, and a knitting needle. Each card has initially a row of holes punched round the edge. The information is recorded on the cards by using the punch to cut V-shaped slots from certain holes to the edge of the card (Fig. 2.2). When a pack of cards has a needle run through any particular hole, and is then gently shaken, all those cards for which that hole is slotted will be free to fall off the needle, while those cards which have the hole left unslotted will be retained. Thus sorting can be done easily and rapidly.

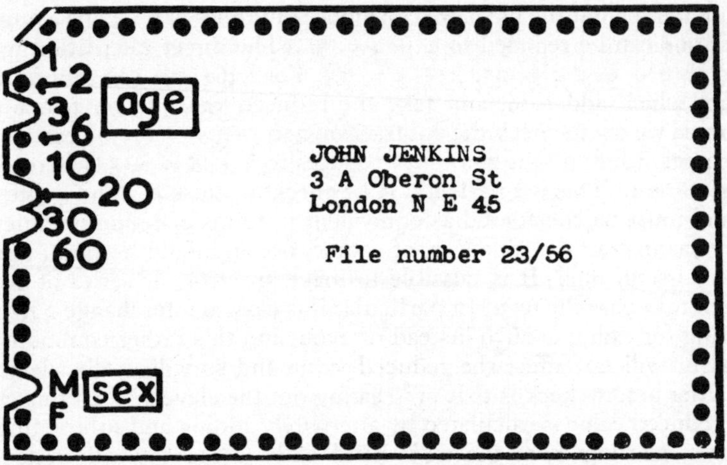

Fig. 2.2—A Cope-Chat card, punched for age 54 and male sex

All kinds of information can be put on a Cope-Chat card: it is only necessary to arrange a method of "coding" by which any particular piece of information is represented by the slotting of certain holes. Such items as sex, place of birth, the first two letters of a surname, can easily be recorded and sorted; we could separate out from a pack such items as "all cards representing males" or "all cards representing females born in London", and these might be counted or sorted further. The easiest way to record numbers is to have holes corresponding to the numbers 1, 2, 3, 6, 10, 20, 30, 60, 100, 200, 300, 600, etc. 12 will then be represented by slotting the holes "2" and "10" and 8 by slotting the holes "2" and "6" (since $2 + 6 = 8$).

Using this system of punching it is an easy matter to find totals. Suppose that a number A_1, say, is punched on the first card, a number A_2 on the second, A_3 on the third, and we wish to find the total

$T = A_1 + A_2 + A_3 \ldots$ We first extract with the needle all cards slotted in the hole "1" and count them: let us say that there are n_1 such cards. Since each contributes 1, together they contribute $1n_1 = n_1$ to the total. These cards are now put back into the pack, and all cards slotted for "2" are taken out, and counted. If there are n_2 such cards, they will contribute $2n_2$ to the total. We replace these cards, take out all cards slotted for "3", and continue in that way. The required total will be

$$T = n_1 + 2n_2 + 3n_3 + 6n_6 + 10n_{10} + 20n_{20} + \ldots$$

If we have two sets of numbers punched on the cards, A_1 and B_1 (say) on the first card, A_2 and B_2 correspondingly on the second, and so on, then we can find (though rather clumsily) the sum of products $S = A_1B_1 + A_2B_2 + A_3B_3 + \ldots$ This as we shall see later is a calculation of importance biometrically, used in calculations of variances, standard deviations, and correlations. To do this we first separate out all cards slotted for "1" in the representation of the number A. By the method already described, we find the total of all the B numbers on the separated cards: let this total be T_1. These cards are then replaced, and those slotted for "2" in the representation of A are separated. The total of the B numbers on these cards is found to be, say, T_2. The cards are replaced, and the process continues. The sum of products

$$\begin{aligned} S &= A_1B_1 + A_2B_2 + A_3B_3 + \ldots \\ &= T_1 + 2T_2 + 3T_3 + 6T_6 + 10T_{10} + \ldots \end{aligned}$$

There is no need for the numbers "B" to be different from the numbers "A": the process can equally well be used to evaluate

$$A_1A_1 + A_2A_2 + A_3A_3 + \ldots = A_1^2 + A_2^2 + A_3^2 + \ldots,$$

where A_1 is a number recorded on the first card, A_2 the corresponding number on the second, A_3 that on the third, and so on.

The Cope-Chat system is very convenient for dealing with data punched on to a moderate number of cards—up to a few hundred. It has the good points that it is simple, cheap, and flexible; and also it is possible to write extra information on the face of the card. Its bad points are that the counting of cards by hand is rather slow, and that only a limited amount of information can be recorded on a single card with about 150 holes.

The other punched card systems use cards containing columns of figures; usually a hole is punched through just one of the figures in each column. There are several makes available. One typical system uses cards with 80 columns; each column contains the ten digits 0 to 9, and also two extra symbols X and Y for occasional use when desired. A number such as 11002 is indicated by punching through the digits

1, 1, 0, 0, 2 in successive columns (Fig. 2.3). These cards can hold in punched form much more information than a Cope-Chat card of comparable size, and can be counted and sorted at the rate of several hundred cards a minute; but special mechanical equipment is required. Some simple arithmetic operations, such as addition and multiplication, can be performed directly with the cards. The numbers on the cards can also be read directly into many types of electronic computer, making complicated calculations possible.

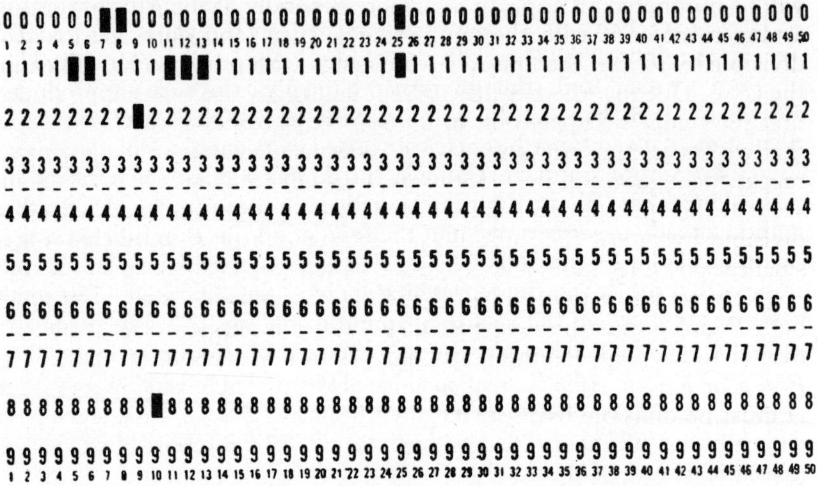

Fig. 2.3—Part of a card punched for ... 110028111 ...

2.9 The uniqueness of factorization

We now give the promised proof that any positive integer, greater than 1, can be split into prime factors in only one way.

This theorem depends on a simpler lemma that if A and B are two positive integers such that the product AB is divisible exactly by a prime p, then at least one of the two integers A and B is divisible by p.

To see why this lemma should be true let us consider the meaning

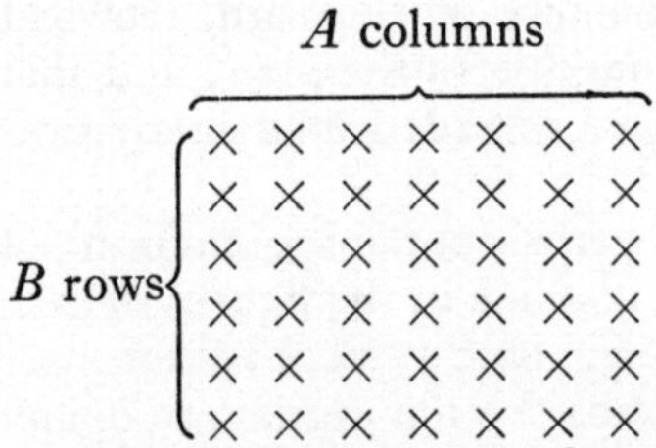

Fig. 2.4—The product AB

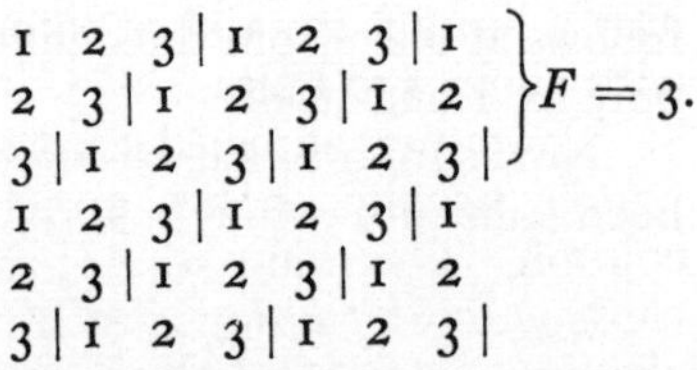

Fig. 2.5—Test for divisibility by p (= 3)

of the product AB. This can be considered as the number of objects arranged rectangularly in A columns and B rows, as in Fig. 2.4 (where $A = 7$, $B = 6$).

Now let us number off these AB objects in order as 1, 2, 3 . . . up to p; 1, 2, 3 . . . up to p; 1, 2, . . . (starting again from 1 at every pth number), and so on till we have exhausted the rectangle (see Fig. 2.5, where p is taken as 3). Then every time we come to p in this counting we shall have numbered off a multiple of p. Since AB is supposed to be exactly divisible by p, the last object (in the lower right-hand corner) will be numbered off as p. In fact, more generally, if the Rth row of this arrangement ends with the numbering p, then (and only then) is AR divisible by p. Let the first row which ends with p be he Fth. Then in the $(F + 1)$th row the numbering repeats exactly the numbering in the first row; and this repetition continues. It is not until the $(2F)$th row that we get another row ending in p. The whole pattern then begins to repeat once again, and the $(3F)$th row is the next that ends in p. We thus see that the only numbers R for which the Rth row ends in p, i.e. AR is divisible by p, are the multiples F, $2F$, $3F$ . . . of F, where F is the smallest possible value of R.

Now we know that both AB and Ap are divisible exactly by p, so both B and p must be multiples of F. But p is supposed to be prime: that is to say, by definition, its only factors are 1 and p. So either $F = 1$ or $F = p$. If $F = 1$, then since AF is divisible by p, we see that A must be divisible by p. If on the other hand $F = p$, then since B is a multiple of F, it must be a multiple of p. This establishes our result that at least one of A and B must be divisible by p.

This result can be readily generalized. If a product $ABCD$ is exactly divisible by a prime number p, then at least one of the factors A, B, C, and D is divisible by p. (This will hold for any number of factors.) For supposing that $ABCD$ is divisible by p, on writing this as $A \,.\, (BCD)$ we see that either A or BCD (or both) is divisible by p. If p divides A we have proved the required result. If not, then p must divide $B \,.\, (CD)$, and therefore B and/or CD is divisible by p. If B is divisible by p, then we have proved the required result: if not, then CD is divisible by p, so that at least one of C and D has p as a factor. This proves that at least one of A, B, C, D is divisible by p.

To prove our main theorem we need one further result, that if p divides P, and both p and P are prime numbers, then $p = P$. This follows at once from the definition of a prime, that P is exactly divisible only by 1 and itself.

Now suppose that x is a positive number (greater than 1) which has been split into prime factors in 2 ways: let us say

$$x = pqrst = PQRSTUV \qquad (2.1)$$

where p, q, etc., are all prime. We then know that $PQRSTUV$ is divisible by p, so that at least one of the prime numbers P, Q, R, S,

T, U, V is divisible by p, and so must be equal to p. For example, suppose R was equal to p. Then we can rewrite equation (2.1) as

$$pqrst = PQpSTUV.$$

Dividing both sides of this equation by p, it becomes

$$qrst = PQSTUV.$$

A similar argument now shows that at least one of the remaining prime numbers P, Q, S, T, U, V must be equal to q; let us say, $U = q$. The equation now becomes

$$qrst = PQSTqV$$

and on dividing both sides by q,

$$rst = PQSTV.$$

We can continue in this way, showing that every prime number on the left-hand side also occurs somewhere on the right. Finally, when we have exhausted all the prime numbers on the left-hand side, which then becomes 1, the right-hand side, being equal to the left, also must become 1, so that we must also have exhausted all the primes on the right. That means that the prime numbers on the right-hand side are simply those on the left arranged in a different order; so that apart from the order the factorization into prime numbers can be done in one way only.

2.10 Brackets

Brackets are used to indicate the order in which different operations are performed: e.g. $(3 + 4) \times 5$ means "add 3 and 4, and multiply the result by 5", giving 35 altogether, whereas in the expression $3 + (4 \times 5)$ the product 4×5 has to be found first, and then added to 3, giving 23. To avoid having too many brackets the convention is made that operations are to be performed in the following order (except where otherwise indicated by suitable brackets): first, multiplications; second, divisions and the taking of logarithms, roots, sines, cosines, etc.; third, additions and subtractions. For example, $3 + 4 \times 5$ means $3 + (4 \times 5)$, not $(3 + 4) \times 5$, while similarly $\log 2x + 3$ means $\log (2x) + 3$, not $\log (2x + 3)$, and $3x/5y = \dfrac{3x}{5y}$. Usually it is quite clear how to interpret a formula, but where there is doubt it is best not to omit any brackets: thus $\log x/2$ might be taken to mean $(\log x)/2$ or $\log (x/2)$.

One point to note (as some authors are careless here) is that the denominator following a division sign only extends up to the next + or − sign. Thus the expression $x + 3/y + 2$ is to be read as $x + \dfrac{3}{y} + 2$, not as $\dfrac{x + 3}{y + 2}$, which should properly be written $(x + 3)/(y + 2)$. And $1/2x$ does *not* mean $\frac{1}{2}x = x/2$.

A dot is sometimes written instead of $\times$, and is read "times" or "into": thus $a \times b = a.b = ab$. In America this dot is written high, as $a \cdot b$, and the decimal point low, as 13.24. The international signs for division are a colon : and /, read as "over" or "by". The English sign $\div$ means "minus" in some countries. Most continental European countries use a comma for a decimal point: thus the English 3002·16 equals the European 3.002,16.

PROBLEMS

(1) Check the following sums by casting out nines and elevens. Which are correct?

(*a*) $7123 + 604 - 6215 = 1152$
(*b*) $30{\cdot}74 + 1{\cdot}38 - 20{\cdot}53 + 40{\cdot}10 - {\cdot}67 = 51{\cdot}02$
(*c*) $123 \times 56 = 6888$
(*d*) $72{\cdot}3 \times 13{\cdot}4 = 966{\cdot}82$
(*e*) $162^2 = 26244$
(*f*) $25{\cdot}3 \times 11{\cdot}1 + 7{\cdot}2 \times 6{\cdot}3 - 5{\cdot}1 \times 42{\cdot}4 = 109{\cdot}95$.

(2) Show that any exact square must have a reduced value 0, 1, 4, or 7 on casting out the nines, and 0, 1, 3, 4, 5, or 9 on casting out the elevens.

(3) Show that any exact square must end in one of the figures 0, 1, 4, 5, 6, 9. Which of the following numbers are exact squares: 62,500; 137,926; 26,451; 41,616; 235,724; 40,903; 1,210; 5,287?

(4) Show that any exact cube must have reduced value 0, 1, or 8 on casting out nines. Which of the following numbers are exact cubes: 12,531; 110,592; 176,024?

(5) If *A*, *B*, . . . *G* are all positive integers, then any positive integer which divides exactly into all of them is called a "common factor". Show that any factor of a common factor is also a common factor.

(6) By splitting the numbers *A*, *B*, . . . *G* into prime factors, show how to find their greatest common factor [G.C.F.]*. Find the greatest common factor of 120, 2100 and 144. Every common factor is a factor of the G.C.F. Why? Find all the common factors of 120, 2100, and 144.

(7) Similarly show that any multiple of a common multiple of *A*, *B*, . . . *G* is also a common multiple, and that every common multiple is a multiple of the least common multiple [L.C.M.].

(8) Let *a*, *b* be two positive integers. Let *c* be the difference between them, that is, $c = a - b$ if *a* is greater than *b*, and $c = b - a$ if *b* is

* Also called the "highest common factor" [H.C.F.].

greater than a. Show that any common factor of a and b is also a common factor of b and c, and conversely; and in particular that the G.C.F. of a and b is equal to the G.C.F. of b and c. What happens if $a = b$? If $c \neq 0$, let d be the difference between b and c. Then G.C.F. (c and d) = G.C.F. (b and c) = G.C.F. (a and b). What happens if $c = d$? Show that by continuing in this way we can find fairly quickly the G.C.F. of any two numbers, even where we cannot conveniently split them into prime factors. Find the G.C.F. of 45 and 84 in this way.

By imagining a and b split into prime factors, show that

$$(\text{G.C.F. of } a \text{ and } b) \times (\text{L.C.M. of } a \text{ and } b) = ab.$$

Hence find the L.C.M. of 45 and 84.

(9) By imagining the positive integers r, s, and t split into prime factors, show that the L.C.M. of r, s, and t is the least common multiple of L and t, where L is the least common multiple of r and s. By referring to the previous question, show that this enables us to find the L.C.M. of any three numbers, even where we cannot in practice split them into prime factors. Extend this to a method for finding the L.C.M. of four or more numbers.

CHAPTER THREE

SOME POINTS IN ALGEBRA

3.1 Constants, variables and equations

We suppose that the reader has a working knowledge of most points in elementary algebra, as for example how to solve an equation like $3x + 8 = 11$ [$3x = 11 - 8$, $x = 1$] or to multiply expressions such as

$$(x + y)^2 = x^2 + 2xy + y^2$$
$$(x - y)^2 = x^2 - 2xy + y^2$$
$$(x + y)(x - y) = x^2 - y^2$$

The usual definition of algebra in elementary textbooks is that " algebra is the science of operating with numbers, where the numbers are represented by letters". However, we can expand this definition to make certain useful distinctions between the different kinds of numbers represented by the symbols.

Some symbols represent fixed unvarying numbers, or "constants". Examples of such constants are π, the ratio of the circumference of a circle to its diameter (derived from the Greek word περίμετρος, perimeter or circumference), e, the "base" of natural logarithms (Section 6.13), J, the number of joules to one gram-calorie, and K, the dissociation constant of any given chemical reaction. But more often a symbol will represent a varying quantity, or "variable". Such a quantity may quite literally be changing from one moment to another, or from one day to another. Examples of such quantities are the population of the world, the weight of a particular man or animal, the amount of chlorophyll produced by a particular plant. Quantities such as g, the attraction of gravity, or h, the height above sea-level of the earth's surface, vary from place to place rather than from time to time; and the temperature and pressure of the air vary, both with the place and time of measurement—and the actual observed result may also vary slightly with the instrument used to make the measurement. This distinction is, of course, a relative one, though none the less useful on that account; if we are concerned with experiments made at various points on the earth's surface, then g, the acceleration due to gravity, will vary slightly from place to place. But if the experiments are all performed in one place then g will be constant.

The equations which express the relations between these quantities will likewise be of several kinds. An equation such as $(x + y)^2 =$

$x^2 + 2xy + y^2$ is a consequence of the laws of arithmetic, and is true for all numbers x and y whether they are fixed numbers, e.g. $[1 + \sqrt{2}]^2 = 1 + 2\sqrt{2} + (\sqrt{2})^2 = 3 + 2\sqrt{2}$, or whether they are variables. An equation of this sort is often called an "identity". Other equations may represent "laws of nature"; e.g. that if m is the mass of a volume V of gas, measured in molecular weight units, P its pressure, and T its temperature, then $PV = mRT$, R being the "gas constant". This holds approximately for most gases, and will be true whether P, V, and T are varying quantities or constants. Finally we may have relations which are found to hold good in particular circumstances only, as for example a single observation in an experiment, "$T = 15{\cdot}7°$ C". In such a particular observation T is, of course, a constant, not a variable: this remark may sound trivial, but, as explained in Section 8.13, neglect of it is a common cause of error in differential calculus, where we are dealing with varying quantities.

3.2 Polynomials

The simplest kinds of algebraical expression are those containing no fractional expressions with letters in the denominator, but only powers and products of powers, as for example, $uv + u + v + 2$, $3 + 2x + x^2$, $P^3 + 3PQ^2 + 3P^2Q + \frac{1}{2}uv$.

Such expressions are called *polynomials* (from the Greek, "many terms") and the parts of the expression which are separated by + and − signs are called *terms*. In $uv + u + v + 2$ the terms are uv, u, v, 2. These polynomials are particularly important because of their great mathematical simplicity. As we have said, we shall suppose that the reader is familiar with the ordinary laws of algebra: but for convenience we may summarize them as follows.

Laws of addition and multiplication

(1) Since algebraical expressions represent numbers, any two can be added or multiplied.

(2) Additions can be done in any order, e.g. $2x^2y + (x + 3y) = (3y + x) + 2x^2y = x + 3y + 2x^2y$.

(3) Multiplications can be done in any order, e.g. $3ab\,(a + b) = 3\,(a + b)\,ba = (3b)\,a\,(a + b)$.

(4a) $A\,(B + C + \ldots + D) = AB + AC + \ldots + AD$ where $A, B, C \ldots D$ may stand for any numbers or algebraical expressions, e.g. $(x + y)\,(u^2 + uv + z) = (x + y)u^2 + (x + y)uv + (x + y)z$.

(4b) $(B + C + \ldots + D)A = BA + CA + \ldots + DA$: for example, $3A + 5A + 10A = (3 + 5 + 10)A = 18A$. This law is equivalent to (4a) by virtue of law 3, that it does not matter in which order we perform a multiplication.

(5) Index laws (i.e., laws involving powers A^n):

$$
\begin{aligned}
&(5a) \quad A^m A^n = A^{m+n} \\
&(5b) \quad (AB)^n = A^n B^n \\
&(5c) \quad (A^m)^n = A^{mn}
\end{aligned}
$$

These are really consequences of law 3, by virtue of the definition of a power A^n as a repeated product $AAA \ldots A$ of n A's.

(6) For any number or expression A

$$
\begin{aligned}
&0A = 0 \\
&A + 0 = 1A = A^1 = A
\end{aligned}
$$

(7) If $ABC \ldots = 0$, then at least one of the factors $A, B, C \ldots$ is zero.

All other operations of algebraic addition and multiplication can be analysed into a repeated application of these seven simple laws, although they will probably have become so familiar that the process is practically instinctive without such an analysis. Thus the operation $A+A=2A$ can be analysed as $A+A=1A+1A$ [law 6] $= (1+1)A$ [law 4b] $= 2A$; and the operation $A.A = A^2$ as $A.A = A^1.A^1$ [law 6] $=$ A^{1+1} [law 5a] $= A^2$.

Similarly

$$
\begin{aligned}
(x+y)^2 &= (x+y)(x+y) \\
&= (x+y)x + (x+y)y && \text{[law 4a]} \\
&= (x.x + y.x) + (x.y + y.y) && \text{[law 4b]} \\
&= (x^2 + xy) + (xy + y^2) && \text{[law 3]} \\
&= x^2 + (xy + xy) + y^2 && \text{[law 2]} \\
&= x^2 + 2xy + y^2
\end{aligned}
$$

Subtraction does not need any further laws; all we need say is that $x - y$ can be written as $x + (-1)y$, from which all the properties of subtraction follow. For example,

$$
\begin{aligned}
x - x &= 1x + (-1)x \text{ [law 6]} = [1 + (-1)]x \text{ [law 4b]} \\
&= 0.x = 0 \text{ [law 6]}.
\end{aligned}
$$

This precise statement of the laws of algebra is not merely of purely theoretical or academic interest; later on we shall find it useful to bring in symbols (such as matrices) which obey fairly similar but slightly different laws, and it is helpful to be clear as to what exactly can or cannot be done with algebraic symbols.

There are a few equations which are of such common occurrence that it is wise to commit them to memory. These are

$$(x+y)^2 = x^2 + 2xy + y^2 \quad \ldots \quad (3.1)$$

$$(x-y)^2 = x^2 - 2xy + y^2 \quad \ldots \quad (3.2)$$

$$(x+y)(x-y) = x^2 - y^2 \quad \ldots \quad (3.3)$$

$$(x+y)^3 = x^3 + 3x^2y + 3xy^2 + y^3 \quad \ldots \quad (3.4)$$

$$x^3 + y^3 = (x+y)(x^2 - xy + y^2) \quad \ldots \quad (3.5)$$

$$x^3 - y^3 = (x-y)(x^2 + xy + y^2) \quad \ldots \quad (3.6)$$

$$x^5 - y^5 = (x-y)(x^4 + x^3y + x^2y^2 + xy^3 + y^4)$$

and in general

$$x^n - y^n = (x - y)(x^{n-1} + x^{n-2}y + x^{n-3}y^2 + \ldots + y^{n-1}) \quad (3.7)$$

All these formulas can be verified by direct multiplication, and they are identities, that is, they are true for all values of x and y without restriction.

3·3 One-variable polynomials

Polynomials containing only one variable x are of special importance and simplicity.

Consider first the following table, derived from a survey made by Karl Pearson and A. Lee, *Biometrika*, **2** (1902), p. 357, on the inheritance of height and other physical characters. x here represents the height of a father (rounded off to the nearest inch), and y the average height of sons of fathers of height x.

Table 3.1—Inheritance of height

Father's height (inches) x	*Average son's height* y	$y - \frac{1}{2}x$
59	64·7	35·2
60	65·6	35·6
61	66·3	35·8
62	65·6	34·6
63	66·7	35·2
64	66·7	34·7
65	67·2	34·7
66	67·6	34·6
67	68·0	34·5
68	69·1	35·1
69	69·4	34·9
70	69·7	34·7
71	70·5	35·0
72	70·9	34·9
73	72·0	35·5
74	71·5	34·5
75	71·7	34·2

We can plot these points graphically in the usual way, measuring x, the father's height, horizontally, and y, the average son's height, vertically. The points we obtain are represented by black circles in Fig. 3.1.

But we notice that if we subtract half the father's height from the son's, obtaining $y - \frac{1}{2}x$, this difference is nearly constant and equal to about 34·8 inches (see Table 3.1). Naturally in a limited sample taken from the general population we should not expect absolute constancy in this difference. There will inevitably be random fluctuations in the values, especially at the ends of the range, where the number of specially tall or specially short fathers will be small. We shall have

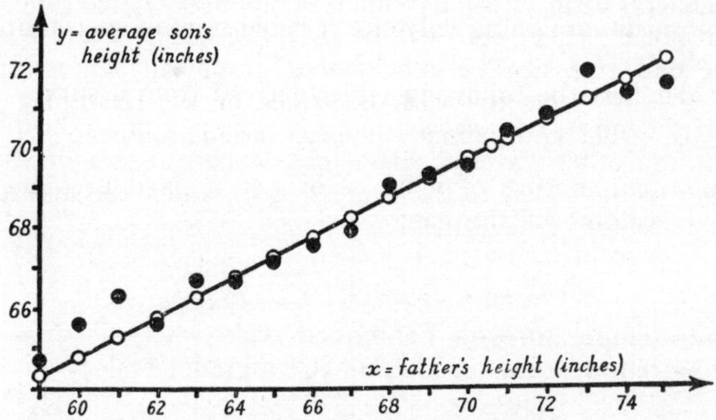

Fig. 3.1—The relation between the height of fathers and the average height of their sons

more to say about this later (Section 21.10). On the whole, however, the relation $y - \frac{1}{2}x = 34{\cdot}8$ is reasonably accurate for all the values of x and y, and we can write this as

$$y = \tfrac{1}{2}x + 34{\cdot}8 \quad . \quad . \quad . \quad . \quad (3.8)$$

We can also plot on the graph the "theoretical" values which would be obtained if the relation $y = \frac{1}{2}x + 34{\cdot}8$ was absolutely accurate. Thus when $x = 60$, $y = 64{\cdot}8$. These values are represented by the white circles, and it will be seen that they lie on a straight line. This is true even for fractional values of x; when $x = 70{\cdot}6$, $y = 70{\cdot}2$, and this also gives a point on the line. This line is a picture or "graph" of the polynomial $\frac{1}{2}x + 34{\cdot}8$, in the sense that every pair of values x and y for which $y = \frac{1}{2}x + 34{\cdot}8$ gives when plotted a point on this line, and all points on the line represent values of x and y for which $y = \frac{1}{2}x + 34{\cdot}8$. From this graph (Greek *graphé*, "writing" or "picture") we can see at a glance how the expression $\frac{1}{2}x + 34{\cdot}8$ behaves for various values of x, and we can use it to predict the probable height y of the son, when we know the height x of the father. The upward slope of the graph as we proceed from left to right shows that the taller the father is, the taller the son will be (on the average). The fact that this is only a mild upward slope shows that the average son's height does not increase as rapidly as that of the father; an increase of 1 inch in the father's height only gives an increase of $\frac{1}{2}$ inch in the son's.

Pearson and Lee showed that a more accurate relation between the height of father and height of son is given by

$$y = 33{\cdot}73 + {\cdot}516\,x.$$

This, however, hardly differs from the formula $y = 34{\cdot}8 + \frac{1}{2}x$ when plotted graphically, at least in the relevant ranges of values of x and y.

The general form of a polynomial of the first degree (one containing x, but not x^2, x^3, x^4, etc.) will be $A + Bx$, where A and B are fixed numbers; and this can be represented graphically by plotting the value of $y = A + Bx$ for various values of x. Examples of such polynomials, either approximate or exact, are as follows:

(i) $x =$ temperature of n moles of gas at fixed pressure P,
$y =$ volume of the gas;
$y = [nR/P]\,x = A + Bx$
where $A = 0$ and $B = nR/P$.

(ii) $x =$ temperature on Fahrenheit scale,
$y =$ temperature on Celsius (Centigrade) scale;

$$y = -\frac{160}{9} + \frac{5}{9}x = A + Bx \text{ where } A = -\frac{160}{9},\ B = \frac{5}{9}$$

(iii) $x =$ year between 1891 and 1901 inclusive,
$y =$ population of England and Wales in year x;
$y = -\,632{,}850{,}000 + 350{,}000\,x$ (approx.).

(iv) $x =$ tension in an elastic thread or spring,
$y =$ length of the thread;
$y = A + Bx$ (Hooke's law)
where $A =$ natural length under no tension
$B = A \times$ modulus of elasticity.

All these expressions will be found to give straight-line graphs: for that reason they are known as *linear* expressions (Section 5.11). The word "linear" or "linear in x" applied to an algebraic expression means that it is a polynomial containing x but no higher power of x, such as x^2 or x^3.

If we plot various graphs of the form $y = A + Bx$ we can see the effect of changes in the constants A and B (Fig. 3.2). When B is positive the graph slopes upward to the right; when negative, downwards. If $B = 0$, the graph becomes simply $y = A$, and is horizontal. Different graphs with the same value of B are parallel, as for example $y = \frac{1}{2}x$, $y = 2 + \frac{1}{2}x$, $y = -1 + \frac{1}{2}x$. Thus B determines the *slope* of the graph, and indeed it is usual to consider B as the measure of the slope. (For the present we shall not attempt to give formal proofs of these assertions, but simply explore the subject pictorially.) The number "A"

is even simpler to interpret: it is the value of y when $x = 0$, i.e. the distance from the "origin" or point $x = 0$, $y = 0$, at which the graph cuts the vertical line $x = 0$. This vertical line is known as the "y-axis", and the corresponding horizontal line $y = 0$ is the "x-axis". Thus the effect of changes in A is simply to move the graph as a whole vertically upwards or downwards, as is shown by a comparison of the three graphs $y = \frac{1}{2}x$, $y = 2 + \frac{1}{2}x$, $y = -1 + \frac{1}{2}x$. When $A = 0$ the graph must pass through the origin $x = 0$, $y = 0$.

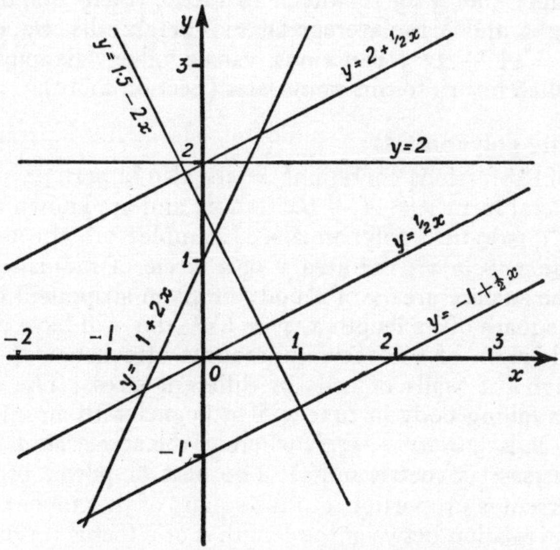

Fig. 3.2—Graphs of the linear functions

(i) $y = \frac{1}{2}x$ **(iv) $y = 2$**
(ii) $y = 2 + \frac{1}{2}x$ **(v) $y = 1 + 2x$**
(iii) $y = -1 + \frac{1}{2}x$ **(vi) $y = 1{\cdot}5 - 2x$**

Besides giving a general picture of the behaviour of the polynomial, graphs are useful in two ways. For each value of x we can read off the corresponding value of y, and for each y the value of x. This reading by eye is not usually very accurate, but quite often it may be good enough, and more convenient than numerical calculation. But it is quite an easy matter with a linear relation to find the value of x for any given y: this simply amounts to solving a linear equation. Thus if $y = A + Bx$, we have by transferring x to the left-hand side and y to the right, $-Bx = A - y$; on dividing by $-B$, $x = (-A/B) + (1/B)y$ which is again a linear relation. Thus our relation between father's height x and average son's height y gives $y = \frac{1}{2}x + 34{\cdot}8$, $x = 2y - 69{\cdot}6$. Here, however, a rather unexpected word of warning is needed. In this relation x is the father's height, and y the average of the heights of sons of such fathers. When we write the relation as $x = 2y - 69{\cdot}6$, there is a definite temptation to suppose that it will not matter much if we take y as the actual son's height, and x as the average height of fathers having such sons. If the relation between height of father and height of son was an exact one, so that the son's height depended only on the father's and on nothing else, we could change the meaning of our symbols in this way without much harm. But, as is common knowledge, one father can have sons of very different heights, and it is only when we average the heights throughout a large population that we get a simple and exact relation. In fact, it turns out that if Y is the son's height, and X the average father's height, the relation is much more like $X = \frac{1}{2}Y + 33{\cdot}3$ than $x = 2y - 69{\cdot}6$, which is a very different graph. We shall return to this point later (Section 21.10).

3.4 Quadratic polynomials

Graphs of expressions containing x^2 are also important. These will be of the general form $y = A + Bx + Cx^2$, and are known as "second degree" or "quadratic" polynomials. Examples are the area y of a square of side x, $y = x^2$; the area y of a circle of radius x, $y = \pi x^2$; in general, the surface area y of a body of given shape will be proportional to the square of its length x, $y = Cx^2$: this will have importance in the consideration of the varying amounts of substances which can diffuse through the walls of cells of different sizes. The distance y travelled by a falling body in time x, if it begins with an initial downward velocity u, is $y = ux + \frac{1}{2}gx^2$, where g, the acceleration of gravity, is $9{\cdot}81$ metres/sec^2 in metric units. The heat or power produced by an electric current is proportional to the square of the current. Another example is the relation between the length x of a foetus (in metres) and its age y (in months) which is $y = 2{\cdot}30 + 9{\cdot}0\,x + 12{\cdot}8\,x^2$ over a certain period, according to Scammon and Calkins (*Growth of the human body in the pre-natal period*, 1929, Univ. Press, Minnesota).

Fig. 3.3 shows the graphs of a number of such polynomials. The equation (i), $y = 29{\cdot}43x - 4{\cdot}905x^2$, represents the height y (in metres) of a body projected upwards with velocity 29·43 metres per second, x being the time in seconds after the body leaves the ground. (This is a particular case of the equation $y = ux + \frac{1}{2}gx^2$ referred to above. However, here y refers to the distance measured upwards, not downwards, so u is the initial upward velocity, 29·43 m/sec, and $\frac{1}{2}g = -4{\cdot}905$, the minus sign showing a downward acceleration.) Equation (ii), $y = 10 + 10x - 4{\cdot}905x^2$, gives the height of a body thrown up with velocity 10 metres/sec and starting from a point 10 metres above the ground. It will be seen that all these graphs are of a very similar shape, called a *parabola*. If the constant C in the formula $y = A + Bx + Cx^2$ is positive, they have somewhere a lowest or *minimum* point, and the graph curves upwards symmetrically on either side of this point. For example $y = 18 - 8x + 2x^2$ has a minimum when $x = 2$.

If C is negative the graph has a peak or *maximum* point, and curves downwards on either side. For example, the graph $y = 29{\cdot}43x - 4{\cdot}905x^2$ has a maximum at $x = 3$; this means that a body projected upwards with velocity 29·43 metres/sec will reach its greatest height after 3 seconds, and will then begin to fall.

We know that if x is large, x^2 is very much larger: when $x = 100$, $x^2 = 10{,}000$, and when $x = -2000$, $x^2 = 4{,}000{,}000$. This shows us at once that if x is very large, either positively or negatively, then in the formula $y = A + Bx + Cx^2$ the values of the terms $A + Bx$ will

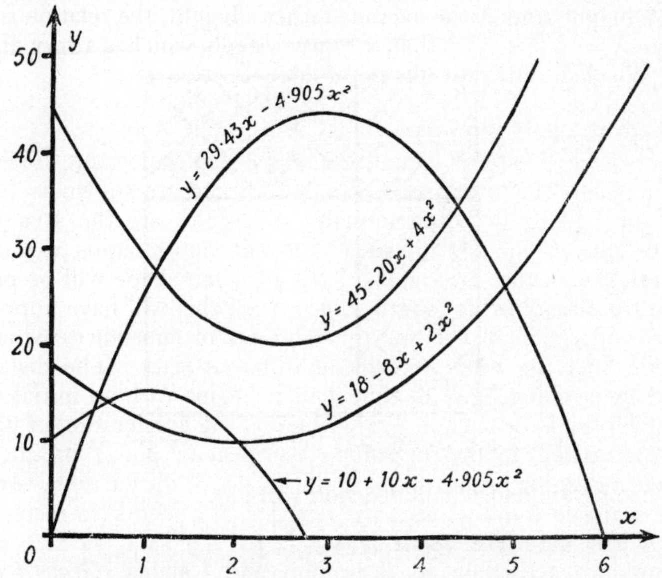

Fig. 3.3—Graphs of the quadratic functions

(i) $y = 29{\cdot}43x - 4{\cdot}905x^2$ (iii) $y = 18 - 8x + 2x^2$
(ii) $y = 10 + 10x - 4{\cdot}905x^2$ (iv) $y = 45 - 20x + 4x^2$

become small in comparison with the term Cx^2. (This would not be so if C was exactly zero, but then we would be dealing with the linear expression $A + Bx$ which we have already discussed, and not with a quadratic one.) The terms $A + Bx$ may actually be large in themselves, but relatively speaking they will be negligible, and in that sense we may say that $y = Cx^2$ approximately for large x. But x^2 must always be positive, so that y must be positive for large values of x if C is positive, and negative if C is negative. This explains the behaviour of the graph at the left- and right-hand ends of the scale. To study the behaviour of the graph at intermediate points a special trick is used, that of "completing the square". Suppose we wanted to investigate the behaviour of the expression $y = 4x + x^2$. This can be written as $y = 2x + 2x + x^2$, and can be pictured as in Fig. 3.4 where x^2 is represented as the area of a square of side x, and the two products $2x$ as rectangles of sides 2 and x. This figure suggests that if we add the unshaded square of area $2^2 = 4$ we shall get a square of side $(2 + x)$; that is confirmed by the identity $(2 + x)^2 = 4 + (4x + x^2)$ which we get by direct multiplication. The original quantity $y=4x+x^2$ can therefore be written as $y=(2+x)^2-4$. Now $(2 + x)^2$ being a perfect square is always positive, except when $x = -2$ when it is zero; and therefore y is greater than -4 except when $x = -2$, $y = -4$. Thus this expression has the minimum value -4 and the minimum occurs when $x = -2$.

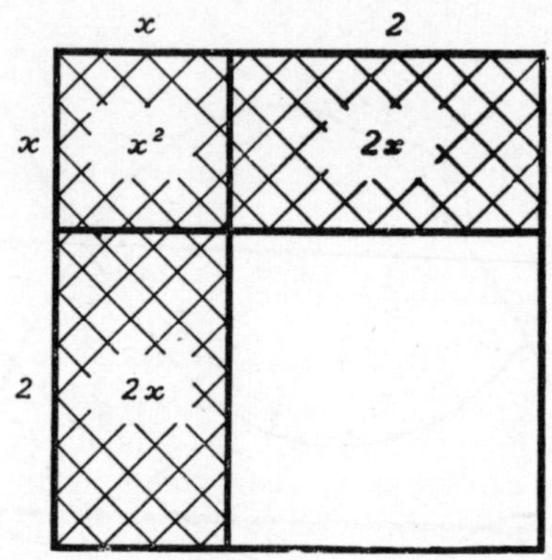

Fig. 3.4—"Completing the square" for $4x + x^2$

The general quadratic expression will be

$$y = A + Bx + Cx^2 = C\left(\frac{A}{C} + \frac{B}{C}x + x^2\right)$$

where A, B and C are constants. A similar argument suggests that this should be compared with the square

$$\left(x + \frac{B}{2C}\right)^2 = \left(\frac{B}{2C} + x\right)^2 = \frac{B^2}{4C^2} + \frac{B}{C}x + x^2$$

for when this is multiplied by C the terms in x and x^2 agree exactly with those in the expression for y. The constant term is different; thus

$$y = C\left(x + \frac{B}{2C}\right)^2 - C\frac{B^2}{4C^2} + C\frac{A}{C}$$

On bringing the last two fractions to a common denominator $4C^2$ (by multiplying numerator and denominator by $4C$) this becomes

$$y = C\left(x + \frac{B}{2C}\right)^2 - C\left(\frac{B^2 - 4AC}{4C^2}\right)$$

$$= C(x - \alpha)^2 - C\beta$$
$$= C[(x - \alpha)^2 - \beta] \quad . \quad . \quad . \quad (3{\cdot}9)$$

where we have written α for $-B/2C$ and β for $(B^2 - 4AC)/4C^2$.

Now $(x - \alpha)^2$ is a perfect square, and is therefore positive except when $x = \alpha$, when it is zero. Thus $C(x - \alpha)^2$ is positive when C is positive, and negative when C is negative, but zero when $x = \alpha$. It follows that if C is positive $y = C(x - \alpha)^2 - C\beta$ is always greater than $- C\beta$ except when $x = \alpha$, so that y has the minimum value $- C\beta$ and takes this minimum when $x = \alpha$. Similarly if C is negative y takes the maximum value $-C\beta$ when $x = \alpha$.

PROBLEMS

Find algebraically the maximum or minimum values of the following expressions, and the values of x at which they occur: (1) $3 + x^2$, (2) $18 - 8x + 2x^2$, (3) $45 - 20x + 4x^2$, (4) $29{\cdot}43x - 4{\cdot}905x^2$.

The expression $y = C(x - \alpha)^2 - C\beta$ also shows us what the effect is on the graph of altering α, β and C. If we alter $C\beta$ we merely add a certain quantity on to y, i.e. we move the whole graph bodily upwards or downwards. If we change α we change the position of the maximum or minimum $x = \alpha$, and simply shift the graph horizontally. On the other hand if we change C we change the shape of the graph: the greater C is, positively or negatively, the more rapidly the graph curves away from its maximum or minimum, and the "sharper" is its tip.

This trick of completing the square also enables us to find the value of x when we are given the value of y. For to solve the equation

$$y = C(x - \alpha)^2 - C\beta$$

we write it as

$$C(x - \alpha)^2 = y + C\beta,$$

or, after division by C,

$$(x - \alpha)^2 = y/C + \beta.$$

There are now three cases to consider. If $(y/C + \beta)$ is negative this is impossible, as $(x - \alpha)^2$ cannot be negative, so that the graph cannot take this value of y anywhere. This will happen if y is greater than the maximum value, or less than the minimum. If $(y/C + \beta)$ is positive then $(x - \alpha)^2 = (y/C + \beta)$ if $(x - \alpha)$ is either the positive or negative square root of $(y/C + \beta)$; i.e. $(x - \alpha) = \pm \sqrt{(y/C + \beta)}$. (The mathematical sign $\sqrt{u}$ is usually interpreted to mean the *positive* square root, the negative one being denoted by $-\sqrt{u}$, e.g. $\sqrt{4} = 2$, but both 2^2 and

$(-2)^2$ are equal to 4.) This gives us two possible values of x, say $x' = \alpha + \sqrt{(y/C + \beta)}$, and $x'' = \alpha - \sqrt{(y/C + \beta)}$. As before, $\alpha = -\frac{1}{2}B/C$ and $\beta = (B^2 - 4AC)/4C^2$, so that the values x', x'' are

$$x', x'' = [-B \pm \sqrt{(4yC + B^2 - 4AC)}]/2C \qquad (3.10)$$

These are the values of x for which $y = A + Bx + Cx^2$. ($\pm$ means "plus or minus".) Thus a ball thrown in the air attains a given height y at times x'', x', once when rising, once when falling.

The last case occurs when $(y/C + \beta) = 0$, i.e. at a maximum or minimum point. The equation $(x - \alpha)^2 = (y/C + \beta) = 0$ then has only one solution, $x = \alpha = -\frac{1}{2}B/C$. This is in fact covered by our general formula $x = [-B \pm \sqrt{(4yC + B^2 - 4AC)}]/2C$; it is the special case when the square root is zero. If we throw a ball in the air it will be at its highest point for only one instant of time.

PROBLEMS

Find the values of x for which $y = 40$ when (5) $y = 3 + x^2$, (6) $y = 18 - 8x + 2x^2$, (7) $y = 45 - 20x + 4x^2$, (8) $y = 29{\cdot}43x - 4{\cdot}905x^2$.

The device of completing the square often helps us to factorize $A + Bx + Cx^2$ by using the formula $u^2 - v^2 = (u + v)(u - v)$: for example

$$\begin{aligned} x^2 + 6x + 5 &= (x + 3)^2 - 4 \\ &= (x + 3)^2 - 2^2 \\ &= (x + 3 + 2)(x + 3 - 2) \\ &= (x + 5)(x + 1) \end{aligned}$$

If follows that $x^2 + 6x + 5 = 0$ when either $x + 5 = 0$ or $x + 1 = 0$, but not otherwise [law (7), p. 35], i.e. the roots of the equation $x^2 + 6x + 5 = 0$ are $x = -5$ and -1. In general, by (3.9),

$$\begin{aligned} A + Bx + Cx^2 &= C[(x - \alpha)^2 - (\sqrt{\beta})^2] \\ &= C(x - \alpha + \sqrt{\beta})(x - \alpha - \sqrt{\beta}) \end{aligned} \qquad (3.11)$$

Therefore the equation

$$A + Bx + Cx^2 = 0$$

has roots

$$x = \alpha \pm \sqrt{\beta} = \frac{-B \pm \sqrt{B^2 - 4AC}}{2C} \quad . \quad . \quad (3.12)$$

which can also be got by putting $y = 0$ in (3.10). If β is negative the method fails; the equation then has no roots, for as we have shown above, $A + Bx + Cx^2 = y$ has no roots when $y/C + \beta$ is negative (put $y = 0$). Also $A + Bx + Cx^2$ has then no factors, for if, say,

$x - h$ was a factor, giving x the value h would make this factor zero, and hence also $A + Bx + Cx^2 = 0$, i.e. h would be a root of this equation, which is impossible.

PROBLEMS

Find the factors, if any, of (9) $-4 + x^2$, (10) $6 + 5x + x^2$, (11) $3 + x^2$, (12) $18 - 8x + 2x^2$, (13) $45 - 20x + 4x^2$, (14) $3x - 5x^2$.

3.5 Cubic polynomials

Polynomials containing x^3, known as "third degree" or "cubic" expressions, may also occur. For example, the mass of a cube of density ρ and side x will be ρx^3. In general if we have a number of objects of similar shape and made of the same material, then the mass y will be

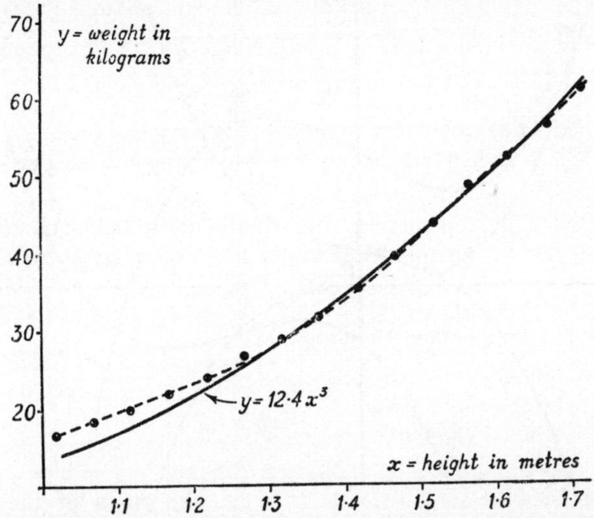

Fig. 3.5—The relation between height and weight for London schoolgirls

Continuous line: $y = 12{\cdot}4x^3$

Dotted line: $y = -670{\cdot}2 + 2136x - 2492x^2 + 1280x^3 - 238x^4$

proportional to the cube of the length x, i.e. $y = Dx^3$. It is of interest to ask how far this relation holds among human beings.

Fig. 3.5 shows the average weight (in kilograms) of a sample of London schoolgirls plotted against their height in metres (*L.C.C. report on the heights and weights of school pupils in the County of London in 1949*). More precisely the children have been divided into groups, of heights from ·995 to 1·045, 1·045 to 1·095, and so on, and the weights of all children in each group averaged and plotted against the central height of the group, i.e. 1·02, 1·07, etc. This grouping will slightly distort the curve, but that does not greatly matter in the present connection. The continuous line is the curve $y = 12{\cdot}4x^3$, and it will be seen that it fits the data well above a height of 1·3 metres, but is rather low below that height. Some such effect might indeed be expected, since the shape and relative proportions of bones, muscles, and other constituents change during growth. We shall explain later, using the theory of statistics, how a more accurate fit can be obtained; here we merely demonstrate that the dotted line, which represents the fourth degree polynomial, $y = -670{\cdot}2 + 2136x - 2492x^2 + 1280x^3 - 238x^4$ gives a much better approximation.

There are three morals which we can draw from this. The first is the danger of extrapolation. If we had only the portion of the curve above 1·3 metres, we might suppose that height and weight were always related by the equation $y = 12{\cdot}4x^3$, and extend this curve to

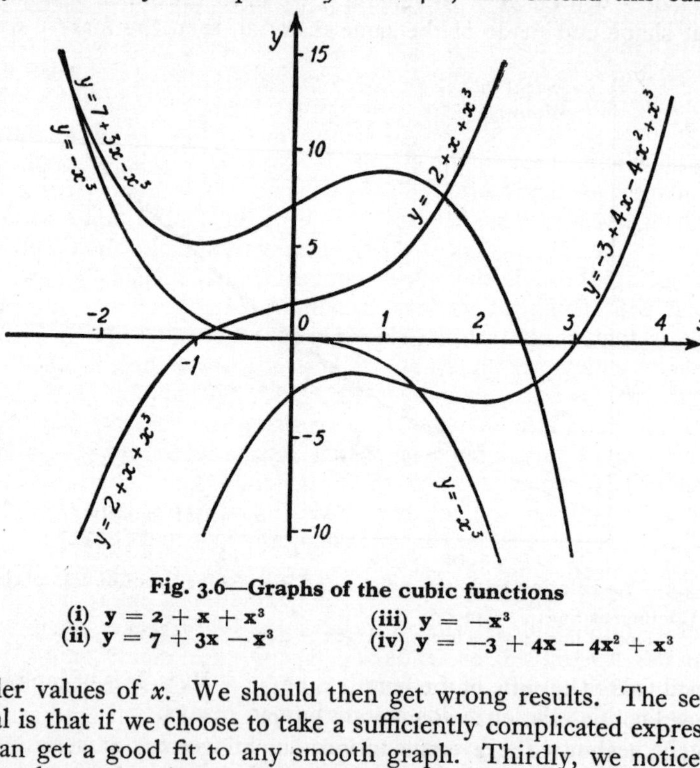

Fig. 3.6—Graphs of the cubic functions

(i) $\mathbf{y = 2 + x + x^3}$ **(iii)** $\mathbf{y = -x^3}$
(ii) $\mathbf{y = 7 + 3x - x^3}$ **(iv)** $\mathbf{y = -3 + 4x - 4x^2 + x^3}$

smaller values of x. We should then get wrong results. The second moral is that if we choose to take a sufficiently complicated expression, we can get a good fit to any smooth graph. Thirdly, we notice that while the expression $-670{\cdot}2 + 2136x - 2492x^2 + 1280x^3 - 238x^4$ represents the weight pretty accurately, at least in our range of heights, it has no very simple or obvious interpretation, unlike the less accurate relation $y = 12{\cdot}4x^3$. Thus, although purely mathematical accuracy undoubtedly has its uses, it must be sought after with discretion, as otherwise it may sometimes obscure the real issues.

The general form of a cubic polynomial will be $y = A + Bx + Cx^2 + Dx^3$. We show the graphs of certain cubics in Fig. 3.6. They are all sinuous curves with two bends, rather like the letter *S* placed on its side. Some of them rise steadily throughout, like $y = 2 + x + x^3$, others rise up to a point, then fall for a time, and then rise again, like $y = -3 + 4x - 4x^2 + x^3$. Some fall throughout, and others fall, rise, and fall. Since for large positive or negative values of x the quantity x^3 is much larger in magnitude than x or x^2, we can say that for large x the curve $y = A + Bx + Cx^2 + Dx^3$ will behave in a similar way to $y = Dx^3$. If D is positive, then y will be negative for large negative values of x, and positive for large positive x. If D is negative the reverse will be true. Thus, unlike a quadratic, a cubic can never have an ***absolute*** maximum or minimum. It can, however, have peaks and valleys such that the value of y is greater at the peak and smaller at the bottom of the valley than it is for any nearby value of x. For example, $y = -3 + 4x - 4x^2 + x^3$ has a peak when $x = \frac{2}{3}$, $y = -\frac{49}{27}$, and a valley at $x = 2$, $y = -3$. Such points are called ***local*** maxima and minima.

Such local maxima are not to be despised; they may for example have some significance in the theory of evolution. A species may be better adapted to its environment in its natural condition than if any small change was made. A large change might on the other hand be of benefit to the species, but it may not be able to make such a change without going through the disadvantageous intermediate conditions: and so it will remain at its local maximum of adaptation. A simple example is that of the rabbit: this is well adapted to life in Britain, and even better adapted to regions of Australia where there is a plentiful supply of food and few enemies. This degree of adaptation could be measured in various ways—the simplest being the density, or number of rabbits per square kilometre, which results or would result from colonization by rabbits. Thus this adaptation has local maxima in Britain and Australia. The species also inhabits Europe and Asia, with doubtless various points of maximum and minimum success. But the Promised Land of Australia is separated from other colonized countries by sea, where, of course, the adaptation is zero, and which presents an impassable barrier to the species. Only by the intervention of man has Australia been colonized, and it presumably would have remained free of rabbits had man not interfered: i.e. the species would have been confined to those regions to which it was locally rather than absolutely best fitted. A similar example, in which the variation is genetic rather than geographical, occurs in the dachshund breed of dogs with long jaws. The dachshund type is well adapted, if only because its odd shape arouses curiosity and interest among human beings and keeps the breed going. The normal types of dog also prosper, partly for similar reasons. But it seems that the lower and upper jaws are lengthened by different independent genes, so that an intermediate breed in which some but not all the dogs had long jaws would be sure to contain some dogs who had the genes to lengthen the lower jaw but not the upper one, and vice versa, and such dogs would be in a very sad state. Thus both the normal and dachshund types represent local maxima of adaptation, with a minimum somewhere between. In view of this it is rather a puzzle to know how the dachshund type ever arose in the first place: it would appear to require a very happy accident for the mutations necessary for the lengthening of both upper and lower jaws to have appeared simultaneously.

There is no simple device like that of completing the square which can be used to find the position of the maximum and minimum of a cubic curve, the factors, or the value of x for a given value of y: so that discussion of these points will have to be deferred until Chapter 12 where we consider more powerful methods of tackling these problems. We shall, however, note that for certain values of y the equation $y = A + Bx + Cx^2 + Dx^3$ may be satisfied for three distinct values of x; that is, the graph may attain the same height y at three distinct points. This will be seen from Fig. 3.6 to be true of the curve $y = 7 + 3x - x^3$ for the value $y = 7$. In fact, in this case it is easy to determine the particular values of x, for if $7 + 3x - x^3 = 7$, then $3x - x^3 = 0$, or in factors $x(3 - x^2) = x(\sqrt{3} - x)(\sqrt{3} + x) = 0$. One of these factors must be zero, i.e. either $x = 0$, or $x = \sqrt{3}$, or $x = -\sqrt{3}$.

From cubic curves we may go on to fourth degree and even higher ones. These can be of even more complicated shape: for example, suppose we wish to investigate the curve

$$y = 10 + 12x - 5x^2 - \tfrac{3}{2}x^3 + \tfrac{1}{2}x^4 \quad . \quad . \quad (3.13)$$

For small values of x we calculate y directly, obtaining the table:

$x =$	−6	−5	−4	−3	−2	−1	0	1	2	3	4	5	6
$y =$	730	325	106	10	−14	−5	10	16	10	1	10	70	226

(*Note.* A quick way of calculating $y = A + Bx + Cx^2 + Dx^3 + Ex^4$ is to multiply E by x, add D, multiply again by x, add C, multiply by x, add B, multiply by x and add A. Thus to find y in (3.13) when $x = 2$, calculate in turn $\frac{1}{2} \times 2 - 3 = -\frac{1}{2}, (-\frac{1}{2}) \times 2 - 5 = -6, (-6) \times 2 + 12 = 0, 0 \times 2 + 10 = 10 = y$.)

The table shows "valleys" near $x = -2$ and $x = -3$ and a "peak" near $x = 1$. Since x^4 grows much more rapidly than x, x^2 or x^3 when x becomes large, we see that for large positive or negative values of x the term $\frac{1}{2}x^4$ dominates: the lower degree terms are small in comparison, even if large in absolute value. For example, if x is 7, we have $\frac{1}{2}x^4 = 1200{\cdot}5$, whereas $-\frac{3}{2}x^3 = -514{\cdot}5$ and $-5x^2 = -245$. When $x = 20$, $\frac{1}{2}x^4 = 80{,}000$, whereas $-\frac{3}{2}x^3 = -12{,}000$: the bigger x is, the more $\frac{1}{2}x^4$ dominates. Thus beyond the table given above the curve will resemble $y = \frac{1}{2}x^4$, i.e. y will be positive and increase rapidly as x increases. The complete curve will be rather like a distorted letter W.

PROBLEMS

Investigate the following curves, drawing rough graphs:

(1) $y = 1 + x^3$, (5) $y = x^2 + x^4$,
(2) $y = x^2 + x^3$, (6) $y = 4x^2 - x^4$,
(3) $y = x^2 - x^3$, (7) $y = x + x^2 + x^3 + x^4$,
(4) $y = 3x - x^3$, (8) $y = -20 + 64x - 20x^3 + x^5$.

3.6 Functions

At this point it is convenient to introduce a new name and notation, which we shall frequently use later on. If x and y are two variable quantities such that the value of y depends on the value of x, then y is said to be a *function* of x. Thus the yield y of wheat from a given field depends on the amount of rainfall x, the weight of an animal depends on the amount of food it eats, and the speed of a chemical reaction depends on the temperature. The cases with which we are mostly concerned will be those in which y is a function of x only, and not of any other variables—or at any rate those cases in which the effect of any other variable quantities can be neglected for the purposes of the problem in hand. For example, the volume of a cube, y, depends only on the length of its side, x, and on nothing else: given x we know the value of y exactly. If the temperature is kept fixed, then the osmotic pressure of a solution depends only on its concentration, and not on the shape of the vessel containing it, or any other such factor. The osmotic pressure, we say, is a function of the concentration. If we study one particular animal, then its weight y is a function of its age x, for if we specify any particular age then the weight is also determined. Henceforth, whenever we use the word "function" without qualification we shall mean a relationship of this type; that is to say, whenever we know the value of x we can find y. This of course is a very different meaning of the word "function" from its use in physiology.

Many functional relationships can be represented by mathematical formulas. If x represents the side of a cube, and y its volume, then $y = x^3$. If x represents the concentration of a solution, and y its osmotic pressure, then for sufficiently small concentrations y is proportional to x, so that $y = Bx$ where B is constant. If $y = \frac{1}{2}x + \sqrt{1 + x^2}$ then y is a function of x. This is also true if y is a polynomial in x, such as $y = 2 + 15x + 3x^2$.

It is customary to indicate that y is a function of x by writing "$y = f(x)$". [This is read as "y equals fx" or "y equals f of x".] The statement "$y = f(x)$" means therefore that the value of y depends on that of x, but that we are not concerned with writing out in detail the exact relationship: we represent it simply by the letter f. Thus instead of writing out a long expression such as $y = [1 + 3x + \sqrt{(x^2 - 1)}]x^3$ every time it occurs, we can agree to write it simply as $y = f(x)$.

Another relationship, such as $y = 29{\cdot}43x - 4{\cdot}905x^2$, could be written $y = F(x)$, and still further functions as $y = g(x)$, or $y = \phi(x)$ and so on.

Any function $y = F(x)$ occurring in practice, whether it has an algebraic form or not, can be represented graphically by plotting the value of y for each value of x. Thus the relation

$$y = F(x) = 29{\cdot}43x - 4{\cdot}905x^2 \quad . \quad . \quad . \quad (3.14)$$

means that y is the height at time x of a cricket ball thrown up at a speed of 29·43 m/sec. Its graph is shown in Fig. 3.3. The height of the ball after 2 seconds, i.e. the value of $y = F(x)$ when $x = 2$, is written $F(2)$: by putting $x = 2$ in (3.14) we get

$$F(2) = 29{\cdot}43 \times 2 - 4{\cdot}905 \times 2^2 = 39{\cdot}24$$

Similarly the height after $(z^2 + 1)$ seconds is written

$$F(z^2 + 1) = 29{\cdot}43(z^2 + 1) - 4{\cdot}905(z^2 + 1)^2$$

and is got by replacing x in (3.14) by $(z^2 + 1)$.

The above notation $y = F(x)$ is the standard and customary one, and will be found in most books. But it seems worth noting that it can be slightly simplified. In "$F(x)$" the left-hand bracket often sticks to the letter F closer than a brother, whatever we may put inside the bracket, e.g. $F(a^2b^2c^2)$. It therefore does no work which cannot equally well be done by the letter F itself, and could easily and conveniently be dropped as superfluous. Thus it would be perfectly possible to write "$y = Fx)$" for "y is the function F of x"; or, even more simply, we may on many occasions reduce the bracket to a mere point, writing "$y = Fx.$". Similarly, $F2)$ or $F2.$ will stand for what is customarily written as $F(2)$, and $Fx^2)$ or $Fx^2.$ for the function of x^2. This slight simplification enables us to dispense with not only one bracket, but often with several brackets at a time, making the formulas considerably more legible. Thus just as the square of y is written y^2, so the square of $Fx)$ can be written $Fx)^2$, whereas in the conventional notation it has to be written $[F(x)]^2$, involving three more brackets. As the notation $Fx)$ or $Fx.$ is rather unusual in appearance we shall keep to the standard usage $F(x)$ in this book: but we invite readers to give serious consideration to the merits of "$Fx.$".

We can also have functions of two or more variables. The phrase "z is a function of x and y", or in symbols, "$z = F(x,y)$" or "$z = Fx,y)$" means that when we know the values of x and y we can find that of z. Thus the distance to which a cricket ball is thrown is a function of both the velocity and inclination of the throw: if projected either horizontally or vertically it will not go very far. The volume of a given quantity of gas is a function of both pressure and temperature. If $z = x^2 + xy + y^2$, then z is a function of x and y.

It is harder to represent such a two-variable function graphically. If $z = F(x,y)$, then ideally we can represent x and y by two distances or "co-ordinates" measured in perpendicular directions in a horizontal plane and z as the height above this plane (Fig. 3.7). The values of z for varying x and y will then form some kind of surface, which will be a three-dimensional representation of the function F. Unfortunately such a surface is difficult to construct and cumbersome to use, and it is

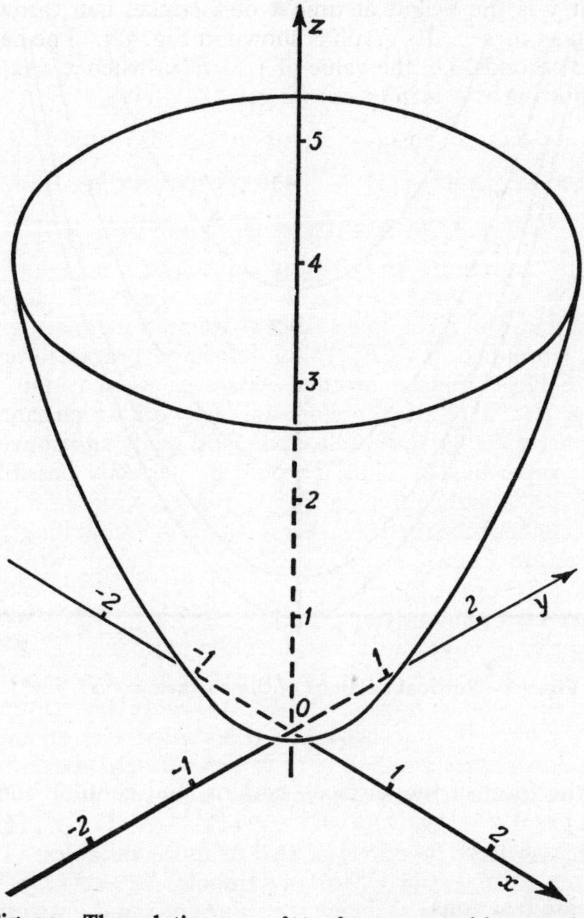

Fig. 3.7—The relation $z = x^2 + y^2$ represented by a surface

usual to draw instead on a single diagram a number of plane sections of this surface. One way is to select a certain number of values of y, and for each of these values draw the graph of z against x (Fig. 3.8). This corresponds to taking a number of parallel vertical sections through the surface. Another method is to take horizontal sections, plotting the curves in the horizontal plane at which z has a given value (Fig. 3.9).

This is precisely what we do when we plot the contour lines of a mountain. Unfortunately even this device fails for a function of three variables, such as $u = F(x, y, z)$. Considerable ingenuity may be needed to provide a really suitable diagrammatic representation of such a function.

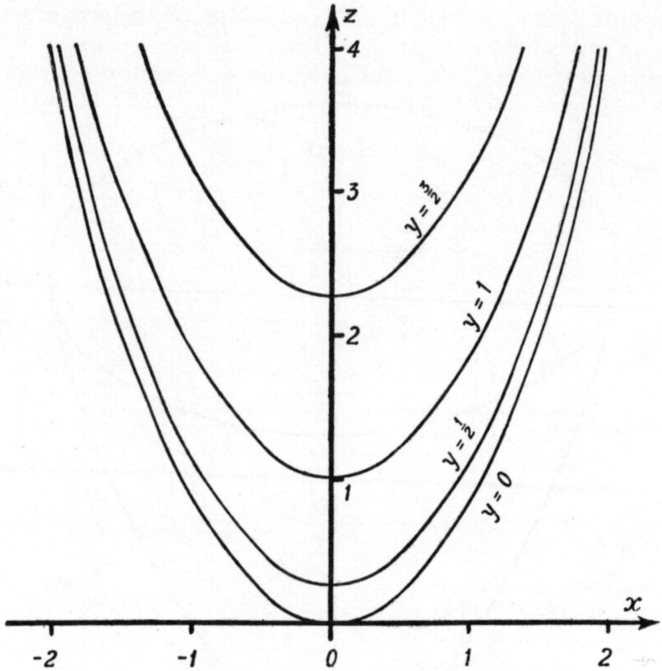

Fig. 3.8—Vertical sections of the surface $z = x^2 + y^2$

PROBLEMS

Sketch the form of the surface, sections, and contour lines for the functions (1) $z = x + 2y$, (2) $z = x - y$, (3) $z = x^2 - y^2$, (4) $z = xy$, (5) $z = x + 1/x - y^3$.

3.7 Algebraic fractions

Besides polynomials we may of course have many other types of dependence. Since algebraic symbols represent ordinary numbers, we can have fractions such as $y = (x + 1)/(x - 1)$, $y = (x^2 + 1)/x$, and so on. Thus when $x = 2$, $(x + 1)/(x - 1) = 3/1 = 3$. The one caution which must be observed is that no fraction may have zero denominator: $(x + 1)/(x - 1)$ is not defined when $x = 1$.

EXAMPLES

(1) Boyle's law: for a gas, if $x =$ the volume and $p =$ the pressure at a fixed temperature, then xp constant $= K$ (say), or $p = K/x$. The graph $p = 1/x$ is shown in Fig. 3.10; the curve is known as a "rectangular hyperbola". It will be observed that as x approaches 0, p increases in magnitude without any limit, or "bound" (to use the proper technical phrase). Thus when $x = \cdot 01$, $p = 100$, when $x = \cdot 0001$, $p = 10{,}000$ and when $x = \cdot 000001$, $p = 1{,}000{,}000$. If $x = -\cdot 001$, $p = -1000$, and

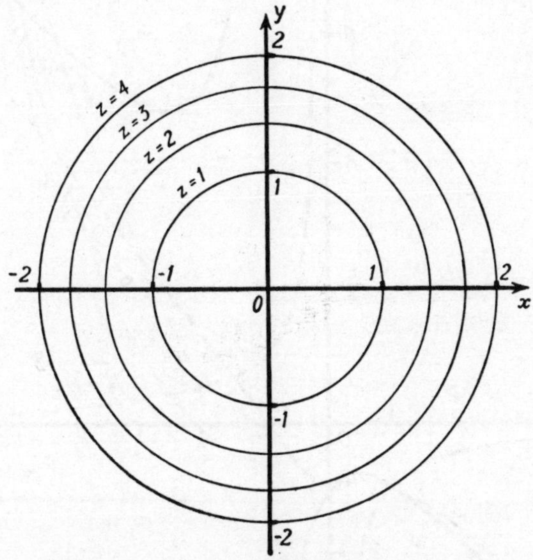

Fig. 3.9—Contours of the surface $z = x^2 + y^2$

so on. We say that p becomes "infinity" (symbol ∞) when x becomes zero: but that is merely to be taken as a shorthand way of saying that as x nears 0, p becomes large, as large as we like. The interpretation of this would be that by sufficiently increasing the pressure on a given mass of gas we could make it contract to as small a volume as we liked. At the other end of the scale if x becomes large, p becomes very small: this is in sharp contrast to the behaviour of a polynomial function, which must always become large for large x.

We see also that Boyle's law cannot hold exactly for very great pressures, for if it did then we could reduce the volume x to as near zero as we wish, and that we know to be impossible. In fact, by sufficient compression the gas will be turned into a liquid—the change may be either discontinuous, accompanied by condensation, or else if the temperature is above the critical point it will be continuous with no precise point dividing the gaseous and liquid states.

(2) Baggally's theory of psychological forces. According to a theory put forward by Baggally (*Inter. J. Psycho-Anal.*, **28**, 1947), if we have two pleasures of a similar kind, measured by quantities x and y, then

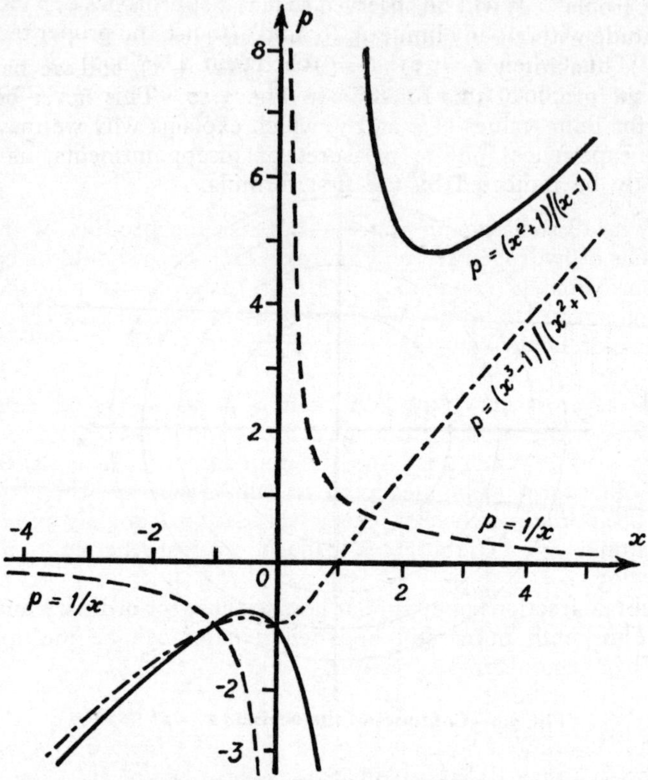

Fig. 3.10—Graphs of the rational functions
(i) $p = 1/x$
(ii) $p = (x^2 + 1)/(x - 1)$
(iii) $p = (x^3 - 1)/(x^2 + 1)$

they combine together to give a resultant pleasure $p = (x^2 + y^2)/(x + y)$. Unfortunately it is far from clear how we can measure the intensity of a pleasure, so that the quantities x and y must remain at present rather hypothetical. But we can investigate the consequences of the theory in a general way.

Let us put $y = -1$, i.e. a unit amount of distaste, and plot the graph of $p = (x^2 + 1)/(x - 1)$ (Fig. 3.10). When x is just slightly greater than 1, this becomes very large and positive, whereas when x

is just below 1, p is large and negative: i.e. we should get the greatest amount of pleasure or unpleasure under the influence of two opposite desires which nearly balance. Now this does often seem to happen; it may be related to the fact that we get a great deal of satisfaction from solving a problem or confronting a situation which we can master, but not too easily, while the bitterest disappointments are those where we are just beaten. If the desires x and y are of *different* kinds, then Baggally supposes the resultant pleasure to be given by $p = (x^3 + y^3) \div (x^2 + y^2)$. Taking $y = -1$, $p = (x^3 - 1)/(x^2 + 1)$, and we have also plotted the graph of this function in Fig. 3.10. This never becomes infinite for finite values of x and y, which explains why we may never actually experience infinite pleasures or disappointments, as would apparently be indicated by the first formula.

(3) The dissociation of water. In water the product of the concentrations of hydrogen H^+ and hydroxyl OH^- ions should be constant (and approximately equal to $1/10^{14}$). If therefore we plot one concentration against the other we should obtain a rectangular hyperbola, exactly as for Boyle's law.

It is easy to predict the behaviour of a fractional expression for large values of the variable. Consider, for example, $y = (2+3x+5x^2) \div (1 + 72x + x^3)$. When x is large, the numerator will be approximately $5x^2$, the other terms being negligible in comparison, and the denominator will be approximately x^3. Thus y will be approximately $5x^2/x^3 = 5/x$. For example, if $x = 100$, $y = \cdot 049943$, against the approximation $5/100 = \cdot 05$.

Algebraic fractions obey similar laws to those for ordinary numerical ones. Thus both numerator and denominator can be multiplied or divided by any non-zero expression:

$$\frac{1}{x+1} = \frac{5}{5x+5} = \frac{(x-1)}{(x-1)(x+1)} = \frac{x-1}{x^2-1}$$

To add or subtract fractions, bring them to a common denominator. Thus suppose we wish to evaluate

$$\frac{3}{(x+1)^2} + \frac{1}{x-2} - \frac{5}{x(x+1)}$$

The lowest common denominator can be written as $x(x+1)^2(x-2)$, since it must contain $(x+1)^2$, $(x-2)$ and $x(x+1)$ as factors. The given expression can therefore be written

$$\frac{3x(x-2) + x(x+1)^2 - 5(x+1)(x-2)}{x(x+1)^2(x-2)} = \frac{x^3+10}{x(x+1)^2(x-2)}$$

[Theoretically according to our above rule we might run into trouble if the common denominator becomes zero. *Query:* when does this happen, and what occurs then?]

If we wish to solve an equation containing fractions it must be reduced to a common denominator: thus the equation

$$3/(x+1)^2 + 1/(x-2) - 5/x(x+1) = 0$$

becomes

$$(x^3+10)/x(x+1)^2(x-2) = 0.$$

Multiplying both sides of this equation by $x(x+1)^2(x-2)$, we get $x^3 + 10 = 0$, or $x = -\sqrt[3]{10}$.

3.8 Algebraic division

We have already seen that it is possible to add, subtract and multiply algebraic expressions, obtaining new algebraic expressions. For example

$$(2x^2+3y+1) + (4xy-5y+4) = 2x^2+4xy-2y+5$$
$$2^x(x+2) \times 2^y(x-2) = 2^{x+y}(x^2-4)$$

and so on. In particular, if we add, subtract, or multiply polynomials, we obtain new polynomials.

We now consider division, but only for polynomials. This is important from the practical point of view in helping to simplify expressions. If the answer to a problem is

$$(6x^3+x^2-8x+6)/(3x^2-4x+2),$$

it is not immediately obvious that this is the same as $(2x+3)$. But there are also theoretical reasons for studying division; for example, it helps to construct a theory of interpolation.

Compare the following division sums:

$$\begin{array}{r} 212 \\ 231\overline{)48995} \\ \underline{462} \\ 279 \\ \underline{231} \\ 485 \\ \underline{462} \\ \underline{\underline{23}} \end{array} \qquad \begin{array}{r} 2x^2 + x + 2 \\ 2x^2+3x+1\overline{)4x^4+8x^3+9x^2+9x+5} \\ \underline{4x^4+6x^3+2x^2} \\ 2x^3+7x^2+9x \\ \underline{2x^3+3x^2+x} \\ 4x^2+8x+5 \\ \underline{4x^2+6x+2} \\ \underline{\underline{2x+3}} \end{array}$$

The first is an ordinary division of 48995 by 231, and the second an algebraic division of one polynomial, $4x^4+8x^3+9x^2+9x+5$, by another, $2x^2+3x+1$. They are put side by side to emphasize the essential similarity of the procedure. (In this particular case the similarity is so close that precisely the same figures occur throughout the two sums, e.g. the divisor is **231** on the left, and $\mathbf{2x^2 + 3x + 1}$ on the right, and so on. That will not be true in general.) There is, however, an important difference between the objects of these two sums. In the arithmetic division the process consists in subtracting the divisor, 231, from the "dividend" or number 48995 to be divided, as many times as possible, so as to leave the least positive (or zero) remainder. In this case we can take 231 away 212 times, leaving remainder 23, so that

$$48995 - 212 \times 231 = 23$$

or

$$48995 = 212 \times 231 + 23,$$

which is an example of the general rule

$$\text{Number divided} = \text{quotient} \times \text{divisor} + \text{remainder} \qquad (3.15)$$

In an algebraic division we aim to reduce the remainder to the smallest *degree* possible (not to the smallest magnitude; for if the remainder contains x we will not know how large it is numerically until we know the value of x). Thus in the first step of the algebraic division we try to remove the term in x^4, at the second step the term in x^3, and at the last step the term in x^2, leaving in the remainder only the term involving x and the constant term. To remove the leading term $4x^4$ we multiply the divisor by $2x^2$ (which is the first term in the quotient) giving $2x^2 \times (2x^2 + 3x + 1) = 4x^4 + 6x^3 + 2x^2$, and subtract this from the dividend. This leaves an expression beginning $2x^3 + \ldots$; to eliminate the $2x^3$ we multiply the divisor by x (the second term in the quotient), getting $2x^3 + 3x^2 + x$, and subtract this. Proceeding in this way, we end up with the quotient $2x^2 + x + 2$ and remainder $2x + 3$. The rule (3.15) still holds, so that

$$4x^4 + 8x^3 + 9x^2 + 9x + 5 = (2x^2 + x + 2)(2x^2 + 3x + 1) + (2x + 3).$$

That is to say, if the right-hand side is multiplied out and added together we get the algebraic expression on the left. Another example is the following:

$$\begin{array}{r|l}
 & \text{quotient} = \qquad 3x^2 - 2x - 1 \\
\hline
x^3 - 2x^2 - x + 2) & 3x^5 - 8x^4 \qquad\quad + 12x^2 - 2x - 9 \\
 & 3x^5 - 6x^4 - 3x^3 + 6x^2 \\
\cline{2-2}
 & \quad - 2x^4 + 3x^3 + 6x^2 - 2x \\
 & \quad - 2x^4 + 4x^3 + 2x^2 - 4x \\
\cline{2-2}
 & \qquad\quad - x^3 + 4x^2 + 2x - 9 \\
 & \qquad\quad - x^3 + 2x^2 + x - 2 \\
\cline{2-2}
 & \text{remainder} = \qquad 2x^2 + x - 7
\end{array}$$

In general, if $F(x)$ and $G(x)$ are two polynomials of respective degrees n and m, we can always divide $F(x)$ by $G(x)$, finding a quotient polynomial $Q(x)$ and a remainder $R(x)$ such that

$$F(x) = Q(x)G(x) + R(x) \quad . \quad . \quad . \quad (3.16)$$

The degree of the remainder $R(x)$ will be less than m, the degree of the divisor $G(x)$. Furthermore, provided that m is not greater than n, we begin the division operation by trying to divide the highest degree term in $F(x)$, or x^n term, by the highest degree term in $G(x)$, or x^m term. The first term in the quotient $Q(x)$ will therefore contain x^{n-m}.

PROBLEMS

(1) Divide $2x^2 + 5x + 9$ by $2x + 1$.

(2) Divide $6x^2 + 4$ by $2x + 3$.

(3) Divide $2x^3 + 7x^2 + 3x - 1$ by $x^2 - x + 1$.

(4) Divide $3x^4 + 7x^2 + 1$ by $2x^2 - 5x^2 + 1$.

(5) If n is less than m in the argument above, show that $Q(x) = 0$, $R(x) = F(x)$.

(6) What polynomial gives quotient $(x^2 + 1)$ and remainder $(2x + 1)$ when divided by $(x^2 - 1)$?

(7) A polynomial $x^3 + ax^2 + 2x + b$ is zero when $x = 1$, and divides exactly by $x - 2$ (with zero remainder). Find a and b.

(8) Show that the division of a polynomial by $(x - 3)$ involves in the numerical calculation only multiplications by 3 and additions. Generalize.

(9) What is the remainder when $Cx^2 + Bx + A$ is divided by $x - 3$?

(10) What is the remainder when $x^2 + 5x + 6$ is divided by $x - k$?

(11) When does $x - k$ divide into $Cx^2 + Bx + A$ without remainder?

(12) Show that the algebraic division of a polynomial by $x^2 - 3x - 2$ requires only arithmetic multiplication and addition. Generalize.

3·9 Some properties of polynomials

A number of important consequences follow from (3.16). Let $F(x)$ be any polynomial of degree not greater than n.

$$F(x) = A + Bx + Cx^2 + \ldots + Hx^n \quad . \quad . \quad (3.17)$$

($H = 0$ if the degree is less than n.) Choose arbitrarily a set of n numbers, which we call $a_1, a_2, \ldots a_n$.

(*Note on notation.* A letter with a suffix, like a_1, a_2, or a_n, is considered as a single symbol. Compare "Henry Jones", "Paul Jones", "Bill Jones", etc., naming different individuals in one family, or "Gallus gallus", "Gallus domesticus", etc., naming different species of fowl. "a_2" can be read as "a subscript 2", or "a sub 2", or simply "a 2" if there is no danger of confusion.)

We now prove—

Theorem 3.1. *We can find a set of numbers* α, β, γ, ... *such that*

$$F(x) = \alpha + \beta(x - a_1) + \gamma(x - a_1)(x - a_2) + \dots + \lambda(x - a_1)(x - a_2)\dots(x - a_n) \quad (3.18)$$

in the sense that when the right-hand side is multiplied out it becomes identical with the original polynomial $F(x) = A + Bx + \dots + Hx^n$.

Proof. Take as an example $n = 3$, and $F(x)$ is of degree 3; the general case will follow similarly. We begin by dividing $F(x)$ by $(x - a_1)(x - a_2)(x - a_3)$ which is also a polynomial of degree 3. The quotient is degree $3 - 3 = 0$, i.e. it is a constant δ. The remainder $R_1(x)$ is at most of degree 2. Hence by (3.15)

$$F(x) = \delta(x - a_1)(x - a_2)(x - a_3) + R_1(x) \quad . \quad (3.19)$$

Now divide $R_1(x)$ by $(x - a_1)(x - a_2)$; the quotient will be a constant γ, and the remainder $R_2(x)$ will be at most of the first degree: that is,

$$R_1(x) = \gamma(x - a_1)(x - a_2) + R_2(x) \quad . \quad . \quad (3.20)$$

Divide $R_2(x)$ by $(x - a_1)$; the quotient will be a constant, β, and the remainder being of degree less than 1 will also be a constant α, so that

$$R_2(x) = \beta(x - a_1) + \alpha \quad . \quad . \quad . \quad (3.21)$$

Now substitute the value of $R_2(x)$ given by (3.21) into (3.20), and the value of $R_1(x)$ given by (3.20) into (3.19); we get finally

$$F(x) = \alpha + \beta(x - a_1) + \gamma(x - a_1)(x - a_2) + \delta(x - a_1)(x - a_2)(x - a_3),$$

which is the equation (3.18) for the particular case $n = 3$. For example, suppose we take $F(x) = 2x^2 + x - 7$, and $a_1 = 2$, $a_2 = 1$. Then a division of $2x^2 + x - 7$ by $(x - a_1)(x - a_2) = x^2 - 3x + 2$ gives quotient $\gamma = 2$ and remainder $7x - 11$; and a division of $7x - 11$ by $x - a_1 = x - 2$ gives quotient $\beta = 7$ and remainder $\alpha = 3$. Hence

$$\begin{aligned} 2x^2 + x - 7 &= \alpha + \beta(x - a_1) + \gamma(x - a_1)(x - a_2) \\ &= 3 + 7(x - 2) + 2(x - 2)(x - 1). \end{aligned}$$

PROBLEMS

(1). Express x^2 in the form $\alpha + \beta x + \gamma x(x - 1)$.

(2) Express x^2 in the form $\alpha + \beta(x + 1) + \gamma(x + 1)(x + 2)$.

(3) Express $x^3 + 3x$ in the form $\alpha + \beta(x + 2) + \gamma x(x + 2) + \delta x(x + 2)(x - 2)$.

(4) If $x^3 + 3x + A$ is divisible by $x + 2$, what is A?

(5) Show by division that any polynomial of the fifth degree can be expressed in the form

$$(\alpha + \beta x) + (\gamma + \delta x)(x^2 + 3x + 6) + (\varepsilon + \zeta x)(x^2 + 3x + 6)^2$$

by a suitable choice of α, β, γ, δ, ε, ζ.

(6) Similarly show that, given any quadratic expression of the form $(x^2 + hx + k)$, any polynomial can be expressed in the form

$$(\alpha + \beta x) + (\gamma + \delta x)(x^2 + hx + k) + (\varepsilon + \zeta x)(x^2 + hx + k)^2 + \ldots$$

where enough terms are taken to give a polynomial of the correct degree.

3.10 Interpolation

An important function, such as $y = \log x$, $y = \sin x$, $y = \sqrt{x}$ or $y = x^2$ is often put into the form of a table, so that, given x, we read off the value of the function at once, without calculation. Much work can thus be saved. In the case of an empirical function such as the height y of a man at age x, there need be no simple relationship between y and x, and the easiest way of representing the function is by a table of measured heights at different ages.

Suppose we want to know the value of a function to about 4-figure accuracy. It would be possible to tabulate the value of y for all four-figure values of x within a suitable range. But such a table would be extremely long. It is more practical to use the fact that (for most functions) the graph is a smooth curve. If we have only a selection of values of y and x, corresponding to a selection of points on the graph, we may be able to deduce intermediate points with considerable confidence, since they must lie on a smooth curve through the known points.

The simplest method is "linear interpolation", or the use of simple proportion. Consider the problem of finding the common logarithm of 2·004, given a four-figure log table with the following values.

$x =$ 2·000	2·010	2·020	2·030	2·040	...
$y = \log x =$ ·3010	·3032	·3054	·3075	·3096	...

Each time x increases by ·01, its logarithm y increases by very nearly the same amount, for

$$\log 2{\cdot}010 - \log 2{\cdot}000 = {\cdot}3032 - {\cdot}3010 = {\cdot}0022,$$
$$\log 2{\cdot}020 - \log 2{\cdot}010 = {\cdot}3054 - {\cdot}3032 = {\cdot}0022,$$

$$\log 2{\cdot}030 - \log 2{\cdot}020 = {\cdot}3075 - {\cdot}3054 = {\cdot}0021,$$
$$\log 2{\cdot}040 - \log 2{\cdot}030 = {\cdot}3096 - {\cdot}3075 = {\cdot}0021.$$

The slight variation in the last figure is in part due to the limitation of the table to four-place accuracy. In short, within this range of values, any change in the value of x produces a proportional change in $y = \log x$.

To find $\log 2{\cdot}004$, argue as follows. When x increases by ·01 (from 2·000 to 2·010), y increases by ·0022 (from ·3010 to ·3032). Hence if x increases from 2·000 to 2·004, this increase ·004 is only a fraction $\cdot 004/\cdot 010 = \cdot 4$ of the increase ·01 in the table. The increase in the value of y should also be ·4 times the increase given in the table, i.e. $\cdot 4 \times \cdot 0022 = \cdot 0009$. Hence $\log 2{\cdot}004 = \log 2{\cdot}000 + \cdot 0009 = \cdot 3010 + \cdot 0009 = \cdot 3019$.

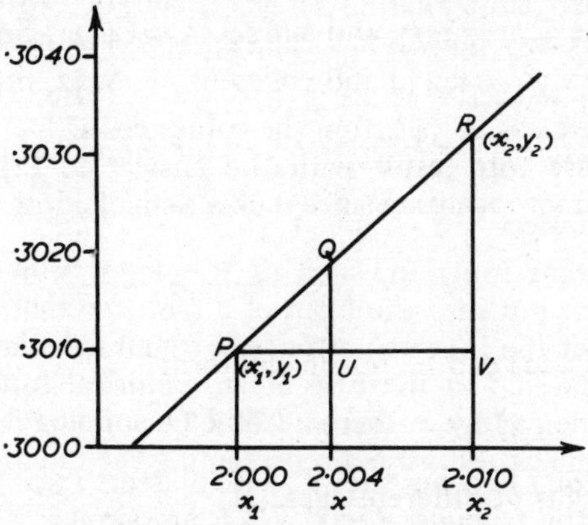

Fig. 3.11—Linear interpolation

Similarly, if we are given that $\sin 45{\cdot}1^\circ = \cdot 7083$ and $\sin 45{\cdot}2^\circ = \cdot 7096$, and wish to find $\sin 45{\cdot}12^\circ$, we begin by noting that as x increases by $\cdot 1^\circ$, $\sin x$ increases by $\cdot 7096 - \cdot 7083 = \cdot 0013$. Hence if x increases by $\cdot 02^\circ$, we expect $\sin x$ to increase by $\cdot 2 \times \cdot 0013 = \cdot 0003$. Therefore $\sin 45{\cdot}12^\circ = \sin 45{\cdot}10^\circ + \cdot 0003 = \cdot 7085$.

This "rule of proportional parts" can be expressed by a formula. Denote two particular values of x by the symbols x_1 and x_2, and the corresponding values of $y = F(x)$ by $y_1 = F(x_1)$ and $y_2 = F(x_2)$. This situation is illustrated in Fig. 3.11, where $x_1 = 2{\cdot}000$, $x_2 = 2{\cdot}010$, $y_1 = \log x_1 = \cdot 3010$, $y_2 = \log x_2 = \cdot 3033$. P and R are the corresponding points (x_1, y_1), (x_2, y_2) on the graph of $y = F(x)$. We wish to find $y = F(x)$ where, say, $x = \cdot 2004$ (between x_1 and x_2). Q is the corresponding point (x, y).

Suppose that the increase $(x - x_1)$ in x is a fraction f of the tabulated increase $(x_2 - x_1)$, i.e.

$$f = (x - x_1)/(x_2 - x_1) = PU/PV \text{ in Fig. 3.11.}$$

In our example, $f = (2{\cdot}004 - 2{\cdot}000)/(2{\cdot}010 - 2{\cdot}000) = {\cdot}004/{\cdot}010 = {\cdot}4$. Then we expect the change $y - y_1 = UQ$ to be a fraction f of the change $y_2 - y_1 = VR$, i.e.

$$y - y_1 = f(y_2 - y_1) \quad . \quad . \quad . \quad (3.22)$$

whence

$$y = y_1 + f(y_2 - y_1) \quad . \quad . \quad . \quad (3.23)$$
$$= fy_1 + (1 - f)\, y_2 \quad . \quad . \quad . \quad (3.24)$$

Formula (3.23) is most convenient when calculating by hand, and (3.24) when using a machine.

PROBLEMS

(1) If $\sin 24{\cdot}3^\circ = {\cdot}4115$ and $\sin 24{\cdot}4^\circ = {\cdot}4131$, find $\sin 24{\cdot}37^\circ$.

(2) If $\cos 22{\cdot}5^\circ = {\cdot}9239$ and $\cos 22{\cdot}6^\circ = {\cdot}9232$, find $\cos 22{\cdot}54^\circ$.

(3) Calculate $\cos 22{\cdot}54^\circ$ from the values $\cos 22^\circ = {\cdot}9272$, $\cos 23^\circ = {\cdot}9205$. Compare this result with the answer to (2), which may be expected to be more accurate since it uses a smaller interval of tabulation.

Note that (3.22) can be put in the form

$$(y - y_1)/(y_2 - y_1) = VR/UQ = f = PU/PV$$

(by definition of f). This shows that to assert (3.22) is equivalent to asserting that the triangles QUP, RVP are similar, so that the angles $\angle UPQ$, $\angle VPR$ are equal, or, in other words, PQR is a straight line. Thus interpolation on the basis of simple proportion is valid when the small part PQR of the graph $y = F(x)$ is nearly enough a straight line (whence the name "linear interpolation").

If the graph $y = F(x)$ shows appreciable curvature, linear interpolation will not be quite accurate. However, it is possible that the graph may be approximated to sufficiently well by a parabolic curve $y = A + Bx + Cx^2$, or by a cubic curve $y = A + Bx + Cx^2 + Dx^3$. The more terms we take, the greater variety of curves we have to choose from, and hence the better fit we can make to the graph. (In the same way, the more measurements a tailor makes, the better fitting suit he will produce.) However, increasing the number of terms also increases the labour of computation.

We now proceed to show that we can fit a quadratic curve (parabola) to pass through any three points on the graph of a function, and a cubic curve $y = A + Bx + Cx^2 + Dx^3$ to pass through any four points, and so on. More exactly, we illustrate the procedure on the particular case of 4 points. The generalization to n points should then be clear.

Suppose that we have 4 measurements y_1, y_2, y_3, y_4 of the weight of an animal (in mg) taken at the respective ages of x_1, x_2, x_3 and x_4 days.

Age (days)	$x_1 = 2$	$x_2 = 3$	$x_3 = 4$	$x_4 = 6$
Weight (mg)	$y_1 = 228$	$y_2 = 335$	$y_3 = 460$	$y_4 = 716$

We wish to estimate the weight y at $x = 5$ days, assuming that the graph of weight against age is smooth. By using Theorem 3.1 (Section 3.9) we see that any cubic curve can be put in the form

$$y = \alpha + \beta(x - x_1) + \gamma(x - x_1)(x - x_2) + \delta(x - x_1)(x - x_2)(x - x_3) \quad (3.25)$$

that is, in our case

$$y = \alpha + \beta(x - 2) + \gamma(x - 2)(x - 3) + \delta(x - 2)(x - 3)(x - 4) \quad (3.26)$$

We have to make this pass through the four given points by choosing appropriate values for α, β, γ and δ.

Now when $x = x_1 = 2$, we have $y = y_1 = 228$. Since the curve has to pass through the point (x_1, y_1), we can substitute these values in (3.26), obtaining

$$228 = \alpha + 0 + 0 + 0,$$

that is, $\alpha = 228$. The next point on the curve is $x = x_2 = 3$, $y = y_2 = 335$. Substituting these values in (3.26) gives

$$\begin{aligned} 335 &= \alpha + \beta \times 1 + 0 + 0 \\ &= 228 + \beta, \end{aligned}$$

whence $\beta = 107$. Now substitute $x_3 = 4$, $y_3 = 460$, getting

$$\begin{aligned} 460 &= \alpha + \beta \times 2 + \gamma \times 2 \times 1 + 0 \\ &= 228 + 107 \times 2 + 2\gamma, \end{aligned}$$

whence $\gamma = 9$. Finally substitute $x_4 = 6$, $y_4 = 716$, getting

$$\begin{aligned} 716 &= \alpha + \beta \times 4 + \gamma \times 4 \times 3 + \delta \times 4 \times 3 \times 2 \\ &= 228 + 107 \times 4 + 9 \times 4 \times 3 + \delta \times 4 \times 3 \times 2 \end{aligned}$$

whence $\delta = -2$. The constants α, β, γ, δ have thus been successively determined, and (3.26) becomes

$$y = 228 + 107(x - 2) + 9(x - 2)(x - 3) - 2(x - 2)(x - 3)(x - 4) \quad (3.27)$$

This is the cubic curve passing through the four given points. To find the value of y when $x = 5$, we substitute $x = 5$ in (3.27), assuming that that represents the growth curve as nearly as possible. A direct substitution in (3.27) gives

$$y = 228 + 221 + 54 - 12 = 591.$$

An alternative way of calculating y in terms of x, which is a little simpler than direct substitution, is as follows. On reading (3.27) from right to left, it can be put into the alternative form

$$y = [\{-2(x-4)+9\}(x-3)+107](x-2)+228,$$

whence, if $x = 5$,

$$\begin{aligned} y &= [\{-2 \times 1 + 9\} \times 2 + 107] \times 3 + 228 \\ &= [7 \times 2 + 107] \times 3 + 228 \\ &= 121 \times 3 + 128 \\ &= 591. \end{aligned}$$

An advantage of this method of calculation is that the coefficients $\alpha, \beta, \gamma, \delta, \ldots$ in the interpolation formula (3.25) can be found successively. If we use only α and β, we have linear interpolation; α, β, and γ give quadratic interpolation, and so on. Usually successive coefficients $\alpha, \beta, \gamma, \ldots$ decrease rapidly in value; when one has become negligibly small, the polynomial (3.25) can be terminated at that point.

Note carefully that this process fits a polynomial *exactly* to a set of values. Quite often the problem arises of fitting an approximate smooth curve to a set of empirical values which are subject to random error. It would then be foolish to try to make the curve pass exactly through all the points—the result would be a very irregular and meaningless curve. This approximate curve fitting, known as "regression analysis", is discussed in Chapter 21, Vol. 2 (Sections 21.14 and 21.15). Examples of such curves occur in Fig. 3.1 and 3.5, above.

PROBLEMS

(4) According to the Highway Code, a vehicle travelling at 20 miles per hour can normally come to a sudden stop in 40 feet; one travelling at 30 or 40 miles per hour will require 75 or 120 feet respectively. How far will be needed for vehicles travelling at respectively 25 and 35 m.p.h.?

(5) Given that $\sqrt{81} = 9$, $\sqrt{100} = 10$, and $\sqrt{121} = 11$, find by interpolation $\sqrt{90}$ and $\sqrt{110}$ to 3 places of decimals.

(6) Given that tan $87{\cdot}5° = 22{\cdot}904$, tan $87{\cdot}6° = 23{\cdot}859$, tan $87{\cdot}7° = 24{\cdot}898$, tan $87{\cdot}8° = 26{\cdot}031$, find tan $87{\cdot}64°$ to 3 places of decimals.

This discussion has important theoretical consequences. We have shown that we can fit one and only one cubic curve

$$y = \alpha + \beta(x-x_1) + \gamma(x-x_1)(x-x_2) + \delta(x-x_1)(x-x_2)(x-x_3)$$

to pass through four specified points (x_1, y_1), (x_2, y_2), (x_3, y_3) and (x_4, y_4) (provided that x_1, x_2, x_3 and x_4 are all different, i.e. no two of the points lie in the same vertical line). For the process described above then determines the values of α, β, γ and δ successively. A similar argument shows that there is just one polynomial curve $y = F(x)$ of degree $n - 1$ which passes through n given points (no two on the same vertical line).

This leads to the following result:

Theorem 3.2. An equation of the nth degree

$$F(x) = A + Bx + Cx^2 + \ldots + Hx^n = 0$$

cannot have more than n roots.

For suppose, if possible, that it had $n + 1$ roots, say $\rho_1, \rho_2, \ldots, \rho_{n+1}$. Then the two curves $y = F(x)$ and $y = \frac{1}{2}F(x)$ would both be of the nth degree, and both pass through all the points $(\rho_1, 0)$, $(\rho_2, 0)$, ..., $(\rho_{n+1}, 0)$, contrary to the assertion that there is only one polynomial curve of the nth degree through $(n + 1)$ points. Hence we have a contradiction, showing that the equation cannot have more than n roots.

3.11 Tabulation at equally spaced values

Most tables of functions $F(x)$ are calculated for equally spaced values of x. For example, the values of $\log x$ are usually given for $x = 1{\cdot}00, 1{\cdot}01, 1{\cdot}02, \ldots$, and so on by regular intervals of $\cdot 01$ up to $x = 9{\cdot}99$. [This set of values of x can be expressed concisely as $x = 1{\cdot}00(\cdot 01)9{\cdot}99$.] In such cases interpolation formulas become simplified. We do not discuss the general case here: details can be found in a number of books, such as *Interpolation and Allied Tables* (H.M. Stationery Office). We merely mention a specially simple and useful formula for quadratic interpolation.

Let x_0, x_1, x_2, x_3 be four consecutive values of x in the tables of $F(x)$, at equal intervals h, so that $x_1 = x_0 + h$, $x_2 = x_1 + h$, and $x_3 = x_2 + h$. Suppose we wish to find $y = F(x)$, where x lies between x_1 and x_2. Let, as before, $f = (x - x_1)/h$. Then, very nearly

$$y = F(x) = y_1 + f(y_2 - y_1) + \tfrac{1}{4}f(1 - f)\Delta \quad . \quad (3.28)$$

where $y_0 = F(x_0)$, $y_1 = F(x_1)$, $y_2 = F(x_2)$, $y_3 = F(x_3)$ are the tabulated values and

$$\Delta = -y_0 + y_1 + y_2 - y_3.$$

Suppose, for example, that we wish to calculate $\tan 66{\cdot}14^\circ$ from a 5-figure table of tangents. We have

$$x_0 = 66{\cdot}0^\circ \quad y_0 = \tan 66{\cdot}0^\circ = 2{\cdot}24604$$
$$x_1 = 66{\cdot}1^\circ \quad y_1 = \tan 66{\cdot}1^\circ = 2{\cdot}25663$$
$$x_2 = 66{\cdot}2^\circ \quad y_2 = \tan 66{\cdot}2^\circ = 2{\cdot}26730$$
$$x_3 = 66{\cdot}3^\circ \quad y_3 = \tan 66{\cdot}3^\circ = 2{\cdot}27806$$

whence $y_2 - y_1 = \cdot 01067$ and $\Delta = -y_0 + y_1 + y_2 - y_3 = -\cdot 00017$.

Also $\qquad f = (66{\cdot}14 - 66{\cdot}10)/{\cdot}1 = {\cdot}4$. Hence

$$y = \tan 66{\cdot}14 = 2{\cdot}25663 + {\cdot}4({\cdot}01067) + \tfrac{1}{4}({\cdot}4)({\cdot}6)(-{\cdot}00017)$$
$$= 2{\cdot}26089.$$

The first two terms represent linear interpolation. The last term is the correction for curvature in the graph, and is very small, amounting to only -1 in the fifth place of decimals. This correction can be neglected, and the linear formula (3.24) used, if Δ does not exceed 16 times the maximum permissible error in the final answer.

Sometimes a table gives the value of a function of two or more variables, say $z = F(x, y)$, for equidistant values $x_0, x_1, x_2, \ldots$ of x and for equidistant values $y_0, y_1, y_2, \ldots$ of y. Such a table is known as one of "double entry". If we wish to calculate $F(x, y)$ for values of x and y between the tabulated values, the simplest procedure is as follows. First keeping y fixed at each of its values $y_0, y_1, y_2, \ldots$ in turn, we interpolate for the desired value of x, finding $F(x, y_0)$, $F(x, y_1)$, etc. Then, keeping x fixed at its chosen value, we interpolate for y, finding $F(x, y)$. In particular, for linear interpolation in x and y we have the formula—

$$z = F(x,y) = (1 - f)(1 - \phi)F(x_0, y_0) + f(1 - \phi)F(x_1, y_0)$$
$$+ (1 - f)\phi F(x_0, y_1) + f\phi F(x_1, y_1) \qquad . \quad . \quad . \quad (3.29)$$

where $f = (x - x_0)/(x_1 - x_0)$ and $\phi = (y - y_0)/(y_1 - y_0)$.

PROBLEMS

(1) Find tan 66·18° by linear and by quadratic interpolation.

(2) Given that $\sqrt{4{\cdot}0} = 2$, $\sqrt{4{\cdot}1} = 2{\cdot}0248$, $\sqrt{4{\cdot}2} = 2{\cdot}0494$ and $\sqrt{4{\cdot}3} = 2{\cdot}0736$, find $\sqrt{4{\cdot}123}$ to 4 decimal places.

3.12 Factors

The polynomial $x^3 - 1$ divides exactly by $(x - 1)$ without remainder, for $(x^3 - 1) = (x - 1)(x^2 + x + 1)$. Hence $(x - 1)$ is said to be a factor of $(x^3 - 1)$. In general a polynomial $G(x)$ is a factor of a polynomial $F(x)$ if it divides exactly into $F(x)$. In particular, $(x - a)$ is a factor of $F(x)$ if

$$F(x) = (x - a)Q(x) \qquad . \quad . \quad . \quad (3.30)$$

where $Q(x)$ is the quotient. Hence $F(a) = (a - a)Q(a) = 0$; that is, if $(x - a)$ is a factor of $F(x)$, then a must be a root of the equation $F(x) = 0$. Conversely, suppose that $F(a) = 0$. We know that $F(x)$ can be expressed in the form (Theorem 3.1)

$$F(x) = \alpha + \beta(x - a) + \gamma(x - a)x + \delta(x - a)x^2 + \ldots$$

Since $F(a) = 0$, it follows that $\alpha = 0$; so

$$F(x) = (x - a)[\beta + \gamma x + \delta x^2 + \ldots]$$

showing that $F(x)$ has the factor $(x - a)$. That is,

Theorem 3.3. *$(x - a)$ is a factor of a polynomial $F(x)$ if and only if $F(a) = 0$.*

This is a useful test for factors. For example, since when $x = 1$ we obviously have $x^3 - 1 = 0$, it follows that $(x - 1)$ is a factor of $(x^3 - 1)$. Since $x^2 + 5x + 4 = 0$ when $x = -1$, it follows that $(x + 1)$ is a factor of $x^2 + 5x + 4$. A similar proof shows that if $a_1, a_2, a_3, \ldots$ are different, then $F(x)$ divides by $(x - a_1)(x - a_2)(x - a_3)\ldots$ if and only if, $F(a_1) = 0$, $F(a_2) = 0$, $F(a_3) = 0, \ldots$ In particular, we have

Theorem 3.4. *Suppose that $F(x)$ is a polynomial of degree n, and the equation $F(x) = 0$ has n distinct roots. Then*

$$F(x) = \lambda(x - a_1)(x - a_2)\ldots(x - a_n)$$

where λ is a constant.

Proof. Since $F(a_1) = 0$, $F(a_2) = 0, \ldots F(a_n) = 0$, $F(x)$ must divide by the expression $(x - a_1)(x - a_2)\ldots(x - a_n)$, which is of degree n. Since $F(x)$ is also of degree n, the quotient must be of degree $n - n = 0$, i.e. a constant λ.

It follows from Theorem 3.4, on multiplying out the right-hand side, that the term of highest degree in $F(x)$ is λx^n.

PROBLEMS

(1) Show from Theorem 3.3 that $x^n - 1$ is always divisible by $x - 1$, and if n is even it is also divisible by $x + 1$.

(2) Show that $x^{2n+1} + 1$ is divisible by $x + 1$, but $x^{2n} + 1$ is not.

(3) Factorize $x^2 + 6x - 7$.

(4) Factorize $x^3 + 6x^2 + 3x - 10$.

(5) Factorize $x^3 - 3x^2 + 6x - 4$.

(6) $x^4 + Cx^2 + A$ is divisible by $(x - 1)(x + 2)$. Find C, A, and the other factors.

(7) $F(x) = A + Bx + Cx^2 + \ldots + Gx^{n-1} + Hx^n$. There are n distinct roots, $a_1, a_2, \ldots a_n$ of the equation $F(x) = 0$, so that Theorem (3.4) applies. Show that

$$\lambda = H$$
$$\lambda(a_1 + a_2 + \ldots + a_n) = -G$$
$$\lambda a_1 a_2 \ldots a_n = A.$$

3.13 Ratios and proportions

Many ratios are biologically important; e.g. the sex-ratio of males to females in a population, mendelian ratios in genetics, and so on. Ratios may help to indicate causal relationships. There is a higher proportion of ulcer patients among individuals of blood-group O than among those of group A. This seems to show some influence of blood-group on disease, although the situation is still obscure. For a further discussion of this kind of association, see Section 21.1, Vol. 2.

If four numbers a, b, A, B are such that $a/b = A/B$, they are said to be "in proportion". On multiplying through by bB, the condition becomes

$$aB = Ab \quad . \quad . \quad . \quad (3.31)$$

For the present, we shall suppose all the numbers to be positive. We can then illustrate the relationship geometrically, as in Fig. 3.12.

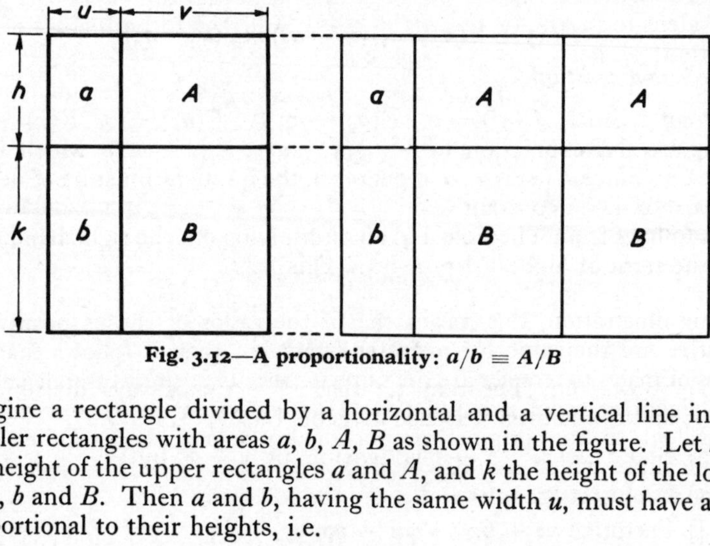

Fig. 3.12—A proportionality: $a/b \equiv A/B$

Imagine a rectangle divided by a horizontal and a vertical line into 4 smaller rectangles with areas a, b, A, B as shown in the figure. Let h be the height of the upper rectangles a and A, and k the height of the lower ones, b and B. Then a and b, having the same width u, must have areas proportional to their heights, i.e.

$$a/b = h/k.$$

Similarly $A/B = h/k$, and therefore $a/b = A/B$, as required.

One interpretation of this diagram could be as follows. The rectangle represents a population. The vertical line divides it into two parts according to one characteristic, e.g. males on the left, females on the right, in the ratio $u : v$. The horizontal division corresponds to an independent character, e.g. rhesus positive above the line, rhesus negative below. The relation $a : b = A : B$ would then show that the ratio of rhesus positive to negative is the same in the two sexes, as appears to be true (at least approximately) in actual populations.

Various consequences arise from this geometric representation. On the right of Fig. 3.12 we see that the rectangles $a + A$, $A + A$ $(= 2A)$ and $a + A + A$ $(= a + 2A)$ all have height h, and the corresponding rectangles $b + B$, $B + B$ and $b + B + B$ all have height k. Since rectangles of equal width have areas proportional to their widths, we have

$$\frac{h}{k} = \frac{a}{b} = \frac{A}{B} = \frac{a + A}{b + B} = \frac{2A}{2B} = \frac{a + 2A}{b + 2B} = \text{etc.}$$

For example, the relation $a : b = A : B = (a + A) : (b + B)$ can be interpreted as follows: if the ratio of rhesus positive to negative is the same in males as in females, then it is the same in the population as a whole (with both sexes put together).

More generally, we can assert that if $a/b = A/B$, then both these ratios are equal to $(\lambda a + \mu A)/(\lambda b + \mu B)$ for any λ, μ (not both zero). This can also be shown algebraically. The relation $a/b = A/B$ is equivalent to (3.31), or to $\mu aB = \mu Ab$. Add λab to both sides of this equation: we get

$$a(\lambda b + \mu B) = (\lambda a + \mu A)b$$

or

$$a/b = (\lambda a + \mu A)/(\lambda b + \mu B).$$

Fig. 3.12 can also be looked at vertically instead of horizontally, so we can deduce a new proportionality

$$a/A = b/B = u/v \quad . \quad . \quad . \quad . \quad (3.32)$$

In our illustration, this means that if the ratios of rhesus positive to negative are the same for males and females, then it follows that the ratios of males to females are the same for rhesus positives and negatives. The algebraic proof is simple; (3.32) is equivalent to (3.31), which is equivalent to the original proportionality $a/b = A/B$. Other proportionalities which can also be deduced from $a/b = A/B$, and which are fairly obvious geometrically, are the following:

$$b/a = B/A$$
$$2a/b = 2A/B$$
$$a/(a + b) = A/(A + B).$$

The last type is particularly important; it means that we can add the numerator to the denominator and preserve proportionality. More generally, we can add any fixed multiple of the numerators to the respective denominators, or any fixed multiple of the denominators to the respective numerators, and proportionality will be preserved.

EXAMPLES

(1) The ratio of males to females in Britain is 106 : 100. What fraction of the population is male?

Let M denote the total number of males, F that of females; then $M/F = 106/100$. Add the numerator to the denominator on each side: we get

$$M/(M + F) = 106/(106 + 100) = \cdot 515;$$

that is, males form a fraction $\cdot 515 = 51 \cdot 5$ per cent of all individuals.

(2) Solve the equation $2x/(2 - x) = 6$.

We have $$2x/(2 - x) = 6/1.$$

Add half the numerator to the denominator on each side. We get

$$2x/(x + 2 - x) = 6/(3 + 1)$$

i.e. $$x = 6/4 = 1 \cdot 5.$$

(3) Solve the equation

$$\frac{1 + x}{1 - x} = \frac{5}{4}.$$

Add numerator to denominator on each side: we get

$$(1 + x)/2 = 5/9$$

Subtract half the denominator from the numerator on each side:

$$x/2 = \tfrac{1}{2}/9 = 1/18.$$

Multiply both sides by 2; we get $x = \frac{1}{9}$.

So far we have restricted ourselves to positive values of the quantities involved, as this gives a simple geometric interpretation. There is no difficulty in principle in using negative numbers as well, since the propositions can be proved algebraically, although the geometric interpretation will be less obvious. However, if any of the denominators become zero we could run into trouble. The simplest way to get round this difficulty is to take (3.31) to be the *definition* of proportionality; this is equivalent to the relation $a/b = A/B$, provided that b and B are neither zero. It will also be satisfied under various other conditions, e.g. if $b = B = 0$, which is simply a particular case of Fig. 3.12 with $k = 0$; or $a = b = 0$, which is simply a particular case of Fig. 3.12 with $u = 0$. In fact, the equation (3.31) seems to give a natural definition of proportionality whether a, b, A, B are positive, negative, or zero; and as it involves no fractions, it can give no trouble from zero denominators.

PROBLEMS

Solve the following equations for x:

(1) $x/(2 - x) = 3$.

(2) $3x/(x + 1) = 2$.

(3) $(x + 1)/(x - 3) = 3$.

(4) $(x + 1)/(x - 1) = (1 - a)/(1 + a)$.

If $a/b = A/B$, show algebraically and geometrically that

(5) $b/a = (B - b)/(A - a)$.

(6) $A/a = (B + 2A)/(b + 2a)$.

(7) $a/(a + A) = (b - a)/(b - a + B - A)$.

We can also have several numbers in proportion: for example, Fig. 3.13 illustrates the proportion

$$a : b : c = A : B : C = \alpha : \beta : \gamma$$

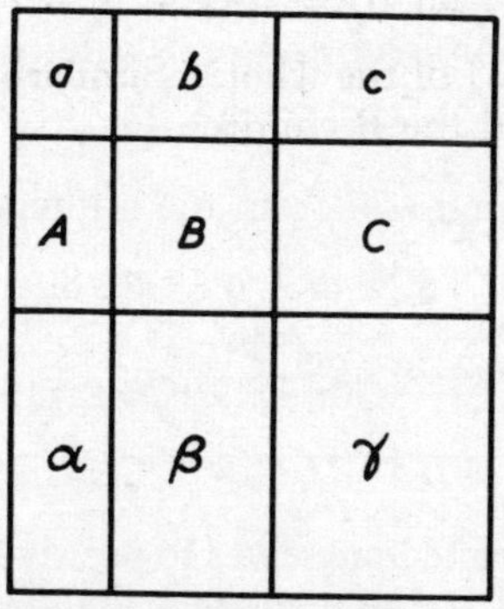

Fig. 3.13—Triple proportionality: $a : b : c = A : B : C = \alpha : \beta : \gamma$

This is interpreted as meaning that all 2×2 proportions extracted from this in the obvious way hold good; e.g. $a/b = A/B = \alpha/\beta$; $a/c = A/C = \alpha/\gamma$, etc. Such continued proportions obey laws very similar to the 2×2 type $a/b = A/B$; thus, without altering proportionality, one can perform the following operations on Fig. 3.13:

(i) multiply any row throughout by a constant: e.g. $2a : 2b : 2c = a : b : c = A : B : C = \alpha : \beta : \gamma$.

(ii) add or subtract any row from any other,

e.g. $$(a + A) : (b + B) : (c + C) = a : b : c = \text{etc.}$$

(iii) multiply any column by a constant and/or add or subtract it to any other column.

(iv) interchange rows and columns (i.e. rotate the whole figure through 90°).

Proofs are left to the reader.

EXAMPLE

Matings in which both parents have blood-group AB produce children of the three groups A, AB, and B in the ratios 1 : 2 : 1 (in the long run). What proportion does each blood-group form of the total number of children?

Let $N(\text{A})$ denote the number of children of group A, and similarly $N(\text{AB})$, $N(\text{B})$. Then we have the proportion

$$\begin{aligned} & N(\text{A}) : N(\text{AB}) : N(\text{B}) : N\,(\text{Total}) \\ = {} & 1 \quad : \quad 2 \quad : \quad 1 \quad :(1+2+1) \end{aligned}$$

in which we have added an extra column, which is the sum of all the other columns. From this it follows at once that

$$N(\text{A})/N\,(\text{Total}) = 1/(1+2+1) = 1/4,$$

i.e. the A children form $\frac{1}{4}$ of the whole. Similarly the AB children form $\frac{2}{4} = \frac{1}{2}$ of the whole, and the B children $\frac{1}{4}$.

CHAPTER FOUR

COMPARISONS OF MAGNITUDES

4.1 Inequalities

In elementary algebra we spend much time considering equations like $1 + 4x + 3x^2 = 0$, $x + y = 2$, and discussing how to solve them. But in practice it is important not only to be able to say when two quantities are equal, but also to compare them and say which is greater and which is less. For example, we may find it convenient to use an approximate formula: if we know that the error in this formula is *never greater than* ·0001 we may be perfectly satisfied with the formula, without asking exactly what the error is. Comparisons of magnitude such as this are known as *inequalities*.

We have already come across a number of such inequalities, as when we say that the function $x^2 + 1$ has a minimum value 1, and can never be less than 1: the reason for that is that x^2 is never negative. We must now bring in strict definitions and rules for dealing with inequalities, so that we can state them with precision and manipulate them with ease.

The first thing to notice is that we have actually been using the words "greater" and "less" in two different senses—although that should not have caused any confusion, as so far the context should have made clear which sense was meant. The ambiguity in question occurs in comparing negative numbers: which is greater, -4000 or -4? In one sense of the word "greater", -4000 is obviously a much bigger or greater number than -4.

An electrified wire at a potential of -4000 volts is very dangerous to touch; one at a potential of -4 volts is completely harmless. A pressure of -4000 kilograms weight, i.e. a tension of 4000 kilograms, will tear most common objects apart, whereas a tension of 4 kilograms is comparatively harmless. On the other hand the sea bed at a height of -4000 feet is clearly lower than at a height of -4 feet; a man with £(-4000) in the bank, i.e. an overdraft of £4000 is worse off than one with £(-4), and a temperature of $-200°$ is colder than $-2°$.

Now as has been said we can to some extent make this distinction clear in ordinary language. If we say that a number is "large and negative" then we think of something like -4000 rather than -4. But the best thing to do is to invent a notation which expresses our ideas without this ambiguity.

4.2 Symbols for comparisons

The sense we shall first deal with is the one which is natural in comparing heights, or bank balances, in which £(−4000) is considered as smaller than £(−4). We shall say that "A is greater than a", or in symbols $A > a$, if the difference $(A - a)$ is positive. (Notice that the wide end of the symbol "$>$" is next to the greater number, and the narrow end next to the smaller, so that the symbol is a picture of the relationship.) Thus $4000 > 4 > -4 > -4000$; for all x, $x^2 > -4$ (because $x^2 + 4$) is always positive); and $x > 0$ is another way of writing "x is positive". In the same way we shall say that "a is less than A", or $a < A$, if $(a - A)$ is negative.

The principal rules of operation with these symbols are as follows:

(A) The statements $A > a$ and $a < A$ are equivalent, i.e. have the same meaning. For if $(A - a)$ is positive, then $(a - A)$ is negative, and conversely.

(B) If $A > a$, then whatever x may be, $A + x > a + x$. In words, "we may add or subtract the same quantity from both sides of an inequality" (subtraction of x being equivalent to addition of $-x$). The proof follows from the identity $(A + x) - (a + x) = A - a$; we are supposing that $A > a$, i.e. $A - a$ is positive, and so $(A + x) - (a + x)$ is positive.

(C) If $A > a$ and $B > b$ then $A + B > a + b$, i.e. "we may add together the left-hand sides (L.H.S.) of two inequalities to give a new L.H.S., and the two R.H.S's to give a new R.H.S., provided that the two inequality signs point in the same direction". This follows from the identity $(A + B) - (a + b) = (A - a) + (B - b) > 0$. [*Query*: Is $(A - B)$ necessarily greater than $(a - b)$? If so, why? If not, why not?]

(D) If $A > a$ and $k > 0$ then $kA > ka$. "We can multiply or divide both sides of an inequality by the same *positive* number k." (Division is included since division by k is equivalent to multiplication by $1/k$.) For $kA - ka = k(A - a)$, and k and $(A - a)$ are both positive. If, however, k is negative then the inequality is reversed: $A > a$ implies $kA < ka$ (for example, $-A < -a$).

(E) If A and a are both *positive* and $A > a$, then $A^2 > a^2$ and $\sqrt{A} > \sqrt{a}$, but $1/A < 1/a$. "The greater a positive number is, the greater is its square, the greater its square root, and the smaller its reciprocal." The proofs follow from $(A^2 - a^2) = (A - a)(A + a)$, $(\sqrt{A} - \sqrt{a}) = (A - a)/(\sqrt{A} + \sqrt{a})$, and $1/A - 1/a = (a - A)/aA$, respectively.

(F) If $A > B$ and $B > C$ then $A > C$.

It is also convenient to have a composite symbol "$A \geq a$" or "$A \geqslant a$" for "A is greater than or equal to a", and similarly "$A \leq a$" or "$A \leqslant a$" means "A is less than or equal to a". The symbol

"$A \ngtr a$", which is sometimes used, means "A is not greater than a", and is exactly equivalent to "$A \leqslant a$".

Theorems concerning inequalities can vary from very simple and obvious results to very difficult and subtle ones. There are three main lines of attack on the simpler cases. We can use the definition that $A > a$ if $(A - a)$ is positive; we can combine known inequalities using the above rules (A) to (F); and we can rely on the fact that a square is always positive or zero; e.g. since $(x - 1)^2 = x^2 - 2x + 1 \geqslant 0$ we see that $x^2 + 1 \geqslant 2x$ for all values of x.

PROBLEMS

(1) Show that the rules (A) to (F) remain true if "$>$" is replaced throughout by "$\geqslant$", and "$<$" by "$\leqslant$".

(2) What happens to rule (C) if $A > a$ and $B \geqslant b$?

(3) If $A > a$, then $A^3 > a^3$, and $A^3 + A^2 + A > a^3 + a^2 + a$. But why does $A^2 > a^2$ not necessarily hold?

(4) For all values of x, show that $x^2 + 4x \geqslant -4$. When does the equality hold?

(5) For all positive x, $x + 1/x \geqslant 2$. What happens for negative x?

(6) What happens to rule (E) if A and a are not necessarily positive?

(7) If A, a, B, b, are all positive, and $A > a$, $B > b$, then $AB > ab$.

(8) For all values of x, X, y, Y show that $(x^2 + y^2)(X^2 + Y^2) \geqslant (xX + yY)^2$. When does equality occur?

(9) If x and y are positive, and $x > y$, what relation holds between $1/\sqrt{x}$ and $1/\sqrt{y}$? What relation holds between $1/(x^2+1)$ and $1/(y^2+1)$? And between $(x^2 + 1)/(y^3 + 2)$ and $(y^2 + 1)/(x^3 + 2)$?

(10) If $a + b + c \geqslant 0$, then $a^3 + b^3 + c^3 \geqslant 3abc$.

4·3 Comparisons of absolute magnitudes

The "modulus" or "absolute value" of a number x is simply the number x with its sign made positive: for example, the modulus* of -4 is $+4$, while the modulus of $+4$ is also $+4$. The modulus of x is denoted by the symbol $|x|$ (spoken as "mod x"). Thus, speaking precisely, we can define $|x|$ as follows:

$$\text{If } x \text{ is positive or zero, } |x| = x,$$
$$\text{if } x \text{ is negative, } |x| = -x.$$

This notation enables us to express in unambiguous fashion the second meaning of the phrase "A is greater than a" in untechnical

* The reader must be warned that mathematicians use the word "modulus" in several quite different senses. A "modulus of elasticity", for example, measures the relationship between stress and strain. But the notation $|x|$ is never used except for the "absolute value" of x, either in the sense used above or some closely related sense.

language. That is, that A is greater in magnitude than a irrespective of their positive or negative signs: or simply that A has a greater modulus (or absolute value) than a. In symbols $|A| > |a|$. Thus a potential of -4000 volts is greater in its absolute value than one of $+4$ volts, and is more dangerous to approach. This is shown by the relation $|-4000| = 4000 > |4|$. On the other hand -4000 volts is still the lower potential, in the sense that an electric current will flow *from* the higher potential of 4 volts *to* the lower potential of -4000 volts: and that is shown by the inequality $-4000 < 4$ without the modulus sign.

In order to manipulate these relations it is necessary to know the properties of the function $|x|$.

The chief properties are:

$$\text{For all values of } x \text{ and } y, \quad |xy| = |x|\,|y| \qquad (4.1)$$

$$\text{and provided that } y \neq 0, \quad |x/y| = |x|/|y| \qquad (4.2)$$

These properties follow from the usual rules for multiplication and division of numbers with signs. In multiplying, say, $+3 \times (-4)$, to get -12, we first multiply the positive numbers 3 and 4, i.e. $|+3|$ and $|-4|$, to give 12, the absolute value of the product, and then attach the sign, here minus. Thus $|(+3)(-4)| = 12 = |+3|.|-4|$.

$$\text{For all } x, \ |x| = |-x| = \sqrt{x^2} \geqslant x \qquad (4.3)$$

This follows immediately from the definition of $|x|$.

$$\text{For all } x \text{ and } y, \quad |x + y| \leqslant |x| + |y| \qquad (4.4)$$

This is almost obvious: if x and y were forces acting (in the same or opposite directions) on a body, then $(x + y)$ would be the resultant force. This would mean that "the magnitude of the resultant force is not greater than the sum of the magnitudes of the component forces". A formal proof is as follows. If $(x + y)$ is positive or zero it equals $|x + y|$ by definition; hence by the use of (4.3) $|x| \geqslant x$, $|y| \geqslant y$, and so by (C) of Section 4.2,

$$|x| + |y| \geqslant x + y = |x + y|.$$

If $(x + y)$ is negative, $-(x + y) = (-x) + (-y)$ is positive, and by the previous argument, $|-x| + |-y| \geqslant |-(x + y)|$, whence by (4.3),

$$|x| + |y| \geqslant |x + y|.$$

A similar argument shows that

$$|x - y| \leqslant |x| + |y| \qquad (4.5)$$

PROBLEMS

(1) Draw the graphs
$$y = |x|,\ y = |5 + x| + |5 - x|,\ y = |x + 2| - |x|.$$
(2) Show that $|x^n| = |x|^n$.

(3) If $|A| > |a|$, and $|B| > |b|$, show that $|AB| > |ab|$.

(4) If $|A| > |a|$ is it always true that $|A + x| > |a + x|$? If so, why? If not, why not?

(5) Prove that $|1/x| = 1/|x|$.

(6) If $|A| > |a|$ is it always true that $|1/A| < |1/a|$ (provided that $a \neq 0$)? If so, why? If not, why not?

(7) When does $|x + y| = |x| + |y|$?

(8) Prove that $|x + y + z| \leqslant |x| + |y| + |z|$.
[Hint: apply (4.4) twice.]

We can also use this notation to prove formally many of the relations we have already stated in words. For example, we have already said that we can make $1/x$ as large as we like in magnitude by making x small enough—suppose we were challenged to make it larger than 1,000,000. We could reply:

"You want $1000000 < |1/x| = 1/|x|$, do you? Well, let's take $|x|$ to be less than ·000001. On multiplying by 1000000 we have, by rule (D), $1000000|x| < 1$, and on dividing by $|x|$, $1000000 < 1/|x|$. For example, $x = \cdot 0000005$ would satisfy you, wouldn't it?"

But here our challenger replies that he would really like $1/x$ to be greater in magnitude than 1,000,000,000,000,000.

Retort: It's difficult to see why you should, but it's only necessary to take $|x|$ to be less than ·000000000000001.

We have also said that an expression like $x^2 + 2x + 3$ can be taken to be x^2 without appreciable percentage error, provided that $|x|$ is large enough.

Challenger: Well, suppose you say that $|x| > 1000$; what percentage error are you likely to commit?

Retort: The actual error in taking $x^2 + 2x + 3$ as x^2 is, in magnitude, $|2x + 3|$; the percent error is therefore in magnitude

$$E = 100\,|2x + 3|/|x^2|\,\%$$

(taking the error as a percentage of the assumed value x^2. If we take the error as a percentage of the true value $(x^2 + 2x + 3)$ we do not significantly alter the results, but the calculation is rather more complicated.) Now we can rewrite E as

$$E = \frac{100}{|x|}\frac{|2x + 3|}{|x|} = \frac{100}{|x|}\left|2 \pm \frac{3}{|x|}\right|\%$$

Since $|x| > 1000$, $1/|x| < \cdot 001$

and
$$\left|2 \pm \frac{3}{|x|}\right| \leqslant |2| + \left|\frac{3}{|x|}\right| \quad \text{[by (4.4) and (4.5)]}$$
$$= 2 + 3/|x| \quad \text{[by (4.2)]}$$
$$\leqslant 2\cdot 003 \quad \text{[rule (E)]}$$

Therefore
$$E < 100 \times \cdot 001 \times 2\cdot 003\%$$
$$< \cdot 2003\%$$

i.e. the error in taking $x^2 + 2x + 3$ to be equal to x^2 cannot exceed $\cdot 2003\%$ (or practically 1 part in 500) when $|x| > 1000$.

4.4 Bounded quantities

We have seen that by making x small enough in modulus we can make $1/x$ as large as we like. Similarly by making x large enough, we can make x^2 as large as we like. The quantity $1/(x^2 + 1)$ does not, however, behave in this way: whatever x may be, $1/(x^2 + 1)$ always lies between 0 and 1. For since x^2 is always positive, $1/(x^2 + 1)$ is always positive, i.e. > 0. And $(x^2 + 1) \geqslant 1$, so that by rule (E) $1/(x^2 + 1) \leqslant 1$. A quantity such as $1/(x^2 + 1)$ is said to be "bounded" (i.e. confined within bounds. This corresponds exactly to the word "limited" in ordinary language, but unfortunately "limit" has a special technical sense in mathematics, which we shall encounter later). More precisely, we say that a fixed or variable quantity β is "bounded by a positive number B" if $|\beta|$ is never greater than B, i.e. $|\beta| \leqslant B$. Thus $1/(x^2 + 1)$ is bounded by 1: we leave it to the reader to show that $x/(x^2 + 1)$ is bounded by $\frac{1}{2}$. On the other hand, if x can vary without restriction, then x^2 is not bounded. But if we restrict x to a certain range, such as $|x| < 100$, then x^2 is bounded (here by 10,000). Natural quantities directly measured are necessarily bounded, so that if a mathematical formula for one indicates an unbounded answer there must be a serious discrepancy between the theoretical mathematical model and the actual phenomena near the point at which the formula gives absurdly large results. The formula may, however, fit well over the rest of the range.

The rules for dealing with bounded quantities are very simple.

Theorem 4.1. Any constant quantity is bounded (by its own modulus). This follows at once from the definition.

Theorem 4.2. The sum, difference, and product of two bounded quantities are all bounded. For let β_1, β_2 be the two bounded quantities, $|\beta_1| \leqslant B_1$, $|-\beta_2| = |\beta_2| \leqslant B_2$. Then $|\beta_1 + \beta_2| \leqslant |\beta_1| + |\beta_2| \leqslant B_1 + B_2$; *similarly* $|\beta_1 - \beta_2| \leqslant B_1 + B_2$, *and* $|\beta_1 \beta_2| \leqslant B_1 B_2$.

Theorem 4.3. The sum or product of any number of bounded quantities is bounded.

This is just a repeated application of Theorem 4.2.

CHAPTER FIVE

SHAPES AND NUMBERS

5.1 The use of algebra to solve geometric problems

In Chapter 3 we saw how an algebraic relation, such as $y = 2x + 3$, or $y = x^2 + 2x^3$, can be represented pictorially by a plane curve or graph. But this procedure can equally well be reversed: if we have a suitable curve, we may be able to express it as an algebraic equation, and so reduce the study of geometry, i.e. relations between shapes, to algebra, i.e. relations between numbers.

This idea of using algebraical methods systematically to solve geometric problems was first conceived by Descartes (1596–1650), who called it "analytical geometry". To a large extent the usual fumbling and searching for correct construction lines can thus be replaced by straightforward algebraic manipulation. By consistent use of this method we can often avoid fallacies arising from a badly drawn figure. For example, take a triangle ABC (Fig. 5.1) in which $AB \neq AC$, and

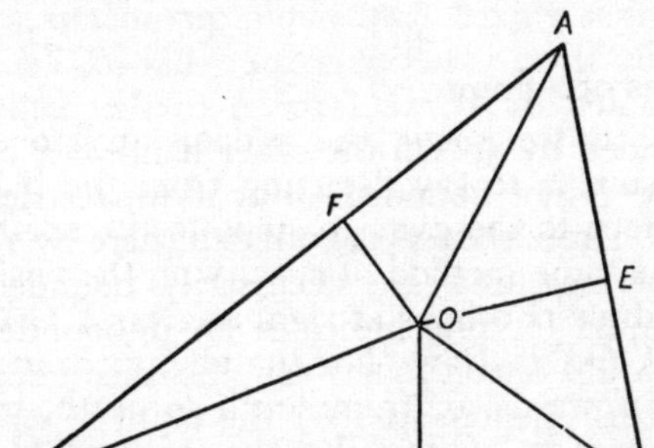

Fig. 5.1—A proof that any triangle ABC has equal sides AB and AC

let the perpendicular bisector of BC meet BC at D and the bisector of the angle A at O. Join OB, OC, and draw the perpendiculars OE, OF from O onto the sides AC, AB respectively. Then since $BD = DC$, and the $\angle BDO = \angle CDO$, the triangles ODB, ODC are congruent, so that $OB = OC$. Also the triangles OAF, OAE are congruent, since they have two angles the same and the side OA in common, so that $AF = AE$ and $OF = OE$. Thus the triangles BOF, COE have two sides and a right-angle the same, and therefore are congruent, and so $BF = CE$. It follows that $AB = BF + FA = CE + EA = AC$, contrary to the supposition that $AB \neq AC$. The fallacy?—it lies simply in the plausible but inaccurate drawing of Fig. 5.1; the points do not lie in the correct relative positions. If the reader will take the trouble to re-draw the figure accurately, he will find that there is no cause for alarm.

Just as ordinary Euclidean geometry can be developed from the beginning by using certain "axioms" about points, lines, angles, etc., so in the same way we can make analytic geometry stand entirely on its own feet by using suitable definitions of points, lines, etc., in terms of numbers. But although aesthetically it may be rather more satisfying to be completely consistent in this way, as well as helping to avoid possible fallacies (as we have explained above), nevertheless it hardly falls within the proper scope of this book. We shall content ourselves here with using whatever method may be the simplest and most suitable for each problem. We shall also assume that the reader is familiar with simple geometric theorems, such as that the sum of the three angles of a triangle is 180°; that two triangles are "similar", or of the same shape, if the three angles of the one are equal to the three angles of the other; and that in such pairs of similar triangles corresponding sides are of proportional length, i.e. one triangle is simply a magnification or reduction of the other in a fixed ratio. We shall also mainly confine our attention to the geometry of the plane for the present.

5.2 Co-ordinates of a point

Traditionally, as we know, the proper way to explain where to find buried treasure is to say "starting from the blasted tree on the heath, walk 40 steps to the east, 30 steps to the north, and then dig". We can adopt the same method of specifying the position of a point P in a plane. We draw two lines at right angles, $X'OX$ $Y'OY$, meeting at O. (Usually $X'OX$ is drawn horizontally, or from west to east, and $Y'OY$ vertically upwards, or from south to north, but that is merely a convenient convention. O is called the *origin*, $X'OX$ the *x-axis*, and $Y'OY$ the *y-axis*.) If we draw the perpendicular PU from P onto the x-axis (Fig. 5.2), then if the distance OU (say) $= x$, and $UP = y$, measured in suitable units, we can reach P from O by going a distance x "east" and a distance y "north". x and y are the *cartesian co-ordinates* of P with respect to the reference axes $X'OX$ and $Y'OY$, x being the x-co-ordinate (or *abscissa*, from the Latin "cut off") and y the y-co-ordinate (or *ordinate*, from the Latin "set up"): and we write for short $P = (x, y)$. If the point lies on the left of the y-axis, or to the "west", then its x-co-ordinate is naturally counted as negative. If it lies below ("south" of) the x-axis, then its y-co-ordinate is negative. Thus in Fig. 5.2, Q has a negative x-co-ordinate, but positive y, while R has both co-ordinates negative. In this way each point of the plane can be specified by two co-ordinates; this is exactly the system used in plotting graphs. In using the system for geometric purposes, however, we always suppose that x and y are measured in the same units, whereas in drawing a graph it may be helpful to use different scales on the two axes. (The name *cartesian* for this system of co-ordinates is, of course, in honour of Descartes, its inventor.)

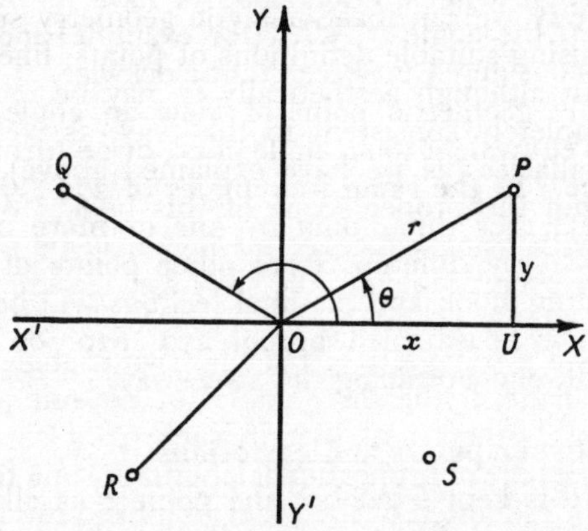

Fig. 5.2—Cartesian and polar co-ordinates of a point P

Another perfectly good way of specifying the position of the treasure is to say how far to go in a direct line, and in what direction to go—e.g. we could say "walk 50 steps at an angle of 36·9° north of east". In general we can use the *polar co-ordinates* of P, namely $r =$ the distance OP, $\theta =$ the angle $\angle XOP$, and we shall write $P = \{r, \theta\}$ (using brace brackets). r is here always taken to be positive.

5·3 Measurement of angles

An angle is essentially measured as an amount of turning: the angle $\angle XOP$ is the turn needed to bring the line OX into coincidence with OP. The natural unit of angle is the complete turn: this is customarily divided into 360 degrees (°). The ancient sexagesimal system divides each degree into 60 minutes ($'$), and each minute into 60 seconds ($''$) so that we may speak of an angle of $19° 36' 36''$, for example. But nowadays we more often simply divide the degree decimally, writing the angle $19° 36' 36''$ as $19{\cdot}61°$. The conversion is quite simple, since $6' = {\cdot}1°$, and $36'' = {\cdot}01°$.

In certain parts of Europe another system is used: the complete turn is divided into 400 "grades", which are divided decimally. It is difficult to see much advantage in this, while it makes such familiar angles as 30° and 60° difficult to write: but it is well to be aware of its existence.

In a plane, rotation may be either clockwise or anticlockwise. We usually take the anticlockwise direction or "sense" of rotation as the positive one—that is, the sense of rotation from the positive x-axis OX to the positive y-axis OY. Thus in Fig. 5.2 the angle θ or "polar angle" of the point Q may be taken as 150°, while that for R can be taken as 225°, or else as −135°, according to whether we turn anticlockwise (positively) or clockwise (negatively) from OX. In future therefore every angle will be understood to have its own proper sign associated with it, the angle $\angle ABC$ (for example) meaning the turn *from BA to BC*, considered positive if anticlockwise.

From a purely geometric point of view an angle of 360°, or a complete turn, is equivalent to an angle of 0°, or no turn at all: the final result is the same. In the same way angles of 270°, 630°, and −90° are all equivalent: they differ only by one or more complete turns. But they may be very different from other points of view: place a person on a rotating stool, and his final feelings will be very different according to whether he is turned through 270°, 630°, or −90°, although in all cases he will end up facing the same way.

5.4 Relation between polars and cartesians

If the angle θ is kept fixed but the point P is allowed to move, then although the values of x, y, and r will alter, their ratios will remain fixed. Thus if $P' = (x', y') = \{r', \theta\}$ i.e. P' is on OP (Fig. 5.3), and

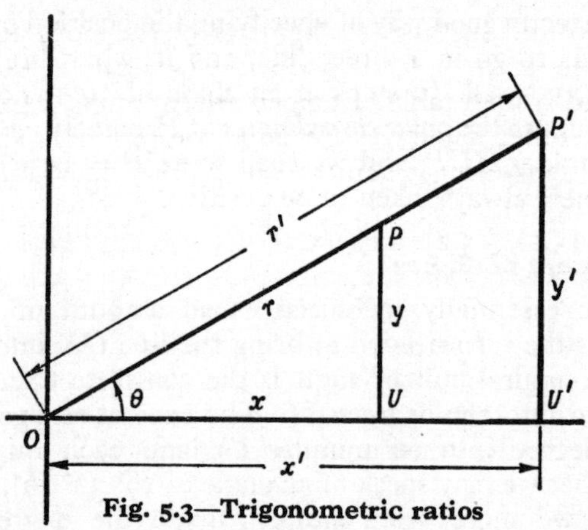

Fig. 5.3—Trigonometric ratios

$P'U'$ is the perpendicular from P' onto OU, so that U' has coordinates $(x', 0)$ the triangles OUP and $OU'P'$ are similar (having two angles, θ and 90°, the same). Accordingly $x/y = x'/y'$, $x/r = x'/r'$, and so on. This shows that these ratios depend only on the angle θ and on nothing else. There are six such ratios. We call x/r the *cosine* of the angle θ, or $\cos\theta$; y/r is the *sine* of θ, or $\sin\theta$ (pronounced "sine", not "sin"); y/x is the *tangent*, $\tan\theta$, x/y the *cotangent*, $\cot\theta$, r/x the *secant*, $\sec\theta$, and r/y the *cosecant*, $\operatorname{cosec}\theta$ or $\csc\theta$. These six ratios are known as the *trigonometric ratios* or *circular functions* of the angle θ, and their values for values of θ ranging from 0° to 90° will be found in most books of mathematical tables. We shall shortly see how they may be calculated.

Note that since x and y are cartesian co-ordinates, they may be either positive or negative, according to the position of P. As drawn in Fig. 5.3, both x and y happen to be positive, but if (for example) P was to the left of O (when θ lies between 90° and 270°) x would be negative, and if P lay below O, y would be negative. But the radius r is by definition always taken to be positive.

Although the trigonometric functions are defined as ratios, their chief use is to find the other two sides of a right-angled triangle when one side is known, together with the angle θ. For from their definitions we have at once:

Given $x = OU$, then

$$y = UP = x \tan\theta, \quad r = OP = x \sec\theta \quad . \quad . \quad (5.1)$$

Given $y = UP$, then

$$x = OU = y \cot\theta, \quad r = OP = y \operatorname{cosec}\theta \quad . \quad (5.2)$$

Given $r = OP$, then

$$x = OU = r \cos\theta, \quad y = UP = r \sin\theta \quad . \quad . \quad (5.3)$$

We can also find the angles when two of the sides are known. If, for example, we know x and y, then $y/x = \tan\theta$, is also known, and we can find from a table of tangents what angle θ gives this value of $\tan\theta$. The other angle $\angle OPU$ must be $90° - \theta$, since the three angles of the triangle must sum to 180°. Also, knowing x and θ we can find r from (5.1).

Equations (5.1), (5.2) and (5.3) accordingly give the relations between the cartesian co-ordinates (x, y) and the polar co-ordinates $\{r, \theta\}$. They may, however, be used for many other purposes. As we have seen, they give the complete solution of problems involving right-angled triangles. Other figures can usually be cut into right-angled triangles and thereby solved (whence the word "trigonometric", from the Greek for the "measurement of triangles").

EXAMPLES

(1) The roof OP of a building slopes upwards at an angle of 30°. The highest point P is 1·5 metres above the lowest point O. What is the length OP of the roof?

Complete the triangle OUP, as in Fig. 5.4. Then $OP = UP \operatorname{cosec} 30°$: from tables, $\operatorname{cosec} 30° = 2$, and UP is given to be 1·5 metres. Therefore $OP = 3$ metres.

We can see that cosec $30° = 2$ from the following construction. Draw OP', of the same length as OP, downwards from O at an angle of $-30°$. Then the triangle POP' is isosceles (with two equal sides

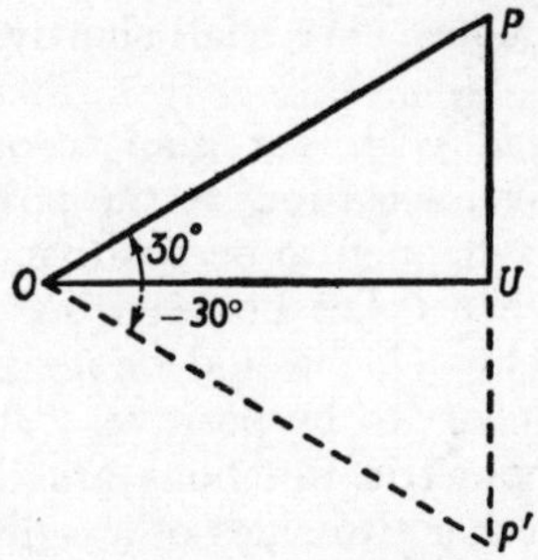

Fig. 5.4—Problem of the length of a roof

$OP = OP'$) and OU bisects PP' at U. But since the $\angle P'OP = 60°$, the triangle POP' is in fact equilateral (all three sides equal) and $OP = P'P = 2\ UP$. Thus $\sin 30° = UP/OP = \frac{1}{2}$, and $\text{cosec } 30° = OP/UP = 2$.

(2) A man wishes to find the height of a flagpole. He walks 4·5 metres away from it, and then finds that the tip P of the pole appears to be 45° above the horizon. If the man's eyes are 1·5 metres above the ground, what is the height of the pole?

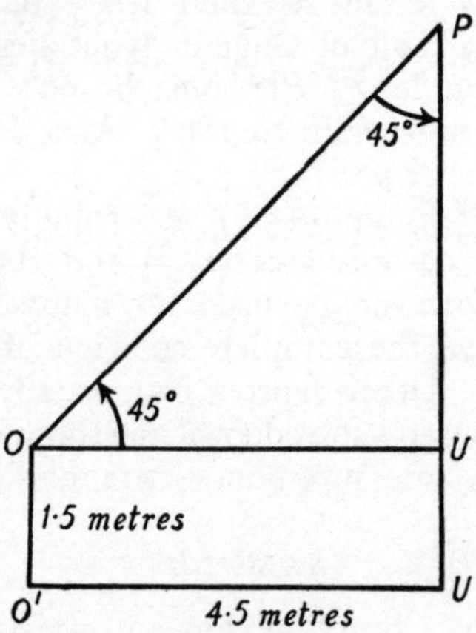

Fig. 5.5—Problem concerning the height of a flagpole

Let O be the man's eye, O' his foot, U' the foot of the pole, and U the point on the pole at the same height as O. Then $U'U = O'O = 1{\cdot}5$ m, and $UP = OU \tan 45° = 4{\cdot}5$ m, since $\tan 45° = 1$ (Fig. 5.5).

Thus the total height of the pole is $U'U + UP = 6$ metres. We can see that $\tan 45° = 1$ from Fig. 5.5: for since the three angles of the triangle OPU sum to 180°, we must have $\angle OPU = 45° = \angle UOP$ whence $OU = UP$, $\tan 45° = UP/OU = 1$.

(3) A surveyor wishes to find the height of a hill. He finds that from a point M in a level plain not far from the hill the peak P appears to be 10° above the horizontal, while from a point N 1 kilometre further away from the hill the elevation is only 5° (Fig. 5.6). What is the height y of the hill?

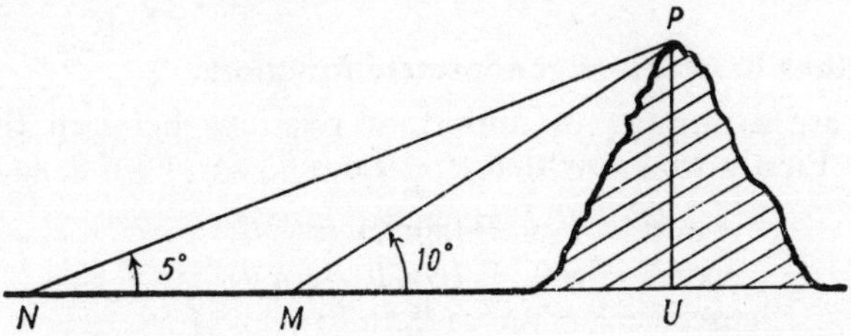

Fig. 5.6—A method of finding the height of a hill

Solution: referring to Fig. 5.6, $MU = y \cot 10°$, $NU = y \cot 5°$, and therefore $NM = NU - MU = y (\cot 5° - \cot 10°) = 5{\cdot}859\ y$ (from tables). But $NM = 1$ km $= 1000$ m, so that $y = 1000/5{\cdot}859 = 170{\cdot}6$ metres.

(4) In a triangle ABC, to find a relation between the angles and the opposite sides. Draw AD perpendicular to BC (Fig. 5.7). We shall

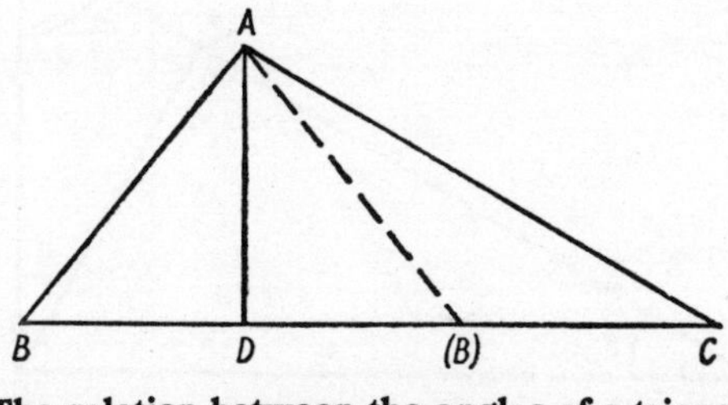

Fig. 5.7—The relation between the angles of a triangle and the opposite sides

call the angle $\angle CBA$, "$\angle B$" for short. Then $AD = AB \sin \angle B = AC \sin \angle C$; or dividing through by $\sin \angle B \sin \angle C$

$$AB/\sin \angle C = AC/\sin \angle B \quad . \quad . \quad . \quad (5{\cdot}4)$$
$$= BC/\sin \angle A \text{ by a similar argument.}$$

If we are given any side and opposite angle of a triangle, and also any one other measurement (of a side or angle), we can find all the remaining dimensions of the triangle using this formula. For example, suppose $AB = 2$, $AC = 3$, $\angle C = 30°$. Thus $2/\sin 30° = 3/\sin \angle B = BC/\sin \angle A$ or, since $\sin 30° = \frac{1}{2}$,

$$\sin \angle B = 3/4 \quad \text{and} \quad BC = 4 \sin \angle A.$$

From tables, $\angle B = 48{\cdot}6°$ or $131{\cdot}4°$, $\angle A = 180° - \angle B - \angle C = 101{\cdot}4°$ or $18{\cdot}6°$, and $BC = 3{\cdot}93$ or $1{\cdot}28$ respectively. (Thus there are two possible triangles satisfying the given conditions, one with $\angle B = 48{\cdot}6°$, $\angle A = 101{\cdot}4°$, $BC = 3{\cdot}93$, and the other with $\angle B = 131{\cdot}4°$, $\angle A = 18{\cdot}6°$, $BC = 1{\cdot}28$, shown as (*B*) in Fig. 5.7.)

5·5 Relations between trigonometric functions

There are a number of important relations between the circular functions. Firstly we know that $x = r \cos \theta$, $y = r \sin \theta$, so that

$$\left.\begin{aligned} \tan \theta &= y/x = (\sin \theta)/(\cos \theta)^* \\ \cot \theta &= x/y = (\cos \theta)/(\sin \theta) \\ \operatorname{cosec} \theta &= r/y = 1/(\sin \theta) \\ \sec \theta &= r/x = 1/(\cos \theta) \end{aligned}\right\} \quad . \quad . \quad . \quad (5{\cdot}5)$$

so that all the six functions can be expressed in terms of $\sin \theta$ and $\cos \theta$ alone.

Now in the triangle *OUP* (where $OU = x$, $UP = y$, $OP = r$ and $\angle UOP = \theta$) draw the perpendicular *UZ* from *U* onto *OP*. Then $\angle OPU = 90° - \theta$, and therefore $\angle PUZ = \theta$. Thus (Fig. 5.8)

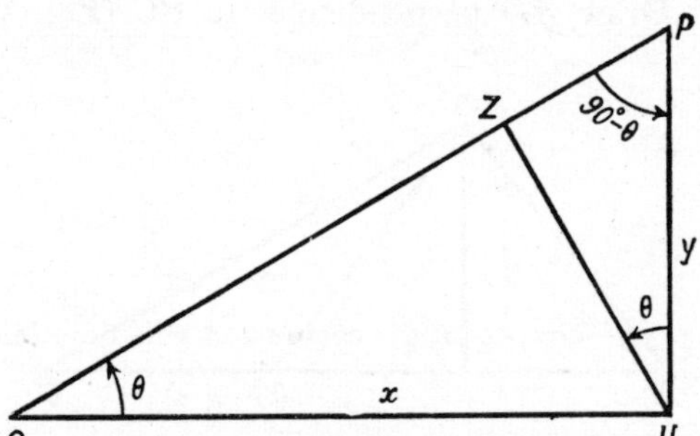

Fig. 5.8—Pythagoras's theorem

$$OZ = x \cos \theta = r \cos \theta . \cos \theta = r(\cos \theta)^2$$
$$ZP = y \sin \theta = r \sin \theta . \sin \theta = r(\sin \theta)^2$$

It is usual to write $(\cos \theta)^2$ as $\cos^2 \theta$, $(\sin \theta)^3$ as $\sin^3 \theta$, and so on; this is a simple abbreviation, easily understood, which avoids brackets.

* Expressions like $(\sin \theta)/(\cos \theta)$, $(\sin A)(\cos B)$ are often written without brackets, as $\sin \theta/\cos \theta$, $\sin A \cos B$, although this is not strictly consistent with the conventions of Section 2.10.

Thus $OZ = r\cos^2\theta$, $ZP = r\sin^2\theta$. But $OP = OZ + ZP$, i.e. $r = r(\cos\theta)^2 + r(\sin\theta)^2$. If we multiply this equation by r, we get

$$r^2 = (r\cos\theta)^2 + (r\sin\theta)^2 = x^2 + y^2 \quad . \quad . \quad (5.6)$$

which is Pythagoras's theorem connecting r, x and y. On the other hand, if we divide by r we get

$$(\cos\theta)^2 + (\sin\theta)^2 = 1, \quad . \quad . \quad . \quad (5.7)$$

a relation between $\cos\theta$ and $\sin\theta$ which can alternatively be written

$$\cos\theta = \pm\sqrt{1 - (\sin\theta)^2}$$
$$\sin\theta = \pm\sqrt{1 - (\cos\theta)^2}$$

so that when the value of $\sin\theta$ is known, that of $\cos\theta$ is fixed apart from its sign, and vice versa.

We can see what this means by considering the way in which $\cos\theta$ and $\sin\theta$ vary as θ changes, as shown by their graphs (Fig. 5.9). If

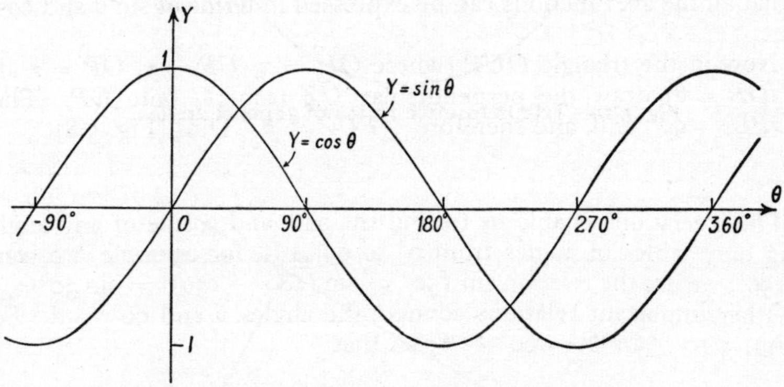

Fig. 5.9—Graphs of the cosine and sine functions

$r = 1$ then $\cos\theta = x$ and $\sin\theta = y$, i.e. $(\cos\theta, \sin\theta)$ are the co-ordinates of the point on the circle $r = 1$ in direction θ (Fig. 5.10). As θ increases from 0° to 90°, $x = \cos\theta$ decreases from 1 to 0, and $y = \sin\theta$ increases from 0 to 1. When θ increases from 90° to 180°, $\cos\theta$ becomes negative, decreasing from 0 to -1, while $\sin\theta$ remains positive, decreasing from 1 to 0. For $180° < \theta < 270°$ we have both sine and cosine negative, while for $270° < \theta < 360°$ only the sine is negative. We have now reached the initial position, and the whole sequence repeats. Thus any combination of signs of $\sin\theta$ and $\cos\theta$ is possible: in fact it is not difficult to see that

$$\cos(180° - \theta) = -\cos\theta$$
$$\sin(180° - \theta) = \sin\theta$$

$$\cos(180° + \theta) = -\cos\theta$$
$$\sin(180° + \theta) = -\sin\theta$$

$$\cos(360° - \theta) = \cos(-\theta) = \cos\theta$$
$$\sin(360° - \theta) = \sin(-\theta) = -\sin\theta \quad . \quad . \quad (5.8)$$

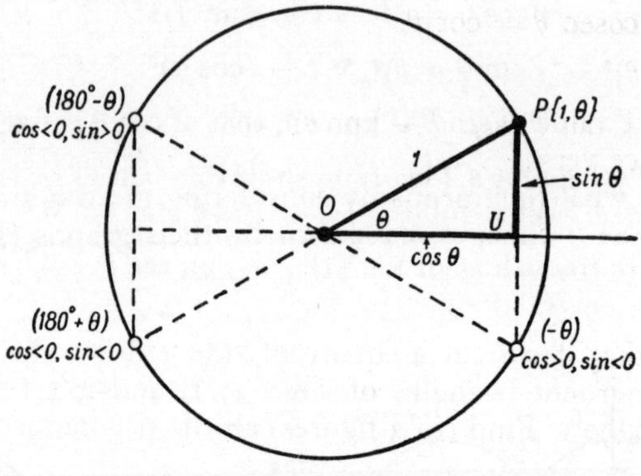

Fig. 5.10—Trigonometric ratios of general angles

These relations enable us to find the sine and cosine of any angle, using only tables of angles from 0° to 90°. If, for example, we want sin 150°, we use the relation $\sin 150° = \sin(180 - 150)° = \sin 30° = \frac{1}{2}$.

Other important relations connect the angles θ and $90° - \theta$. For in Fig. 5.10 $\angle OPU = 90° - \theta$; so that

$$\sin(90° - \theta) = OU/OP = \cos\theta$$
$$\cos(90° - \theta) = UP/OP = \sin\theta$$

$$\tan(90° - \theta) = OU/UP = \cot\theta$$
$$\cot(90° - \theta) = UP/OU = \tan\theta$$

$$\text{cosec}(90° - \theta) = OP/OU = \sec\theta$$
$$\sec(90° - \theta) = OP/UP = \text{cosec}\,\theta \quad . \quad . \quad (5.9)$$

As the triangle is drawn in Fig. 5.10 θ lies between 0° and 90°; but a little investigation will show that these relations hold for all values of θ. Fortunately they are easy to express in words: "by changing θ into $90° - \theta$, the sine is changed into the COsine, and vice versa, the tangent into the COtangent, and vice versa, and the secant into the

COsecant, and vice versa: that is, a 'co-' is either added on or taken off". In addition, the relation $\sin\theta = \cos(90° - \theta)$ has the practical use that it enables sines to be obtained from cosine tables.

Similarly the relations $\cot\theta = \tan(90° - \theta)$ and $\operatorname{cosec}\theta = \sec(90° - \theta)$ dispense with the need for special tables of cotangents and cosecants.

PROBLEMS

(1) Prove that $\tan\theta \,.\, \operatorname{cosec}\theta = \sec\theta$.

(2) $\cos\theta \,.\, \operatorname{cosec}\theta = \cot\theta$.

(3) $(\operatorname{cosec}\theta)^2 = 1 + (\cot\theta)^2$.

(4) $(\sec\theta + \tan\theta)(\sec\theta - \tan\theta) = 1$.

(5) Using Pythagoras's Theorem, show that $|\sin\theta| \leqslant 1$, $|\cos\theta| \leqslant 1$, for all values of θ. What follows for $\operatorname{cosec}\theta$ and $\sec\theta$?

(6) What are the values of $\tan(180° + \theta)$, $\sec(-\theta)$, $\sin(90° + \theta)$, $\cot(180° - \theta)$, $\cos(270° + \theta)$?

(7) Show that if we cut a square of side 1 down the diagonal we obtain two congruent triangles of sides 1, 1, and $\sqrt{2}$ ($= 1{\cdot}414$) and angles 45°, 45°, 90°. Find (to 3 figures) all the trigonometric functions of 45°.

(8) Show that by bisecting an equilateral triangle of side 2 we obtain a triangle of sides 1, $\sqrt{3}$ ($= 1{\cdot}732$), and 2, and angles 30°, 60°, 90°. Deduce the values of the trigonometric functions of 30° and 60°.

(9) Given that $\sin 15° = {\cdot}259$ and $\cos 15° = {\cdot}966$ (to 3 figures), calculate tan 15°, cot 15°, sec 15°, cosec 15°. What are the trigonometric functions of 75°?

(10) Use the results of the three previous questions to draw rough graphs of $\operatorname{cosec}\theta$, $\sec\theta$, $\tan\theta$, $\cot\theta$, and $(\sin\theta)^2$ for values of θ from 0° to 360°.

(11) From a boat sailing due east it is noticed that a buoy appears to be 15° east of north. After sailing 1 kilometre, the buoy then appears to be 30° west of north. How far was the boat from the buoy (*a*) at the beginning of the kilometre run, (*b*) at the end, (*c*) at its nearest approach?

(12) A cyclist is going up a hill at an incline of 1° at a speed of 18 kilometres per hour. Given that $\sin 1° = {\cdot}0175$, $\cos 1° = {\cdot}9998$, at what rate (in metres per second) is his height increasing? Assuming that the cycle and cyclist together weigh 70 kilograms, and that to raise 1 kilogram through 1 metre in 1 second is to do work at the rate of $g = 9{\cdot}81$ watts, how many watts is he using in overcoming the pull of gravity, apart from the friction of the cycle and air resistance?

5.6 Projection and the addition formulas

Definition: If P is any point, and PU the perpendicular from P onto the x-axis OX, then U is called the *projection of P on OX* (Fig. 5.11). If Q is another point, with projection W, then the segment UW is called the *projection of PQ on OX*. If PQ makes an angle ψ_{PQ} with the direction OX, then $UW = PQ \cos \psi_{PQ}$. In this formula UW is to be taken with the proper sign, positive if W is to the right of U,

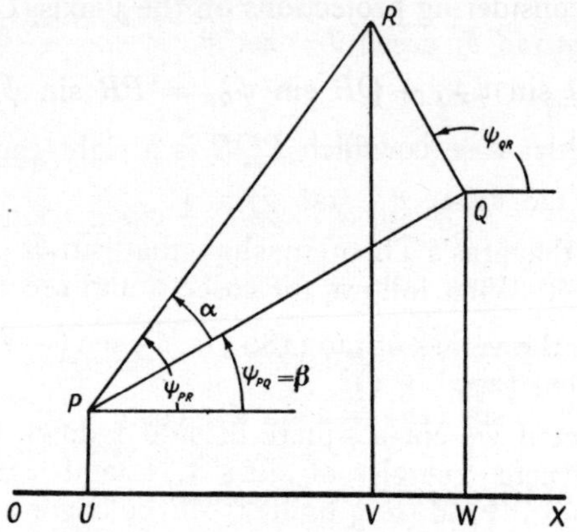

Fig. 5.11—Projection on the line OX

and negative if to the left. In other terms, if P has cartesian co-ordinates (x', y'), and Q has co-ordinates (x'', y''), then since $x' = OU$, by definition, $x'' = OW$, then $UW = OW - OU = x'' - x'$.

If R is another point (x''', y''') with projection V, and QR makes an angle ψ_{QR} with OX, then the projection of QR is

$$WV = QR \cos \psi_{QR} = x''' - x''.$$

Now the projection of PR is $UV = x''' - x'$; and if UW, WV, UV are taken with their proper signs

$$UW + WV = UV$$

[or $(x'' - x') + (x''' - x'') = (x''' - x')$]; so that if PR makes an angle ψ_{PR} with OX,

$$PQ \cos \psi_{PQ} + QR \cos \psi_{QR} = PR \cos \psi_{PR} . \qquad . \ (5.10)$$

This equation holds for *any* triangle PQR. But a particularly important case occurs when the angle at Q is a right-angle, and $PR = 1$. If then $\angle QPR = \alpha$, $PQ = \cos \alpha$ and $QR = \sin \alpha$. Also in this case, if $\psi_{PQ} = \beta$, then $\psi_{QR} = \beta + 90°$ and $\psi_{PR} = \alpha + \beta$. Thus formula (5.10) becomes

$$\cos \alpha \,.\, \cos \beta + \sin \alpha \,.\, \cos (90° + \beta) = \cos (\alpha + \beta).$$

Now $\cos(90° + \beta) = \cos[90° - (-\beta)]$

$$= \sin(-\beta) \text{ [by equation (5.9)]}$$

$$= -\sin\beta \text{ [by (5.8)], whence}$$

$$\cos(\alpha + \beta) = \cos\alpha . \cos\beta - \sin\alpha . \sin\beta \quad . \quad . \quad (5.11)$$

This is an identity, true whatever the values of the angles α and β may be.

Similarly by considering projections on the y-axis, OY, one obtains in general

$$PQ \sin\psi_{PQ} + QR \sin\psi_{QR} = PR \sin\psi_{PR}$$

and in the particular case in which PQR is a right-angled triangle

$$\sin(\alpha + \beta) = \sin\alpha . \cos\beta + \cos\alpha . \sin\beta \quad . \quad (5.12)$$

Particularly important cases of these formulas occur when $\alpha = \beta$; we then have

$$\cos(2\alpha) = (\cos\alpha)^2 - (\sin\alpha)^2$$

$$\sin(2\alpha) = 2\sin\alpha . \cos\alpha \quad . \quad . \quad . \quad (5.13)$$

5.7 Calculation of trigonometric functions

We have mentioned above the existence of tables giving the values of the trigonometric functions of angles. The reader may have wondered how these were computed in the first place. We are now in a position to explain one possible method of computation; this is a very simple method, though not the most efficient one. We have already seen from equations (5.5) that all other functions can be expressed in terms of the sine and cosine, while $\sin\theta = \cos(90° - \theta)$. Thus it is only necessary to compute a table of $\cos\theta$ in the first place; all other tables can be derived from this.

The key formula is (5.13). Since $(\sin\alpha)^2 = 1 - (\cos\alpha)^2$ [by equation (5.6)], this can be written $\cos 2\alpha = 2(\cos\alpha)^2 - 1$, or $(\cos\alpha)^2 = \frac{1}{2}(1 + \cos 2\alpha)$. Putting $2\alpha = \beta$ and taking square roots gives

$$\cos\tfrac{1}{2}\beta = \pm\sqrt{\frac{1 + \cos\beta}{2}} \quad . \quad . \quad . \quad (5.14)$$

We now begin by taking some simple and convenient value for β, say $\beta = 90°$, $\cos\beta = 0$. This gives us $\cos 45° = 1/\sqrt{2}$ (the + sign is taken, since $\cos 45°$ must be positive). From $\cos 45°$ we get at once $\cos 135° = \cos(180° - 45°) = -\cos 45° = -1/\sqrt{2}$. We now substitute in (5.14) the values $\beta = 45°$ and $\beta = 135°$, obtaining $\cos 22{\cdot}5° = \sqrt{\left[\frac{1}{2} + \frac{1}{\sqrt{8}}\right]}$ and $\cos 67{\cdot}5° = \sqrt{\left[\frac{1}{2} - \frac{1}{\sqrt{8}}\right]}$ From these we get at once cos 157·5° = cos (180° − 22·5°) and cos 112·5°. Again by substituting in formula (5.14) the values β = 22·5°, 67·5°, 112·5° and 157·5° we find the cosines of half these angles, i.e. of 11·25°, 33·75°, 56·25° and 78·75° (i.e. of 1/32, 3/32, 5/32 and 7/32 of a complete turn of 360°). Proceeding in this way we find the cosines of more and more angles—the next step gives us the cosines of all angles which are multiples of 360°/64, the step after that all multiples of 360°/128, halving each time. In time we shall get as near as we like to any given angle; and when we have enough values we can fill in the gaps in the table by interpolation.

5.8 The use of tables

The values of the sines and cosines computed as in the preceding section, or by any other convenient method, will usually be compactly presented in a table of columns of which a small part is shown in Table 5.1.

Table 5.1—Table of sines

θ	·0	·1	·2	·3	·4	·5	·6	·7	·8	·9
23°	·3907	·3923	·3939	·3955	·3971	·3987	·4003	·4019	·4035	·4051
24°	·4067	·4083	·4099	·4115	·4131	·4147	·4163	·4179	·4195	·4210

This table is to be read as sin 23·0° = ·3907, sin 23·1° = ·3923, sin 23·2° = ·3939, etc. Now suppose we were given the angle to 2 places of decimals, say as 23·13°, we would then use linear interpolation (formula 3.12). Since sin 23·1° = ·3923, sin 23·2° = ·3939, we see that as θ increases by ·1°, sin θ increases by ·0016. Therefore as θ increases by ·03°, sin θ increases by ·0016 × ·3 = ·0005 (to 4 figures). We can say that "the difference in sin θ for ·03° is ·0005", and so sin 23·13° = sin 23·1° + ·0005 = ·3928.

Now if we calculate sin 23·23° in a similar way, we find that the correction for the added ·03° is again ·0005, the same as before. The reason is that the difference between sin 23·2° and 23·3° is again ·0016, the same as between sin 23·1° and sin 23·2°. The same is true for all angles between 23° and 24°, so that we can give a table of differences applicable along the whole row.

Difference in angle	·01	·02	·03	·04	·05
Difference in sine	·0002	·0003	·0005	·0006	·0008

For compactness, however, we usually express the differences in terms of the last figure tabulated, i.e. we call ·0005 a difference of 5, it being understood that this is to be added to the last figure. Thus we find, as a rule, at the right-hand side of the table 9 "mean differences", showing the amounts which have to be added to allow for a second place of decimals in the angle.

Table 5.2—Mean differences

θ	1	2	3	4	5	6	7	8	9
23°	2	3	5	6	8	10	11	13	14
24°	2	3	5	6	8	10	11	13	14

Example. To find sin 23·77°

$$\begin{aligned} \sin 23{\cdot}7^\circ &= {\cdot}4019 \\ \text{difference for } 7 &= \phantom{{\cdot}40}11 \\ \hline \sin 23{\cdot}77^\circ &= {\cdot}4030 \end{aligned}$$

One pitfall must be noted. The use of "mean differences" as above is justified by its great simplicity and convenience, but *only if no appreciable accuracy is being lost by using the same differences for all entries in the given row of the table.* Fortunately it is easy to see at a glance when this is so: the mean differences in successive rows should not differ by more than 2 if we wish to be reasonably sure of not committing an error exceeding 1 in the last place. In the example given above the mean differences for 23° and 24° turn out to be identical, and so their use is completely justified by the above criterion. The accuracy can be increased, when using mean differences, by using differences positively or negatively up to 5 in the last place (see Section 22.10), e.g. an angle 23·37° would be considered as 23·40° − ·03°; sin 23·4° = ·3971, difference for 3 to be *subtracted* = ·0005, whence sin 23·37° = ·3966.

If the mean differences in successive rows differ by more than 2 they should not be used, but the actual correction must be calculated from the individual differences, as we have done above in evaluating sin 23·13°. The printing of mean differences in a book of tables is *not* always a sufficient guarantee that it is safe to use them!

Another difficulty with certain tables is that differences are negative, but are often printed positive, with only the most casual of warnings that they should be subtracted and not added. This occurs for example in many tables of cosines and reciprocals. In finding cos 22·63°, the usual four-figure tables give $\cos 22{\cdot}6^\circ = {\cdot}9232$, difference for $3 = 2$. This really means -2, and so $\cos 22{\cdot}63^\circ = {\cdot}9232 + (-{\cdot}0002) = {\cdot}9230$. Cosine and cosecant tables also have negative differences.

We can also use these tables in the inverse manner: given the sine or cosine, to find the angle. For example, if $\sin\theta = {\cdot}3945$, we see that $\sin 23{\cdot}2^\circ = {\cdot}3939$, the difference being ${\cdot}3945 - {\cdot}3939 = {\cdot}0006$. Now a mean difference of 6 in the sines corresponds to a difference ${\cdot}04^\circ$ in θ (from the mean difference table), so that $\theta = 23{\cdot}2 + {\cdot}04 = 23{\cdot}24^\circ$.

5·9 Trigonometric identities

The addition formulas (5.11) and (5.12) give rise to a large number of identities. Since (5.11) and (5.12) are true for *all* values of α and β, they will remain true if we replace β by $-\beta$, thus obtaining the difference formulas

$$\left.\begin{aligned}\cos(\alpha-\beta) &= \cos\alpha\,.\,\cos\beta + \sin\alpha\,.\,\sin\beta\\ \sin(\alpha-\beta) &= \sin\alpha\,.\,\cos\beta - \cos\alpha\,.\,\sin\beta\end{aligned}\right\}\quad . \qquad (5.15)$$

The formulas for angles such as $(90^\circ - \theta)$, $(180^\circ + \theta)$ are simply special cases of the general sum and difference formulas. The cosine formulas are peculiarly perverse in having a + sign in the formula for $\cos(\alpha-\beta)$ and a $-$ sign in the formula for $\cos(\alpha+\beta)$. Since $\tan\theta = (\sin\theta)/(\cos\theta)$ we have, using (5.11) and (5.12),

$$\tan(\alpha+\beta) = \frac{\sin(\alpha+\beta)}{\cos(\alpha+\beta)} = \frac{\sin\alpha.\cos\beta + \cos\alpha.\sin\beta}{\cos\alpha.\cos\beta - \sin\alpha.\sin\beta}$$

or, on dividing numerator and denominator by $\cos\alpha\,.\,\cos\beta$,

$$\tan(\alpha+\beta) = (\tan\alpha + \tan\beta)/(1 - \tan\alpha\,.\,\tan\beta)\quad . \quad . \quad (5.16)$$

In the same way

$$\tan(\alpha-\beta) = (\tan\alpha - \tan\beta)/(1 + \tan\alpha\,.\,\tan\beta)\quad . \quad (5.17)$$

PROBLEMS

(1) Prove that $\tan 2\alpha = \dfrac{2\tan\alpha}{1 - (\tan\alpha)^2}$

(2) $\cot(\alpha+\beta) = (\cot\alpha\,.\,\cot\beta - 1)/(\cot\alpha + \cot\beta)$.

(3) $\cos 3\alpha = 4(\cos\alpha)^3 - 3\cos\alpha$.

(4) $\sin 3\alpha = 3\sin\alpha - 4(\sin\alpha)^3$.

(5) $2\sin\alpha\,.\,\cos\beta = \sin(\alpha+\beta) + \sin(\alpha-\beta)$.

(6) $\sin\theta + \sin\phi = 2\sin\frac{1}{2}(\theta+\phi)\,.\,\cos\frac{1}{2}(\theta-\phi)$.

(7) $\sin(\alpha+\beta)\,.\,\sin(\alpha-\beta) = (\sin\alpha + \sin\beta)(\sin\alpha - \sin\beta)$.

(8) $\tan 3\alpha = [3\tan\alpha - (\tan\alpha)^3]/[1 - 3(\tan\alpha)^2]$.

5.10 The distance between two points*

We now proceed to apply these formulas to the study of geometric curves and shapes. First of all, however, we have to work out the formulas for such basic ideas as the distance between two points, or the angle between two lines, when expressed algebraically.

Let $P = (x, y)$ and $Q = (x', y')$ be two points in the plane. What is the distance $\delta = PQ$?

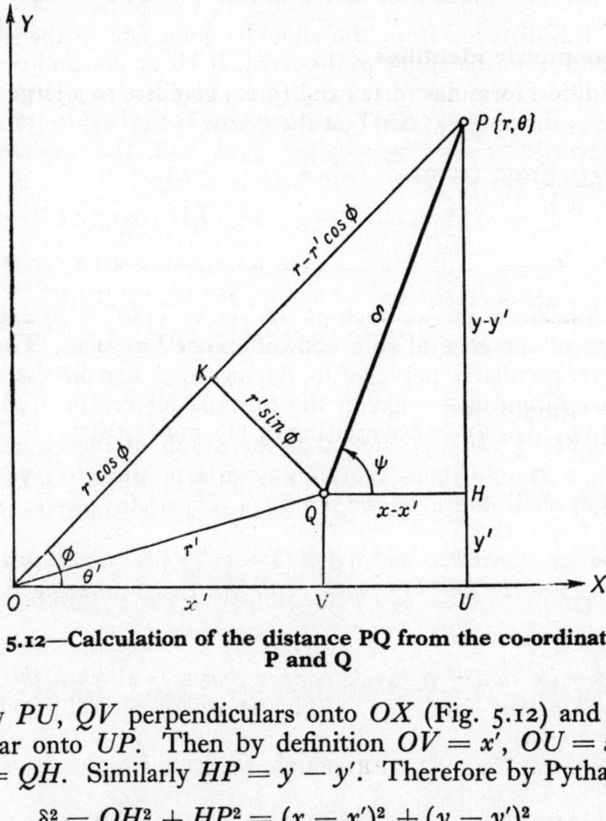

Fig. 5.12—Calculation of the distance PQ from the co-ordinates of P and Q

Draw PU, QV perpendiculars onto OX (Fig. 5.12) and QH perpendicular onto UP. Then by definition $OV = x'$, $OU = x$, $VU = x - x' = QH$. Similarly $HP = y - y'$. Therefore by Pythagoras

$$\delta^2 = QH^2 + HP^2 = (x - x')^2 + (y - y')^2 \quad . \quad . \quad (5.18)$$

If the points are given in polar co-ordinates, $P = \{r, \theta\}$, $Q = \{r', \theta'\}$. Draw QK perpendicular to OP. Let $\angle QOP = \theta - \theta' = \phi$. Then $QK = r' \sin \phi$, $OK = r' \cos \phi$, $KP = OP - OK = r - r' \cos \phi$, and

$$\begin{aligned}\delta^2 &= QK^2 + KP^2 \\ &= (r' \sin \phi)^2 + (r - r' \cos \phi)^2 \\ &= r'^2 (\sin \phi)^2 + r^2 - 2rr' \cos \phi + r'^2 (\cos \phi)^2 \\ &= r^2 + r'^2 - 2rr' \cos \phi \quad \text{[by (5.6)]} \\ &= OP^2 + OQ^2 - 2 . OP . OQ . \cos \angle QOP \quad . \quad . \quad (5.19)\end{aligned}$$

* The remainder of Chapter 5 is more specialized, and may be omitted on first reading.

This formula is known as the extension of Pythagoras. It gives the length of the third side PQ of a triangle when two sides, OP, OQ, and the angle $\angle QOP = \phi$ between them are known. If $\phi = 90°$ it becomes Pythagoras's theorem. The formula (5.19) can also be used to find an angle when the three sides of a triangle are known: it covers those cases of the solution of triangles not provided for by formula (5.4).

EXAMPLE

(1) If the distance from the shoulder joint (S) to the elbow (E) is 35 cm, and from the elbow to the wrist (W) is 25 cm, and if when the elbow is fully flexed the distance from shoulder to wrist is 20 cm, what then is the angle $\angle SEW$ at the elbow?

Solution: From (5.19),

$$SW^2 = SE^2 + EW^2 - 2 \,.\, SE \,.\, EW \cos \angle SEW$$

We have

$$20^2 = 35^2 + 25^2 - 2 \,.\, 35 \,.\, 25 \,.\, \cos \angle SEW$$

i.e.
$$\cos \angle SEW = \cdot 829, \quad \angle SEW = 34°.$$

5.11 The straight line

In Chapter 3 we suggested that the graph of the equation $y = A + Bx$ is a straight line. We are now in a position to give a formal proof. First of all, however, we require a very useful result:

Theorem 5.1. *Let* $P = (x, y)$ *and* $Q = (x', y')$ *be two distinct points in the plane. Then the line PQ makes with the x-axis an angle* ψ *given by*

$$\tan \psi = (y - y')/(x - x'),$$

with the proviso that if it should happen that $x = x'$, *then this is interpreted to mean that* $\psi = \pm 90°$. [This is a very natural proviso, since $\tan(\pm 90°) = \infty$.]

A glance at Fig. 5.12 is enough to establish the truth of this theorem.

Returning now to the graph $y = A + Bx$, let $P = (x, y)$ be *any* point on this graph. Let Q be the point at which the graph cuts the y-axis ($x = 0$). That is to say, $Q = (x', y') = (0, A)$. The equation $y = A + Bx$ can therefore be written as $(y - y') = B(x - x')$. We now have two possibilities to consider. Either $x = x'$, in which case $P = Q$, or we can divide through by $(x - x')$, giving $B = (y - y') \div (x - x') = \tan \psi$, where ψ is the inclination of PQ to the x-axis. If therefore we find ψ from the equation $\tan \psi = B$, and draw the straight line L through Q at an angle ψ with the x-axis, then the equation $y = A + Bx$ states that P must lie on L, i.e. the graph of $y = A + Bx$ is simply the straight line L (since P was defined as ***any*** point satisfying that equation).

Actually we have jumped a step in the above argument. The equation $\tan \psi = B$ has not one but *two* solutions: if one solution is Ψ, the other is $\Psi + 180°$. But this does not really matter—each solution simply covers one of the two halves into which Q splits the line L, so that the final conclusion is unimpaired.

By reversing the argument we can easily show that any straight line which is not parallel to the y-axis has a graph of the form $y = A + Bx$. The constant $B = \tan \psi$ is called the ***gradient*** or ***slope*** of the line, and we shall call the angle ψ the ***inclination***.

Having found in this way the general equation $y = A + Bx$ of a straight line L we can perform the following calculations, which correspond to the constructions we can do with rulers, set-squares and protractors in ordinary practical geometry.

(I) To find where L meets the x and y axes. On the y-axis $x = 0$, whence on substitution in the equation $y = A + Bx$, $y = A$, that is, L meets the y-axis at $(0, A)$. Similarly, provided $B \neq 0$, L meets the x-axis at $(-A/B, 0)$. For example, $y = 2 + 3x$ meets the y-axis at $(0, 2)$ and the x-axis at $(-2/3, 0)$.

(II) Given the equation $y = A + Bx$, to draw the line L (say on squared paper). The simplest method is to find two points on the line, such as $(0, A)$ and $(-A/B, 0)$, plot them, and join with a ruler.

(III) To find the line through two given points Q (x', y') and R (x'', y''). Let the line be $y = A + Bx$; then since Q and R lie on it, $y' = A + Bx'$, $y'' = A + Bx''$. We have to solve these equations for A and B. By subtraction we find that $y' - y'' = Bx' - Bx''$, whence $B = (y' - y'')/(x' - x'')$. It follows that $A = (x'y'' - x''y')/(x' - x'')$. For example, the line joining $(1, 2)$ and $(3, 5)$ is $y = \frac{1}{2} + \frac{3}{2}x$. If $x' = x''$ this construction fails, the line is then parallel to the y-axis, and has the equation $x = x'$.

(IV) To find if 3 points Q (x', y'), R (x'', y'') and S (x''', y''') lie on a straight line. If $x' \neq x''$ then the line joining Q and R has the equation $y = A + Bx$, where A and B have the values given above. The condition that S should also lie on this line is $y''' = A + Bx'''$; when this is written out at length it becomes

$$y''' = [(x'y'' - x''y') + x'''(y' - y'')]/(x' - x''),$$

and on multiplication by $(x' - x'')$ becomes

$$y'(x'' - x''') + y''(x''' - x') + y'''(x' - x'') = 0 \quad . \quad (5.20)$$

For example $(1, 2)$, $(3, 5)$ and $(7, 11)$ all lie on a straight line.

PROBLEM

(1) Show that even when $x' = x''$, equation (5.20) gives the condition for Q, R, and S to lie on a straight line.

(V) To find the equation of a line through a given point $Q\ (x', y')$ and of given direction (specified by its inclination ψ or gradient $B = \tan \psi$). From Fig. 5.12 or Theorem 5.1 we see that, if P is any point (x, y) on the line, $(y - y') = (x - x') \tan \psi = B(x - x')$, or $y = (y' - Bx') + Bx = A + Bx$ where $A = y' - Bx'$.

Example. The line of gradient $\frac{3}{2}$ through (1, 2) is $y = \frac{1}{2} + \frac{3}{2}x$.

(VI) To find the angle between two given lines L' (with equation $y = A' + B'x$) and L'' (with equation $y = A'' + B''x$). We know that L' has inclination ψ', where $\tan \psi' = B'$, and L'' has inclination ψ'', where $\tan \psi'' = B''$. The angle between L' and L'', measured as the turn from L' to L'', is therefore $\alpha = \psi'' - \psi'$. Thus

$$\begin{aligned}\tan \alpha &= \tan (\psi'' - \psi') \\ &= (\tan \psi'' - \tan \psi')/(1 + \tan \psi'' . \tan \psi') \\ &= (B'' - B')/(1 + B'B'') \quad . \quad . \quad . \quad (5.21)\end{aligned}$$

In particular the two lines are parallel, i.e. $\alpha = 0$, if $B'' = B'$. They are perpendicular, i.e. $\alpha = 90°$, if $\tan \alpha = \infty$, i.e. $1 + B'B'' = 0$, i.e. $B'B'' = -1$.

Example. $y = 3 + 2x$ is parallel to $y = 1 + 2x$ and perpendicular to $y = 3 - \frac{1}{2}x$.

(VII) To find the point of intersection of two lines L' $(y = A' + B'x)$ and L'' $(y = A'' + B''x)$. If $Q = (X, Y)$ is the point of intersection, then both equations $Y = A' + B'X$ and $Y = A'' + B''X$ must be true. By subtraction $(A' - A'') + (B' - B'')X = 0$, i.e. $X = (A'' - A') \div (B' - B'')$. By substitution in the equation $Y = A' + B'X$ we find $Y = (A''B' - A'B'')/(B' - B'')$. This method fails if $B' = B''$, when the lines L' and L'' are parallel; if $A' \neq A''$ they have then no point of intersection [since there is then no solution of the equation $(A' - A'') + (B - B'')X = 0$], while if $A' = A''$ the lines L' and L'' are then identical, and any point on one necessarily lies on the other.

Example. The lines $y = 1 + 2x$ and $y = 5 - 2x$ meet at (1, 3).

(VIII) To find if three lines L' $(y = A' + B'x)$, L'' and L''' meet in a point ("are concurrent"). If so, and L' and L'' do not coincide, then the point $Q\ (X, Y)$ of intersection of L' and L'' must lie on L''', so that $Y = A''' + B'''X$. On simplification this becomes

$$(A' - A'')\,B''' + (A'' - A''')\,B' + (A''' - A')\,B'' = 0 \quad (5.22)$$

Example. The lines $y = 1 + 2x$, $y = 7 - 4x$, and $y = 4 - x$ are concurrent.

PROBLEM

(2) Show that (5.22) is still the correct condition for concurrence even if $B' = B''$, but $B' \neq B'''$. What happens if $B' = B'' = B'''$?

(IX) To find the distance from a point $Q = (x', y')$ to a straight line L $(y = A + Bx)$.

Let QH be the perpendicular from Q onto L, and let P be an arbitrary point of L (Fig. 5.13). Then the distance QP obeys Pytha-

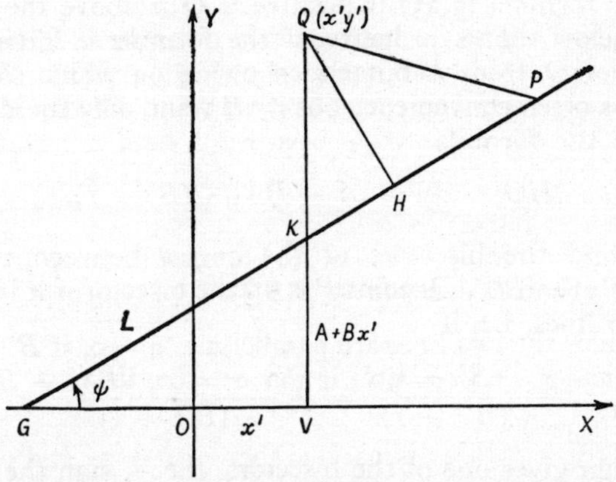

Fig. 5.13—The distance of a point Q from a line L

goras's theorem, $QP^2 = HQ^2 + HP^2$. Thus if we imagine P to move up and down the line L, the distance QP will be a minimum when P is at H, and will then be equal to the perpendicular HQ. This minimum distance is called the "distance of the point Q from the line L".

To find HQ in terms of known quantities, let L meet the x-axis at G; and let QV be the perpendicular from Q onto the x-axis, meeting L at K. Then $OV = x'$, $VQ = y'$, by definition, and since K is on L, $VK = A + Bx'$. Therefore $KQ = (y' - A - Bx')$. Moreover, since $\angle VGK = \psi$, $\angle GKV = 90° - \psi = \angle HKQ$, so that $\angle KQH = \psi$. It follows that $HQ = KQ \cos \angle KQH = (y' - A - Bx') \cos \psi$. It remains only to express $\cos \psi$ in terms of

$$B = \tan \psi = \frac{\sin \psi}{\cos \psi} = \frac{\sqrt{1 - (\cos \psi)^2}}{\cos \psi} = \sqrt{\frac{1}{(\cos \psi)^2} - 1}$$

Solving this equation, we have $B^2 = 1/(\cos \psi)^2 - 1$, or $(\cos \psi)^2 = 1/(B^2 + 1)$, or $\cos \psi = \pm 1/\sqrt{(B^2 + 1)}$. Which sign we take here depends on how we measure the angle ψ between the x-axis and the line L. There are two possible values of ψ, differing by 180°, and one of them will always lie between $-90°$ and $90°$. (ψ could not be equal to $\pm 90°$, since then the line L would not have an equation of the form $y = A + Bx$.) If we take this value of ψ, then $\cos \psi$ must be positive, and accordingly $= 1/\sqrt{(B^2 + 1)}$, so that

$$HQ = (y' - A - Bx')/\sqrt{(B^2 + 1)} \quad . \quad . \quad (5.23)$$

Example. The distance of the point (4, 9) from $y = 1 + \frac{3}{4}x$ is $(9 - 1 - 3)/\sqrt{(\frac{9}{16} + 1)} = 5/\frac{5}{4} = 4$.

Note on sign. The above expression for HQ is not necessarily positive, but has the same sign as $(y' - A - Bx')$, i.e. as KQ. That is, HQ given by formula (5.23) is positive if Q is above the line L, and negative if below. This property of the formula of giving not only the distance of Q from L but also showing on which side of L the point Q lies is often convenient; but if we want only the distance then we must use the formula

$$|HQ| = |y' - A - Bx'|/\sqrt{(B^2 + 1)}.$$

(X) To find the bisectors of the angles between the lines L' $(y = A' + B'x)$ and L''. A point P is on the bisector if it is equidistant from the two lines, i.e. if

$$\frac{y - A' - B'x}{\sqrt{(B'^2 + 1)}} = \pm \frac{y - A'' - B''x}{\sqrt{(B''^2 + 1)}}$$

The + sign gives one of the bisectors, the − sign the other.

PROBLEM

(3) Find the bisectors of the angles between the lines $y = 1 + \frac{3}{4}x$ and $y = 2 - \frac{4}{3}x$.

In this way we can repeat in algebraic form any construction we can do with ruler, set-square, and protractor. However, it is a great nuisance to always have to make an exception of lines parallel to the y-axis, which cannot be written in the form $y = A + Bx$. To overcome that difficulty we can write the equation of the straight line in the general form $lx + my = n$. This includes *all* straight lines: if $m \neq 0$ then it can be written as $y = n/m - lx/m$, which is in the standard form $y = A + Bx$ with $A = n/m$ and $B = -l/m$. If $m = 0$, then the equation becomes $x = n/l$, i.e. is a line parallel to the y-axis. (The case when both $l = 0$ and $m = 0$ must be excluded, since the equation becomes then simply $n = 0$, which gives no information about the point (x, y) and has accordingly no geometric meaning). All the above calculations can be repeated without difficulty in terms of the $lx + my = n$ notation, instead of the notation $y = A + Bx$.

FURTHER PROBLEMS

(4) The London suburb of Croyhurst has two main streets, High Street running south to north, and East Street running west to east. These streets intersect at Ox Cross. There are four stations. The Underground station, Croyhurst High Street, lies 300 metres south of Ox Cross, and East Croyhurst Underground is 800 metres east of Ox Cross. The two British Railways stations are Croyhurst North, on the High Street 400 metres north of Ox Cross, and Croyhurst Central on East Street, 200 metres east of the cross. Assuming that both railway lines are dead straight, find the equations which represent them. (The obvious co-ordinate axes are East Street and High Street, with distances measured in kilometres.) At what point does the British Railways line cross the Underground? What is the angle between them? If a straight road from Ox Cross passes over the railway intersection, what is its equation? How far is it from Ox Cross to the intersection? What is the nearest distance each of the railways comes to Ox Cross?

(5) What is the angle between the lines $l'x + m'y = n'$ and $l''x + m''y = n''$? What is the condition (*a*) that they should be parallel, (*b*) that they should cut at right-angles?

(6) In Fig. 5.13 P is the point $(x, A + Bx)$. Write down directly an expression for the square of the distance PQ, and find its minimum value as x varies, i.e. as P moves along L. Thus verify equation (5.23).

(7) Prove algebraically that the two bisectors of the angles between two straight lines L' and L'' cut at right angles.

(8) Prove that the three bisectors of the angles of a triangle meet in a point.

(9) Show that the equation of a straight line in polar co-ordinates is of the form $r = p \operatorname{cosec}(\theta - \psi)$. What is the meaning of "p" (i.e. what quantity does it measure?)

5.12 The division of a line

A particularly important problem is the following one. $Q = (x', y')$ and $R = (x'', y'')$ are two given points. We want to find the point $P = (X, Y)$ on the line QR such that the ratio QP/PR takes a given value, say λ/μ (Fig. 5.14). We could use the properties of straight lines and distances we have developed above, but the simplest method is the following: Draw QV, PU, RW perpendicularly onto the x-axis: then $VU = (X - x')$, $UW = (x'' - X)$. Now if QR has inclination ψ, $VU = QP \cos\psi$, $UW = PR \cos\psi$, so that $VU/UW = QP/PR = \lambda/\mu$ by hypothesis: that is

$$(X - x')/(x'' - X) = \lambda/\mu.$$

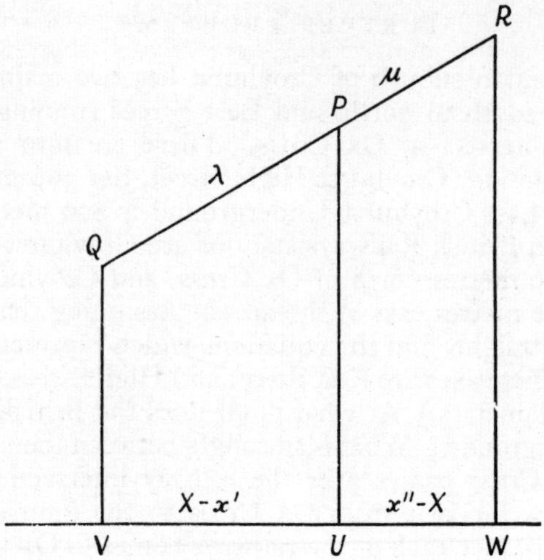

Fig. 5.14—The division of a line QR in the ratio λ : μ

Solving this equation for X we find

$$X = (\mu x' + \lambda x'')/(\mu + \lambda)$$

and by a similar argument

$$Y = (\mu y' + \lambda y'')/(\mu + \lambda) \qquad (5.24)$$

(Joachimsthal's formula). For example, $\lambda = 1$, $\mu = 1$ gives the mid-point of QR with co-ordinates $(\frac{1}{2}[x' + x''], \frac{1}{2}[y' + y''])$.

5.13 The parabola

We have already named the curve represented by the equation $y = Cx^2$, where C is a constant, a "parabola". More generally the curve $y = A + Bx + Cx^2$ has been called a parabola (Section 3.4), and it was there shown that this can be alternatively written $y + C\beta = C(x - \alpha)^2$. By shifting the origin of co-ordinates to the point $(\alpha, -C\beta)$, i.e. the maximum or minimum point, this becomes simply $y = Cx^2$. But this change of origin and axes of co-ordinates is merely an algebraic convenience, which does not affect the curve itself. We shall therefore consider only the form of equation $y = Cx^2$. It is then possible to deduce certain properties of the curve from this equation.

The origin (0, 0) is called the "vertex" of the parabola. If $C > 0$ it is the lowest, or minimum, point; if $C < 0$ it is the maximum. The y-axis $x = 0$ is the "axis" of the curve, and since the values of y at $+x$ and $-x$ are the same the curve is symmetrically placed around the axis (Fig. 5.15). The point F $(0, 1/4C)$ is called the "focus" of the parabola, and the line $y = -1/4C$ is the "directrix". The directrix meets the axis at the point D $(0, -1/4C)$ such that $FO = OD$.

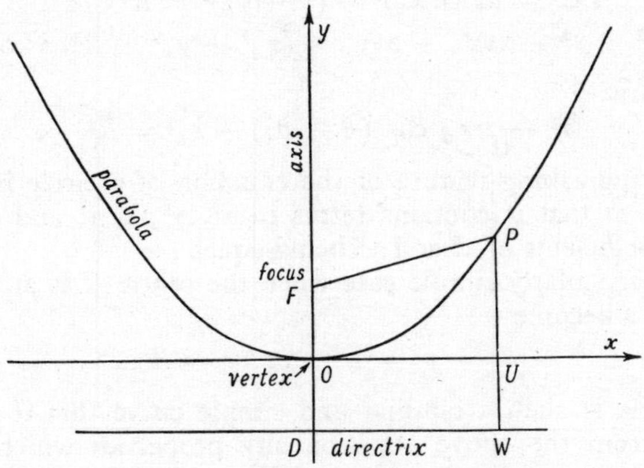

Fig. 5.15—Properties of a parabola

Now let P be any point (x, y), and let δ be the distance of P from the directrix: then

$$\delta = |y + 1/4C|$$

Also $PF^2 = x^2 + (y - 1/4C)^2$, so that

$$\begin{aligned}\delta^2 - PF^2 &= (y + 1/4C)^2 - (y - 1/4C)^2 - x^2 \\ &= (y - Cx^2)/C\end{aligned}$$

It follows that $\delta = PF$, or equivalently $\delta^2 = PF^2$ (since both δ and PF are positive), if and only if $y = Cx^2$, and P lies on the parabola. Thus a parabola may alternatively be defined as the path of a point P which moves in such a way that $PF = \delta$ where PF is the distance of P from a certain fixed point F (the "focus") and δ is the distance of P from a certain fixed line (the "directrix").

Finally we may determine the points of intersection of the parabola $y = Cx^2$ and an arbitrary straight line $y = A + Bx$. At any point of intersection (X, Y), both equations $Y = CX^2$ and $Y = A + BX$ must hold, and so by subtraction $A + BX - CX^2 = 0$. This is a quadratic equation: it may have two roots (if $B^2 + 4AC > 0$), no roots (if $B^2 + 4AC < 0$), or exceptionally a single root $(B^2 + 4AC = 0)$. Thus we see that a straight line usually meets a parabola either in two points or none. Exceptionally it may meet the curve in a single point.

5.14 The circle

A circle is the path of a moving point P whose distance PC from a given fixed point C (the "centre") has a constant value R (the "radius"). Accordingly we can readily write down the equation of a circle from the equations (5.18) and (5.19). If $C = (x_0, y_0) = \{r_0, \theta_0\}$ and $P = (x, y) = \{r, \theta\}$ the equation becomes in cartesians

$$PC^2 = (x - x_0)^2 + (y - y_0)^2 = R^2 \quad \text{or}$$

$$x^2 + y^2 - 2xx_0 - 2yy_0 + (x_0^2 + y_0^2 - R^2) = 0 \quad . \quad (5.25)$$

and in polars

$$r^2 - 2rr_0 \cos(\theta - \theta_0) + r_0^2 = R^2 \quad . \quad . \quad (5.26)$$

The distinguishing feature of the equation of a circle in cartesian co-ordinates is that it contains terms in x^2, y^2, x, y, and a constant term, the coefficients of x^2 and y^2 being equal.

In the particularly simple case when the centre C is at the origin, the equations become

$$x^2 + y^2 = R^2, \quad \text{or} \quad r = R \quad . \quad . \quad . \quad (5.27)$$

The circle is such a familiar and simple curve that it is difficult to deduce from the above equation any properties which are both useful and not immediately obvious. We may note one property of equation (5.27) of interest. Up to this point most equations have been given explicitly in the form "$y =$ a function of x"; a form which is most convenient for plotting a graph. But equation (5.27) gives a relation between x and y which confines the point P (x, y) to a curve just as effectively as an explicit relation. y is then said to be an "implicit function" of x. It is true that we can in this case readily solve (5.27) to give $y = \pm \sqrt{[R^2 - x^2]}$, but this does not affect the general principle that almost any relation between x and y will define a curve.

Incidentally another moral to be drawn from the equation of the circle $y = \pm \sqrt{[R^2 - x^2]}$ is the necessity for the use of unending decimals, as explained in Chapter 2. If we limited ourselves to exact fractional values for x and y (such as 1/4, 27/43) then for most values of x there would be no value of y, since $[R^2 - x^2]$ would usually have no exact square root. It would no longer be possible to think of a circle as a continuous curve. It might be possible to overcome this difficulty by sufficiently ingenious dodges—but on the whole the use of unending decimals gains much in simplicity and flexibility, even if they are really only a convenient fiction.

PROBLEMS

(1) What is the equation of a circle with radius 25 and centre $(0, -15)$? What are the points of intersection of this circle with the x and y axes, and with the lines $x = 7$, $x = 3y$, $2x = y + 25$?

(2) Show that in general a circle and straight line cut in two or no points, exceptionally in one.

(3) A circle passes through the three points (0, 0), (5, −5), and (8, 4). By substituting these in the general equation $(x - x_0)^2 +$

$(y - y_0)^2 = R^2$ find its centre (x_0, y_0) and its radius R. Where does the circle cut the lines $y = \frac{1}{3}x$, $4x + 3y = 20$?

(4) A circle passes through the three points (4·4, 12·3), (10·4, ·3), (−10·4, −10·1). What is its centre and its radius?

(5) Show that the points $Q = (a, b)$ and $R = (a, -b)$ lie on a circle with centre at O. Find the equation of this circle. If $P = (x, y)$ is any point on this circle, find tan $\angle QPR$, and show that this does not change as P moves round the circle.

5·15 The ellipse

If a circle is squashed it becomes an oval curve called an "ellipse". More precisely let us take a circle with centre O and radius a, and imagine it compressed in a vertical direction, so that the distance of each point from the x-axis is reduced to $1/c$th part of its value, while the distance from the y-axis is unaltered: the resulting curve is an ellipse (Fig. 5·16).

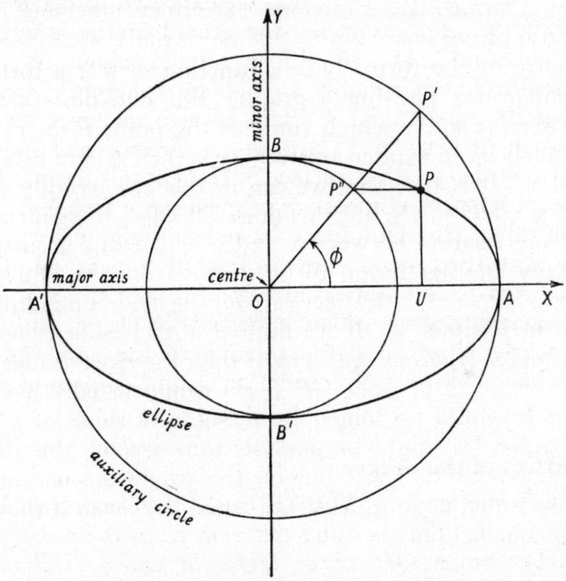

Fig. 5·16—Properties of an ellipse

If $P = (x, y)$ is a point on the ellipse, and P' the point on the circle from which it was derived by compression, then by definition $P' = (x, cy)$, or $UP = UP'/c$ in Fig. 5·16. Since P' lies on the circle of radius a, x and y must satisfy the condition $x^2 + (cy)^2 = a^2$, or on dividing through by a^2,

$$x^2/a^2 + y^2/b^2 = 1 \quad . \quad . \quad . \quad . \quad (5.28)$$

where

$$b = a/c \quad . \quad . \quad . \quad . \quad (5.29)$$

(5.28) is therefore the equation of the ellipse in cartesian co-ordinates. In polar co-ordinates, since $x = r \cos \theta$, $y = r \sin \theta$, it becomes

$$\left(\frac{r \cos \theta}{a}\right)^2 + \left(\frac{r \sin \theta}{b}\right)^2 = 1$$

From the equation (5.28) it is clear that if (x, y) is a point on the ellipse, so that $x^2/a^2 + y^2/b^2 = 1$, then $(-x, y)$, $(x, -y)$, and $(-x, -y)$ are also on the ellipse. In other words the curve is symmetrical about both the x-axis, or "major axis", and y-axis, or "minor axis". It intersects the x-axis at the points $A = (a, 0)$ and $A' = (-a, 0)$; $AA' = 2a$ is called the "(length of the) major axis". Similarly the curve cuts the minor axis at $B = (0, b)$ and $B' = (0, -b)$.

From the symmetry of the equation we deduce that we can equally well look upon an ellipse as a circle of radius b expanded in the ratio $c = a/b$ in the direction of the x-axis. (We always take the major axis to be the longer axis, so that $a > b$, and $c > 1$.)

A curve of this simple form may be expected to occur fairly often in natural objects. For example, the gravid uterus is approximately a "prolate spheroid", i.e. an ellipse rotated about its major axis, whereas a sea-urchin is roughly an "oblate spheroid", i.e. an ellipse rotated about its minor axis. Certain bivalve molluscs have elliptical lines of growth (d'Arcy W. Thompson's *Growth and Form*, p. 583) and muscles may also be elliptical (see S. Haughton's *Animal Mechanics*). In the physical sciences ellipses are of even more frequent occurrence. A penny or other circular object looked at obliquely appears to be elliptical; a disturbed pendulum swings in a near ellipse; bridges and tunnels often have elliptical arches. Most of the planets and the sun are, in varying degree, oblate spheroids in shape, and the planets have very nearly elliptical paths relative to the sun. In biological statistics we have "ellipses of correlation" and "ellipses of discrimination".

5.16 Properties of the ellipse

In Fig. 5.16 the angle $\angle XOP'$ is called the "eccentric angle" ϕ of the point P. (N.B. This is quite different from the polar co-ordinate $\theta = \angle XOP$.) Since $OP' = a$, $OU = a \cos \phi$, $UP = UP'/c = (a \sin \phi)/c = b \sin \phi$, and so $P = (a \cos \phi, b \sin \phi)$. If P'' is the point on OP' for which $OP'' = b$, then P'' has y-co-ordinate $b \sin \phi$, the same as for P, so that $P''P$ is parallel to OX. This gives a construction for the point P with eccentric angle ϕ: draw the line $OP''P'$ at inclination ϕ, cutting the circles of radius b and a at P'' and P' respectively, and draw $P''P$, PP' parallel to the axes to meet at P. By repeating this construction we can find as many points of the ellipse as we wish.

PROBLEM

(1) Verify the following alternative construction for an ellipse. Take a straight ruler HK of length $a + b$, and divided at the point P such that $HP = a$, $PK = b$. Let the end H be constrained to move along the y-axis, and K along the x-axis. Show that when $\angle HKO = \phi$, P is the point $(a \cos \phi, b \sin \phi)$, and so as the ruler HK moves, P describes an ellipse.

This construction shows us incidentally yet another way of representing a curve. An ellipse is the path described by a point $P = (a \cos \phi, b \sin \phi)$ which moves as ϕ is allowed to vary. This is called the "parametric form" of the curve, in contrast to the "implicit" and "explicit" forms of the equation, $x^2/a^2 + y^2/b^2 = 1$ and $y = \pm b \sqrt{(1 - x^2/a^2)}$ respectively. ϕ is the "parameter".

The property that an ellipse is a circle "pulled out" along its major axis makes it almost obvious that the section of a circular cylinder by

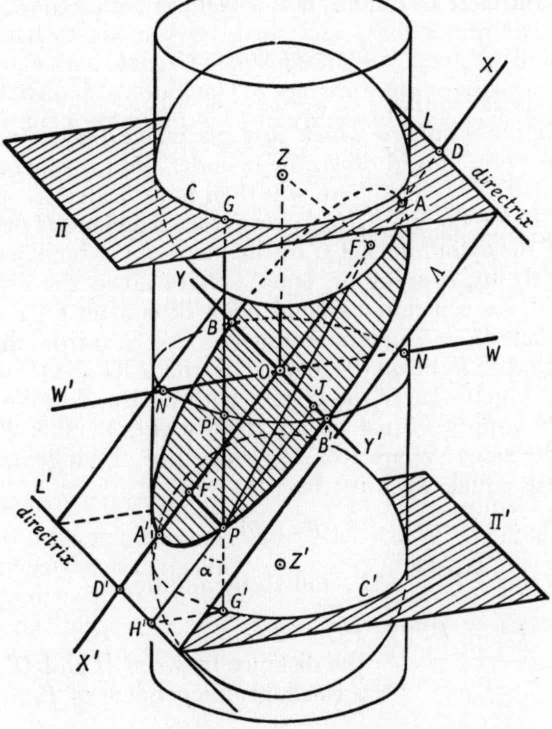

Fig. 5.17—An ellipse as a section of a cylinder

a plane Λ (not at right angles or parallel to the axis) is an ellipse. Consider a cylinder of radius b intersected by this plane Λ, which passes through a point O on the axis of the cylinder and makes an angle α with the axis (Fig. 5.17). Complete the construction as follows: a plane ($W'YWY'$) perpendicular to the axis meets the section plane Λ in the line $Y'OY$, and $W'OW$ is the line perpendicular to $Y'OY$ in this plane. $X'OX$ is perpendicular to $Y'OY$ in the cutting plane Λ. P is an arbitrary point on the curve of intersection of Λ and the cylinder, PJ the perpendicular from P onto $Y'OY$. The intersection of the plane $W'YWY'$ and the cylinder is a circle of radius b. Let P' be the point on this circle such that PP' is parallel to the axis of the cylinder.

Then referred to axes $X'OX$, $Y'OY$ the co-ordinates of P are $x = JP, y = OJ$, while referred to axes $W'OW$, $Y'OY$ the co-ordinates of P' are JP' and $OJ = y$ respectively. But $\angle P'PJ = \alpha$, so that

$$x = JP = JP' \operatorname{cosec} \alpha.$$

Since P' lies on a circle, it follows that the curve of intersection on which P lies is a circle of radius b expanded in the ratio cosec α in the direction of the x-axis OX, i.e. it is an ellipse with major axis

$$2a = 2b \operatorname{cosec} \alpha \quad . \quad . \quad . \quad (5.30)$$

and minor axis $2b$ [so that $c = a/b = \operatorname{cosec} \alpha$].

Now imagine a sphere which just fits inside the cylinder (and so has radius b) to be moved down the cylinder until it touches the plane of the ellipse at F. The sphere will then have centre Z, say, and will touch the cylinder along a circle C which lies in a plane Π perpendicular to the axis of the cylinder. Let G be the point of C which lies on the line PP'. Similarly by running an equal sphere down the cylinder from the other end, we can find a second point of contact F', a centre Z', a circle C', a plane Π' and an intersection G'. It is clear from the symmetry of Fig. 5.17 that F, F' lie on the major axis, that $F'O = OF$, $Z'O = OZ$, and that PG touches at G the sphere with centre Z. (We shall leave the reader to supply formal proofs if he wishes.) Now PF is also a tangent to the same sphere from P; and since all tangents from P to the sphere are equal, we must have

$$PF = PG.$$

In the same way $PF' = PG'$, and so by addition

$$\begin{aligned} PF + PF' &= PG + PG' \\ &= \text{the distance between } \Pi \text{ and } \Pi' \\ &= \text{a constant independent of } P. \end{aligned}$$

We can readily find the value of this constant by moving P along the ellipse until it coincides with A, the end of the major axis. Then

$$\begin{aligned} PF + PF' &= AF + AF' \\ &= AF + A'F \text{ (by symmetry)} \\ &= AA' = 2a \qquad (5{\cdot}31) \end{aligned}$$

This gives us a new definition of an ellipse: "an ellipse is the path of a point P which moves in such a way that the sum of its distances from two fixed points F and F' is constant". F is called a "focus" of the ellipse; and thus it has two foci, F and F'. We can readily find the distance OF, for in the right-angled triangle OFZ, $\angle FOZ = \alpha$ and $FZ = b$, so that

$$OF = b \cot \alpha \qquad (5{\cdot}32)$$

We can use this focal property of the ellipse for an alternative geometric construction. If we place two drawing pins at the two foci, F, F', and run round them a loop of string PFF' of appropriate length, then by keeping this loop taut by a pencil at P we can make P describe an ellipse.

Now the plane Π will cut the plane of the ellipse in a line L perpendicular to OX and cutting it at a point D. This line L is called the "directrix" of the ellipse corresponding to the focus F. Similarly there will be a second directrix L' corresponding to the focus F' (Fig. 5.17).

Now produce JP to meet L' at right angles at a point H' (JP is parallel to OX'). Then $\angle H'G'P = 90°$ and $\angle G'PH' = \alpha$, so that

$$PF' = PG' = PH' \cos \alpha \qquad (5{\cdot}33)$$

This gives yet another definition of the ellipse. PF' is the distance of P from the focus F', and PH' the distance from the directrix L'. Accordingly "an ellipse is the path of a point P which moves in such a way that its distance PF' from a fixed point F' bears a constant ratio to the distance PH' from a fixed line L',

$$PF'/PH' = \epsilon = \text{const.}"$$

Since $\epsilon = \cos \alpha$ we must add the condition that $\epsilon < 1$. If $\epsilon = 1$ we have seen in Section 5.13 that this definition gives us a parabola, not an ellipse. The constant ϵ is called the "eccentricity" of the curve.

Since from (5·30) $\sin \alpha = b/a$, we must have the connecting equation

$$\epsilon^2 + (b/a)^2 = \epsilon^2 + 1/c^2 = 1 \qquad (5{\cdot}34)$$

FURTHER PROBLEMS

(2) Draw the ellipse with semi-axes $a = 4$, $b = 3$, and verify both from the figure and from its equation that it passes through the points $(\pm 2{\cdot}4, \pm 2{\cdot}4)$ and $(\pm 3{\cdot}2, \pm 1{\cdot}8)$.

(3) Show that the distance from F to $(0, b)$ is a. Where are the foci of the ellipse of problem (2)?

(4) Find the intersections K and K' of the ellipse $x^2/a^2 + y^2/b^2 = 1$ and the line $y = l + mx$. Find the mid-point S of the line KK', and show that it lies on the line $y = -b^2x/a^2m$.

Prove that in Fig. 5.17

(5) $OF = OF' = a\epsilon$.

(6) $OD = OD' = a/\epsilon$.

(7) $OZ = OZ' = a$.

(8) Prove the focus–directrix property algebraically. Take the focus F to be the point $(a\epsilon, 0)$, the directrix L to be $x = a/\epsilon$, the point P to be (x, y) and PH to be the perpendicular from P onto L. Write down the equation $PF^2 = \epsilon^2 PH^2$ in terms of these co-ordinates, and show that it reduces to $x^2/a^2 + y^2/b^2 = 1$.

(9) Show that the focus–directrix property implies the relation $PF + PF' = 2a$.

5.17 Conic sections

Now let us consider a plane section not of a cylinder but of a circular cone. A circular cone is defined in the following way. Take a fixed straight line VZZ' called the "axis", and a fixed point V on the axis. Let a variable line VGG' through V move in such a way that the angle $\angle GVZ$ remains fixed (at, say, β). Then the line VGG' traces out a cone (Fig. 5.18 overleaf).

Now consider the intersection of the cone with a plane Λ (making an angle α greater than β with the axis of the cone) (Fig. 5.18). Let P be any point on the curve of intersection. As in the discussion for a cylinder we can imagine two spheres touching both cone and plane Λ; one has, say, centre at Z, touches the cone in a circle C and the plane Λ at a point F, and the other has centre Z', touches the cone in C' and the plane at F'. Let the line VP cut C in G, C' in G'. Then $PF = PG$, since both are tangents to the sphere; $PF' = PG'$, and $PF + PF' = PG + PG' = GG' =$ a constant. Thus by equation (5.31) the curve traced out by P, i.e. the intersection of Λ and the cone, is an ellipse with foci at F and F'. Furthermore let L be the intersection of Λ and the plane Π containing the circle C. Then $PF = PG$

$$= (\text{distance of } P \text{ from } \Pi)(\sec\beta)$$
$$= (\text{distance of } P \text{ from } L)(\sec\beta . \cos\alpha).$$

Thus L is the directrix corresponding to the focus F, and the eccentricity $\epsilon = \sec\beta . \cos\alpha$.

So far we have obtained the same curve—an ellipse—whether it is a cylinder or a cone that we cut through. But by allowing the angle α

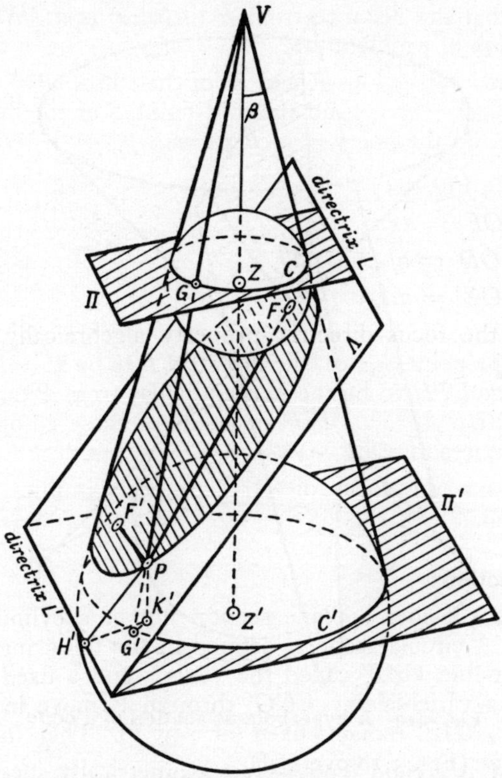

Fig. 5.18—An ellipse as section of a cone by a plane Λ

The angle between Λ and the axis VZZ' is α

between Λ and the axis of the cone to decrease we can obtain new curves, and by the sphere construction we can see that they must all have the focus-directrix property with eccentricity $\epsilon = \sec\beta . \cos\alpha$. (The change in shape can be watched by shining a conical beam of light from a torch onto a wall Λ.) As α decreases the ellipse becomes longer and the second focus F' recedes into the distance so that when $\alpha = \beta$ there is only one focus F, the eccentricity $\epsilon = \sec\beta . \cos\alpha = 1$, and the curve has therefore become a parabola. When α decreases still further we get a new curve of eccentricity $\sec\beta . \cos\alpha$, greater than 1. This curve is known as a *hyperbola*.

Now in geometry a straight line is imagined as extending indefinitely in *both* directions, so that the complete cone, defined as the figure swept out by a line through the vertex V making a fixed angle β with the axis, is really a "double cone" consisting of two ordinary or "single cones" meeting at the vertex V (Fig. 5.19). When we talk of a cone we usually think of a single cone, which can be easily constructed in

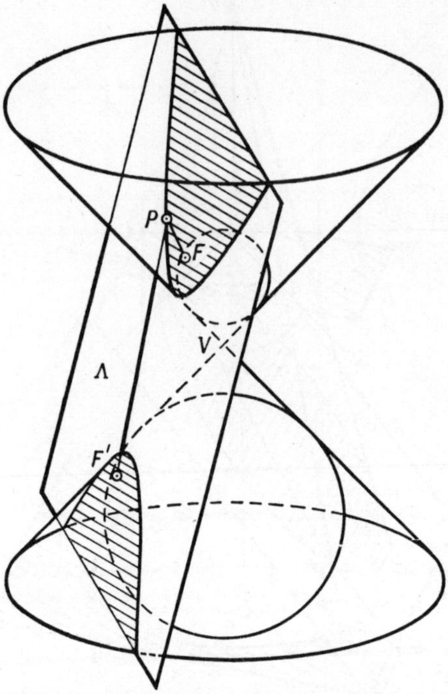

Fig. 5.19—A hyperbola as section of a cone

paper or wood as a rigid model—but geometrically speaking it is the double cone which is the complete figure. Now when $\alpha < \beta$ the cutting plane Λ intersects both halves of the double cone, so that the hyperbola splits into two parts which are not joined together. We can fit a sphere touching Λ into each single cone (Fig. 5.19), so that there are two foci F and F', one inside each half of the curve. By considering tangents from a point P on the curve to these spheres, we can show that

$$PF - PF' = \pm 2a \quad . \quad . \quad . \quad (5.35)$$

where $2a$ is a constant (the distance between the circles of contact of the two spheres measured along the cone).

We can find the equation of the hyperbola as follows (Fig. 5.20). Let F be a focus, L the corresponding directrix. Draw FD perpendicularly from F onto L and produce it to O where $OF = \epsilon^2 OD$, i.e. $OD = DF/(\epsilon^2 - 1)$. Take OF as x-axis, and a perpendicular line OY as y-axis. Denote the distance $\epsilon \times OD$ by a, so that F has co-ordinates $(a\epsilon, 0)$ and L has equation $x = a/\epsilon$. If P is any point on the curve, then $|PH|$, the perpendicular distance of P from L, is given by

$$|PH| = |x - (a/\epsilon)|,$$

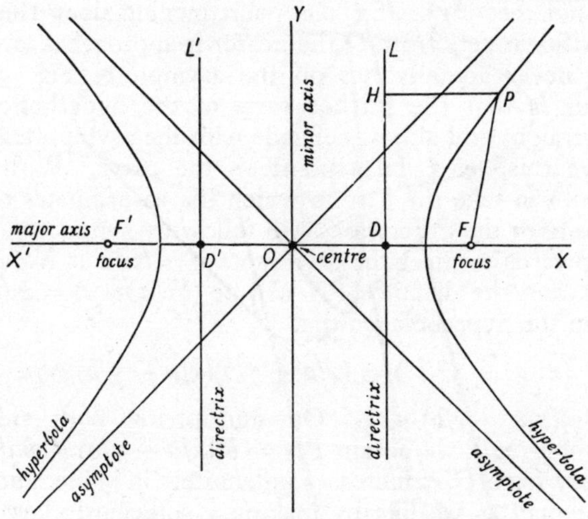

Fig. 5.20—Properties of a hyperbola

while $PF^2 = (x - a\epsilon)^2 + y^2$. But the focus–directrix property shows that $PF^2 = \epsilon^2 PH^2$, i.e. $(x - a\epsilon)^2 + y^2 = \epsilon^2(x - a/\epsilon)^2$, which reduces to

$$x^2/a^2 - y^2/b^2 = 1 \quad . \quad . \quad . \quad (5.36)$$

where $b = a\sqrt{(\epsilon^2 - 1)}$.

From this equation we see that if (x, y) lies on the curve so also do $(-x, y)$, $(x, -y)$ and $(-x, -y)$; i.e. the curve is symmetrical about both the x-axis, or "major axis", and y-axis, or "minor axis". The point O is called the "centre". The curve meets the major axis at the "vertices" $(0, \pm a)$, but does not cut the minor axis $x = 0$ at all.

A peculiarity of the hyperbola which is not shared by the ellipse or parabola is the existence of "asymptotes". The two lines $y = bx/a$, $y = -bx/a$ are called "asymptotes" (from the Greek *asýmptōtos*, not

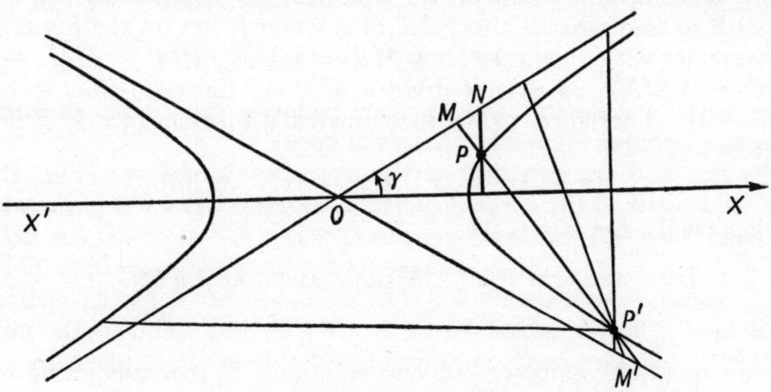

Fig. 5.21—A geometrical construction for a hyperbola

coincident, not meeting). If P is a point moving along the hyperbola, then the further it gets from O the nearer it approaches an asymptote, although it never actually lies on the asymptote (Fig. 5.21). The effect of this is that the further parts of the hyperbolic curve are practically straight and almost coincide with the asymptotes.

To prove this, let P be a point on the curve. Without loss of generality we can take the case in which the co-ordinates of P, (x, y), are both positive: the other cases will follow by symmetry. Draw PN vertically upwards to meet the asymptote $y = bx/a$ at N. N will then be $(x, bx/a)$, and the distance PN will be $(bx/a - y) = b\,(x/a - y/b)$. But P lies on the hyperbola, so that

$$(x^2/a^2 - y^2/b^2) = (x/a + y/b)(x/a - y/b) = 1$$

whence $(x/a)(x/a - y/b) \leqslant 1$. On multiplying both sides by the positive number ba/x we obtain $PN = b\,(x/a - y/b) \leqslant ba/x$. Now as P moves away from O, x increases indefinitely in value, and (ba/x) can be made as small as we like by making x sufficiently large. That is, PN can be made as small as we choose by going far enough away along the hyperbola. On the other hand the asymptote does not intersect the hyperbola, for on the asymptotes $x^2/a^2 - y^2/b^2 = 0$, whereas on the hyperbola $x^2/a^2 - y^2/b^2 = 1$. We can relate the slope of the asymptotes to the eccentricity ϵ of the hyperbola. If $\gamma = \angle XON$ (Fig. 5.21) is the angle of inclination of the asymptote $y = bx/a$, then $\tan\gamma = b/a$. But $b^2 = a^2(\epsilon^2 - 1)$, i.e. $(\tan\gamma)^2 = (\sin\gamma)^2/(\cos\gamma)^2 = \epsilon^2 - 1$. Therefore $\epsilon^2 = [(\sin\gamma)^2 + (\cos\gamma)^2]\,/\,(\cos\gamma)^2 = 1/(\cos\gamma)^2 = (\sec\gamma)^2$, and

$$\epsilon = \sec\gamma \qquad . \quad . \quad . \quad . \quad (5.37)$$

Another useful property of the asymptotes is the following one. Let M be a point on one asymptote ($y = bx/a$), and M' a point on the other ($y = -bx/a$). Let MM' intersect the hyperbola in P and P': then $MP = P'M'$. To prove this, let $M = (x, bx/a)$, $M' = (x', -bx'/a)$. Let R be a movable point on the line MM', and put $MR/MM' = \lambda$. We wish to find what is the value of λ when R lies on the hyperbola, i.e. coincides with P or P'. Now $MR = \lambda.MM'$, $RM' = MM' - MR = (1 - \lambda)\,MM'$, so that R divides MM' in the ratio λ/μ, where $\mu = (1 - \lambda)$. Therefore, using Joachimsthal's formula (5.24), R is the point

$$([x - \lambda x + \lambda x'],\ b[x - \lambda x - \lambda x']/a)$$

and lies on the hyperbola if

$$[x - \lambda x + \lambda x']^2/a^2 - b^2\,[x - \lambda x - \lambda x']^2/a^2b^2 = 1,$$

i.e. if $(4xx'/a^2)\,(\lambda - \lambda^2) = 1$.

This quadratic equation has two solutions, λ_1 (corresponding to P) and λ_2 (corresponding to P')—

$$\lambda_1 = \tfrac{1}{2} - \sqrt{\frac{xx' - a^2}{4xx'}};\ MP = \lambda_1 MM',$$

$$\lambda_2 = \tfrac{1}{2} + \sqrt{\frac{xx' - a^2}{4xx'}};\ MP' = \lambda_2 MM'.$$

But $\lambda_1 + \lambda_2 = 1$, so that $MP + MP' = MM'$, and

$$MP = MM' - MP' = P'M'.$$

This property can be used to construct a hyperbola, if its asymptotes and one point P' on the curve are given.

We draw any line through P' cutting the asymptotes at M and M', and find the point P where $MP = P'M'$. P is then also on the curve. By repeating the construction with different lines $MP'M'$ we can find as many points as we like.

PROBLEM

(1) Draw the hyperbola with asymptotes $y = \pm \tfrac{1}{2}x$ passing through (1, 0). Also find its equation, and show that it passes through $(\pm \tfrac{5}{4}, \pm \tfrac{3}{8})$ and $(\pm \tfrac{5}{3}, \pm \tfrac{2}{3})$. What are its foci, eccentricity, and directrices?

5.18 The rectangular hyperbola

If $a = b$ the asymptotes of the hyperbola are the lines $y = \pm x$, which are at right angles. The curve is then known as a "rectangular hyperbola" and has equation $x^2 - y^2 = a^2$, or in polar co-ordinates

$$r^2 (\cos\ \theta)^2 - r^2 (\sin\ \theta)^2 = a^2,$$

that is, by (5.13), $r^2 \cos 2\theta = a^2$. This equation takes a specially simple form if we take new co-ordinate axes parallel to the asymptotes, that is, at an angle of 45° to the old axes. This change of axes is most easily done in polar co-ordinates, where it amounts simply to increasing θ by 45°. Thus if Θ is the new value of the polar angle, $\Theta = \theta + 45°$, or $\theta = \Theta - 45°$. So the equation of the hyperbola $r^2 \cos 2\theta = a^2$ becomes $r^2 \cos 2(\Theta - 45°) = r^2 \cos (2\Theta - 90°) = r^2 \sin 2\Theta = 2r \sin \Theta \,.\, r \cos \Theta = a^2$; or if (X, Y) are the co-ordinates in terms of the new axes, $XY = \tfrac{1}{2}a^2 =$ constant. It is in this form that we previously encountered the equation (Section 3.9).

We have seen above that the circle, ellipse, parabola and hyperbola can all be considered as sections of a cone, and classified according to their "eccentricity" ϵ. There remains one other possible form of a conic section: if the cutting plane passes through the vertex we obtain a pair of straight lines, or possibly a single straight line or single point. All these are represented by second degree equations; see Section 18.19 for some further discussion.

CHAPTER SIX

LOGARITHMS

6.1 Powers and logarithms

Powers of 10 are easy to multiply and divide, because of the rules that $10^m \times 10^n = 10^{m+n}$ and $10^m/10^n = 10^{m-n}$. For example, $1000 \times 100 = 10^3 \times 10^2 = 10^5 = 100000$ and $1000/100 = 10^3/10^2 = 10^1 = 10$. These rules are immediate consequences of the definition of the power 10^m as m tens multiplied together. Since $10^3 = 10 \times 10 \times 10$ and $10^2 = 10 \times 10$, on multiplying these we get $10 \times 10 \times 10 \times 10 \times 10$ with $3 + 2 = 5$ tens altogether; hence $10^3 \times 10^2 = 10^5$. The index m is usually called the *common logarithm* (or Briggsian logarithm) of the power 10^m; e.g. $\log_{10} 1000 = 3$, because $1000 = 10^3$. Multiplication of numbers corresponds to addition of their logarithms; e.g. the multiplication $1000 \times 100 = 100000$ becomes in terms of the logarithms, $3 + 2 = 5$. Similarly the division $1000/100 = 10$ corresponds to a subtraction $3 - 2 = 1$ of the logarithms. Common logarithms are also known as "logarithms to the base 10". If $m = \log_{10} x$, we write $x = \text{antilog}_{10} m$. Since $m = \log x$ means by definition that $x = 10^m$, the symbol "$\text{antilog}_{10} m$" is merely another way of writing 10^m.

The same device can be used with any other series of powers, such as $2^1 = 2$, $2^2 = 4$, $2^3 = 8$, $2^4 = 16$, . . . which are said to have the respective logarithms 1, 2, 3, 4, . . . to the base 2. Thus the equation $16 \times 4 = 2^4 \times 2^2 = 2^6 = 64$ can be put in the form

$$\begin{aligned} \log_2 (16 \times 4) &= \log_2 16 + \log_2 4 \\ &= 4 \times 2 = 6, \end{aligned}$$

so that $16 \times 4 = \text{antilog}_2 6 = 2^6 = 64$. That is, to multiply numbers we add their logarithms, and take the antilog of the sum. Similarly, the relation $16/4 = 2^4/2^2 = 2^{4-2} = 2^2$ becomes

$$\log_2 (16/4) = \log_2 16 - \log_2 4 = 4 - 2 = 2,$$

whence $16/4 = \text{antilog}_2 2 = 2^2 = 4$. To divide one number by another, subtract the logarithm of the second from the logarithm of the first, and take the antilog of the difference.

Pursuing this line of thought, we say in general that if $x = b^m$, then m is the logarithm of x to base b, and x is the antilogarithm of m, or in symbols

$$(m = \log_b x) \text{ means } (x = b^m = \text{antilog}_b m)$$

It follows by a similar argument that for any base b

$$\left.\begin{aligned}\log_b(xy) &= \log_b x + \log_b y\\ \log_b(x/y) &= \log_b x - \log_b y\end{aligned}\right\} \quad . \quad . \quad (6.1)$$

provided that the logarithms exist. Also, since $\text{antilog}_b m$ is simply another way of writing b^m, we have

$$\left.\begin{aligned}\text{antilog}_b(m+n) &= (\text{antilog}_b m)(\text{antilog}_b n)\\ \text{antilog}_b(m-n) &= (\text{antilog}_b m)/(\text{antilog}_b n)\end{aligned}\right\} \quad . \quad (6.2)$$

As addition and subtraction are much simpler than multiplication and division, it would be most convenient if all multiplications or divisions could be done using logarithms, using equations (6.1) and (6.2). Unfortunately the definition, as given here, allows logarithms to be defined only for exact powers $b^1, b^2, b^3, \ldots$ of the base b, but not for numbers lying between them. Can we interpolate between these numbers, so that $\log_b x$ should be defined for all x? To put the question in another way, can we think of a new definition of logarithms, which will agree with the previous definition whenever x is a power of b, but which will also be applicable to other values of x? Such a definition will still have to satisfy (6.1) and (6.2) if it is to be practically useful.

This kind of generalization is very common in mathematics. When a technique has been found for solving one class of problems, it is natural to try to extend it to solve a wider class of problems of a similar type. For example, "division" when first invented meant only exact division, such as $6/3 = 2$: it was only later on that fractions, such as $4/3$, which are not whole numbers were thought of. Such fractions represent a useful generalization of the idea of division. Negative numbers are another example. They allow us to do subtractions such as $3 - 4 = -1$, which seem impossible when only positive numbers are allowed. We meet further examples later in the book. Often much ingenuity is needed to find a suitable method of generalization in any particular case.

6.2 The equiangular spiral

One possible method of generalizing a logarithm is suggested by the growth of bacteria. If a colony of bacteria is left on its own, with sufficient food, it will double in numbers every half-hour. If therefore we take a colony of unit size, after one half-hour it will have grown to $2 = 2^1$ units, after two half-hours to $2 \times 2^1 = 2^2$ units, and so on; after n half-hours its size will be 2^n units, provided that n is an integer. It is therefore natural to define 2^n to mean the size x of the colony after n half-hours, whether n is an integer or not, and to say that $n = \log_2 x$. This is in principle a valid method of definition, but it involves ideas about rates of growth which we only discuss later on in the book. We shall therefore adopt here instead a geometrical approach, which, we will show later, readily connects with the ideas of rates of growth.

The method is suggested by the following construction (Fig. 6.1). Let b be any positive number. Draw a line OI of unit length, and a

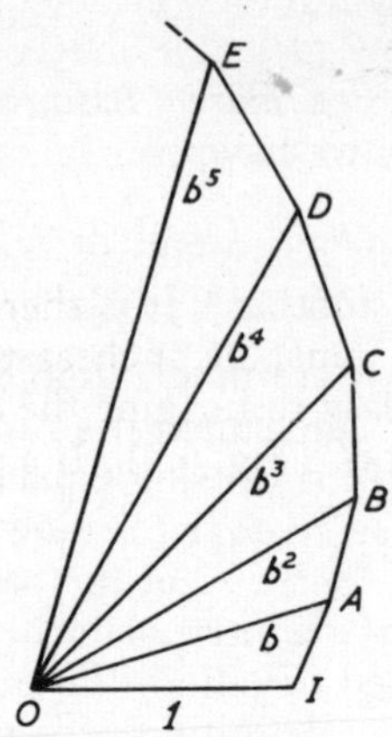

Fig. 6.1—Geometric representation of the powers of *b*

second line OA of length b. On OA draw a triangle AOB similar to IOA, and continue in the same way with further triangles BOC, COD, DOE, etc. By similarity we have

$$OA/OI = b = OB/OA = OC/OB = OD/OC = \dots$$

Thus
$$OB = b.OA = b^2$$
$$OC = b.OB = b^3$$
$$OD = b.OC = b^4$$

and so on: the lengths of the successive radii OA, OB, OC, . . . are the successive powers b^1, b^2, b^3, . . . Also, the angles $\angle IOA$, $\angle AOB$, $\angle BOC$, etc., are all equal. Now we can use any angle we wish as unit of angle, measuring others in terms of that one. We are accustomed to use a right angle or a degree as unit, but here it is more convenient to take $\angle IOA$ as the unit of angle. In that case

$$\angle IOA = 1 = \log_b b^1 = \log_b OA$$
$$\angle IOB = 2 = \log_b b^2 = \log_b OB$$
$$\angle IOC = 3 = \log_b b^3 = \log_b OC$$

and so on. The angles $\angle IOA$, $\angle IOB$, $\angle IOC$, are the respective logarithms of the corresponding lengths OA, OB, OC. Looked at in another way, if $\{r, \theta\}$ are the polar co-ordinates of the point A, then by definition $r = OA$, $\theta = \angle IOA$, and hence $\theta = \log_b r$. The same applies for all the other points B, C, D, . . .; for each of them, provided that the angular co-ordinate θ is measured in the agreed units, with $\angle IOA = 1$, then $\theta = \log_b r$, and $r = b^\theta$. The points A, B, C, . . . look as though they fall on some kind of continuous smooth curve. If we imagine such a curve passing through these points it is tempting to suppose that as $P = \{r, \theta\}$ moves along this curve, the angle θ can be called the logarithm of the radius r, and this definition will obey the fundamental relations (6.1) and (6.2). This seems plausible, but we have yet to find such a curve, and prove that it is the one we require.

Now in Fig. 6.1 we have an approximation to such a continuous curve by the series of straight lines IA, AB, BC, . . . We notice that our method of construction has made all the angles $\angle OAI$, $\angle OBA$, $\angle OCB$, . . . equal to one another. It is therefore plausible to assume that the continuous curve must be such as to cut every radius at the same angle, ϕ. Now it is easy to imagine the existence of such a curve, called an "equiangular spiral" (Fig. 6.2). If a pole is fixed in the ground

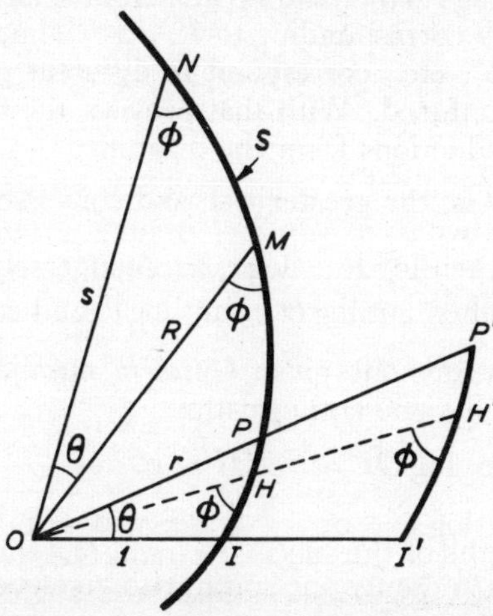

Fig. 6.2—An equiangular spiral *S*, showing the addition law for logarithms

at a centre-point O, and we start walking from a point N in such a way as to keep the point O always at a fixed angle ϕ on our right, our path will be an equiangular spiral. Some insects fly towards a light in this way, and equiangular spiral markings are often found on shells. Of course, to imagine the existence of the spiral in this way does not constitute a completely logical proof that it does exist. Such an argument would be rather detailed and complicated, so we rely here on the general reasonableness of the geometric picture. A completely rigorous proof of the existence and properties of logarithms is to be found in textbooks on the theory of functions.

Imagine, therefore, an equiangular spiral drawn as in Fig. 6.2 around a "centre" O, in the sense that every radius, or line through O, cuts the spiral at the same angle ϕ. Let OI be a radius of unit length. If P is any other point on the curve, we say that the angle $\angle IOP = \theta$, measured in suitable units, is the logarithm of the radius $OP = r$, and that OP is the antilogarithm of $\angle IOP$, i.e.

$$\theta = \log r \qquad r = \text{antilog}\,\theta.$$

Before going on to consider whether this is in fact a suitable generalization of the idea of logarithms, we first make a few observations. We shall usually suppose that ϕ lies between $0°$ and $90°$. If $\phi = 0°$, the spiral becomes a straight line, and if $\phi = 90°$, it becomes a circle around O. Both of these are useless for our purpose. Furthermore, if we imagine P moving outward along the spiral, starting from I, as it moves both θ and r will increase steadily. It may happen that θ will eventually reach $360°$, i.e. P has made one complete revolution about O along the spiral. As P has been getting further and further from O all the time, it will not return to the point I corresponding to $\theta = 0°$. Thus the angles $\theta = 0°$, $\theta = 360°$, $\theta = 720°$, etc., correspond to different points on the spiral and are not to be confused. With that proviso, the following assertions seem reasonable deductions from the diagram:

(i) the greater θ is, the greater r is, and conversely, i.e.

$$\left.\begin{array}{l}\text{if } R > r, \text{ then } \log R > \log r, \text{ and conversely} \\ \text{if } \Theta > \theta, \text{ then antilog}\,\Theta > \text{antilog}\,\theta, \text{ and conversely}\end{array}\right\} \quad . \quad (6.3)$$

It follows that, on any one given spiral, a number has at most one logarithm and at most one antilogarithm.

(ii) by definition, $\log OI = \angle IOI = 0$, i.e.

$$\log 1 = 0; \qquad \text{antilog}\, 0 = 1.$$

(iii) there is one and only one spiral with given centre O and angle ϕ through any given point I.

Another property which is almost obvious is that if we draw two further radii OM, ON, such that $\angle MON = \angle IOP$ (see Fig. 6.2), the triangles IOP, MON (with two sides straight and one curved) are similar. We can prove this as follows. Imagine the triangle IOP enlarged in the ratio $R : 1$, where $R = OM$, but keeping O fixed. This triangle IOP then enlarges into $I'OP'$. We now have $OI' = R = OM$. Since angles are unaltered by this magnification, any radius OH is magnified into a radius OH', cutting the enlarged curve $I'P'$ at the angle ϕ, i.e. $I'P'$ is according to the definition part of an equiangular spiral of centre O and angle ϕ. Now rotate the triangle $I'OP'$ about O until OI' falls on OM. By property (iii) above, the arcs $I'P'$ and MN will overlap, and because $\angle MON = \theta = \angle IOP = \angle I'OP'$, the triangle MON will coincide with $I'OP'$, which is similar to IOP.

Because of this similarity, $ON/OM = OP/OI = r$, so that

$$ON = r\,.\,OM = rR = OP\,.\,OM \quad . \quad . \quad (6.4)$$

Also

$$\begin{aligned}\log(rR) = \log ON &= \angle ION \text{ (by definition)}\\ &= \angle IOM + \angle MON\\ &= \angle IOM + \angle IOP\\ &= \log R + \log r \quad . \quad . \quad . \quad (6.5)\end{aligned}$$

Thus this definition of a logarithm does satisfy the fundamental rule for multiplication. If we put $rR = s$, so that $R = s/r$, it becomes

$$\log s = \log(s/r) + \log r,$$

whence by a slight rearrangement,

$$\log(s/r) = \log s - \log r \quad . \quad . \quad . \quad (6.6)$$

so that the division rule is also true. Thus this new definition of logarithms does obey the addition and subtraction laws as we required it to do.

EXAMPLE

By using a table of common logarithms, which we can think of as derived from a suitable spiral as above, we find

$$\log 1{\cdot}6 + \log 1{\cdot}5 = {\cdot}2041 + {\cdot}1761 = {\cdot}3802 = \log 2{\cdot}4,$$

whence $1{\cdot}6 \times 1{\cdot}5 = 2{\cdot}4$. Inversely written,

$$\log 2{\cdot}4 - \log 1{\cdot}6 = {\cdot}3802 - {\cdot}2041 = {\cdot}1761 = \log 1{\cdot}5,$$

whence $2{\cdot}4/1{\cdot}6 = 1{\cdot}5$.

Another way of writing this relation arises from (6.4). In Fig. 6.2 put $\angle IOP = \theta$, $\angle IOM = \Theta$, so that $\angle ION = \psi$ (say) $= \theta + \Theta$. Then $OP = \text{antilog}\,\theta$, $OM = \text{antilog}\,\Theta$, and $ON = \text{antilog}\,(\theta + \Theta)$, so that

$$\text{antilog}\,(\theta + \Theta) = (\text{antilog}\,\theta)\,(\text{antilog}\,\Theta). \quad . \quad . \quad (6.7)$$

Similarly, since $OM = ON/OP$, we have

$$\text{antilog}\,(\psi - \theta) = (\text{antilog}\,\psi)/(\text{antilog}\,\theta). \quad . \quad . \quad (6.8)$$

Incidentally, we see also that we can define an antilogarithm in the following way. Draw two radii OM, ON to the spiral making an angle $\angle MON = \theta$. Then $\text{antilog}\,\theta = ON/OM$. This presents the antilogarithm in the form of a ratio dependent on an angle θ, which thus partly resembles the ordinary trigonometric ratios $\sin\theta$, $\cos\theta$, etc. In fact, we will see later that there are "hyperbolic functions" $\sinh\theta$, $\cosh\theta$, etc., defined in terms of antilogarithms, which have properties which closely parallel those of the trigonometric ratios.

The analogy between this new definition of logarithms and the old one can be shown by taking an equiangular spiral with centre O, drawing a radius OI of unit length, and other radii OA, OB, OC, OD, . . . such that the angles $\angle IOA$, $\angle AOB$, $\angle BOC$, . . . are all equal (Fig. 6.3).

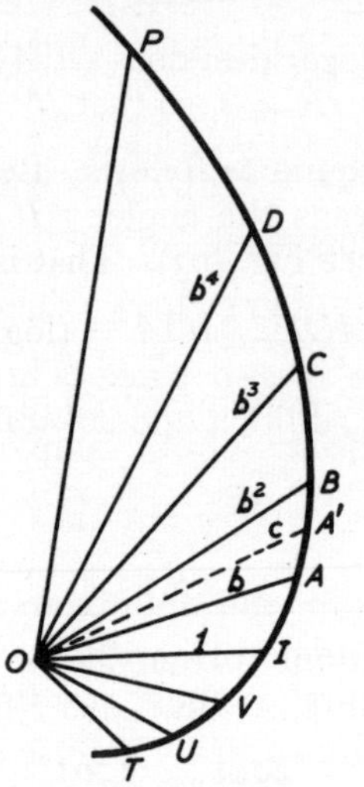

Fig. 6.3—The relation between logarithms and powers

The triangles IOA, AOB, BOC, . . . are therefore all similar, so that if $OA = b$, we have by proportion $OB = b^2$, $OC = b^3$, and so on. Thus (using the definition of logarithms as angles),

$$\left.\begin{aligned} \log b^2 &= \angle IOB = 2\,\angle IOA = 2\log b \\ \log b^3 &= \angle IOC = 3\,\angle IOA = 3\log b \end{aligned}\right\} \quad . \quad . \quad (6.9)$$

and so on (compare with Fig. 6.1).

So far we have said nothing about the unit of measurement for the angle θ. If we take $\angle IOA$ as unit (as we did in Fig. 6.1) these equations will become $\log b = 1$, $\log b^2 = 2$, $\log b^3 = 3$, etc., so that the new definition agrees exactly with the former definition of a logarithm to base b, according to which $\log b^m = m$ when m is an integer. The new definition of $\log x$ applies, however, even when x is not an exact power of b, so it is the kind of generalization we were looking for.

6.3 Change of base

If we choose any other unit of angle, say $\angle IOA'$ in Fig. 6.3, where $OA' = c$, then we now have by definition $\log c = \log OA' = \angle IOA' = 1$, i.e. the logarithms are now to base c. Thus the difference between different systems of logarithms is no more than the difference between different units of angle. Many properties of logarithms do not depend on the particular unit chosen: for example, equations (6.5) and (6.6) really state the relation between angles in Fig. 6.2,

$$\angle ION = \angle IOP + \angle IOM,$$

and this is true whatever unit we use. Thus we usually need not specify the unit in our formulas, i.e. the logarithms can be taken to any convenient base. Where numerical values are required, we shall always use common logarithms, unless otherwise stated. The values of logarithms to any other base can be found as follows. Let $r = OP$, where P is on the spiral: then $\log_b r$ means the angle $\angle IOP$ measured in units of $\angle IOA$, where $OA = b$ (see Fig. 6.3). That is,

$$\log_b r = \angle IOP / \angle IOA = (\log r)/(\log b) \quad . \quad . \ (6.10)$$

where the last two logarithms are measured in any scale. They could be taken to be common logarithms. For example,

$$\log_2 3 = (\log 3)/(\log 2) = (\cdot 4771)/(\cdot 3010) = 1\cdot 585.$$

6.4 The use of log tables for multiplication and division

In order to use logarithms in practice, it is necessary to have log tables. For notes on the use of these, see Section 5.9. Thus, to find $1\cdot 534 \times 1\cdot 618$ to four-figure accuracy we use the relation

$$\log(1\cdot 534 \times 1\cdot 618) = \log 1\cdot 534 + \log 1\cdot 618.$$

From four-figure tables of common logarithms (Appendix Table 2) we have

$$\begin{aligned} \log 1\cdot 530 &= \cdot 1847 \\ \text{difference for } 4 &= 11 \\ \hline \log 1\cdot 534 &= \cdot 1858 \end{aligned} \qquad \begin{aligned} \log 1\cdot 610 &= \cdot 2068 \\ \text{difference for } 8 &= 21 \\ \hline \log 1\cdot 618 &= \cdot 2089 \end{aligned}$$

However, $1\cdot 618$ is nearer to $1\cdot 620$ than to $1\cdot 610$, and it is simpler and more accurate to proceed thus:

$$\begin{aligned} \log 1\cdot 620 &= \cdot 2095 \\ \textit{subtract} \text{ difference for } 2 &= -5 \\ \hline \log 1\cdot 618 &= \cdot 2090 \end{aligned}$$

Hence $\quad \log(1\cdot 534 \times 1\cdot 618) = \cdot 1858 + \cdot 2090 = \cdot 3948,$

i.e. $1\cdot 534 \times 1\cdot 618 = \text{antilog} \cdot 3948$. This can be found in two ways. We have, directly from a table of antilogarithms, antilog $\cdot 3948 = 2\cdot 482$. Alternatively, we can search through a table of logarithms to find the number whose logarithm is $\cdot 3948$. The nearest tabular entries are $\log 2\cdot 48 = \cdot 3945$, and $\log 2\cdot 49 = \cdot 3962$; hence the antilog lies between $2\cdot 48$ and $2\cdot 49$, and is nearer to $2\cdot 48$. Suppose it is $2\cdot 48q$, where q stands here for the next figure (we do not mean the product $2\cdot 48$ times q). If we knew what q was, we would find $\log 2{\cdot}48q$ from the tables by looking up first $\log 2{\cdot}48 = {\cdot}3945$, and then adding on the mean difference for q; the result must come to ·3948. Hence the mean difference for q must be $\cdot 3948 - \cdot 3945 = \cdot 0003$, or 3 in the last place tabulated. We therefore search the mean difference entries in the row beginning 2·4, to see if we can find the entry "3". The nearest one is "4" in the difference column headed "2". Hence q, the last figure of the antilog, is 2, nearly enough, and the answer required is 2·482. This procedure may seem clumsy when explained at length, but in practice it is almost as quick as the use of a table of antilogs.

Logarithms can also be used for a sequence of multiplications and divisions. By repeated use of (6.5) and (6.6)

$$\log(AB/CD) = \log A + \log B - \log C - \log D.$$

Thus

$$\log[7{\cdot}2 \times 3{\cdot}5/(2{\cdot}1 \times 1{\cdot}5)] = {\cdot}8573 + {\cdot}5441 - {\cdot}3222 - {\cdot}1761 = {\cdot}9031,$$

whence $\quad 7{\cdot}2 \times 3{\cdot}5/(2{\cdot}1 \times 1{\cdot}5) = \text{antilog}\, {\cdot}9031 = 8{\cdot}000.$

PROBLEMS

(1) Find, using logarithms, $2{\cdot}41 \times 3{\cdot}42$, $2{\cdot}318 \times 1{\cdot}713$, $8{\cdot}713/2{\cdot}517$, $2{\cdot}327\pi/1{\cdot}609$, $1{\cdot}82 \times 5{\cdot}79/(3{\cdot}24 \times 1{\cdot}73)$.

(2) Using only log tables, find the antilogs of the following: ·4036, ·8485, ·9975, ·5405, ·6791.

(3) Find $\log_3 5$, $\log_3 \pi$, $\log_4 8$.

(4) Prove that $\text{antilog}_b\, x = \text{antilog}\,(x \log b)$.

Since in common logarithms $\log 10 = 1$, $\log 100 = 2$, etc., we can readily find the logarithm of any number, however great, by using a table of logarithms for numbers between 1 and 10:

$$\log 72{\cdot}1 = \log(10 \times 7{\cdot}21) = \log 10 + \log 7{\cdot}21 = 1{\cdot}8579$$
$$\log 3282 = \log(1000 \times 3{\cdot}282) = \log 1000 + \log 3{\cdot}282 = 3{\cdot}5162.$$

In short, the integral part of the logarithm (or "characteristic") of x is one less than the number of figures to the left of the decimal point in x; the fractional part of $\log x$ (or "mantissa") is found directly from the table of logs, ignoring the decimal point. In the same way, it is only necessary to tabulate antilog θ for values of θ between 0 and 1, since, for example, by (6.7),

$$\text{antilog}\, 3{\cdot}5162 = (\text{antilog}\, 3)(\text{antilog}\, {\cdot}5162) = 1000 \times 3{\cdot}282 = 3282.$$

FURTHER PROBLEMS

(5) Find the logarithms of 24·1, 342, 231·8, 3400.

(6) Find, to four-figure accuracy, the values of $24{\cdot}1 \times 342$, $231{\cdot}8 \times 3400$, $232{\cdot}7\pi/16{\cdot}09$.

(7) Find antilog 3·42, antilog 2·51, antilog 4·1872.

(8) Find $\log_5 243$, $\log_2 27$.

6.5 Logarithms of powers

From (6.9), or from an inspection of Fig. 6.3, we see that in general

$$\log b^n = n \log b. \qquad (6.11)$$

This enables one to calculate expressions such as $A^m B^n / C^r D^s$ using logarithms, for

$$\begin{aligned}\log (A^m B^n / C^r D^s) &= \log A^m + \log B^n - \log C^r - \log D^s \\ &= m \log A + n \log B - r \log C - s \log D.\end{aligned}$$

For example

$$\begin{aligned}\log [7{\cdot}2 \times 3{\cdot}5^2/(2{\cdot}1^2 \times 1{\cdot}5^3)] &= {\cdot}8753 + 2({\cdot}5441) - 2({\cdot}3222) - 3({\cdot}1761) \\ &= {\cdot}7908,\end{aligned}$$

whence

$$7{\cdot}2 \times 3{\cdot}5^2/(2{\cdot}1 \times 1{\cdot}5^3) = \text{antilog } {\cdot}7908 = 6{\cdot}177.$$

Note that (6.11) can also be written equivalently

$$b^n = \text{antilog}\,(n \log b). \qquad (6.12)$$

One important use of this is to generalize the idea of "power". So far, b^n has been defined only when n is a positive integer. However, the right-hand side of equation (6.12) exists for fractional as well as integral values of n; thus antilog$({\cdot}2 \log 2)$ = antilog$({\cdot}2 \times {\cdot}3010) = 1{\cdot}149$. Now if an expression exists for all positive values of n, and is equal to b^n when n is an integer, it is reasonable to regard it as a "generalized value of b^n". Thus in future we shall interpret the symbol "b^n" (when b is positive) to mean "antilog$(n \log b)$", or, equivalently, that b^n is defined by (6.11), whatever the value of n. There is no harm in using this extended definition, since we know that it gives $b^2 = \text{antilog}\,(2 \log b) = bb$, $b^3 = \text{antilog}\,(3 \log b) = bbb$, etc., according to the more familiar definition of powers. But it permits us also to write, for example,

$$2^{\cdot 2} = \text{antilog}\,({\cdot}2 \log 2) = 1{\cdot}149.$$

This definition has a simple geometric interpretation (Fig. 6.4). Draw radii OI, OA to an equiangular spiral, with respective lengths 1 and b. Draw the radius OQ making the angle $\angle IOQ = n \angle IOA$; then $OQ = b^n$. Thus in Fig. 6.3 it is clear that $OB = b^2$, $OC = b^3$, etc., according to the new definition as well as the old. If we draw the radius OS bisecting $\angle IOA$, then $OS = b^{\frac{1}{2}}$ (Fig. 6.4).

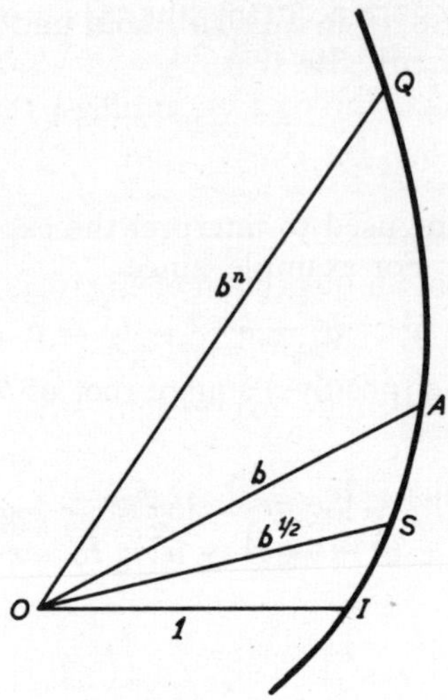

$\angle IOQ = n\angle IOA$

Fig. 6.4—Definition of b^n for general n

Do the usual laws of powers apply under this new definition? Is it still true that $b^m b^n = b^{m+n}$, that $(b^m)^n = b^{mn}$, and that $a^m b^m = (ab)^m$? Consider the first of these relations. We know that

$$m \log b + n \log b = (m + n) \log b.$$

By (6.11) this can be written

$$\log b^m + \log b^n = \log b^{m+n}$$

and by (6.5) this is equivalent to

$$b^m b^n = b^{m+n}. \quad . \quad . \quad . \quad . \quad . \quad (6.13)$$

By applying (6.11) repeatedly we have

$$\log (b^m)^n = n \log b^m = n (m \log b) = mn \log b = \log b^{mn},$$

so that

$$(b^m)^n = b^{mn}. \quad . \quad . \quad . \quad . \quad . \quad (6.14)$$

Furthermore, since $\log ab = \log a + \log b$, we have

$$m \log ab = m \log a + m \log b,$$

which by (6.11) is equivalent to

$$\log (ab)^m = \log a^m + \log b^m,$$

i.e. by (6.5) to

$$(ab)^m = a^m b^m. \qquad (6.15)$$

Thus these familiar laws of indices still hold under the new definition. The reader is left to verify also that

$$b^m/b^n = b^{m-n} \qquad (6.16)$$

$$(a/b)^m = a^m/b^m. \qquad (6.17)$$

These properties can be used to interpret the expression b^n for certain particular values of b. For example, since

$$b^{\frac{1}{2}} \times b^{\frac{1}{2}} = b^{\frac{1}{2}+\frac{1}{2}} = b^1 = b$$

it follows that $b^{\frac{1}{2}}$ is the (positive) square root of b. Hence logs can be used to find square roots:

$$\sqrt{7} = 7^{\frac{1}{2}} = \text{antilog}\,(\tfrac{1}{2}\log 7) = \text{antilog}\,(\tfrac{1}{2} \times \cdot 8451) = 2{\cdot}646.$$

Similarly, since $(b^{\frac{1}{3}})^3 = b^{\frac{1}{3}\times 3} = b^1 = b$, it follows that $b^{\frac{1}{3}}$ is the cube root of b:

$$\sqrt[3]{7} = 7^{\frac{1}{3}} = \text{antilog}\,(\tfrac{1}{3} \times \cdot 8451) = 1{\cdot}827.$$

By (6.14), $b^{\frac{3}{2}} = (b^{\frac{1}{2}})^3 = (\sqrt{b})^3$; and $b^{\frac{3}{2}} = (b^3)^{\frac{1}{2}} = \sqrt{b^3}$. Similarly $b^{p/q}$ can be interpreted as the pth power of the qth root of b. More complicated expressions can be dealt with as follows:

$$\begin{aligned}\log\,(3^{\frac{1}{3}}\pi^2/2^{\frac{3}{2}}) &= \log\,(3^{\frac{1}{3}}) + \log\,(\pi^2) - \log\,(2^{\frac{3}{2}}) \\ &= \tfrac{1}{3}\log 3 + 2\log\pi - \tfrac{3}{2}\log 2 \\ &= \cdot 1590 + \cdot 9944 - \cdot 4515 \\ &= \cdot 7019\end{aligned}$$

whence $3^{\frac{1}{3}}\pi^2/(\sqrt{2})^3 = \text{antilog}\cdot 7019 = 5{\cdot}033$.

We also have from the definition

$$b^0 = \text{antilog}\,(0 \log b) = \text{antilog}\,0 = 1. \qquad (6.18)$$

Most people find this surprising. But it fits in with Fig. 6.4: when $n = 0$ the radius OQ lies along OI. It also agrees with the rule (6.16); if we put $m = n$, we get $b^m/b^m = b^{m-m} = b^0$, which implies that $b^0 = 1$. There is a temptation, which we all feel, to suppose that an expression like b^0 which contains a 0 ought to be zero; but it is emotional rather than logical.

PROBLEMS

(1) Find $\sqrt{3}$, $\sqrt{6}$, $\sqrt[3]{9}$ using logarithms.

(2) Prove that $\sqrt{a}.\sqrt{b} = \sqrt{(ab)}$.

(3) Prove that $(\sqrt{a})/(\sqrt{b}) = \sqrt{(a/b)}$.

(4) Prove that $\text{antilog}_b\, n = b^n$ for all n.

(5) Find $(2{\cdot}63)^2(\sqrt{7{\cdot}25})/13{\cdot}3$.

6.6 Logarithms of fractions

So far we have considered only that part of the equiangular spiral lying above the line OI, i.e. with the angle θ positive. But the spiral does not begin at O; there is an arc $IVUT$ . . . curving inwards, for which the radius r is less than 1, and the angle θ is negative. All our arguments apply equally well to this part of the spiral. Its existence shows that if $r < 1$, $\theta = \log r$ is negative, and conversely that the antilogarithm of a negative number θ is less than 1.

Powers b^n can readily be extended to negative n from the definition (6.11) or (6.12). However, we can also look at the matter graphically. In Fig. 6.3 draw additional radii OV, OU, OT, where

$$\angle IOV = \angle VOU = \angle UOT = \ldots = -\angle IOA.$$

Then $\log OV = \angle IOV = -\angle IOA = (-1)\log b$, i.e. by definition, $OV = b^{-1}$. Similarly $OU = b^{-2}$, $OT = b^{-3}$. There is now a regular progression of radii . . . OD, OC, OB, OA, OI, OV, OU, OT, . . . with equal angles between successive radii, and the lengths of the radii are respectively b^4, b^3, b^2, b^1, $b^0\ (= 1)$, b^{-1}, b^{-2}, b^{-3} . . . Now the triangles DOC, COB, BOA, AOI, IOV, . . . are all similar, so the ratios OD/OC, OC/OB, OB/OA, OA/OI, OI/OV, . . . are all equal to b. Each number in this progression is therefore equal to the previous one divided by b; $OA = OB/b = b^2/b = b$. $OI = OA/b = b/b = 1$. $OV = OI/b = 1/b$. $OU = OV/b = 1/b^2$. Thus $b^{-1} = OV$ means $1/b$, and $b^{-2} = OU$ means $1/b^2$, and in general b^{-m} means $1/b^m$. This interpretation agrees with the usual rules for operating with powers:

$$1/b^m = b^0/b^m = b^{0-m} = b^{-m}.$$
$$b^3/b^2 = b^3b^{-2} = b^{3-2} = b^1 = b.$$

In the same way, $b^{-\frac{1}{2}}$ means $1/b^{\frac{1}{2}} = 1/\sqrt{b}$, and $b^{-\frac{3}{2}}$ means $1/\sqrt{b^3}$.

(*Note.* As a negative number has no logarithm, we cannot use (6.11) or (6.12) to define b^n when b is negative. However, it is usual to take b^{-m} to mean $1/b^m$ when m is an integer, even when b is negative. Note that a quantity like $(-2)^{\frac{1}{2}}$ is not defined, since (-2) has no square root. This restriction will, however, be overcome in Chapter 14.)

Numerically, we could calculate the logarithm of a fraction thus:

$$\log \cdot 5 = \log \tfrac{1}{2} = -\log 2 = -\cdot 3010.$$

It is, however, usual to write this differently. We have already shown that the logarithms of 5, 50, 500, etc., do not differ in their fractional part. They are $\log 5 = \cdot 6990$, $\log 50 = 1\cdot 6990$, $\log 500 = 2\cdot 6990$. If we wish to write $\log \cdot 05$ with the same fractional part, we must say $\log \cdot 05 = \log \frac{5}{100} = -2 + \log 5 = -2 + \cdot 6990 = \bar{2}\cdot 6990$ using the symbol $\bar{2}$ to mean -2. In the same way, $\log \cdot 005 = \bar{3}\cdot 6990$*

* These could also be written as ꝛ·6990 and Ɛ·6990, using the inverted digits ꝛ and Ɛ for -2 and -3 respectively, according to the notation suggested in Chapter 23.

written with positive fractional part, and negative integral part, as is the custom. By writing logarithms in this way we can keep the convention that the fractional part of the logarithm can be derived directly from tables, independently of the position of the decimal point in the original number: e.g.

$$\begin{aligned} \log 5{\cdot}16 &= \ {\cdot}7126 \text{ (from tables)},\\ \text{whence } \log 51{\cdot}6 &= 1{\cdot}7126\\ \log {\cdot}516 &= \bar{1}{\cdot}7126\\ \log {\cdot}0516 &= \bar{2}{\cdot}7126 \end{aligned}$$

If the first non-zero figure is the *first* after the point, then the integral part of the logarithm is $\bar{1}$; if the first non-zero figure is the *second* after the point, then the integral part is $\bar{2}$, and so on. Such a notation can be easily converted into the form $\log {\cdot}516 = -{\cdot}2874$, if desired, for $\bar{1}{\cdot}7126 = -1 + {\cdot}7126 = -(1 - {\cdot}7126) = -{\cdot}2874$. Similarly $\log {\cdot}0516 = \bar{2}{\cdot}7126 = -1{\cdot}2874$. This transformation is, however, rarely needed.

EXAMPLES

(1) Find ${\cdot}123 \times {\cdot}0234$ by logs.

We have

$$\begin{aligned} \log {\cdot}123 &= \bar{1}{\cdot}0899\\ \log {\cdot}0234 &= \bar{2}{\cdot}3692\\ \hline \log({\cdot}123 \times {\cdot}0234) &= \bar{3}{\cdot}4591 \end{aligned}$$

${\cdot}123 \times {\cdot}0234 = \text{antilog } \bar{3}{\cdot}4591 = {\cdot}002878$. In doing the addition the fractions are added as usual, and the integral parts thus, $\bar{1} + \bar{2} = \bar{3}$.

(2) Find ${\cdot}0123/{\cdot}234$ by logs.

We have

$$\begin{aligned} \log {\cdot}0123 &= \bar{2}{\cdot}0899\\ \log {\cdot}234 &= \bar{1}{\cdot}3692\\ \hline \text{By subtraction, } \log({\cdot}0123/{\cdot}234) &= \bar{2}{\cdot}7207 \end{aligned}$$

whence ${\cdot}0123/{\cdot}234 = {\cdot}05256$. In doing the subtraction we first deal with the parts following the decimal point in the usual way, obtaining difference ${\cdot}7207$ with $\bar{1}$ to carry. For the integral part to the left of the decimal point, $\bar{1}$ (carry) $+ \bar{2} - \bar{1} = \bar{2}$.

(3) Find $\sqrt{{\cdot}123}$ by logs. We have $\log {\cdot}123 = \bar{1}{\cdot}0899$, whence $\log \sqrt{{\cdot}123} = \frac{1}{2} \times \bar{1}{\cdot}0899$. The simplest way of halving $\bar{1}{\cdot}0899$ is to write it as $(\bar{2} + 1{\cdot}0899)$, whence $\frac{1}{2} \times \bar{1}{\cdot}0899 = \frac{1}{2} \times \bar{2} + \frac{1}{2} \times 1{\cdot}0899 = \bar{1} + {\cdot}5449 = \bar{1}{\cdot}5449$, and taking the antilog, $\sqrt{{\cdot}123} = {\cdot}3507$.

These operations with numbers which are partly negative, partly positive, can be considerably simplified by the more radical method of using Colson's notation (Chapter 23).

6.7 Logarithmic comparisons

In adult human beings height ranges roughly between 1·3 and 1·9 metres, an overall range of ·6 metres, or 60 cm. In flying-fish, however, the range of length in adults is roughly between 7 and 12 cm, a difference of 5 cm. [A. Fr. Bruun, *Flying Fishes* (*Exocoetidae*) *of the Atlantic*, Dana Report No. 6 (1935), Carlsberg Foundation.] Thus the variation in men is much greater than the total length of a flying fish. This, however, is scarcely a fair method of comparison. A more reasonable measure would be the percentage variation—or let us say the ratio of the greatest and least values. For man, the ratio of the greatest to the least height is $\frac{1 \cdot 9}{1 \cdot 3} = 1 \cdot 46$; for the fish it is $\frac{12}{7} = 1 \cdot 71$. We can indeed reasonably expect most species to have ranges of variation of this order of magnitude, when expressed as a percentage or ratio—although when expressed in absolute rather than relative units the ranges will be very different.

Note: such an idea of a "range of variation", though very convenient, is difficult to express precisely. If we limit ourselves to the normal adult population, then by far the greater proportion of heights will be between 1·3 and 1·9 metres. But in a sufficiently large population there will certainly be a few giants and dwarfs outside these limits. Is it really fair to include these exceptional individuals who occur so very rarely? If one does not, where is the line to be drawn between those included and those excluded? If all individuals are supposed included, how can one be sure that some exceptionally short or exceptionally tall person has not been overlooked? We shall return to this point later (Section 20.4); for the present a fairly vague definition of the range as including almost all the population will be sufficiently adequate.

This idea of "proportional range" can readily be expressed in logarithmic form. For if x is (say) the least value of some measured quantity in a population, and X its greatest value, then $\log (X/x) = \log X - \log x$. If X/x is, say, 1·5—that is, the greatest value exceeds the least by 50 per cent, then $\log X - \log x = \log 1 \cdot 5 = \cdot 176$. But $\log X$ is the greatest value of the logarithm of the measurement in question, and $\log x$ the least: so that $\log X - \log x$ is the range of the logarithm. For example among human beings very few exceed a height of 1·9 metres, which has a logarithm 1·28, and very few are shorter than 1·3 metres, with logarithm 1·11. The range of log heights is accordingly about ·17.

We can expect a very similar range for other characters both in man and in most other species—although there will be a few special cases where the variation is exceptionally wide or exceptionally narrow.

There are a number of other cases in which it is better and simpler to consider the logarithm of a measurement or number instead of the measurement itself. Take, for example, a colony of bacteria growing in a medium which provides them with an abundance of food. Then each bacterium will split into two in its reproduction time of, say, ½ hour. In one hour the number of bacteria will be multiplied by 4, in 1½ hours by 8, in 2 hours by 16, and so on. The growth will continue in this extremely rapid fashion—a "geometric series"—until overcrowding or

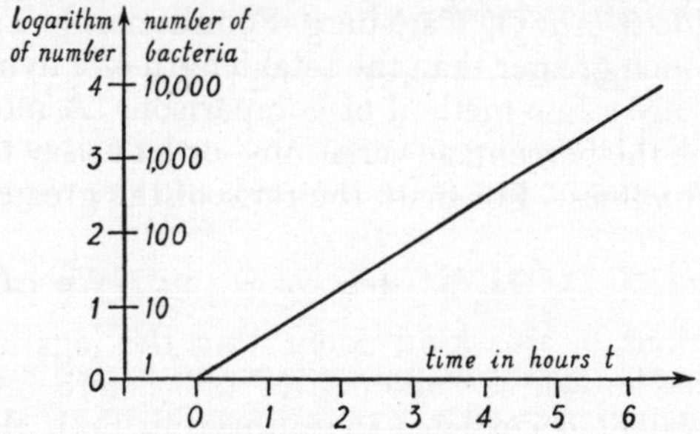

Fig. 6.5—The theoretical growth curve for a bacterial population, plotted logarithmically

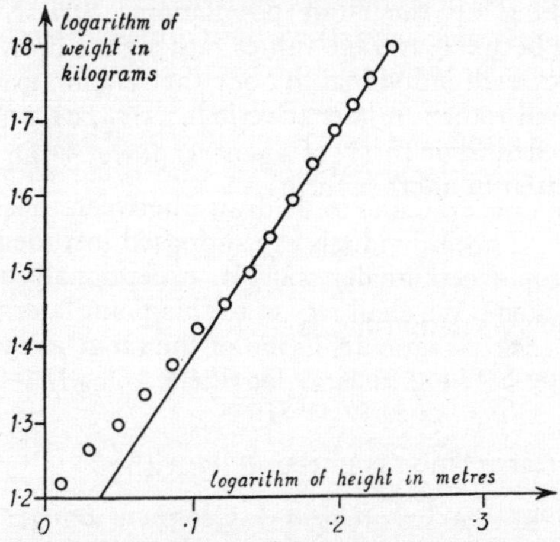

Fig. 6.6—The relation between heights and weights of schoolgirls, plotted logarithmically

exhaustion of food causes it to slow down. It is difficult to form a clear mental picture of such a rapid growth, which multiplies the colony a thousand-fold in less than 5 hours, and a million-fold in under 10 hours: and it is still more difficult to represent it adequately by a graph. But the logarithm of the number of bacteria has a steady rate of increase of log 2 = ·3010 every half-hour, or ·6020 per hour: and plotted against the time it will give a straight-line graph (Fig. 6.5).

A graph can often be made straight by plotting the logarithm of one measurement against the logarithm of the other, instead of using the original measurements as co-ordinates. Thus if we plot the logarithm of the weights of schoolgirls against their heights, using the data of Section 3.5, we obtain the graph of Fig. 6.6, which is practically straight except at the lower end.

Other examples of a linear relation between logarithms are:

(1) Dreyer's relationship between the weight W of a person (in kilograms) and the sitting height λ (in metres):

$$\log W = 1{\cdot}9532 + 3{\cdot}135 \log \lambda$$

(2) Du Bois's relationship between the surface area S of the body (in square metres), the height H (in metres), and the weight W (in kilograms):

$$\log S = -{\cdot}6937 + {\cdot}425 \log W + {\cdot}725 \log H$$

Linear equations of this kind connecting the logarithms of body measurements are known as "allometric" or "heterogonic" relations. They are of frequent occurrence in connecting both the changes in body measurements as an organism grows to maturity, and also in relating the different values of these measurements among the separate individuals making up the adult population. Strictly speaking the relations will be between average values rather than individual values—we cannot expect each individual to obey the relation exactly: but that is a point we shall return to later (Sections 21.14, 21.15, Vol. 2).

By using the definition (6.11) of a general power b^m Dreyer's relationship can be written in another form:

$$\log W = \log 89{\cdot}77 + \log(\lambda^{3\cdot135})$$

i.e.

$$W = 89{\cdot}77\,\lambda^{3\cdot135}$$

Similarly Du Bois's relationship is

$$\log S = \log {\cdot}2024 + \log W^{\cdot425} + \log H^{\cdot725}$$

or

$$S = {\cdot}2024\, W^{\cdot425} H^{\cdot725}.$$

In general, a linear relation connecting logarithms

$$\log w = A + B \log x + C \log y + D \log z$$

can equally well be written

$$w = K x^B y^C z^D$$

where $K = \operatorname{antilog} A$. The two forms are equivalent.

EXAMPLES

(1) The formula $C = \cdot9705\ \lambda^{1\cdot144}$ relating the chest circumference C and sitting height λ, both measured in metres, is a modification of a formula given by Dreyer. If $\lambda = 1\cdot05$, what is the value of C given by this formula? The formula is equivalent to

$$\log C = \log \cdot9705 + 1\cdot144 \log \lambda = \bar{1}\cdot9870 + 1\cdot144 \log \lambda\ . \quad (6.19)$$

If $\lambda = 1\cdot05$, then $\log \lambda = \cdot0212$. Therefore $\log C = \bar{1}\cdot9870 + \cdot0243 = \cdot0113$, $C = 1\cdot027$ metres.

(2) Invert the formula of example (1) to express λ in terms of C.

This is best done by using the formula in its logarithmic form (6.19). We then have
$$\begin{aligned} 1\cdot144 \log \lambda &= \log C - \bar{1}\cdot9870 \\ &= \log C + 1 - \cdot9870 \\ &= \log C + \cdot0130 \end{aligned}$$

Dividing through by $1\cdot144$ we have

$$\log \lambda = \cdot8741 \log C + \cdot0114$$

or

$$\begin{aligned} \lambda &= (\text{antilog } \cdot0114)\ C^{\cdot8741} \\ &= 1\cdot027\ C^{\cdot8741} \end{aligned}$$

(3) Find $(\cdot1024)^{-4\cdot1}$.

$\log(\cdot1024) = \bar{1}\cdot0103$ and therefore $\log(\cdot1024^{-4\cdot1}) = -4\cdot1 \times \bar{1}\cdot0103$. This multiplication can be most readily performed as follows:

$$\begin{aligned} -4\cdot1 \times \bar{1} \qquad &= +4\cdot1 \\ -4\cdot1 \times \ \cdot0103 &= -\ \cdot0422 \\ \hline \text{Total} = -4\cdot1 \times \bar{1}\cdot0103 &= +4\cdot0578 \\ \text{Therefore } \cdot1024^{-4\cdot1} &= \text{antilog } 4\cdot0578 \\ &= 1\cdot142 \times 10^4 \end{aligned}$$

PROBLEMS

(1) Meeh's formula for the surface area S of the human body, in square metres, is $S = KW^{\frac{2}{3}}$ where W is the weight in kilograms, and K is a constant which for children is $\cdot103$. It has been found by Benedict and Talbot that if L is the length of an infant (in centimetres) the amount of heat produced by it in 24 hours is $\cdot1265\ LKW^{\frac{2}{3}}$ calories. How much heat should an infant produce whose weight is 3·63 kilograms, and whose length is 52 cm.?

(2) Dreyer found the following relationship between W (weight in grams) and λ (sitting height in centimetres) of a person: $W^{\cdot319} = \cdot3803\lambda$. Find the sitting height of a person weighing 89780 grams.

6.8 Logarithms of negative numbers

Often we wish to do calculations with negative numbers which would be very well suited for the use of logs, were it not for the difficulty that negative numbers have no logarithms. We can overcome this difficulty by this device: we use for the negative number $(-x)$ the same logarithm as for the positive number x, but mark it with an "N" to show that it is the "logarithm" of a negative number. Thus, since 2 has the logarithm ·3010, − 2 will have the "logarithm" N·3010, according to this convention, and − 200 will have the logarithm N2·3010.

If then we wish to multiply $12{\cdot}3 \times (-2{\cdot}34) \times (-4{\cdot}32) \times (-32{\cdot}1)$ we could set out the work thus:

$$\begin{aligned} \log\ 12{\cdot}3 &= 1{\cdot}0899 \\ \log(-2{\cdot}34) &= \text{N}\ {\cdot}3692 \\ \log(-4{\cdot}32) &= \text{N}\ {\cdot}6355 \\ \log(-32{\cdot}1) &= \text{N}1{\cdot}5065 \\ \hline \text{Total} &= \text{N}3{\cdot}6011 \\ \text{Antilog} &= -3991 \end{aligned}$$

In adding the logarithms we are really multiplying the *positive* numbers 12·3, 2·34, 4·32, and 3·21 to get the product 3991. But since three of the logarithms are marked with an N this shows that three of the numbers we wish to multiply are actually negative, so that the product of them is also negative. The total logarithm is therefore marked with an N, and when we take its antilog we give it a minus sign. In general, if we add or subtract an odd number of N-logarithms, the result will be an N-logarithm: but if two, or any even number of N-logarithms are added or subtracted the N's will cancel and the result will be an ordinary logarithm.

EXAMPLE

(1) Multiply $(-12{\cdot}3) \times (-{\cdot}234)$

$$\begin{aligned} \log(-12{\cdot}3) &= \text{N}1{\cdot}0899 \\ \log(-{\cdot}234) &= \text{N}\bar{1}{\cdot}3692 \\ \hline \text{Total} &= {\cdot}4591 \\ \text{Antilog} &= 2{\cdot}878 \end{aligned}$$

The product 2·878 is positive since it is the product of two negative numbers: and therefore its logarithm is not marked with an N—the two N's cancel on addition.

In the same way an N-logarithm multiplied by an odd number remains an N-logarithm; but if multiplied by an even number it becomes an ordinary logarithm.

(2) Find $(-1{\cdot}23)^4$.

This is positive, since the fourth power of any number is positive.

$$\begin{aligned} \text{Now } \log(-1{\cdot}23) &= \text{N}{\cdot}0899 \\ &\quad\;\; 4 \\ \hline \log(-1{\cdot}23^4) &= \;\;{\cdot}3596 \\ \text{Taking antilogs, } (-1{\cdot}23)^4 &= \;\;2{\cdot}389 \end{aligned}$$

Finally we must notice that since a negative number has no square root we cannot halve an N-logarithm.

6.9 Calculation of a logarithm

We have not yet explained how a table of logarithms can be constructed. The values of common logarithms can be found by the following method. Suppose we require an accuracy of four decimal places. We know that $\log x^{10000} = 10000 \log x$, so that

$$\log x = \frac{1}{10000} \log(x^{10000})$$

Hence, if we know $\log(x^{10000})$ to the nearest integer, this gives $\log x$ to four decimal places. Now x^{10000} can be found by repeated multiplication. This is much less arduous than it sounds, since it is enough to work out the products to about 5-figure accuracy to get the final answer correct to 4 figures. Thus, suppose we want the value of log 2. We have $2^2 = 4$, whence $2^4 = (2^2)^2 = 16$, and $2^8 = (2^4)^2 = 16^2 = 256$. Thus $2^{10} = 2^2 \times 2^8 = 1024 = 1{\cdot}024 \times 10^3$. We now find successively 2^{20}, 2^{40} and 2^{80} by successive squaring. $2^{20} = (2^{10})^2 = 1{\cdot}0486 \times 10^6$, $2^{40} = (2^{20})^2 = 1{\cdot}0996 \times 10^{12}$, and $2^{80} = (2^{40})^2 = 1{\cdot}2091 \times 10^{24}$. It follows that $2^{100} = 2^{20} \times 2^{80} = 1{\cdot}2674 \times 10^{30}$. We continue similarly, finding 2^{200}, 2^{400}, 2^{800} by repeated squares, and

$$2^{1000} = 2^{200} \times 2^{800} = 1{\cdot}0737 \times 10^{30}.$$

Again we can find 2^{2000}, 2^{4000}, 2^{8000} by repeated squaring, and

$$2^{10000} = 2^{2000} \times 2^{8000} = 2{\cdot}0358 \times 10^{3010}.$$

Hence $\log 2^{10000}$ lies between 3010 and 3011 (and is nearer 3010), and so log 2 lies between ·3010 and ·3011 (and is nearer ·3010):

$${\cdot}3010 < \log 2 < {\cdot}3011. \quad . \quad . \quad . \quad (6.20)$$

This process can evidently be used to calculate logarithms to any desired accuracy. It can also be used to increase the accuracy of existing log tables. Thus, since $2^{10} = 1024$, if we have a four-figure log table we can find log 2 to five figures as

$$\log 2 = \tfrac{1}{10} \log 2^{10} = \tfrac{1}{10} \log 1024 = {\cdot}30103.$$

6.10 The length of a circular arc

If A and B are two points on a circle of centre O, then the length of the arc AB depends on the angle $\angle AOB = \theta$ "subtended" by the arc AB at the centre O, and also on the radius $r = OA$ of the circle (Fig. 6.7). We can readily find a formula for the arc length in the following way.

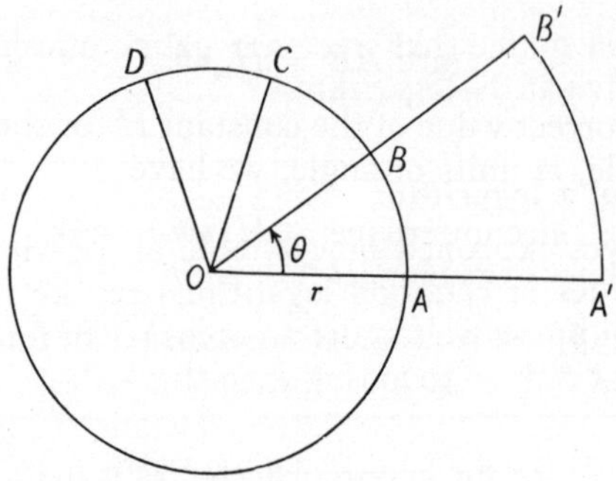

Fig. 6.7—Method of finding the length of an arc

Let C, D, be points on the circle such that $\angle BOC = \angle COD = \theta$. Then since the sectors AOB, BOC and COD are all congruent, the arcs AB, BC and CD must be all equal. Now $\angle AOC = 2\theta$, arc $AC =$ 2 . arc AB; $\angle AOD = 3\theta$, arc $AD = 3$. arc AB. If we double the angle θ we double the arc, and if we treble the angle, we treble the arc. In general it is easy to see in this way that the *arc is proportional to the subtended angle* θ. Again if we double the radius of the circle, obtaining new points A', B', where $OA' = 2r = 2.OA$, $OB' = 2r = 2.OB$, then the sectors AOB, $A'OB'$ are similar, and the arc $A'B'$ is double the arc AB. In general *the length of the arc is proportional to the radius.*

By combining these two facts we have the formula

$$\text{arc } AB = Hr\theta \qquad (6.21)$$

where H is a constant of proportionality. The value of H must depend on the units in which the angle θ is measured. If θ is in degrees, then we shall denote the constant by H_{360}. We shall show in the next section how we can calculate its value; to 6-figure accuracy it turns out that

$$H_{360} = \cdot 0174533$$

From this formula we can find the circumference of a complete circle, for that subtends an angle of 360° at the centre, whence

$$\begin{aligned} \text{circumference} &= H_{360} \cdot r \cdot 360 \\ &= 6{\cdot}28319r \end{aligned}$$

It is usual to call the circumference $2\pi r$, so that it follows that $\pi = 3{\cdot}14159 = 180\,H_{360}$. (The "2" in the formula "$2\pi r$" seems to come in merely on traditional grounds: the fundamental number is really 2π, rather than π.)

We can readily adapt these formulas to any unit of angle we like. Suppose we choose as unit $1/m$th part of the complete circle, so that a complete turn $= 360° = m$ units. It follows that the length of an arc AB can be expressed in the form

$$\text{arc } AB = H_m r\theta$$

where H_m is the correct value of the constant H for these units. Taking the complete circle, m units of angle, we have

$$\text{circumference} = H_m rm = 2\pi r$$
$$\text{whence } H_m = 2\pi/m \qquad . \quad . \quad . \quad (6.22)$$

In this formula there is no need for m to be an integer.

Now for many purposes it is convenient to take a special unit of angle, called the "radian", to make the constant H_m equal to 1, and so simplify the formula for the length of an arc. From (6.22) it follows that we can achieve this by putting $m = 2\pi$, i.e. we take as our unit $1/2\pi$ of a complete turn.

$$\begin{aligned} 1 \text{ radian} &= 1/2\pi \text{ turns} \\ &= (360/2\pi) \text{ degrees} \\ &= 57{\cdot}2958° \text{ (i.e. just under } 60°). \\ 1 \text{ complete turn} &= 2\pi \text{ radians} = 6{\cdot}28319 \text{ radians.} \\ 1° &= 2\pi/360 \text{ radians} \\ &= H_{360} \text{ radians} = {\cdot}0174533 \text{ radians.} \end{aligned}$$

It also follows that if the angle θ is measured in radian units, or in "circular measure" as it is often called, then the formula for the length of an arc becomes

$$\text{arc} = r\theta \qquad . \quad . \quad . \quad . \quad (6.23)$$

In particular, if $\theta = 1$ radian, then the arc AB is r. Thus a radian can be defined as the angle subtended at the centre of a circle by an arc equal in length to the radius (Fig. 6.8) (whence the name radian = RADIus ANgle).

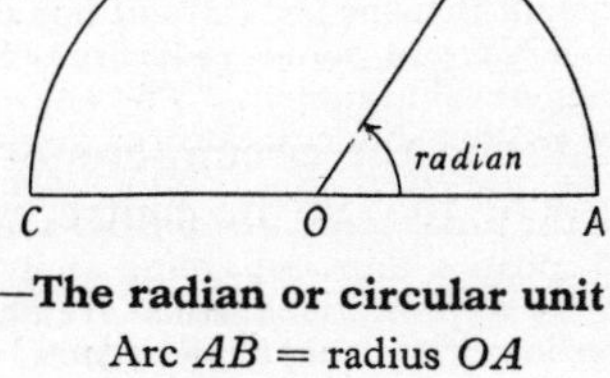

Fig. 6.8—The radian or circular unit of angle

Arc AB = radius OA

This definition however gives very little hint of the importance of a radian as unit of angle, which will become clearer when we come to calculus. We shall simply state here that it is so important that whenever an angle is mentioned without any unit being specified, it is to be understood that it is measured in radians. Thus an angle "π" means π radians or 180°, "$\frac{1}{6}\pi$" = 30°, and "sin θ" means the sine of θ radians unless the contrary is stated.

6.11 Functions of small angles

For many purposes it is important to know the values of the trigonometric functions sin θ, cos θ, tan θ and antilog θ when θ is very small. We shall first consider the functions sin θ, cos θ, and tan θ. Draw the triangle OUP, with $\angle POU = \theta$, and a right angle at U (Fig. 6.9). Let

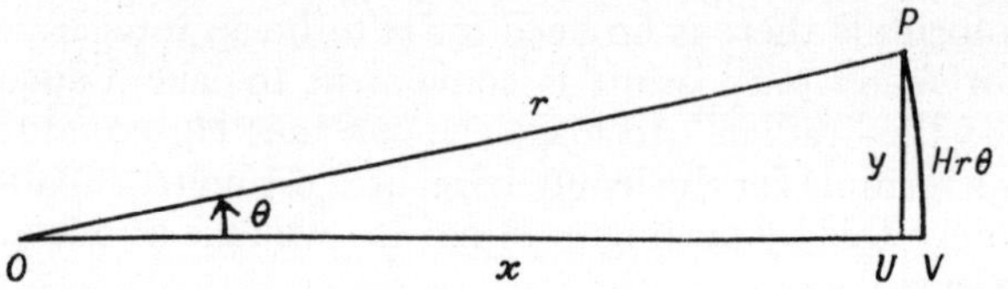

Fig. 6.9—Trigonometric functions of a small angle θ

us call $OP = r$, $OU = x$, $UP = y$ as usual. Let us draw also the arc of the circle PV through P meeting OU produced at V. Then it is fairly evident from the figure that if θ is small enough the straight line UP is practically indistinguishable from the arc VP, which is of length $Hr\theta$ (in general units: $H = H_{360}$ if the angle is measured in degrees, and $H = 1$ if the angle is measured in radians). That is to say,

$$y = UP \simeq Hr\theta \quad . \quad . \quad . \quad (6.24)$$

using the symbol $\simeq$ for "is approximately equal to".* Now sin θ $= y/r$ by definition so that

$$\sin\theta \simeq H\theta \quad . \quad . \quad . \quad (6.25)$$

---

* There does not seem to be any universally accepted symbol for approximate equality: the symbols $\approx$, $\cong$, $\doteqdot$, $\triangleq$, are all used. The symbol $\sim$ is also used in certain cases, but this has a special technical meaning and should not be used indiscriminately.

Although the above argument that the arc VP and line UP are practically identical seems plausible enough from a diagram, some readers may feel that it falls short of the standard required of a mathematical argument. The answer to this objection is that we are here taking ordinary geometric ideas, such as angles and lengths of arcs, for granted, and developing their properties in a common-sense way: and from such a common-sense point of view the proof seems reasonably adequate. To make the proof absolutely watertight would mean a thorough-going analysis of ideas such as angle and length and would bring in complications which are hardly appropriate to a first study of the subject, however important they may be later.

i.e. *the sine of a small angle is nearly equal to that angle multiplied by H.*

We also see that $x = OU \simeq OV = r$, so that

$$\cos\theta = x/r \simeq 1$$
$$\tan\theta = y/x \simeq Hr\theta/r = H\theta \quad . \quad . \quad . \quad (6.26)$$

We have used above the rather vague term "is approximately equal to", and the symbol $\simeq$ to correspond. In fact we can express ourselves rather more precisely. We mean that the *percentage error* in taking $\sin\theta$ or $\tan\theta$ to be $H\theta$ is small and the smaller θ is, the smaller the percentage error will be; so the smaller θ is, the nearer the ratios $(\sin\theta)/\theta$ and $(\tan\theta)/\theta$ will be to H. This is expressed mathematically in the form "$(\sin\theta)/\theta$ tends to H as θ tends to 0", or in symbols

$$(\sin\theta)/\theta \to H \quad \text{as} \quad \theta \to 0\ .$$
$$(\tan\theta)/\theta \to H \quad \text{as} \quad \theta \to 0\ . \quad . \quad . \quad (6.27)$$

This fact can be verified from tables of sines and tangents, and at the same time the values of H_{360} and $\pi = 180H_{360}$ can be determined, as shown in Table 6.1 (derived from 5-figure trigonometric tables).

Table 6.1—Sines and tangents of small angles

θ (degrees)	$\sin\theta$	$(\sin\theta)/\theta$	$\tan\theta$	$(\tan\theta)/\theta$
30	·50000	·01667	·57735	·01924
20	·34202	·01710	·36397	·01820
10	·17365	·01737	·17633	·01763
5	·08716	·01743	·08749	·01750
1	·01745	·01745	·01746	·01746
·5	·00873	·01746	·00873	·01746
·1	·00175	·0175	·00175	·0175

Thus when θ is small, $1°$ or less, $(\sin\theta)/\theta$ and $(\tan\theta)/\theta$ are about ·01746 to 4 places, and this must be the value of H_{360}. By taking more accurate tables we can find $H_{360} \simeq$ ·0174533. In theory we can find H_{360} as accurately as we wish, although this is not actually the most efficient way of calculating it.

Again, in saying that $\cos\theta \simeq 1$, we mean that there is a very small percentage error in taking $\cos\theta$ to be 1. In fact since

$$(1 - \cos\theta)(1 + \cos\theta) = 1 - (\cos\theta)^2 = (\sin\theta)^2$$

we have

$$1 - \cos\theta = \frac{(\sin\theta)^2}{1 + \cos\theta}$$

and since $1 + \cos\theta > 1$, it follows that

$$1 - \cos\theta < (\sin\theta)^2$$

Now $\sin\theta$ is approximately $(H\theta)$, so that $(1 - \cos\theta)$, which is the error in taking $\cos\theta$ to be 1, is less than about $(H\theta)^2$, or $(\sin\theta)^2$: a very small number indeed, much smaller than $\sin\theta$. This is borne out by the tables, which give $\cos 1° = \cdot 99985$, and $\cos \cdot 1° = 1\cdot 00000$ correct to 5 places. Similarly $\sec\theta = 1/\cos\theta$ is nearly 1, and $\operatorname{cosec}\theta$ and $\cot\theta$ are $1/H\theta$ to within a small percentage error.

EXAMPLES

(1) To find the height of a distant object, such as a tree, hill, tower, or chimney. If this has distance x, and subtends an angle θ at the eye, then its height is $y = x\tan\theta$, which is nearly enough $xH\theta$. Thus a tree with an apparent height of $\frac{1}{2}°$ at 1000 metre distance has a height of $\frac{1}{2} \times 1000 \times \cdot 01745 = 8\cdot 73$ metres.

(2) A person of normal sight can read print which has an apparent height of $\frac{1}{12}°$. At 6 metres he can therefore read print of height $600 \times \frac{1}{12} \times \cdot 01745 = \cdot 873$ cm. This is the principle of Snellen's type for testing distant vision. The height of type just legible to a normal person at 6 metres is ·873 cm or ·35 inches. Print legible at 60 metres will accordingly have a height of 8·73 cm, and that legible at 18 metres a height of $3 \times \cdot 873 = 2\cdot 625$ cm. A person who can just read at 6 metres print which he should be able to read at 36 metres is said to have visual acuity 6/36, i.e. he can read print subtending an angle of $\frac{36}{6} \times \frac{1}{12} = \frac{1}{2}°$. If we assume that distinguishing a man at a distance is comparable to reading a letter of type, then he can distinguish a man 1·8 metres tall at a distance x metres such that

$$1\cdot 80 = H_{360} \times \tfrac{1}{2} \times x$$

whence $x \simeq 200$ metres.

If the angle θ is measured in radians our formulas become simply

$$\sin\theta \simeq \tan\theta \simeq \theta \text{ when } \theta \text{ is small.}$$

PROBLEMS

(1) How far away can a normally-sighted person clearly distinguish a tower 10 metres high on a clear day?

(2) If the moon's diameter appears to subtend an angle of $\frac{1}{2}°$, and the distance away is $3\cdot 8 \times 10^8$ metres, what is its actual diameter? What is the size of the smallest object one can expect to distinguish clearly on the moon?

6.12 Natural logarithms

We shall now consider the value of antilog θ for small θ, including in our formulas the general system defined by an equiangular spiral of angle ϕ and centre O. Common antilogarithms will be a special case. Let I be the point on the spiral such that $OI = 1$, and P the point such that $\angle IOP = \theta$ (Fig. 6.10). Then $OP =$ antilog θ by definition. Draw

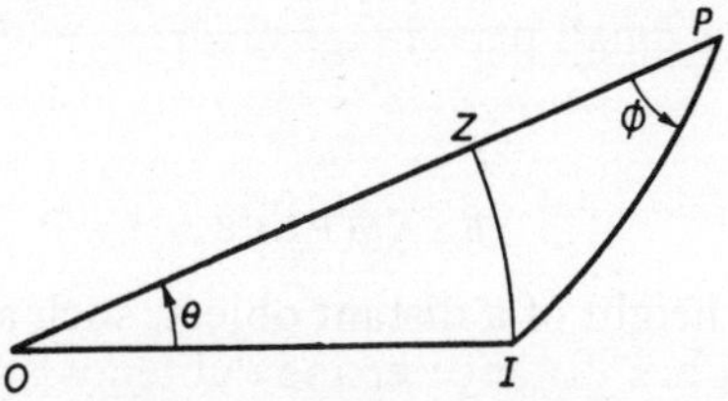

Fig. 6.10—The antilogarithm of a small angle θ

an arc of a circle with centre O and unit radius to meet OP at Z; then $ZP = OP - OZ =$ antilog $\theta - 1$. We therefore have a triangle IZP (with two curved sides) such that the angle at P is ϕ, that at Z is $90°$ and that at I is $90° - \phi$. Also $ZP =$ antilog $\theta - 1$, and the arc $IZ = H\theta$. Now when θ is small this triangle is small and becomes very nearly an ordinary straight-sided triangle. Therefore when θ is small

$$ZP \simeq IZ \cot \phi,$$
$$(\text{antilog } \theta - 1) \simeq H\theta \cot \phi$$

or

$$\text{antilog } \theta \simeq 1 + \theta \,.\, H \cot \phi.$$

It is usual to call $1/(H \cot \phi)$ the "modulus" of the system of logarithms, and denote it by the letter M (or μ). (This is therefore a different use of the word "modulus" from the meaning "absolute value" of Chapter 4.) So

$$\text{antilog } \theta \simeq 1 + \theta/M \quad . \quad . \quad . \quad (6.28)$$

when θ is small.

We can express this in a different way. Let the distance ZP be called u: then $OP = 1 + u =$ antilog θ, so that $\theta = \log(1 + u)$, and $IZ = H\theta = H \log(1 + u)$. Now when $u = ZP$ is small, the triangle IZP is small, and has nearly straight sides and the angle θ is small. The smaller u is the straighter the sides of the triangle, and the nearer IZ/ZP approaches the value $\tan \phi$. Therefore IZ/ZP tends to $\tan \phi$ as u tends to 0, i.e. $H \log(1 + u)/u \to \tan \phi$ as $u \to 0$.

Dividing by H,

$$\frac{\log(1 + u)}{u} \to \frac{\tan \phi}{H} = M \quad \text{as} \quad u \to 0 \quad . \quad . \quad (6.29)$$

Thus the modulus M is the value which $\log (1 + u)/u$ approaches when u becomes very small. We can use this definition to find the value of the modulus for any system of logarithms. For example, if we use a 5-figure table of common logarithms we find the following values:

Table 6.2—Values of log (1 + u)/u

u	$\log (1 + u)$	$\log (1 + u)/u$
1	·30103	·30103
·2	·07918	·3954
·1	·04139	·4139
·05	·02119	·4238
·02	·00860	·430
·01	·00432	·432

Thus $\log (1 + u)/u$ is approaching a value $M \simeq \cdot 43$ as u grows smaller. In fact by using more accurate tables we can find $M = \cdot 43429$ and $M^{-1} = 2\cdot 30259$ to 5 places of decimals. Not only does each system of logarithms have its own modulus M, but also no two different systems can have the same modulus. For if $\log x$ is one system, then any other system is a constant multiple of this, say $k \,.\, \log x$. Now by saying that the first system has modulus M we mean that when u is small, $\log (1 + u)/u$ differs little from M, and in fact that as u approaches 0, $\log (1 + u)/u$ approaches M. The modulus of the second system $k.\log x$ will accordingly be the value approached by $k.\log (1 + u)/u$ as u tends to 0, and accordingly will be simply kM. Since k is not equal to 1 (for otherwise we would be considering only one system of logarithms instead of two different systems) the modulus of the second system must be different from that of the first. In fact it is just the modulus of the first system multiplied by k. In particular if we divide all logarithms of any system by its modulus M, we obtain a new system of modulus 1. This system was discovered by Napier in 1614 and is known as the system of "napierian" (or "naperian"), "natural" or "hyperbolic" logarithms. The symbol $\ln x$ is now frequently used for the natural logarithm of x, and that is the symbol we shall use here. (The reader is warned that this usage is not yet universal, and sometimes the natural logarithm is written simply as $\log x$.) Accordingly this special system of "natural logarithms" is defined by the equation $\ln x = M^{-1} \log x$ or

$$\log x = M \ln x \qquad . \quad . \quad . \quad (6.30)$$

Since the modulus of a system of logarithms is equal to $(\tan \phi)/H$, natural logarithms can also be defined as the system for which $H = 1$ and $\phi = 45°$, i.e. that obtained from a 45° equiangular spiral when the angle θ is measured in radians. The essential property distinguishing natural logarithms from other systems is that when u is small,

$$\ln (1 + u) \simeq u \qquad . \quad . \quad . \ (6.31)$$

(or more precisely that $\ln (1 + u)/u \to 1$ as $u \to 0$).

Query: What follows if we put $x = 10$ in equation (6.30)? In general what is the relation between the base and modulus of a system of logarithms?

The antilogarithms corresponding to natural logs could be written antiln x; but in practice this notation is never used. If $y = \ln x$, we write $x = \exp y$, the *exponential function* of y. $\exp y$ is therefore just a special kind of antilogarithm, the natural antilogarithm: from (6.28) it obeys the law

$$\exp y \simeq 1 + y$$

when y is small. The number

$$e = \exp 1$$

is the number whose natural logarithm is 1, i.e. it is the base of natural logarithms. (Another way of writing $\ln x$ is therefore $\log_e x$.) The letter "e" is usually reserved in mathematical formulas for this particular number, just as π always means the ratio 3·14159 . . . of the circumference of a circle to its diameter. We can readily calculate its value by putting $u = e$ in equation (6.30), giving $\log e = M$, or $e = \text{antilog } M$. Since $M = {\cdot}4343$ . . . we have from tables of common antilogs $e = 2{\cdot}718$. More accurately $e = 2{\cdot}71828182846$ . . .

A further consequence of (6.30) is that

$$\begin{aligned} \log (\exp y) &= M \ln (\exp y) \\ &= My \\ &= y \log e \end{aligned}$$

i.e. $\exp y = e^y$, by definition (6.11) of e^y. This accordingly gives us another way of writing the exponential function, and also a way of calculating its value from a table of common antilogs.

PROBLEMS

(1) Using common logarithms, find ln 2, ln 3, ln 5. Verify that $\ln 10 = \ln 2 + \ln 5 = M^{-1}$.

(2) Using common antilogarithms, find e^2, e^3, e^{10}, e^{-2}, e^{-10}.

(3) Draw rough graphs of $y = \ln x$ (say for $0 < x \leqslant 5$) and $y = e^x$ (say for $-1{\cdot}5 \leqslant x \leqslant 1{\cdot}5$).

To summarize, the natural logarithm and antilogarithm, ln x and exp $x = e^x$, are simply slight modifications of the common logarithm and antilogarithm according to the formulas

$$\ln x = M^{-1} \log x$$
$$e^x = \text{antilog } Mx.$$

But they are nevertheless functions of the greatest importance in every branch of mathematics, both pure and applied to biology, chemistry, physics and engineering. We shall simply remark here that we have already mentioned that functions of the type C^x are important, e.g. in the growth of bacteria. Now C^x is by definition antilog (x log C) in any system of logarithms; using natural logarithms, $C^x = e^{x \ln C} = e^{cx}$ where $c = \ln C$.

The reader should note the general behaviour of the two functions (or of logarithms and antilogs in general). When x is small and positive, ln x is large and negative. The logarithm increases rapidly, until when $x = 1$, $\ln x = 0$, and it thereafter becomes positive. Its rate of increase slows down, and after a time it increases only very slowly: when $x = 10{,}000{,}000{,}000$, ln x is only 23·0. The exponential function behaves very differently. It is always positive, but for values of x which are negative and not too small e^x differs inappreciably from zero. For example, if $x < -5$, $e^x < \cdot 007$, and if $x < -10$, $e^x < \cdot 0001$. When $x = 0$, $e^x = 1$, and thereafter the function increases more and more rapidly; e.g. $e^5 \simeq 148$, $e^{10} \simeq 20000$.

6.13 Hyperbolic functions

We remarked in Section 6.2 that there is a certain analogy between antilogarithms and other trigonometric ratios. This analogy is made even more striking by the use of certain special "hyperbolic functions". The "hyperbolic cosine" of x is defined to be $\frac{1}{2}(e^x + e^{-x})$ and is written cosh x, and the "hyperbolic sine" of x is defined to be $\frac{1}{2}(e^x - e^{-x})$ and is written sinh x.

At first glance the most striking property of these functions is the apparent lack of any resemblance whatever to the ordinary sine and cosine. The graphs of these functions are shown in Fig. 6.11. Unlike the ordinary cosine and sine they have no periodicity or wave form. The cosh curve rather resembles a parabola, but curves upwards more rapidly. It is in fact the shape of a hanging chain, and is sometimes called a "catenary" (from Latin *catēna* = chain). Also unlike cos x, which never exceeds 1, cosh $x \geqslant 1$ for all x, as can be proved in this way. Since $(e^x - 1)^2$ is a perfect square, it is always positive or zero, while e^{-x}, being an antilogarithm, is always positive. Therefore $\frac{1}{2}e^{-x}(e^x - 1)^2$ is always positive or zero, i.e.

$$\tfrac{1}{2}e^{-x}[(e^x)^2 - 2e^x + 1] \geqslant 0$$

But $e^{-x}(e^x)^2 = e^{-x+2x} = e^x$, and $e^{-x}e^x = e^0 = 1$, so that this

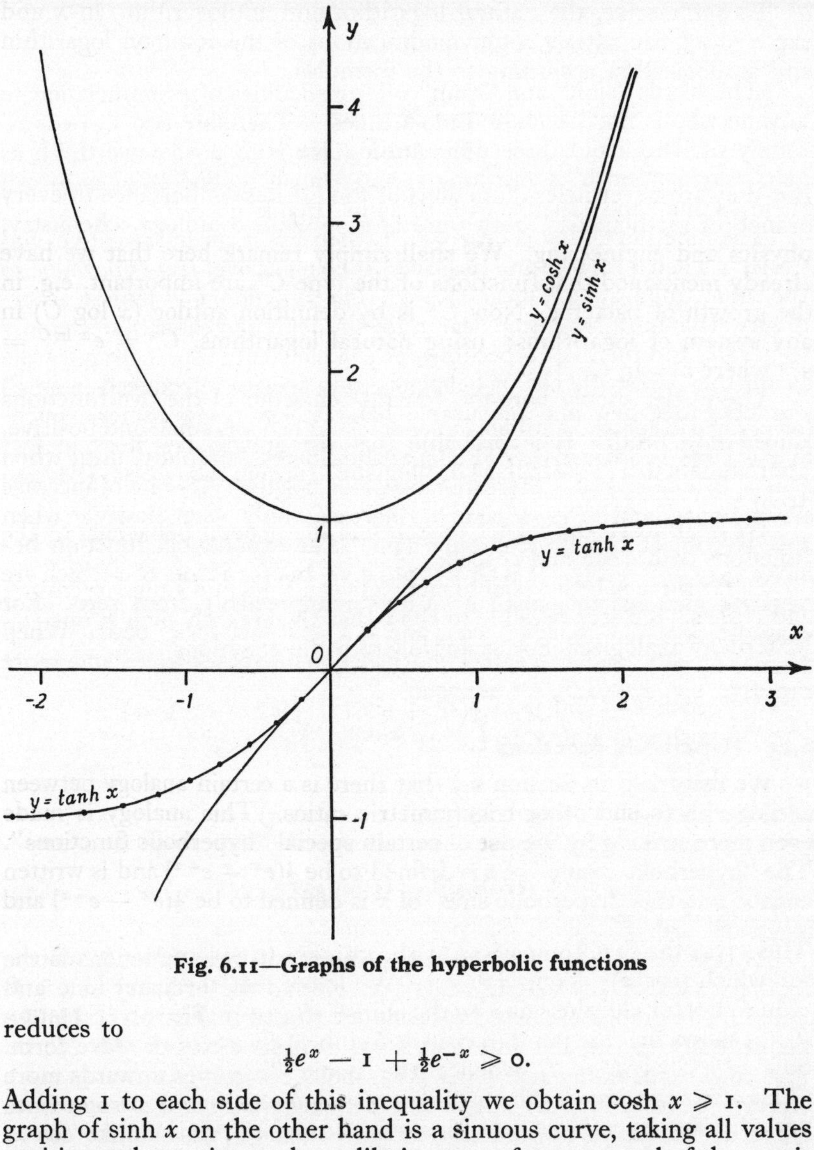

Fig. 6.11—Graphs of the hyperbolic functions

reduces to

$$\tfrac{1}{2}e^{x} - 1 + \tfrac{1}{2}e^{-x} \geqslant 0.$$

Adding 1 to each side of this inequality we obtain $\cosh x \geqslant 1$. The graph of $\sinh x$ on the other hand is a sinuous curve, taking all values positive and negative, and steadily increases from one end of the x-axis to the other.

By analogy with $\tan x = (\sin x)/\cos x$, etc., we can define four further hyperbolic functions

$$\tanh x = (\sinh x)/\cosh x = (e^{x} - e^{-x})/(e^{x} + e^{-x})$$
$$\coth x = (\cosh x)/\sinh x = (e^{x} + e^{-x})/(e^{x} - e^{-x})$$

$$\text{sech } x = 1/\cosh x = 2/(e^x + e^{-x})$$
$$\text{cosech } x = 1/\sinh x = 2/(e^x - e^{-x})$$

[The words "sinh" and "tanh" offer difficulties of pronunciation to anyone who is not Welsh or Indo-Chinese. There are two ways commonly used to avoid these difficulties. One is to pronounce the h as "sh", saying "sinsh" (with long i) and "tansh". The other is to call them "shine" and "than", with th as in "thank".]

The function $\tanh x$ is nearly -1 when x is large and negative, and nearly 1 when x is large and positive. It changes over from one value to the other quite quickly near $x = 0$. The function $1 + \tanh x$ accordingly stays for a long time near zero, suddenly increases quite rapidly to near 2, and then again becomes nearly constant. Apart from a scale factor this is just the law of behaviour of a species introduced in small numbers to a new and favourable habitat, e.g. a few bacteria introduced into broth. It is probable that the growth law does in fact approximate to a $(1 + \tanh x)$ or "logistic" form in many cases (Section 10.8).

We shall leave the reader to examine the general behaviour of the functions coth, sech and cosech.

So far there is little resemblance between sines and sinhs, or cosines and coshes. But it is possible to show that algebraically their properties have many analogies. For example, from the equations

$$\cosh x + \sinh x = \tfrac{1}{2}(e^x + e^{-x}) + \tfrac{1}{2}(e^x - e^{-x}) = e^x$$
$$\cosh x - \sinh x = \tfrac{1}{2}(e^x + e^{-x}) - \tfrac{1}{2}(e^x - e^{-x}) = e^{-x}$$

we see that

$$(\cosh x + \sinh x)(\cosh x - \sinh x) = e^x . e^{-x},$$

or

$$(\cosh x)^2 - (\sinh x)^2 = 1 \quad . \quad . \quad . \quad (6.32)$$

This gives the first comparison with ordinary trigonometric functions, for which $(\cos x)^2 + (\sin x)^2 = 1$. We know that the point $(a \cos \phi, b \sin \phi)$ lies for all values of ϕ on the ellipse $x^2/a^2 + y^2/b^2 = 1$. Equation (6.32) shows that the point $(a \cosh \phi, b \sinh \phi)$ always lies on the hyperbola $x^2/a^2 - y^2/b^2 = 1$, whence the name "hyperbolic functions". (This also suggests why the ordinary trigonometric functions should be periodic, and hyperbolic functions not. For an ellipse is a closed loop: if we go round it once we return to the starting point, and then everything repeats. But the hyperbola is not closed; it stretches away to infinity.) In fact, we can make the analogy a little closer by anticipating some results of the integral calculus. Draw the circle $x^2 + y^2 = 1$ and the rectangular hyperbola $x^2 - y^2 = 1$. Let P be a point on the circle, P' a point on the hyperbola, and I the point (1, 0) (Fig. 6.12). The area of the sector IOP is proportional to the angle $\theta = \angle IOP$; for clearly if we double this angle, we double the area, if we multiply the angle by 3, the area is multiplied by 3, and so on. Thus area $IOP = k\theta$, where k is a constant. Suppose θ is measured in radians; then when $\theta = 2\pi$ the sector IOP becomes the whole circle of unit radius, and has area π (Section 11.13). Thus $\pi = k.2\pi$, or $k = \frac{1}{2}$, or $\theta = 2$ area IOP. This can be taken as an alternative definition of the angle $\theta = \angle IOP$; P is then the point with co-ordinates $x = \cos\theta$, $y = \sin\theta$. By analogy let us call twice the area between the hyperbolic arc IP' and the radii OI, OP' the "hyperbolic angle" θ', then P' can be shown to have co-ordinates $(\cosh\theta', \sinh\theta')$.

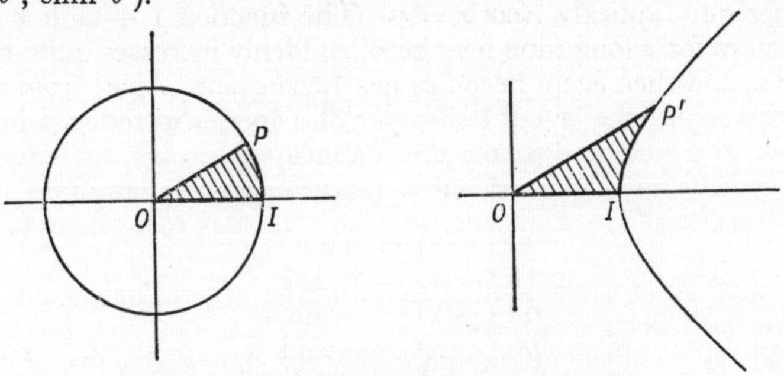

Fig. 6.12—Ordinary and hyperbolic angles

Another analogy comes from the formula for $\cosh(x+y)$:

$$\begin{aligned}\cosh(x+y) &= \tfrac{1}{2}(e^{x+y} + e^{-x-y})\\ &= \tfrac{1}{2}(e^x e^y + e^{-x} e^{-y})\\ &= \tfrac{1}{2}[(\cosh x + \sinh x)(\cosh y + \sinh y)\\ &\qquad + (\cosh x - \sinh x)(\cosh y - \sinh y)]\\ &= \cosh x . \cosh y + \sinh x . \sinh y\end{aligned}$$

on multiplying out and simplifying this expression. Compare this with the trigonometric relation

$$\cos(x+y) = \cos x . \cos y - \sin x . \sin y$$

The hyperbolic formula has the more natural sign + between the two terms; but otherwise the two formulas are alike.

Proceeding in this way we can investigate what formulas connecting hyperbolic functions correspond to the trigonometric formulas of Chapter 5. We find the following remarkable analogies. (The proofs are left as an exercise for the reader.)

$$\cos 0 = \sec 0 = 1$$
$$\cosh 0 = \operatorname{sech} 0 = 1$$

$$\sin 0 = \tan 0 = 0$$
$$\sinh 0 = \tanh 0 = 0$$

If x is small (and measured in radians):

$$\cos x \simeq \sec x \simeq 1$$
$$\cosh x \simeq \operatorname{sech} x \simeq 1$$
$$\sin x \simeq \tan x \simeq x$$
$$\sinh x \simeq \tanh x \simeq x$$
$$\operatorname{cosec} x \simeq \cot x \simeq 1/x$$
$$\operatorname{cosech} x \simeq \coth x \simeq 1/x$$

For all values of x and y (in any units):

$$(\cos x)^2 + (\sin x)^2 = 1$$
$$(\cosh x)^2 - (\sinh x)^2 = 1$$

$$(\sec x)^2 - (\tan x)^2 = 1$$
$$(\operatorname{sech} x)^2 + (\tanh x)^2 = 1$$

$$(\operatorname{cosec} x)^2 - (\cot x)^2 = 1$$
$$(\coth x)^2 - (\operatorname{cosech} x)^2 = 1$$

$$\cos(-x) = \cos x$$
$$\cosh(-x) = \cosh x$$
$$\sin(-x) = -\sin x$$
$$\sinh(-x) = -\sinh x$$
$$\tan(-x) = -\tan x$$
$$\tanh(-x) = -\tanh x$$

$$\cos(x+y) = \cos x \,.\, \cos y - \sin x \,.\, \sin y$$
$$\cosh(x+y) = \cosh x \,.\, \cosh y + \sinh x \,.\, \sinh y$$

$$\sin(x+y) = \sin x \,.\, \cos y + \cos x \,.\, \sin y$$
$$\sinh(x+y) = \sinh x \,.\, \cosh y + \cosh x \,.\, \sinh y$$

$$\tan(x+y) = (\tan x + \tan y)/(1 - \tan x \,.\, \tan y)$$
$$\tanh(x+y) = (\tanh x + \tanh y)/(1 + \tanh x \,.\, \tanh y)$$

$$\cot(x+y) = (\cot x \,.\, \cot y - 1)/(\cot x + \cot y)$$
$$\coth(x+y) = (\coth x \,.\, \coth y + 1)/(\coth x + \coth y)$$

$$\cos 2x = (\cos x)^2 - (\sin x)^2 = 2(\cos x)^2 - 1$$
$$\cosh 2x = (\cosh x)^2 + (\sinh x)^2 = 2(\cosh x)^2 - 1$$

$$\sin 2x = 2 \sin x \,.\, \cos x$$
$$\sinh 2x = 2 \sinh x \,.\, \cosh x$$

$$\tan 2x = \frac{2 \tan x}{1 - (\tan x)^2}$$

$$\tanh 2x = \frac{2 \tanh x}{1 + (\tanh x)^2}$$

$$\cot 2x = \tfrac{1}{2}(\cot x - \tan x)$$
$$\coth 2x = \tfrac{1}{2}(\coth x + \tanh x)$$

$$\cos 3x = 4(\cos x)^3 - 3\cos x$$
$$\cosh 3x = 4(\cosh x)^3 - 3\cosh x$$

$$\sin 3x = 3\sin x - 4(\sin x)^3$$
$$\sinh 3x = 3\sinh x + 4(\sinh x)^3$$

$$\cos(x-y) = \cos x \,.\, \cos y + \sin x \,.\, \sin y$$
$$\cosh(x-y) = \cosh x \,.\, \cosh y - \sinh x \,.\, \sinh y$$

$$\sin(x-y) = \sin x \,.\, \cos y - \cos x \,.\, \sin y$$
$$\sinh(x-y) = \sinh x \,.\, \cosh y - \cosh x \,.\, \sinh y$$

$$\tan(x-y) = (\tan x - \tan y)/(1 + \tan x \,.\, \tan y)$$
$$\tanh(x-y) = (\tanh x - \tanh y)/(1 - \tanh x \,.\, \tanh y)$$

$$\cot(x-y) = (1 + \cot x \,.\, \cot y)/(\cot y - \cot x)$$
$$\coth(x-y) = (1 - \coth x \,.\, \coth y)/(\coth x - \coth y)$$

$$\cos x \cos y = \tfrac{1}{2}\cos(x+y) + \tfrac{1}{2}\cos(x-y)$$
$$\cosh x \cosh y = \tfrac{1}{2}\cosh(x+y) + \tfrac{1}{2}\cosh(x-y)$$

$$\sin x \sin y = -\tfrac{1}{2}\cos(x+y) + \tfrac{1}{2}\cos(x-y)$$
$$\sinh x \sinh y = \tfrac{1}{2}\cosh(x+y) - \tfrac{1}{2}\cosh(x-y)$$

$$\sin x \cos y = \tfrac{1}{2}\sin(x+y) + \tfrac{1}{2}\sin(x-y)$$
$$\sinh x \cosh y = \tfrac{1}{2}\sinh(x+y) + \tfrac{1}{2}\sinh(x-y)$$

If $t = \tan \tfrac{1}{2}x$ then

$$\cos x = (1-t^2)/(1+t^2)$$
$$\sin x = 2t/(1+t^2)$$
$$\tan x = 2t/(1-t^2)$$

If $t = \tanh \tfrac{1}{2}x$ then

$$\cosh x = (1+t^2)/(1-t^2)$$
$$\sinh x = 2t/(1-t^2)$$
$$\tanh x = 2t/(1+t^2)$$

6.14 Inverse functions

It is convenient to have some way of writing "the angle whose cosine (or sine, or tangent, etc.) is y".

Everyday examples of this are quite common. If Paul is the *parent* of Charles, then Charles is a *child* (or *offspring*) of Paul. "Parent" and "child" are inverse relations. If Henry is the *husband* of Winifred, then Winifred is the *wife* of Henry. If John *eats* a sandwich, the sandwich *is eaten by* John: "eats" and "is eaten by" are inverse relations, called by grammarians "active" and "passive". In mathematics we have sometimes special words for inverse relations: as we have seen, if θ is the logarithm of r, we call r the *antilogarithm* of θ. If y is the square of x, then x is the *square root* of y. (It might be called the "anti-square".) But it seems desirable to have a more general notation. For example, if $y = \sin x$, how shall we write x in terms of y? That is, how shall we write an "inverse sine" or "anti-sine"? Now notice that if A is an ordinary number, the inverse of the relation $y = Ax$ is $x = A^{-1}y$. Mathematicians use this as a general pattern; they write the inverse of the relation $y = \sin x$ in the form $x = \sin^{-1} y$. That is to say, the symbol "$\sin^{-1} y$" means simply "the angle whose sine is y", or inverse sine of y. Thus, since $\sin 30^\circ = \frac{1}{2}$, we have $\sin^{-1}\frac{1}{2} = 30^\circ = \frac{1}{6}\pi$ (radians). Similarly, since $\tan 45^\circ = 1$, we have $\tan^{-1} 1 = 45^\circ = \frac{1}{4}\pi$. Of course, in the relation $x = A^{-1}y$, where A is a number, the symbol A^{-1} means another number, namely the reciprocal of A. In the relation "$x = \sin^{-1} y$" the index $^{-1}$ is used merely as a convenient convention to denote the inverse function; it does not mean taking the reciprocal.

The same pattern is used for any function; if $y = F(x)$, we write $x = F^{-1}(y)$.

An alternative notation for "$\sin^{-1} y$" often used is "$\arcsin y$". Similarly $\arctan y = \tan^{-1} y$; $\operatorname{arg\,tanh} y = \tanh^{-1} y$.

Note that even if in the relation $y = F(x)$ for each value of x there is only one value of y it does not follow that for each value of y there is only one value of x. As a rule that is not true: if $y = x^2$, then given y, x can have two possible values, $+\sqrt{y}$ and $-\sqrt{y}$. If $y = \cos x$, then, given y, x can have an infinite number of values—e.g. $\cos^{-1} 0 = 90^\circ$, 270°, 450°, etc.: though it is true that of these values only 90° and 270° represent geometrically distinct angles. All others differ from these by multiples of 360°. In such a case x is said to be a "many-valued" function of y. For example, if $y = x^2$, $x = \pm\sqrt{y}$ is a "two-valued" function of y, the separate values $+\sqrt{y}$ and $-\sqrt{y}$ being two "branches" of the function.

The inverse hyperbolic functions have alternative expressions in terms of natural logarithms. For example, let $x = \cosh^{-1} y$, so that $y = \cosh x = \frac{1}{2}(e^x + e^{-x})$. Multiplying this equation through by $2e^x$, we have $2e^x y = (e^x)^2 + 1$, or $(e^x)^2 - 2ye^x + 1 = 0$. This is a quadratic equation in e^x, with solution

$$e^x = y \pm \sqrt{(y^2 + 1)},$$

i.e.

$$x = \cosh^{-1} y = \ln[y \pm \sqrt{(y^2 - 1)}].$$

In the same way by solving the equation $y = \sinh x = \frac{1}{2}(e^x - e^{-x})$ we find that

$$x = \sinh^{-1} y = \ln[y + \sqrt{(y^2 + 1)}].$$

(We cannot have a minus sign before the square root here because the quantity $y - \sqrt{(y^2 + 1)}$ enclosed in square brackets would then be negative, and so would have no logarithm.)

These formulas agree with the graphs of cosh x and sinh x. If $y > 1$ there are two points on the graph $y = \cosh x$ having this particular value of y. If $y < 1$ there are no such points. The formula

$\cosh^{-1} y = \ln [y \pm \sqrt{(y^2 - 1)}]$ fails to give any value if $-1 < y < 1$ because $(y^2 - 1)$, being negative, has no square root, while if $y \leqslant -1$ the expressions $y \pm \sqrt{(y^2 - 1)}$ are then negative and have no logarithm. On the other hand there is just one point on the curve $y = \sinh x$ which has a given height y, so that the function $\sinh^{-1} y$ exists for all values of y and is a one-valued function, as is shown by the formula

$$\sinh^{-1} y = \ln [y + \sqrt{(y^2 + 1)}].$$

Note: As we mentioned in Section 5.5, it is customary to write $(\sin \theta)^2$ as $\sin^2 \theta$, $(\cos \theta)^3$ as $\cos^3 \theta$, etc. This is a convenient and clear notation, and saves brackets. But one must be careful not to write $(\sin \theta)^{-1}$ as $\sin^{-1} \theta$, as that would conflict with the standard notation for an inverse function: one must write $(\sin \theta)^{-1} = \operatorname{cosec} \theta$, $(\cos \theta)^{-1} = \sec \theta$, etc.

PROBLEMS

(1) Find the values of $\cosh 2$, $\sinh 2$, $\tanh 2$, $\cosh^{-1} 2$, $\sinh^{-1} 2$.

(2) Express $\tanh^{-1} x$ in terms of the natural logarithm function.

(3) Find the inverse function $x = f^{-1}(y)$ for the following functions $y = f(x)$: (i) $y = 2 + 3x$; (ii) $y = 1 + x^2$; (iii) $y = e^x - 1$; (iv) $y = (1 - x)/(1 + x)$; (v) $y = \sqrt{(1 - x^2)}$. Which of these are single-valued?

(4) Show that

$$\sinh^{-1} x + \sinh^{-1} y = \sinh^{-1} [x\sqrt{(1 + y^2)} + y\sqrt{(1 + x^2)}].$$

(5) Show that $\tanh^{-1} x + \tanh^{-1} y = \tanh^{-1} \dfrac{x + y}{1 + xy}$.

CHAPTER SEVEN

GRAPHICAL AIDS TO CALCULATION

7.1 Logarithmic scales

In drawing graphs we have so far always used a uniform scale of measurement along each axis. The actual length chosen as unit of measurement has been allowed to vary from one graph to another: but once the zero point or *origin* and unit have been fixed the whole scale is completely determined.

For many purposes it is better to use scales with non-uniform graduations. As we shall see we can do many complicated calculations, such as the calculation of

$$T = 1724\ W^{\cdot 425}\ H^{\cdot 725}\ C$$

for given values of W, H, and C, without any arithmetic whatever simply by the use of a "nomogram" consisting of suitably graduated lines, together with a straight-edge. (This formula relates the amount of heat T in calories produced by a person in a day with the body weight W in kilograms, the height H in centimetres, and C the number of calories produced per square metre of body surface per hour.)

Let us take a straight line L and choose a point O on it as zero point or "origin". If P is any point on this line we shall call the true distance OP "X": this can be measured by graduating the line OP uniformly. But if we graduate the line in such a way that the mark "x" occurs at the point P such that $OP = X = \log x$, then the line is said to be "logarithmically graduated" (Fig. 7.1). Thus the true distance along the line is then the logarithm of the number marked (e.g. the mark $x = 100$ occurs at distance $X = 2$).

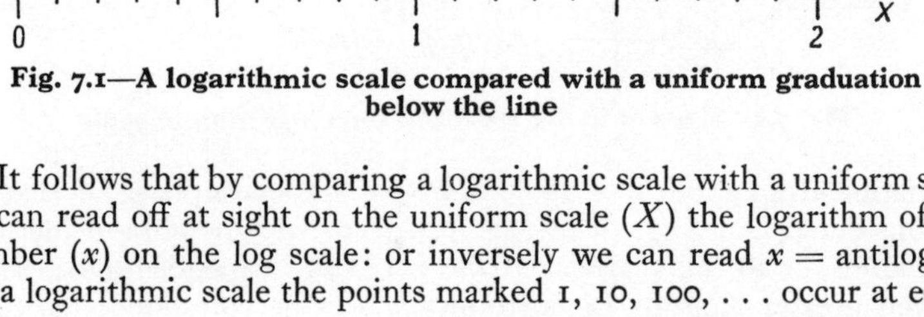

Fig. 7.1—A logarithmic scale compared with a uniform graduation below the line

It follows that by comparing a logarithmic scale with a uniform scale we can read off at sight on the uniform scale (X) the logarithm of the number (x) on the log scale: or inversely we can read $x = \text{antilog}\ X$. On a logarithmic scale the points marked 1, 10, 100, . . . occur at equal distances apart. Hence we can represent an enormous range of values on such a scale. E.g. if we consider lengths, we find in living creatures

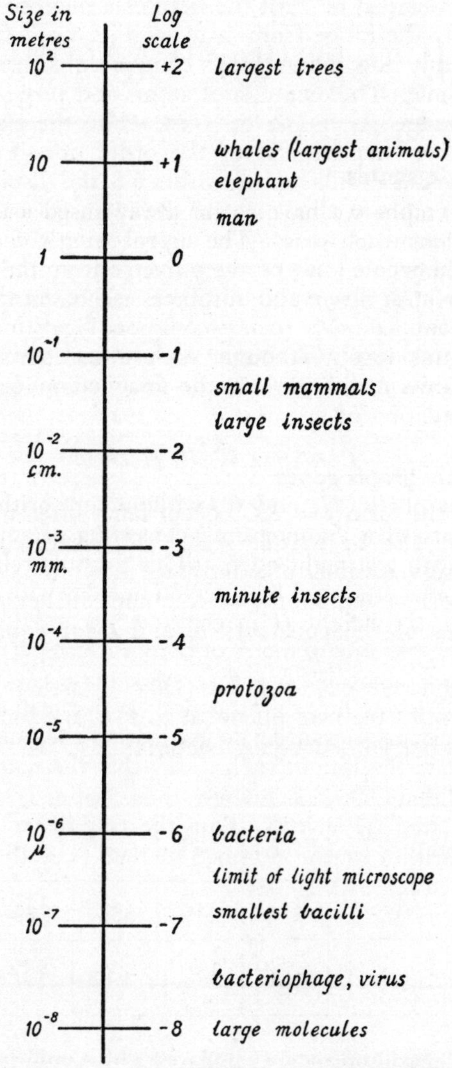

Fig. 7.2—Sizes of living creatures on a logarithmic scale

a variation from about 5×10^{-8} metre for a bacteriophage, if that can be considered as living, to the greatest trees with heights of just over 10^2 metres, or to the seaweed in the Sargasso Sea, sometimes perhaps of even greater length. Logarithmically (using common logarithms) this represents a variation merely from $X = -7\cdot3$ to $X = +2$. Since the limit of vision with a light microscope is about $\frac{1}{5}$ of a micron, or logarithmically $-6\cdot7$, practically the whole of this range is covered by ordinary vision unaided or with the help of a microscope. Adult man occupies roughly the range from ·2 to ·3 (Fig. 7.2). Such a scale also shows more readily how living things compare in size with the physical universe as a whole. The diameter of atoms and very simple molecules comes in the range $X = -9$ to -10, while the size of an atomic nucleus is, rather surprisingly, of the order of -14. In the other direction, the earth's radius has logarithm 6·8, the distance of the sun is 11·2, the nearest stars at about 17 and the extra-galactic nebulas at 22. At somewhere round 25 we may be approaching the farthest limit of vision. Thus the whole scale of the universe from the smallest objects known to the greatest observable distances can be compressed logarithmically into about 44 units (using common logarithms, or just over 100 using natural logs), although such a comparison in ordinary unlogarithmic units utterly defeats the imagination, as it amounts to a ratio of 10^{44} to 1.

7.2 Logarithmic graph paper

Relations of the form $y = KC^x$ occur fairly often in biology. (This relationship may also be written $y = Ke^{cx}$, where $c = \ln C$.) If we have a set of observed values of x and y we may often ask whether these are consistent with an equation $y = Ke^{cx}$, and if they are, what are the numerical values of the constants K and c (or equivalently K and $C = e^c$).

To take a concrete example, Dr (Dame Harriette) Chick treated anthrax spores with 5 per cent phenol at 20·2° C and found after various times x the number y of surviving bacteria:

x (time in hours) ..	0	·5	1·5	2·7	5·95	25·6
y (number of survivors)	434	410	351	331	241	28
$Y = \log y$	2·64	2·61	2·55	2·52	2·38	1·45

Now the equation $y = KC^x$ is equivalent to

$$\log y = \log K + x \log C$$

or

$$Y = A + Bx$$

where

$$Y = \log y, \quad A = \log K, \quad \text{and } B = \log C = Mc,$$

M being the modulus of common logarithms. If therefore we plot the values of $\log y = Y$ against x we should obtain a straight-line graph. We can do this by actually calculating values of Y as in the table above. Alternatively we can use graph paper in which the y axis is logarithmically graduated, and plot the values of y directly. When a point is plotted with co-ordinate y according to the markings on the paper, its actual co-ordinate according to distance measured from the x-axis will be $Y = \log y$, so that we obtain the graph of Y against x, and can verify its straightness (in a rough way) without calculation (Fig. 7.3). The

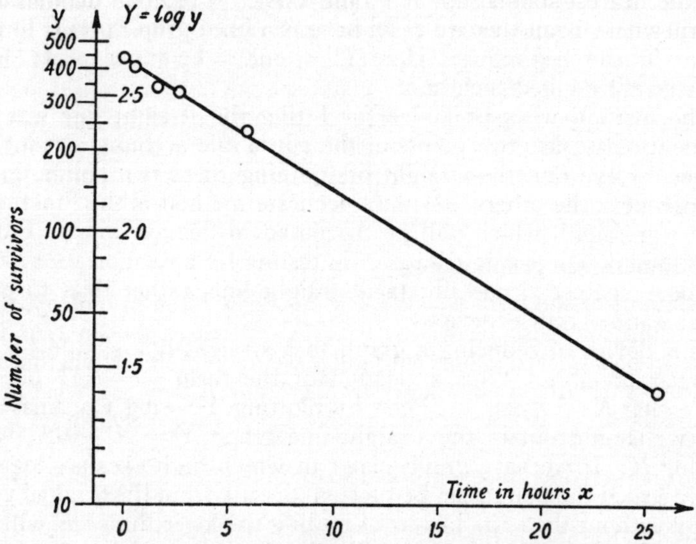

Fig. 7.3—Action of phenol on anthrax spores

graph resulting from Dr Chick's data does in fact look straight, suggesting that apart from experimental error a law $y = KC^x$ does fit the data reasonably well.

It remains to determine the constants A and B in the equation $Y = A + Bx$. A can be interpreted as the value of Y when $x = 0$, and B as the slope of the straight line. The simplest way to determine them is to take two points on the line $Y = A + Bx$ with known co-ordinates x and Y, and use them to provide two equations for the unknowns A and B. Here it is most logical to take the first and last points. At the first point $x = 0$, $Y = 2{\cdot}64$ ($= \log 434$), so that $A = 2{\cdot}64$. At the last point $x = 25{\cdot}6$, $Y = \log 28 = 1{\cdot}45$, so that

$$1{\cdot}45 = 2{\cdot}64 + B \,.\, 25{\cdot}6$$

whence $B = -{\cdot}0465$. Alternatively we can write the relation as

$y = Ke^{cx}$ where $K = \text{antilog } A = 434$ and $c = B/M = -{\cdot}107$. Also $C = \text{antilog } B = {\cdot}898$.

It is clear that the constant K is merely equal to the value of y when $x = 0$, i.e. the number of bacteria at the start of the experiment. It is therefore of no special significance outside the experiment in question. The second constant, whether expressed as B or c or C, is of more interest. It shows how rapidly anthrax bacteria are killed by phenol—at any rate as regards the culture used by Dr Chick, and under her conditions of concentration and temperature. We can interpret C more precisely as follows. At a time x there are KC^x bacteria living; one hour later the number is KC^{x+1}. This is $(KC^x)C$, so that the number of bacteria surviving at the end of an hour is a fixed proportion C of those present at the beginning. Here $C = {\cdot}898 = 89{\cdot}8$ per cent: about 10 per cent are killed each hour.

The method we used above for fitting the straight line was very crude: it only takes two points of the curve into account, although we can see by eye that the straight line joining these two points does in fact pass near the others. A more accurate method is the "method of linear regression" which will be discussed in Section 21.14. But the chief moral is that it is much easier in testing for a relation $y = KC^x$ to plot $\log y$ against x, and obtain a straight line, rather than to plot y against x and obtain a curve.

This device of reducing a graph to a straight line should be used whenever possible. Thus a relation of the form $y = Kx^B$ becomes $\log y = \log K + B \log x$. Then by plotting $Y = \log y$ against $X = \log x$ we should obtain the straight-line graph $Y = A + BX$, where $A = \log K$. If we have graph paper in which both axes are logarithmically scaled the points can be plotted directly from the original values x and y without the extra labour of finding the logarithms. K will then be the value of y when $x = 1$, and B will be the slope of the straight line.

EXAMPLE

(1) The following are the pulse rates y of persons (adult and children) of different heights x (metres).

Height x	·50	·698	·796	·867	·986	1·765
Pulse rate y	134	111	108	104	98	73

On plotting these points logarithmically (Fig. 7.4) we obtain a straight line of slope $-\frac{1}{2}$ which cuts the line $x = 1$ at $y = 94$, i.e. $B = -\frac{1}{2}$, $K = 94$, and the equation is $y = 94x^{-\frac{1}{2}}$.

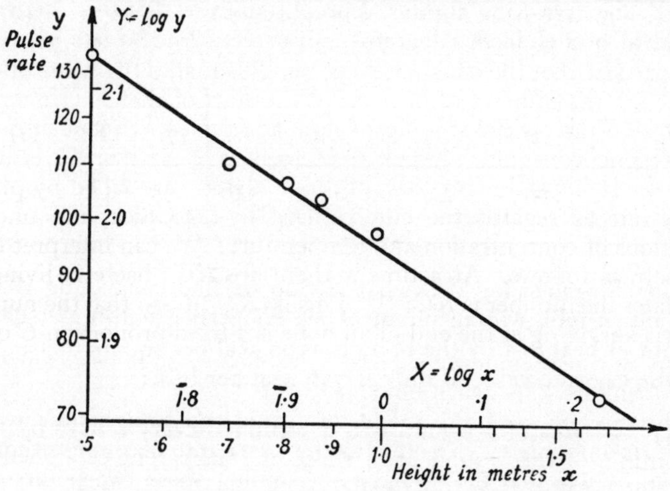

Fig. 7.4—Relation between height and pulse rate

PROBLEMS

(1) Dr Chick tested the bactericidal value of 5 per cent phenol at 33·3° C by mixing a number of bacterial spores with the disinfectant and estimating the number of bacteria in one drop at various intervals of time. She obtained the following results:

Time in hours, t ..	0	·5	1·25	2	3	4·1	5	7
Number of bacteria, n	439	275·5	137·5	46	15·8	5·5	3·5	·5

By plotting log n against t show that the rate of killing obeys the law n = number of survivors = Ke^{ct}. Estimate the constants K and c.

(2) The following values have been found by Feldman and Clark for the rate r of a rabbit's heart at various temperatures t (degrees Centigrade).

Temperature t ..	·4	5·6	6·4	12·8	13·6	14·0	16·0
Heart rate r	5·9	11·7	12·3	25·9	28·4	29·7	37·8

Are they consistent with a law of the form $r = Ke^{ct}$?

(3) The following figures represent the relationship between the weight W of a child in kilograms and its area S in square metres:

W ..	2	3	4	5	6	7	8	9	10
S ..	·163	·214	·260	·301	·340	·377	·412	·446	·479

Plot log S against log W: what law do we find connecting W and S? How much milk will an infant of 2·3 kilograms require per day if the amount of heat lost by the body is 1700 kcal per square metre per day, and the calorific value of milk is 736 kcal per litre?

(4) The area of a wound was determined every 4 days by drawing the outline of the wound on a sheet of transparent cellophane. The following results were obtained:

Time (days) ..	0	4	8	12	16	20	24	28	32	36
Area (cm^2) ..	107	88	74·2	61·8	51	41·6	33·6	26·9	21·3	16·8

Find the law connecting the area with the time, and find how many days should elapse before the wound is reduced to 1 square centimetre.

Note: Carrel, Hartmann, Lecomte du Nouy and others (*J. Exp. Med.*, **24**, 1916, and **27**, 1918) have shown that the logarithm of the area of a wound generally decreases linearly with the time. The rates of decrease of the logarithm, using different antiseptics and different dressings, can thus be compared. A marked deviation from the straight-line law indicates probable infection.

7·3 Other non-uniform scales

Next to the uniform scale, the logarithmic scale is the most convenient and valuable. But clearly it is only one of an immense variety of possible scales. We can have square-root scales, square scales, reciprocal scales, and many others. Many of these will be useful for special problems. In general we shall say that a scale corresponds to a function $F(x)$ if the point P which is marked "x" on the scale is actually at a distance $OP = F(x) = X$ from the origin O. Such scales can be used in very much the same way as logarithmic scales. But, except in a few cases, specially graduated graph paper is not available, and it is necessary to calculate the value of X separately for each observed value of x.

We shall illustrate by an example in which a square root scale is used. Sjöquist studied the course of pepsin digestion by measuring the electrical conductivity y of the protein solution. He found the following values of y for various times x (in hours):

y ..	0	10·5	16·4	19·9	22·7	24·0	27·0	30·4	33·7
x ..	0	2	4	6	8	9	12	16	20
$X = \sqrt{x}$	0	1·41	2	2·45	2·83	3	3·46	4	4·47

If we plot y graphically against $X = \sqrt{x}$ we find that the points lie nearly on a straight line passing through the origin (Fig. 7.5). That is

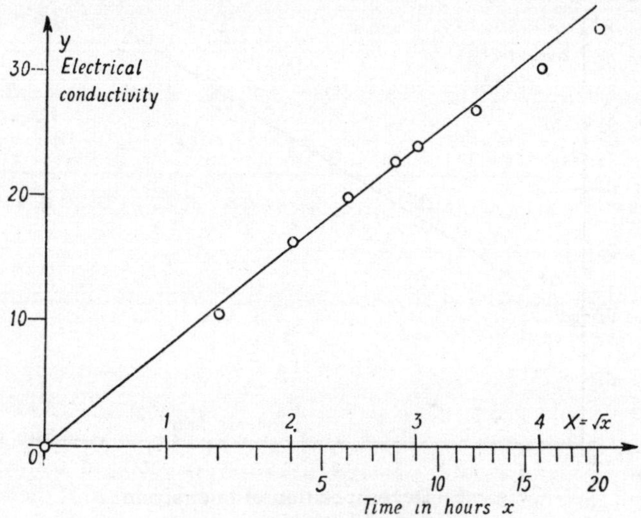

Fig. 7.5—Digestion of pepsin

to say the points obey a law of the form $y = BX = B\sqrt{x}$. The constant B is not far from 8 (when $X = 3$, $y = 24$, $y/X = 8$). This equation $y = B\sqrt{x}$ is the Schütz-Borisoff law.

This example also shows the danger of extrapolating an observed law. Clearly the law $y = B\sqrt{x}$ cannot continue indefinitely, since it would imply that the conductivity y would grow to an indefinitely large value, whereas of course it must approach a certain limiting value L corresponding to complete digestion. The true equation is $kx = L \ln L - L \ln (L - y) - y$, but this approximates to the law $y = B\sqrt{x}$ for small values of x, as can be shown by series methods (Chapter 13).

Each problem will suggest its own most suitable method of scaling.

If we were testing the equation for refraction, $\sin i = \mu \sin r$, where i is the angle of incidence and r the angle of refraction, it would be most natural to plot $\sin i$ against $\sin r$ to see if this gives a straight line. To test the lens formula $1/v - 1/u = 1/f$ where $v =$ distance of object, $u =$ distance of image, and $f =$ focal length, it would be natural to plot $1/v$ against $1/u$.

This technique can be used to find the "order" of an irreversible chemical reaction, i.e. the number of molecules taking part. Let C be the concentration of a reacting substance, t the time, and n the order of the reaction. Then if $n = 1$, i.e. the reaction consists of the decomposition of a single molecule, we obtain a straight-line graph by plotting $\log C$ against t. (See Section 16.23 for a detailed discussion of this point.) If $n > 1$, then we must plot C^{1-n} against t to obtain a straight line. A few trial plots for different values of n will usually determine the correct value.

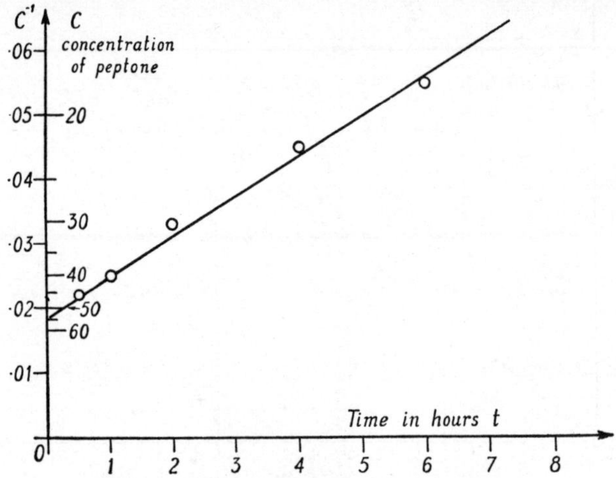

Fig. 7.6—Decomposition of tetanolysin

EXAMPLE

(1) Madsen and Walbum studied the decomposition of tetanolysin by means of peptone, obtaining the following results:

Time in hours t ..	·5	1	2	4	6	8
Concentration of peptone C	47·7	39·7	30·3	22·3	18·1	17·0
C^{-1}	·021	·025	·033	·045	·055	·059

If we plot C^{-1} against t we obtain roughly a straight line (Fig. 7.6) whereas the plot of log C or of C^{-2} or C^{-3} against t is curved. This suggests that the reaction is bimolecular ($C^{1-n} = C^{-1}$ when $n = 2$). [More elaborate investigation indicates the scheme—
2 molecules of tetanolysin + 3 molecules of peptone → lysinpeptone; the lysinpeptone immediately disintegrates with re-formation of peptone. Thus the peptone is not exhausted.]

PROBLEMS

(1) Madsen and Walbum studied the process of tryptic digestion by subjecting 10 grams of casein powder to the action of 100 cc of a 1 per cent solution of trypsin at constant temperature. The amount of casein remaining was found by Kjeldahl's method of nitrogen determination. The following figures represent their results: show that they agree with the supposition that the process is a bimolecular reaction.

Time (hours)	0	·5	2·5	6	11	24	33	48	72
Nitrogen concentration	·110	·108	·102	·100	·096	·076	·070	·060	·049

(2) Madsen and Famulener found the following results for the concentration of vibriolysin at 28° C. Find the order of the reaction.

Time (minutes) ..	0	10	20	30	40	50	60
Concentration ..	100·0	78·3	67·6	59·3	49·8	40·8	34·4

(3) The following values have been found for x and y:

x ..	0	1	2	3	4	5	6	7
y ..	6·29	5·72	5·22	4·78	4·35	4·06	3·75	3·48

By plotting xy against y show that these fit reasonably well to a law $y = a/(x + b)$, and find the values of a and b.

By plotting log y against x show that they fit reasonably well to a law $y = Ke^{Bx}$, and find K and B.

This again points the moral that two quite different formulas can be in good agreement *over a certain range*, though outside this range they may behave very differently. Compare the values of y given by the two formulas when x takes values near $-b$.

7.4 Slide rules

If we have two ordinary uniformly graduated rulers we can readily add and subtract in the following way.

Let O_1 be the zero point on one ruler, and P_1 a typical point "X_1" $= 4$, say, on the scale. This means that the distance O_1P_1 is X_1 units. Now place the second scale alongside the first with its zero point O_2 opposite P_1 (Fig 7.7). Let P_2 be a typical point on the second scale

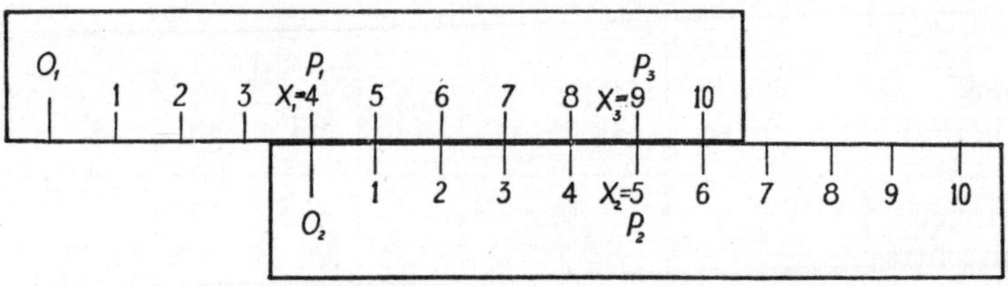

Fig. 7.7—Slide rule for addition and subtraction

"X_2" $= 5$ (in the figure), so that the distance O_2P_2 is X_2 units. It follows that the distance $O_1P_2 = O_1P_1 + O_2P_2 = X_1 + X_2$. Thus opposite P_2 on the lower scale will occur the point $X_1 + X_2$ on the upper scale. If we set P_1 at the point $X_1 = 4$, we see from Fig. 7.7 that $4 + 1 = 5$ (opposite 1 on the lower scale is 5 on the upper), $4 + 2 = 6$, $4 + 3 = 7$, and so on.

Subtraction is equally simple: it is just the inverse of addition. If opposite a point P_3 marked X_3 on the upper scale we place the point P_2 marked X_2 on the lower scale, then opposite O_2 on the lower scale we shall find P_1 on the upper indicating $(X_3 - X_2)$. For example, $9 - 5 = 4$ (Fig. 7.7). For $O_1P_1 = O_1P_3 - O_2P_2 = X_3 - X_2$.

Naturally one would not in practice use such a device merely for addition and subtraction: it would be too cumbersome. But if we graduate the scales non-uniformly then this becomes a very powerful instrument. For example, if we graduate logarithmically, so that the point P_1 is marked x_1 when it is at a distance $O_1P_1 = X_1 = \log x_1$ from O_1, then, instead of adding, the slide rule multiplies, and instead of subtracting it divides (Fig. 7.8). For now P_1 is marked "x_1", where $O_1P_1 = \log x_1$, P_2 is marked "x_2" where $O_2P_2 = \log x_2$, and P_3 opposite P_2 is marked "x_3" ($= 12$) where $O_1P_3 = \log x_3$. Since

$O_1P_3 = O_1P_1 + O_2P_2$, we have $\log x_3 = \log x_1 + \log x_2$, i.e. $x_3 = x_1x_2$, and $x_1 = x_3/x_2$. In Fig. 7.8 the mark "1" on the lower scale is opposite "3" on the upper scale, and the rule is in a position to multiply by 3. Opposite 2 on the lower scale is 6 on the upper, so that $2 \times 3 = 6$; similarly $3 \times 3 = 9$, $4 \times 3 = 12$, and so on. Alternatively the same setting does the division $12/4 = 3$.

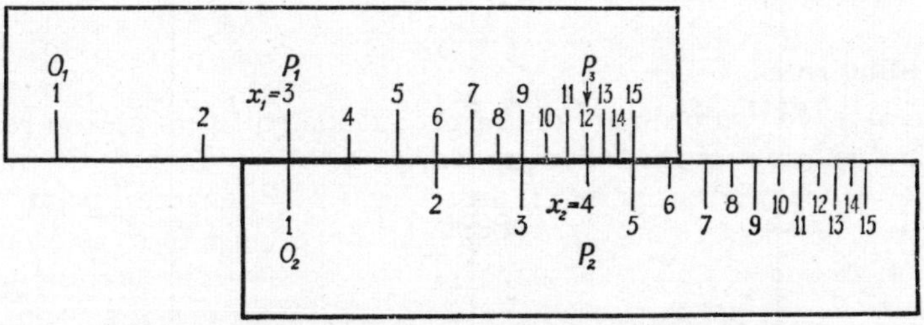

Fig. 7.8—Slide rule for multiplication and division

A pair of logarithmically graduated rules like this forms the simplest form of logarithmic slide rule, and can be used for multiplication and division. It is specially convenient for doing a series of multiplications and divisions, such as $(2{\cdot}73 \times 100{\cdot}6)/(230 \times 56{\cdot}2)$. Most modern slide rules have a number of additional scales which enable one to find square roots, squares, cubes, reciprocals, natural logarithms, exponentials, ordinary logarithms, powers (x^y), sines, cosines, and tangents, with the minimum of trouble. The exact details vary with the make of slide rule, and we shall not discuss them here, since they introduce very little that is new in principle.

It is, however, worth noting that it is a fairly simple matter to make slide rules with graduations on other functional scales besides the familiar logarithmic one. If we use a square graduation the rule will solve $x_3^2 = x_1^2 + x_2^2$, i.e. compute the sides of right-angled triangles, and also calculate "sums of squares", an important statistical operation. In such a graduation the mark x must be placed at a distance x^2 from the zero point. By graduating by reciprocals we can solve the lens equation $1/v - 1/u = 1/f$.

7·5 Nomograms

Instead of using a sliding scale for addition we can often use fixed scales and a straight edge as indicated in Fig. 7.9 overleaf. Here we have three parallel scales which we may call the y_1, y_2 and y_3 scales. They are drawn vertically for convenience in use; and since a vertical scale is usually labelled "y" as a matter of convention, that is our reason for the use of the letter y here. If now the point marked y_1 (=2) on the first scale is joined by a straight line to the point marked y_3 (=5) on the third then this line will cut the central scale at the point $y_1 + y_3$. (A proof of this property will be given in the next section.)

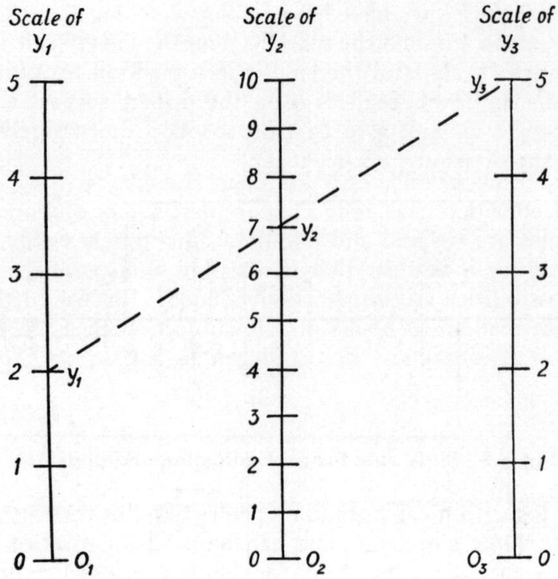

Fig. 7.9—Nomogram for addition

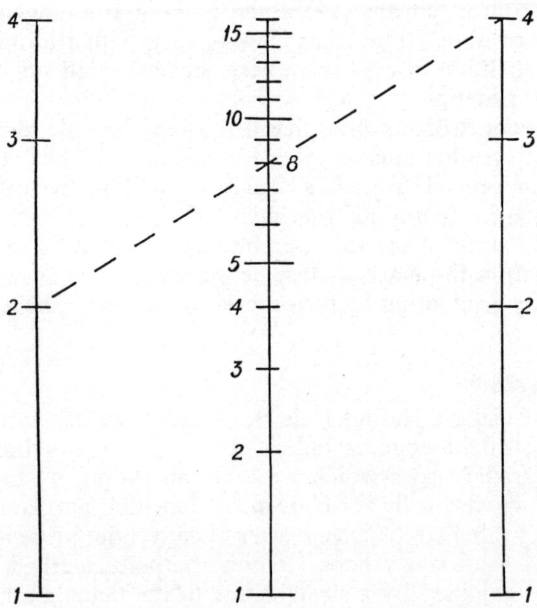

Fig. 7.10—Nomogram for multiplication

The dotted line shows the addition $2 + 5 = 7$, or alternatively the subtraction $7 - 2 = 5$. By replacing the uniformly graduated scales by logarithmic ones we shall get multiplication instead of addition as in Fig. 7.10, where the connecting line shows the multiplication $2 \times 4 = 8$, or alternatively the division $8/2 = 4$. Charts such as these in which a calculation is performed by joining suitable points by straight lines are known as "nomograms" or "alignment charts".

We shall consider here only the simplest kinds of nomograms in which all scales are vertical and parallel. This touches only the fringe of the subject, for it is clear that if we allow non-parallel and curved scales the possibilities are immensely widened. But for further details we must refer readers to books on the subject, such as S. Brodetsky, *First Course in Nomography*, Bell, 1925, or A. S. Levens, *Nomography*, Wiley, 1948.

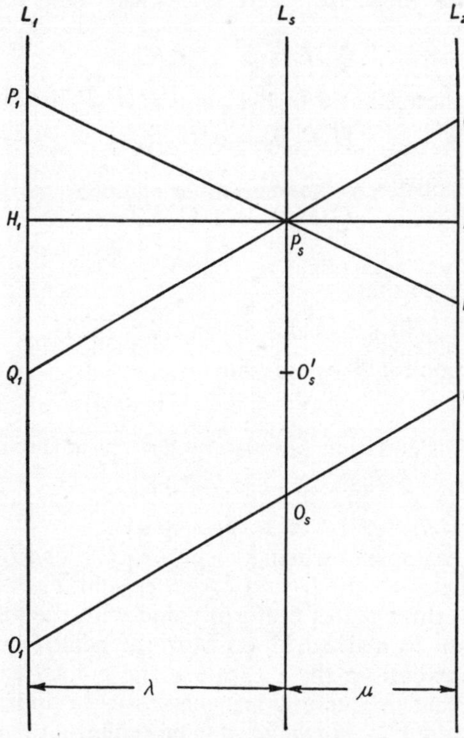

Fig. 7.11—Principle of a parallel-scale nomogram

7.6 Nomograms with three parallel scales

The fundamental theorem of nomography is this.

Let L_1, L_2, and L_s be three parallel straight lines, and O_1, O_2 and O_s be three points on L_1, L_2 and L_s respectively lying on a straight line.

(As a rule O_1, O_2 and O_s will lie in the same horizontal line; but for the sake of generality we shall not assume that to be so.) Let P_1, P_2, P_s be the three points in which a straight line cuts L_1, L_2 and L_s respectively. Call the actual distances O_1P_1, O_2P_2 and O_sP_s, Y_1, Y_2 and Y_s respectively. Furthermore, let the distance between L_1 and L_s (measured positively if L_s is to the right of L_1, negatively otherwise) bear to the distance between L_s and L_2 a ratio $\lambda : \mu$ (Fig. 7.11).
Then

$$Y_s = (\lambda Y_2 + \mu Y_1)/(\lambda + \mu) \qquad . \quad . \quad (7.1)$$

Proof. Draw $Q_1P_sQ_2$ parallel to $O_1O_sO_2$ and $H_1P_sH_2$ horizontally, through P_s, with Q_1 and H_1 on L_1 and Q_2 and H_2 on L_2. Then the triangles $Q_1H_1P_s$ and $Q_2H_2P_s$ are similar, so that $Q_1H_1/H_1P_s = Q_2H_2/H_2P_s$ (using the proper signs, upwards distances being measured positively, and also those from left to right). This relation can be written

$$Q_1H_1/\lambda = -Q_2H_2/\mu$$

In the same way the triangles $H_1P_1P_s$ and $H_2P_2P_s$ are similar, so that

$$H_1P_1/\lambda = -H_2P_2/\mu.$$

Adding these two equations we get

$$Q_1P_1/\lambda = -Q_2P_2/\mu$$

But $Q_1P_1 = O_1P_1 - O_1Q_1 = Y_1 - O_sP_s = Y_1 - Y_s$, and similarly $Q_2P_2 = Y_2 - Y_s$, so that

$$(Y_1 - Y_s)/\lambda = -(Y_2 - Y_s)/\mu$$

Solving this equation for Y_s we obtain

$$Y_s = (\lambda Y_2 + \mu Y_1)/(\lambda + \mu)$$

which is (7.1). This equation can also be written in the form

$$Y_s = \kappa Y_1 + (1 - \kappa)Y_2 \qquad . \quad . \quad . \quad (7.2)$$

where $\kappa = \mu/(\lambda + \mu)$, for $1 - \kappa = \lambda/(\lambda + \mu)$.

The simplest example is when $\lambda = \mu$, $\kappa = \frac{1}{2}$. The L_s scale is then situated midway between the L_1 and L_2 scales, and $Y_s = \frac{1}{2}(Y_1 + Y_2)$. If we graduate all three scales uniformly and with the same unit, then by joining the point P_1 marked Y_1 on L_1 to the point P_2 marked Y_2 on L_2 we obtain a line cutting the L_s scale at the point P_s marked $Y_s = \frac{1}{2}(Y_1 + Y_2)$, so that this nomogram enables us to find the average of two numbers Y_1 and Y_2. However, it is more useful to graduate the L_s scale with unit of graduation half that for the L_1 and L_2 scales, i.e. so that the mark y_s occurs at distance $O_sP_s = Y_s = \frac{1}{2}y_s$ along the scale. We then have

$$\tfrac{1}{2}y_s = \tfrac{1}{2}(Y_1 + Y_2)$$

or

$$y_s = Y_1 + Y_2$$

and the nomogram, which is the one shown in Fig. 7.9, enables us to add the two numbers Y_1 and Y_2.

More generally we can say that if a unit of graduation on the scale L_1 has actual length U_1, then the mark y_1 on the scale will occur at actual distance $O_1P_1 = Y_1 = y_1U_1$ from O_1. Similarly if U_2 is the actual length of the unit of graduation on the L_2 scale, and U_s on the L_s scale, then $Y_2 = y_2U_2$ and $Y_s = y_sU_s$ so that (7.2) becomes

$$U_sy_s = \kappa U_1y_1 + (1 - \kappa) U_2y_2$$

The readings of the points P_1, P_2 and P_s on the scales so graduated will be y_1, y_2 and y_s and will be connected by the relation

$$\begin{aligned} y_s &= (\kappa U_1/U_s)y_1 + ([1 - \kappa] U_2/U_s) y_2 \\ &= B_1y_1 + B_2y_2 \qquad (7.3) \end{aligned}$$

where $B_1 = \kappa U_1/U_s$ and $B_2 = [1 - \kappa] U_2/U_s$. By choosing the quantities κ, U_1, U_2 and U_s suitably we can construct a nomogram to calculate any desired combination of the form: $B_1y_1 + B_2y_2$.

Fortunately it is possible to avoid even this amount of calculation. This is because we can set up and graduate the lines L_1 and L_2 exactly as we wish: the line L_s on which the answer appears will then be fixed both in position and graduation, but is easily determined graphically. (This amounts to saying that we can choose any value we like for U_1 and U_2, but that, given the values of B_1 and B_2, we can then find those of κ and U_s.)

As an example we shall consider the following problem. The calorific values of protein, carbohydrate, and fat are approximately 4, 4, and 9 calories per gram respectively. We wish to construct a nomogram which will enable us to find the calorific value of a diet of given composition. If we let y_1 denote the total number of grams of protein and carbohydrate together, and y_2 the number of grams of fat, then the total calories will be $y_s = 4y_1 + 9y_2$.

Now draw two parallel lines L_1, L_2 at a convenient distance apart to serve as scales for y_1 and y_2, and graduate them on any suitable scale. We may graduate y_1 from 0 to 500 grams, and y_2 from 0 to 300 grams, since the total consumption of fat will in general be less than that of protein and carbohydrate combined (Fig. 7.12 overleaf). We find the position of the y_s scale as follows. Since a diet consisting of 450 grams protein and carbohydrate and no fat has a value of 1800 calories, the line joining the mark 450 on the first scale to 0 on the second must pass through 1800 on the y_s scale. So also must the line joining 0 on the y_1 scale to 200 on the y_2 scale. The point where these two lines intersect must therefore be the point marked 1800 on L_s. So L_s is the line through this point parallel to L_1 and L_2. The zero point on L_s will be its intersection with the line O_1O_2. Having found L_s and the marks y_s for two points on it we can complete the graduation by simple proportion.

This method can also be used, with only slight modification, to construct nomograms for expressions like $y_s = A + B_1y_1 + B_2y_2$, where A, B_1, B_2 are constants. As an example, suppose we want a nomogram for $y_s = 100 + 4y_1 + 9y_2$. We have already shown how to construct one for $y_s = 4y_1 + 9y_2$; the new nomogram will differ only

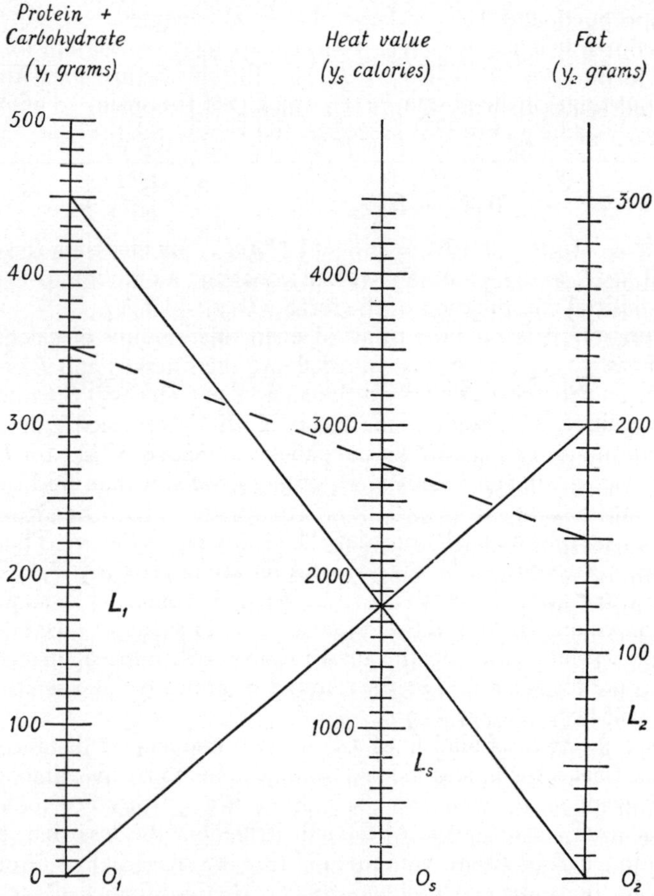

Fig. 7.12—Nomogram for the heat value of a diet

in that all values of y_s will be increased by 100. In other words all that is needed is to take the nomogram for $y_s = 4y_1 + 9y_2$ and add 100 to the values of y_s marked on the line L_s. If we were constructing the nomogram *ab initio* (not having previously obtained one for $y_s = 4y_1 + 9y_2$) then we would proceed at first exactly as before, drawing any two suitable parallel scales L_1 and L_2 for y_1 and y_2 respectively. We now note that the values $y_1 = 450$, $y_2 = 0$ and $y_1 = 0$, $y_2 = 200$

both give $y_s = 1900$, so that the lines joining the points marking $y_1 = 450$ and 0 to the points $y_2 = 0$ and 200 respectively must meet on L_s at $y_s = 1900$. L_s is the line through this point parallel to L_1 and L_2. The line O_1O_2 joining $y_1 = 0$ to $y_2 = 0$ will now meet L_s at $y_s = 100$ (by the formula $y_s = 100 + 4y_1 + 9y_2$) instead of at the point $y_s = 0$. Having found two points on L_s we can graduate the rest of the line by simple proportion.

Of course it is rarely necessary to construct a nomogram for such a simple formula as $A + B_1y_1 + B_2y_2$. But if we use logarithmically graduated scales instead of uniform ones this becomes

$$\log y_s = A + B_1 \log y_1 + B_2 \log y_2$$

or

$$y_s = K y_1^{B_1} y_2^{B_2}$$

where $K = \text{antilog } A$. For example we can construct a nomogram for the Dubois formula $S = \cdot 2025\, W^{\cdot 425} H^{\cdot 725}$, where S is the surface area of the body in square metres, W the weight in kilograms, and H the height in metres. Expressed logarithmically this becomes

$$\log S = -\cdot 6936 + \cdot 425 \log W + \cdot 725 \log H.$$

(since $\log \cdot 2025 = \bar{1}\cdot 3064 = -\cdot 6936$).

Naturally there must be bounds to the values of S, W and H under consideration. Suppose we allow the height H to range from ·5 to 1·8 metres, and weights W from 2 to 70 kilograms. Then the logarithm of H will vary from (roughly) $-\cdot 3$ to $+\cdot 25$, a total range of ·55, and the logarithm of W from ·3 to 1·85, a total range of 1·55.

We must now choose a convenient length for our scales—naturally the longer the scales the greater the accuracy of the nomogram. If we take about 15 cm to be a convenient length, then since the logarithm of H varies by a total amount of ·55, this means that we can graduate the H scale with 3 cm representing a change of ·1 in the logarithm (thus giving an actual total length of $3 \times 5\cdot 5 = 16\cdot 5$ cm). The lowest point in the scale corresponds to a logarithm of about $-\cdot 3$, and therefore the zero point O_1 occurs about $3 \times 3 = 9$ cm up the scale. The first stage in the construction of the nomogram therefore consists in drawing a vertical line L_1, about 17 cm long, and marking the zero point O_1 about 9 cm from the bottom. This point O_1 will actually be marked 1, since $\log 1 = 0$. We can now readily mark off the other points, since the distance of the point marked H from the zero point O_1 is $30 \log H$ cm. (This follows from simple proportion: a difference of ·1 in the log = 3 cm, i.e. a difference of 1 in the log = 30 cm, and a difference $\log H$ = $30 \log H$ cm on the scale.) Thus the mark for 1·1 metres is placed $30 \log 1\cdot 1 = 30 \times \cdot 0414 = 1\cdot 24$ cm above O_1. This is shown (on a reduced scale) in Fig. 7.13.

In the same way on the L_2 scale for weight W, since this is to cover a range 1·55 in logarithms and about 15 cm in actual distance, it will be convenient to take 1 cm = a difference of ·1 in the logarithm, or 10 cm = a difference of 1 in the log. Unfortunately here the logarithm ranges from ·3 to 1·85, so that the zero point does not lie on the part of the scale we are considering. We can overcome this difficulty by using another

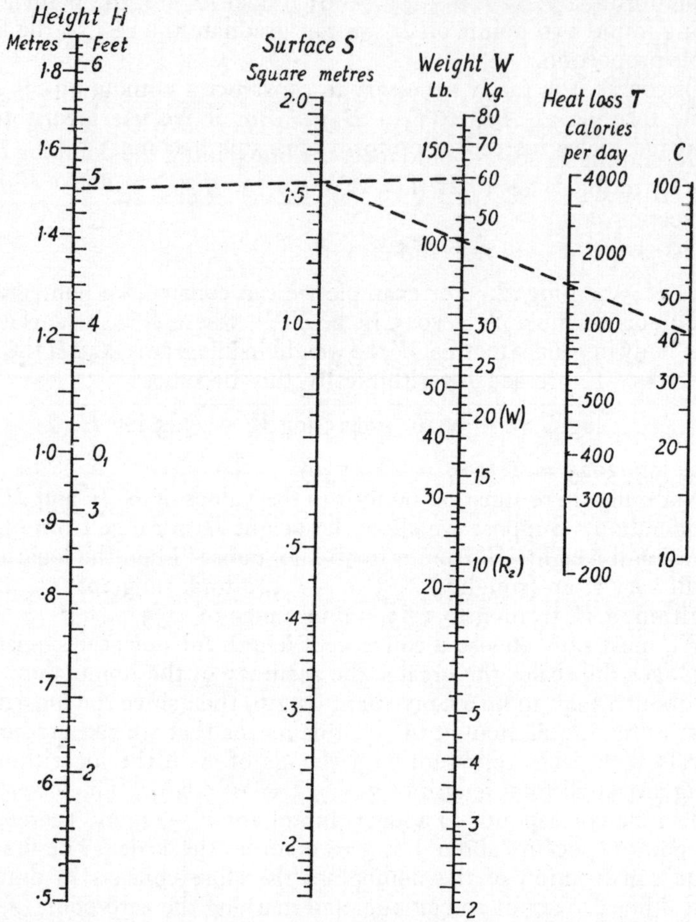

Fig. 7.13—Nomogram for surface area and loss of heat per day

reference point on the scale, say R_2 where $\log W = 1$, and accordingly $W = 10$. Since the bottom of the scale corresponds approximately to $\log W = \cdot 3$, this reference point will come about $10\,(1 - \cdot 3) = 7$ cm up the scale. We therefore draw a second vertical line L_2 at (say) 8 cm distance from L_1, and mark the reference point R_2, $W = 10$. The remainder of the W scale is then readily graduated, the point W, e.g. 20 kg, occurring at a distance $10\,(\log W - 1) = 3{\cdot}01$ cm above R_2.

Having drawn the H (L_1) and W (L_2) scales, we have to find the

S (L_3) scale, both as regards position and graduation, so as to give the equation

$$\log S = -{\cdot}694 + {\cdot}425 \log W + {\cdot}725 \log H$$

when used as a nomogram.

This can be done in two ways. Graphically we do it by noting that if $\log W = 1$, $\log H = {\cdot}425$, then $\log S = -{\cdot}694 + {\cdot}425 + {\cdot}725 \times {\cdot}425 = {\cdot}039$, and if $\log W = 1{\cdot}725$, $\log H = 0$, then $\log S = -{\cdot}694 + 1{\cdot}725 \times {\cdot}425 = {\cdot}039$ again. Thus the line joining the points $\log H = {\cdot}425$ on the H scale ($30 \times {\cdot}425 = 12{\cdot}75$ cm above O_1) to $\log W = 1$ (i.e. R_2) on the W scale should meet the line joining the point $\log H = 0$ (i.e. O_1) to $\log W = 1{\cdot}725$ (i.e. $10 \times {\cdot}725 = 7{\cdot}25$ cm above R_2) at an intersection point $\log S = {\cdot}039$ on the S scale, L_s. We can then draw L_s vertically through this point. The line joining O_1 ($\log H = 0$) to R_2 ($\log W = 1$) will meet the S scale where $\log S = -{\cdot}694 + {\cdot}425 = -{\cdot}269$. Having found two points in this way on the S scale L_s we can graduate it by taking simple proportion in the logarithms.

Another and perhaps simpler way is to use the fact that the equation

$$\log S = A + B_1 \log H + B_2 \log W$$

holds, where

$$B_1 = \kappa U_1/U_s, \quad B_2 = [1 - \kappa]\, U_2/U_s \quad . \quad . \quad . \quad (7{\cdot}4)$$

and U_1 is the length of the L_1 scale corresponding to a difference of 1 in $\log H$, U_2 the length of the L_2 scale corresponding to a difference of 1 in $\log W$, U_s the length along L_s corresponding to a difference of 1 in $\log S$, and κ is the ratio of the distance from L_s to L_2 to the distance from L_1 to L_2. This equation is simply an adaptation of (7.3) to logarithmically graduated scales.

Now we are aiming at the values $A = -{\cdot}694, B_1 = {\cdot}725, B_2 = {\cdot}425$; and we have chosen $U_1 = 30$ cm per unit log, $U_2 = 10$. (These numbers U_1 and U_2 are sometimes called the *moduli* of the scales; this is still another use of the word "modulus".) Thus we have to solve the equations

$$B_1 = \kappa U_1/U_s, \quad \text{i.e.} \quad {\cdot}725 = 30\kappa/U_s$$

and

$$B_2 = [1 - \kappa]\, U_2/U_s \quad \text{i.e.} \quad {\cdot}425 = 10[1 - \kappa]/U_s$$

for the unknowns κ and U_s. This is done thus.

The first equation can be written $B_1/U_1 = \kappa/U_s$, and the second $B_2/U_2 = (1 - \kappa)/U_s$. On addition we get $1/U_s = B_1/U_1 + B_2/U_2$, or

$$\begin{aligned} U_s &= [B_1/U_1 + B_2/U_2]^{-1} \\ &= [{\cdot}0242 + {\cdot}0425]^{-1} \\ &= [{\cdot}0667]^{-1} = 15{\cdot}0 \end{aligned}$$

Now

$$\begin{aligned} \kappa &= U_s B_1 U_1^{-1} = 15{\cdot}0 \times {\cdot}0242 \\ &= {\cdot}363 \end{aligned}$$

and the distance from L_s to L_2

$$= \kappa \times (\text{distance from } L_1 \text{ to } L_2)$$
$$= 8\kappa = 2{\cdot}90 \text{ cm.}$$

We can therefore draw L_s 2·90 cm from L_2, and we know that on this scale a difference of 1 in log S corresponds to a distance of $U_s = 15$ cm. It remains to find a reference point on this scale from which we can plot the others. A convenient point is the zero point O_s, $\log S = 0$, $S = 1$. By our equation this will occur when

$$0 = -{\cdot}694 + {\cdot}425 \log W + {\cdot}725 \log H.$$

We can join the points where $\log H = 0$ (i.e. the point O_1) and where $\log W = {\cdot}694/{\cdot}425 = 1{\cdot}633$ (i.e. a point $10 \times {\cdot}633 = 6{\cdot}33$ cm above R_1). This line will intersect L_s at O_s, and from this the remainder of the scale can easily be graduated. We have now completed the nomogram; a line joining points H and W will intersect L_s at S where $\log S = -{\cdot}694 + {\cdot}725 \log H + {\cdot}425 \log W$, i.e. $S = {\cdot}2025\, H^{\cdot 725}\, W^{\cdot 425}$.

To avoid conversion the H and W scales can also be graduated in British units, S being still read off directly.

7·7 Nomograms with more than three scales

We can readily extend this scheme to construct nomograms for relations such as

$$y_t = y_1 + y_2 + y_3$$
or $$y_t = A + B_1 y_1 + B_2 y_2 + B_3 y_3$$
or $$\log y_t = A + B_1 \log y_1 + B_2 \log y_2 + B_3 \log y_3$$

For we can decompose the equation $y_t = A + B_1 y_1 + B_2 y_2 + B_3 y_3$ into two equations by bringing in an intermediate number y_s in the following way:

$$\left.\begin{aligned} y_s &= A + B_1 y_1 + B_2 y_2 \\ y_t &= y_s + B_3 y_3 \end{aligned}\right\} \quad . \quad . \quad . \quad (7{\cdot}5)$$

We can construct a nomogram which gives y_s from y_1 and y_2, and using the same scale for y_s we can construct a second one giving y_t from y_s and y_3. Thus the procedure in using the nomogram will be to join the points y_1 and y_2 on their respective scales by a line meeting the y_s scale at y_s; this point is then joined to the y_3 scale, and will intersect the y_t scale at the required point $y_t = A + B_1 y_1 + B_2 y_2 + B_3 y_3$. In this process the line L_s bearing the y_s scale is merely brought in as a help in performing the construction, and does not need to be graduated at all. It may, however, be useful to mark two or three points on L_s while drawing the nomogram: these marks can afterwards be erased.

Thus in Fig. 7.13 (p. 158) we have combined the nomogram for surface area $S = {\cdot}2025\, H^{\cdot 725}\, W^{\cdot 425}$ with the product nomogram $T = 24SC$, where C = heat lost in calories per square metre per hour, and T = total heat lost per day. This is a very simple nomogram to construct:

since S is already logarithmically graduated, all that is necessary is to draw the scale for C on a line L_3 parallel to that for S, and graduate it logarithmically. The position and graduation of the T scale is then determined from the equation

$$T = 24\ SC, \quad \text{or}$$
$$\log T = \log 24 + \log S + \log C.$$

In Fig. 7.13 the C scale has been graduated from 10 to 100. The lines joining the points $S = 2$ to $C = 25$ and $S = \cdot 5$ to $C = 100$ must both pass through the point $T = 1200$, which is thus determined. The vertical line through this point will be the T scale, and it will intersect the line joining $S = \cdot 2$ to $C = 25$ at $T = 120$. Now the distance between the points $T = 1200$ and $T = 120$ will correspond to a difference of $\log 1200 - \log 120 = 1$ in $\log T$, and this distance will therefore be the unit U_T of graduation for the T scale. As we have now found two points and the unit, the remaining marks are readily filled in by simple proportion on $\log T$; the point T will be placed at a distance $(\log T - \log 120)\ U_T$ centimetres above the point 120.

As an example of the use of this nomogram we may find the surface area S and the total heat T lost per day for a person of height $H = 1{\cdot}50$ metres and weight $W = 60$ kilograms, losing heat at the rate of $C = 40$ calories per square metre per hour. The line joining the points H and W intersects the S scale at $S = 1{\cdot}55$ square metres: and the line joining this to 40 on the C scale intersects the T scale at 1500 calories per day. If we were only interested in the values of T we could clearly ignore the graduation on the S scale.

PROBLEMS

Construct nomograms for the following:

(1) The respiratory quotient, i.e. the volume of CO_2 expired divided by the volume of O_2 inspired. [The scale for the volume of CO_2 should be placed midway between the scale for respiratory quotient and the scale for volume of O_2. The two external scales should be given the same logarithmic unit, the values for the volume of CO_2 running from 100 to 1000 cc. The intermediate scale for O_2 can be then graduated automatically for volumes of O_2 from 100 to 1000 cc.]

(2) The colour index of blood, viz. the haemoglobin (as per cent of the normal value) divided by the number of erythrocytes (as per cent of normal value).

(3) The gas equation, $V = TR/P$. (If T is the absolute temperature in degrees centigrade, P the pressure in newtons per square metre, i.e. 100 times the pressure in millibars, and V the volume in litres. then $R = 8300$ approximately.)

(4) Pythagoras's theorem, $r^2 = x^2 + y^2$. (Use scales graduated by squares.)

(5) The lens equation, $1/v - 1/u = 1/f$.

(6)
$$y_1 + y_2 + y_3 + y_4 = z.$$

(Place the scales at equal distances apart in the following order. First a scale for y_1, then one for y_2 graduated with half the unit of y_1, and downwards instead of upwards, then one for $y_s = y_1 + y_2$, then one for y_3 graduated upwards with the half unit, one for $y_t = y_s + y_3$, one for y_4 graduated downwards with the half unit, and finally one for z.)

(7) It was stated above that in the nomogram for $y_s = B_1 y_1 + B_2 y_2$ it is possible to choose any parallel scales for y_1 and y_2, the position and unit of graduation of the y_s scale being then determined. This is not quite true: occasionally there may be difficulties. What are these difficulties, and when do they arise?

CHAPTER EIGHT

RATES OF CHANGE

8.1 Movement and change

Everything is (as far as we can tell) always moving and changing. Sometimes the change is slow and minute, as for a mountain which seems the same from year to year, but is nevertheless being worn away and dissolved by wind and rain. Sometimes it is exceedingly rapid, as with the lightning flash, which strikes for an instant and is gone. Living things especially are full of unceasing chemical and physical activity. Even when at rest a plant or animal must constantly breathe; and while some spores and seeds can remain dormant for long periods and still be able to germinate, for most living creatures change is life and the complete cessation of all activity only comes with death.

Changes which occur naturally are of two kinds, continuous and discontinuous. A discontinuous change from X to Y is one in which there are no intermediate states: for example, melting ice undergoes a discontinuous change in state, as there is nothing intermediate between the solid ice and the liquid water. It is true that if we take a block of ice and slowly warm it the change is continuous in the sense that the ice melts gradually, so that the quantity of ice and the quantity of water both change continuously. But all the time there is a clear boundary between the two states, and it is impossible to confuse one with the other. If we warm the water after all the ice has melted it will then undergo continuous changes in temperature and volume, contracting at first until the temperature has reached 4° C, and expanding after that point: and it will go on changing continuously until it reaches the boiling point, when a further discontinuity occurs. A philosophically minded reader may perhaps challenge this distinction, and claim that if we could actually see individual atoms at work we should find that all changes were continuous; or alternatively he could argue that according to quantum theory all changes were abrupt. This point is certainly debatable, but whatever the ultimate truth may be we have in ordinary practice to reckon with both kinds of change.

Now in itself a sudden change can quite easily be specified mathematically: all we have to do is to state the values of the quantities we are considering before and after the change. Thus we can say that ice with a density 0·9 gm/cc changes into water with a density 1·0. A continuous change is a little more subtle: it can take place slowly or rapidly, and it can speed up or slow down. The mathematical theory dealing with the measurement of rates of change is known as the "differential calculus". Fortunately—in spite of the terror which some non-mathematicians feel at the word "calculus"—it is one of the simplest of all branches of mathematics as far as formal manipulation is concerned. Once the definition of "rate of change"— or "derivative" as it is technically named—has been mastered, there is very little more to do except to learn and practise certain simple rules of procedure.

The calculus has the added advantage that it follows very closely our everyday use of language. The simplest kind of continuous change is change of position, or movement: and we speak of the rate of movement, i.e. speed, as "quick" or "slow". These words apply in the strictest sense only to motion: but we are accustomed to carry them over to all kinds of change, as when we say of something that it changes colour rapidly, or that the population of Britain is increasing more slowly than that of India. Indeed this use is so natural that it can hardly be considered as a metaphor. The words "quick" and "slow" are also used in cases where time is not involved at all. For example, we say that the main road out of Brighton rises only slowly for the first mile, then rather more quickly up to the northern edge of the Downs, after which it falls very rapidly down to the Weald. This use of the words "slowly" and "rapidly" might be justified by saying that we imagine ourselves travelling along the road: but that is not an essential part of the situation. What we are talking about is the slope of the ground, which could be expressed alternatively by stating its inclination to the horizontal plane. All these different kinds of rate of change, whether we mean motion, or change of qualities such as colour, or of quantities such as mass, or the more indirect use of the word as a gradient—all these are equally well covered by the differential calculus, and all follow the same rules of calculation.

8.2 Average velocity

If a train travels 30 miles in half an hour, or 60 miles in an hour, or 120 miles in 2 hours, then its speed is 60 miles per hour. More precisely, since it will probably not always be travelling equally quickly, we call the quotient of the distance gone divided by the time taken the "average velocity" over the whole journey: e.g. 20 miles in 20 minutes gives an average velocity of 20 miles/$\frac{1}{3}$ hour $=$ 60 miles per hour.

NOTE: Observe that this is the *definition* of the phrase "average velocity". It has a connection with the ordinary use of the word "average". If a train travels 40 miles between 2 p.m. and 3 p.m., 70 miles between 3 p.m. and 4 p.m., and 25 miles between 4 p.m. and 5 p.m., then the total distance covered is $40 + 70 + 25 = 135$ miles in 3 hours, and the average speed is $135/3 = 45$ miles per hour. This is the average of 40, 70, and 25 in the usual sense. Here we have measured the three velocities of 40, 70 and 25 miles per hour over equal intervals of time of one hour each. If we measure them for *unequal* intervals of time, the average velocity over the whole journey need no longer be the average of the velocities for each part of the journey. For example, if a walker does the first 10 miles at an average velocity of 4 miles per hour, and the next 10 miles at an average velocity of 2 miles per hour, his average for the whole 20 miles is not 3 miles per hour.

Query: What is it? Why?

We can readily express the average velocity by a formula. Suppose we represent the time (measured in hours) by t and the distance which (say) a train has gone (measured in miles) by y. y and t will accordingly be variable quantities, and y will be a certain function of t, that is to say, when we specify a particular time t we can find the corresponding distance y. We can plot a graph of y against t, which will show us how the train is running (the curved line in Fig. 8.1).

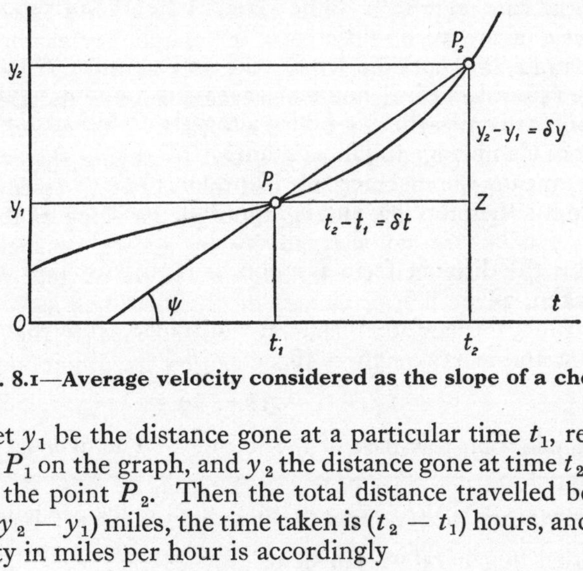

Fig. 8.1—Average velocity considered as the slope of a chord

Now let y_1 be the distance gone at a particular time t_1, represented by a point P_1 on the graph, and y_2 the distance gone at time t_2, as represented by the point P_2. Then the total distance travelled between P_1 and P_2 is $(y_2 - y_1)$ miles, the time taken is $(t_2 - t_1)$ hours, and the average velocity in miles per hour is accordingly

$$v = (y_2 - y_1)/(t_2 - t_1) \quad . \quad . \quad . \quad (8.1)$$

Now if we draw P_1Z horizontally as in Fig. 8.1 to meet the vertical through P_2 at Z, then $P_1Z = t_2 - t_1$, $ZP_2 = y_2 - y_1$, and

$$v = ZP_2/P_1Z = \tan \angle ZP_1P_2 = \tan \psi \quad . \quad . \quad (8.2)$$

where ψ is the inclination of the straight line P_1P_2 to the horizontal. That is to say v is what we have called the "slope" or "gradient" of the line P_1P_2, defined as the tangent of the angle of inclination ψ.

NOTE. We are supposing that the scale of graduation is the same along both axes, i.e. that the same length is used to represent a unit of time t horizontally and a unit of distance y vertically. If not, the relation $v = \tan \psi$ no longer holds without a correction given later in equation (8.5).

We thus have a simple geometrical interpretation of the average velocity v between two times, t_1 and t_2: v is the gradient of the straight line or chord joining the corresponding points P_1 and P_2 on the graph of distance against time.

The quantity $t_2 - t_1$ is the time taken between P_1 and P_2: it may also be considered as the change in the value of t when we go from P_1 to P_2. Similarly $y_2 - y_1$ is the change in distance. It is convenient to use the Greek letter δ to mean "a change", so that we denote the change in t, measured by $(t_2 - t_1)$, by the symbol δt, and the change in the value of y, $(y_2 - y_1)$, by δy. Here δt is to be understood to be a single symbol meaning neither more nor less than $(t_2 - t_1)$: the letter δ does not stand for a number, but for the whole phrase "the change (or difference) in the value of". (The Greek letter δ corresponds to the initial letter d of the word "difference".)

To illustrate this notation let us take an example. A train reaches Cambridge (56 miles) in $1\frac{1}{3}$ hours after leaving London, and it reaches Hunstanton (112 miles) in $3\frac{1}{3}$ hours altogether. What is the average velocity from Cambridge to Hunstanton?

Here y_1 means the distance from London to Cambridge, measured in miles, and is therefore 56, and t_1 represents the time taken, in hours, and is $1\frac{1}{3}$.

y_2 means the distance from London to Hunstanton, or 112, and t_2 the time taken, $3\frac{1}{3}$.

$\delta y = y_2 - y_1$ therefore means the distance from Cambridge to Hunstanton, and is $112 - 56 = 56$.

$$\delta t = t_2 - t_1 = 3\tfrac{1}{3} - 1\tfrac{1}{3} = 2$$

is the time taken on this part of the journey. The average velocity is accordingly

$$v = (y_2 - y_1)/(t_2 - t_1) = 56/2 = 28 \text{ miles per hour.}$$

It follows that in general we can write

$$\text{average velocity } v = (y_2 - y_1)/(t_2 - t_1) = \delta y/\delta t \qquad (8.3)$$

Again, in this formula it must be remembered that δy and δt are convenient abbreviations for $(y_2 - y_1)$ and $(t_2 - t_1)$ respectively and are to be treated as single symbols. We cannot cancel out the δ's in $\delta y/\delta t$ to make it y/t, as we could have done if δ had been an ordinary number.

If we are dealing with two quantities y and t which are connected by a purely empirical relation then we can only calculate the average velocity arithmetically, as we have done above for train times. If, however, there is an algebraic relationship connecting y and t, such as $y = 3t$, or $y = t + t^2$, then we can do better and find a formula for the velocity v.

EXAMPLES

(1) The law $y = 3t$.

To say that the distance y is connected with the time t by the relation $y = 3t$ means that at each particular time (such as t_1) the corresponding distance (y_1) will be 3 times the time, measured in appropriate units, i.e. $y_1 = 3t_1$. In the same way the distance y_2 reached after time t_2 will be $3t_2$, and accordingly $y_2 - y_1 = 3t_2 - 3t_1 = 3(t_2 - t_1)$. Therefore the velocity $v = (y_2 - y_1)/(t_2 - t_1) = 3$. Graphically the relation $y = 3t$ is represented by a straight line, of slope 3, and therefore any chord P_1P_2 coincides with this line and must have slope 3. This relation $y = 3t$ is the law of motion of a point moving with a constant velocity of 3 units.

(2) The law $y = t + t^2$.

If the relation between the position y (measured in metres distance from a fixed point O) and time t (in seconds) is $y = t + t^2$, as might be true of a sphere rolling down a slope, we can still argue in a similar way. We must have $y_1 = t_1 + t_1^2$, $y_2 = t_2 + t_2^2$, and therefore on subtraction

$$\begin{aligned}\delta y = y_2 - y_1 &= (t_2 + t_2^2) - (t_1 + t_1^2)\\ &= (t_2 - t_1) + (t_2^2 - t_1^2)\\ &= (t_2 - t_1) + (t_2 - t_1)(t_2 + t_1)\\ &= (t_2 - t_1)(1 + t_2 + t_1)\end{aligned}$$

(taking out the common factor $t_2 - t_1$). Thus

$$v = (y_2 - y_1)/(t_2 - t_1) = 1 + t_2 + t_1.$$

This formula gives us a quick method of calculating an average velocity: for example the average velocity of the sphere between times $t_1 = 2$ and $t_2 = 4$ would be $1 + t_2 + t_1 = 7$.

Before quoting any further examples there are two points to notice. The first one is that in the formula $v = (y_2 - y_1)/(t_2 - t_1)$, the distances y_1 and y_2 may be measured from any fixed point or origin O on the line of motion of the body: they are not necessarily the distances measured from the point from which the body begins to move. For we are interested only in the difference $\delta y = y_2 - y_1$, or change in position, and this will be the same whatever point we choose to measure y from, provided that we pay proper attention to signs, measuring distances positively in one direction and negatively the other way. Similarly we can measure the times t_1 and t_2 from any convenient instant, such as noon.

Besides this there is a paradox which deserves mention. According to our definition, if a train goes from London to Hunstanton, and then returns to London, the total distance over the whole journey measured by $y_1 - y_2$ is zero, since the initial and final positions are the same. The average velocity is accordingly zero! The reason is that our definition of "average velocity" takes into account the direction as well as the rate of motion. When the train runs from London to Hunstanton it does so with positive velocity: when it returns it has negative velocity, since the distance y from London is continually diminishing. On balance the train has no velocity at all. Now the reader may at first sight think this somewhat unfair, and the engine-driver certainly will. It is customary to meet this objection by defining "average speed" (as distinct from average velocity) as the quotient of the distance travelled, regardless of direction, by the time taken. The "average speed" of the train will not be zero, although its average velocity will. But although this definition may save the engine-driver's honour, it is not on the whole nearly so useful as the one which takes direction into account. If we are standing on the track, and see a train moving at 60 miles per hour, it is a matter of considerable importance to know whether it is approaching or receding. Similarly if we are considering the population of a country, and are told that it is changing by 300,000 inhabitants a year, this information may be quite interesting. But it will be much more valuable if we know whether the change is an increase, i.e. +300,000, or a decrease, −300,000. So except where the contrary is expressly stated, all velocities, and other rates of change, will be considered to have sign as well as magnitude. A positive sign indicates an increase or movement in the positive direction, and a negative sign a decrease or movement in the negative direction.

(3) The law $y = 1/t$.

Here $y_1 = 1/t_1$, $y_2 = 1/t_2$, and

$$y_2 - y_1 = 1/t_2 - 1/t_1 = (t_1 - t_2)/t_1 t_2.$$

The average velocity between times t_1 and t_2 is therefore

$$v = (y_2 - y_1)/(t_2 - t_1) = -1/t_1 t_2.$$

If $t_1 = 1$, $t_2 = 2$, then $v = -\frac{1}{2}$. The negative sign indicates a movement in the negative direction. If for example the distance y was measured north of a fixed point O, then the law of motion $y = 1/t$ indicates a southward movement, towards the point O (for positive t).

So far the quantity y has been a distance, and the quantity $v = \delta y/\delta t = (y_2 - y_1)/(t_2 - t_1)$ a velocity. But a similar definition will apply to any changing quantity. If y_1 is its value at time t_1, and y_2 its value at time t_2, then we call $v = (y_2 - y_1)/(t_2 - t_1) = \delta y/\delta t$ the "average rate of change of y between times t_1 and t_2".

(4) An animal is growing in such a way that its weight y grams is connected with its age t weeks by the relation $y = 2 + 6t^3$. What is the rate of growth between times t_1 and t_2?

If y_1 is its weight at time t_1, and y_2 at time t_2, then $y_1 = 2 + 6t_1^3$, $y_2 = 2 + 6t_2^3$, and

$$y_2 - y_1 = 6t_2^3 - 6t_1^3 = 6(t_2^3 - t_1^3).$$

But $t_2^3 - t_1^3 = (t_2 - t_1)(t_2^2 + t_2t_1 + t_1^2)$ by equation (3.5), so that

$$y_2 - y_1 = 6(t_2 - t_1)(t_2^2 + t_1t_2 + t_1^2),$$

$$v = (y_2 - y_1)/(t_2 - t_1) = 6(t_2^2 + t_1t_2 + t_1^2),$$

measured in grams per week. For example, between $t_1 = 1$ and $t_2 = 1{\cdot}5$ weeks it grows at a rate of $6(t_1^2 + t_1t_2 + t_2^2) = 6(1 + 1{\cdot}5 + 2{\cdot}25) = 28{\cdot}5$ grams per week. This can be verified directly. After 1 week its weight is 8 g, after 1·5 weeks it is 22·25 g. The increase in weight is

$$\delta y = 22{\cdot}25 - 8 = 14{\cdot}25 \text{ g},$$

the time taken

$$\delta t = 1{\cdot}5 - 1 = {\cdot}5 \text{ weeks},$$

and the rate of increase is $\delta y/\delta t = 14{\cdot}25/{\cdot}5 = 28{\cdot}5$ grams per week.

PROBLEMS

Find the formula for the average velocity or rate of change of a quantity y when it is related to the time t by the following laws:

(1) $y = 2 + 3{\cdot}5t$

(2) $y = 1 - t + 1/t$

(3) $y = 3/t^2$

(4) $y = 2/(1 + t)$

(5) $y = (1 + t^2)/(1 + t)$

(6) $y = 1 + t + t^2 + t^3$

(7) $y = 1/(1 - t^2)$

Verify the following formulas for rate of change for the laws stated:

(8) If $y = 1$, $v = \delta y/\delta t = 0$

(9) If $y = t$, $v = 1$

(10) If $y = t^2$, $v = t_1 + t_2$

(11) If $y = t^3$, $v = t_2^2 + t_1 t_2 + t_1^2$

(12) If $y = t^4$, $v = t_2^3 + t_2^2 t_1 + t_2 t_1^2 + t_1^3$

(13) Generalize to the law $y = t^n$ where n is a positive integer.

(14) If $y = \sqrt{t}$, $v = 1/(\sqrt{t_2} + \sqrt{t_1})$

(15) If $y = t^{-1}$, $v = -1/t_1 t_2$

(16) If $y = t^{-2}$, $v = -(t_2 + t_1)/t_1^2 t_2^2$

8.3 Instantaneous velocity

It is very natural to feel that it is sensible to say not only that "this automobile has been travelling at an average speed of 25 miles per hour for the last 5 minutes" but also that "it is moving at this moment at a speed of 23 miles per hour". In other words we feel that there is such a thing as a definite speed at a definite instant, as well as an average speed over an interval of time.

How are we to express this mathematically? We cannot take the definition of average velocity $(y_2 - y_1)/(t_2 - t_1)$ and put $t_2 = t_1$, for that would imply $y_2 = y_1$, and leaves us with a fraction 0/0 which has no meaning. We can, however, imagine the velocity calculated over an interval of time δt which is very short, but not absolutely zero.

To see how this works out let us consider the sphere which we imagined rolling down a slope according to the law $y = t + t^2$. We have already shown that the average velocity of the sphere between time t_1 and time t_2 is $1 + t_2 + t_1$, a formula which will save us some calculation.

Now the average velocity between time $t_1 = 1$ and a time t_2 shortly afterwards is given by the formula $v = 2 + t_2$. Thus the average velocity between times 1 and 1·01 seconds is 3·01 (metres per second); between $t_1 = 1$ and $t_2 = 1·001$ seconds it is 3·001; between times $t_1=1$ and $t_2 = 1·0000001$ seconds the average velocity is 3·0000001. The shorter the interval of time over which the velocity is calculated the nearer it approaches to 3. It is therefore natural to say that the actual or instantaneous velocity at time $t_1 = 1$ is 3 metres/sec. Similarly, if we take $t_1 = 2$, $t_2 = 2·001$ the velocity is 5·001; $t_1 = 2$, $t_2 = 2·0000001$ gives a velocity 5·0000001, and keeping $t_1 = 2$ the nearer t_2 approaches 2 the nearer does the average velocity approach 5. Thus the instantaneous velocity at time $t_1 = 2$ seconds is 5 metres per second. In general we shall say that if as the interval of time $\delta t = t_2 - t_1$ over which it is calculated becomes shorter and shorter the average velocity $v = \delta y/\delta t$ approaches more and more nearly some "limiting" value V, then V is the "instantaneous velocity" at time t_1. In symbols

$$v \to V \text{ as } \delta t \to 0, \quad \text{i.e. as } t_2 \to t_1$$

We can now find a general formula for V in the case of the moving sphere. Such a general formula $v = 1 + t_2 + t_1$ has already been obtained for the average velocity between times t_1 and t_2. Now it is clear that the nearer t_2 approaches t_1 the nearer does v approach $1 + t_1 + t_1 = 1 + 2t_1$. This is therefore the required instantaneous velocity at time t_1. Since there are not now two distinct times t_1 and t_2 to consider, but only a single time t_1, we can drop the suffix $_1$ as superfluous and say that the (instantaneous) velocity at time t is $V = 1 + 2t$. Thus when $t = 1$, $V = 3$; when $t = 2$, $V = 5$, as already found; when $t = 3$, $V = 7$, and so on. Again if y is some measured quantity other than a distance, we have defined $v = \delta y/\delta t = (y_2 - y_1)/(t_2 - t_1)$ to be the average rate of change of the quantity y between the times t_1 and t_2. Again we cannot put $t_1 = t_2$ in this fraction, or we should obtain 0/0. But if as the interval $t_2 - t_1$ becomes very short, so that t_2 tends to t_1, the average v approaches a certain limiting value V, we can speak of V as (by definition) the "instantaneous rate of change" of y at time t_1. Thus, considering the animal of Example (4) above whose weight, y grams, is given by the formula $y = 2 + 6t^3$, where t is its age in weeks, we have already shown that $v = 6(t_2^2 + t_2t_1 + t_1^2)$. The nearer t_2 approaches to t_1 the nearer v will approach to $6(t_1^2 + t_1t_1 + t_1^2) = 18t_1^2$. Dropping the suffix $_1$ as no longer necessary, we see that the instantaneous rate of growth at t weeks is $V = 18t^2$ grams per week.

As a numerical illustration of this let us put $t_1 = 1$. Then if $t_2 = 1{\cdot}01$, the average rate of increase in weight between times t_1 and t_2 will be $v = 6(t_2^2 + t_2t_1 + t_1^2) = 18{\cdot}1806$ grams per week. If we shorten the interval by taking $t_2 = 1{\cdot}001$, we get $v = 18{\cdot}018006$; if we put $t_2 = 1{\cdot}000001$, to get a very short interval indeed of $\cdot 000001$ weeks, then $v = 18{\cdot}000018000006$, a value which differs quite inappreciably from the instantaneous rate $V = 18$ at $t = 1$.

Unfortunately there is a serious defect in the idea of "instantaneous velocity"; it is one which it is utterly impossible to check empirically. In discussing the case of the rolling sphere, we calculated from the formula $y = t + t^2$ that between $t_1 = 1$ and $t_2 = 1{\cdot}01$ seconds the average velocity was $3{\cdot}01$ metres/second; this means that the distance travelled was $\cdot 01 \times 3{\cdot}01 = \cdot 0301$ metres, and that could be verified by careful measurement. But when we shorten the interval of time to $\cdot 0000001$ second and say that the average velocity is then $3{\cdot}0000001$ metres/sec, we are asserting that the distance travelled is $\cdot 00000030000001$ metres in this short time. Although such a statement *must* be true if the law of motion $y = t + t^2$ is accurate, it is nevertheless quite impossible to check in practice.

In the case of the hypothetical growing animal which is supposed to obey the law $y = 2 + 6t^3$ the situation is even more serious. This equation may fit the weights observed very well if we take only daily readings. But a more detailed consideration will show that it simply cannot be true when we consider shorter intervals of time. For it gives a steadily rising weight, whereas in fact the weight will increase very rapidly at feeding time, and fall in between meals. And if the equation $y = 2 + 6t^3$ is untrue so also is the rate of change of weight $V = 18t^2$ derived from it.

Fortunately these objections do not really matter very much in practice. We may be unable to verify that a body is moving with instantaneous velocity $1 + 2t$, but for all practical purposes it behaves as if that was so. We may be actually wrong in supposing that the animal's weight is always increasing at the rate $18t^2$. But in the long run we shall not be very far wrong, and for many purposes it may be considerably easier to take a deliberately simplified model rather than to worry over all the irregularities in detail. Such an idealization is common in many branches of science. We can even treat the population of a country as if it was a continuously varying quantity. We know of course that in fact it can only change by steps of $+1$ for a birth or immigration, or -1 for a death or emigration. But among a population of millions such individual changes are negligible. It is possible to talk of a "continuously varying population" provided that the underlying simplification is not forgotten.

To summarize:

(A) If y is any changing quantity, i.e. in technical language, any function of the time t, then the average rate of change of y between times t_1 and t_2 is defined to be

$$v = \delta y/\delta t = (y_2 - y_1)/(t_2 - t_1)$$

where y_1 is the value of y at time t_1, and y_2 the value at time t_2.

(B) If the relation between y and t can be expressed in mathematical terms, it will frequently happen that when the interval of time $\delta t = t_2 - t_1$ is made shorter and shorter the average velocity v will approach some definite value V. V is then called the "instantaneous" or "actual" velocity at time t_1.

(C) This idea of instantaneous velocity is a purely theoretical device; strictly speaking it can never be completely verified empirically.

(D) However there are many situations, such as the calculation of the velocity of a moving body, where the theory agrees with reality to as close an approximation as our measurements will allow. And there are also many situations which we can treat for all practical purposes as if there was an instantaneous rate of change, even though we know that this is not absolutely true when we consider them in detail.

8.4 Algebraic functions

Once a formula has been obtained for an average velocity v, it is usually quite easy to deduce the value of the instantaneous velocity V. For example, we have already shown that if the law of motion is $y = 1/t$, then the average velocity is $v = -1/t_1t_2$. If t_2 approaches t_1 in value, then v approaches $-1/t_1^2$; or, on dropping the suffix $_1$ as usual, we see that the instantaneous velocity at time t is $V = -1/t^2$.

Again if the law of motion is $y = 1/(1 + t)$, we have

$$\begin{aligned}\delta y = y_2 - y_1 &= 1/(1 + t_2) - 1/(1 + t_1)\\ &= -(t_2 - t_1)/(1 + t_1)(1 + t_2);\end{aligned}$$

$$\text{average velocity } v = \delta y/\delta t = -1/(1 + t_1)(1 + t_2);$$
$$\text{instantaneous velocity} = V = -1/(1 + t)^2.$$

This direct method of finding instantaneous rates of change can be applied to a great variety of expressions—in fact to all algebraic fractions obtained by dividing one polynomial by another. The most important application is to laws of the form $y = t^n$. Here we use the results of problems (8) to (16) of Section 8.2.

Example 1 $y = 1$ for all t

Average velocity $v = 0$

Instantaneous velocity $V = 0$

Example 2 $y = t$
Average velocity $v = 1$
Instantaneous velocity $V = 1$

Example 3 $y = t^2$
Average velocity $v = t_2 + t_1$
Instantaneous velocity $V = 2t$

Example 4 $y = t^3$
Average velocity $v = t_2^2 + t_2 t_1 + t_1^2$
Instantaneous velocity $V = 3t^2$

Example 5 $y = t^4$
Average velocity $v = t_2^3 + t_2^2 t_1 + t_2 t_1^2 + t_1^3$
Instantaneous velocity $V = 4t^3$

These examples suggest the general result: that for the law of motion $y = t^n$ the instantaneous velocity is

$$V = nt^{n-1} \qquad . \quad . \quad . \quad . \qquad (8.4)$$

For the case in which n is a positive integer this general formula can be proved as follows [using equation (3.7)]:

$$\delta y = y_2 - y_1 = t_2^n - t_1^n$$
$$= (t_2 - t_1)(t_2^{n-1} + t_2^{n-2} t_1 + t_2^{n-3} t_1^2 + .. + t_1^{n-1})$$

Average velocity $v = \delta y / \delta t = (t_2^{n-1} + t_2^{n-2} t_1 + \ldots + t_1^{n-1})$.

On the right-hand side there are n terms, and as t_2 approaches t_1 each of these terms approaches t_1^{n-1}, so that their sum tends to nt_1^{n-1}. Therefore $V = nt^{n-1}$.

We have already shown that for the law $y = t^{-1}$ the instantaneous velocity is given by $V = -t^{-2}$; this agrees with (8.4) when $n = -1$.

Example 6 The law $y = t^{-2}$.
Average velocity $v = -(t_2 + t_1)/t_1^2 t_2^2$
Instantaneous velocity $V = -2t/t^4 = -2t^{-3}$

This agrees with (8.4) with $n = -2$. These examples suggest that (8.4) may be true for all values of n, not only positive integers. It will later be proved that that is so.

PROBLEMS

(1) Show that (8.4) is true in general if n is a negative integer. (Put $n = -m$, then $y = 1/t^m$. Find v, and thence V.)

(2) Find V for the following laws of motion; $y = 1/(1 + t^2)$, $y = \sqrt{t}$, $y = 1/\sqrt{t}$. Show that the last two agree with formula (8.4).

8.5 Derivatives and differentiation

Corresponding to any specified law of motion or change, relating a quantity y to the time t, there will be an instantaneous rate of change of y, which we have called V. Technically V is called the "derivative", "derivate", or "differential coefficient" of y with respect to the time t. The operation of finding the instantaneous rate of change is known as "differentiation".

This process can be illustrated graphically (Fig. 8.2). Let us draw the graph of y against the time t, and let P_1 (t_1, y_1) and P_2 (t_2, y_2) be

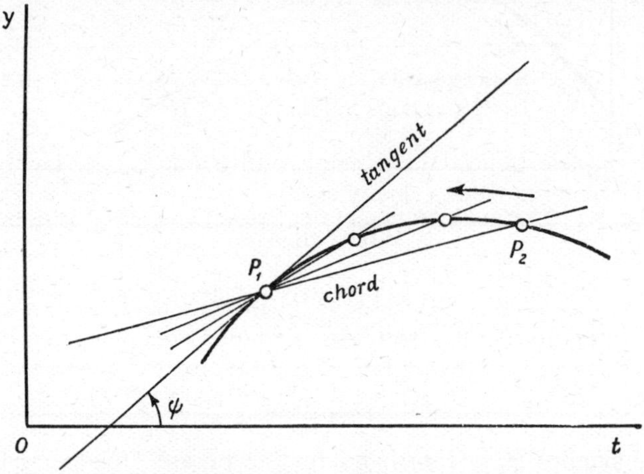

Fig. 8.2—Chord approaching tangent as limiting position

any two points on the graph. Then the slope of the chord P_1P_2 is the average rate of change between times t_1 and t_2. Now imagine P_2 to move along the curve towards P_1. As it does so the chord P_1P_2 will rotate about P_1 and will approach a certain limiting position which we call the "tangent" at P_1 (from Latin *tangēns* = touching. This use of the word "tangent" is, of course, different from its use as the name of a trigonometric ratio, tan θ. However, there is a historical connection between these senses, but one which need not concern us here). As the chord tends to the tangent the slope v of the chord must tend to the slope of the tangent, i.e. the derivative or instantaneous rate of change V = the slope of the tangent = tan ψ, where ψ is the angle between the tangent and the t-axis.

Now this is indeed the sort of relation we would expect. For where y is increasing, the graph will slope upwards (towards the right), ψ will be positive and so will tan ψ (Fig. 8.3). The more rapid the increase, the greater the slope will be. If y is decreasing, the graph is falling; ψ is negative, and so is tan ψ. Where $\psi = 0$, and the tangent is horizontal, y is momentarily stationary, neither increasing nor decreasing.

This relation can be used as a means of roughly estimating the rate V of change by drawing the tangent by eye, measuring ψ and finding $\tan \psi$ from tables.

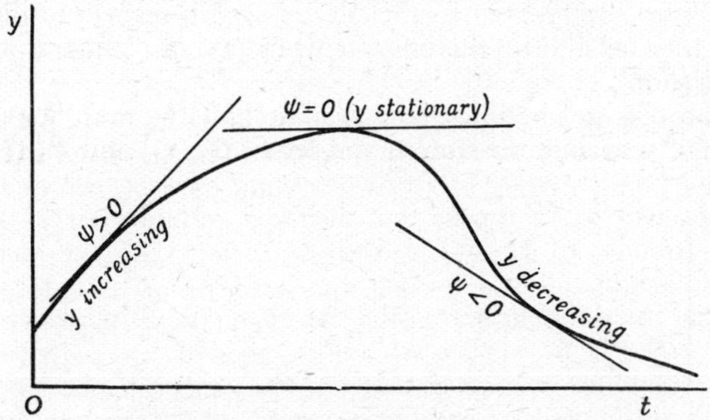

Fig. 8.3—The inclination ψ of the tangent compared with the rate of change

So far we have used the letter V for the rate of change or derivative of y. This notation is rather arbitrary and does not show the relationship between V, t and y. An alternative symbol for the derivative is $D_t y$, i.e.

$$V = D_t y.$$

Here the symbol D_t is shorthand for the phrase "the derivative with respect to t of". It is a symbol which has no meaning standing by itself: it is the complete symbol "$D_t y$" which means something, viz. the rate of change of y. Often the suffix t can be omitted as being understood, and we can write simply "Dy".

This is not the only way of writing a rate of change. Another symbol, even more commonly used, is dy/dt. The reason for such a notation is this: we have already shown that the average velocity v is $\delta y/\delta t$, where δy means "the change in y" and δt "the change in t". To show the relationship between the average velocity v and the instantaneous velocity V the latter is written as dy/dt: but although this looks like a quotient "dy divided by dt" it is to be interpreted as a single symbol; neither dy nor dt are to be taken as having any meaning of their own. (Some books give an interpretation of dy/dt which allows dy and dt to be regarded as actually existing quantities, but it is rather a complicated and artificial interpretation, and will not be considered here.)

Other notations for V which are fairly often met with are y' (read as "y dash" or "y prime"), y_t and $\dot{y}$. (The last is read as "y dot" and is used only when the variable t is the time.) Also if y is written as a function of t in the form $y = f(t)$, then V is written as $V = f'(t)$. We mention these notations because the reader must be prepared to meet them in mathematical books. Sometimes one form may be convenient, sometimes another. In this notation we can write the results so far obtained as

$$D_t t^n = d(t^n)/dt = nt^{n-1}$$
$$D_t(2 + 6t^3) = d(2 + 6t^3)/dt = 18t^2$$
$$D_t[1/(1 + t)] = -1/(1 + t)^2$$

and so on.

We can now generalize our ideas to include the more metaphorical cases where time is not involved. Suppose that the height (h metres) and weight (w kilograms) of a growing child have been found to obey the law $w = 13\ h^3$. Suppose that during a certain interval of time h increases from h_1 to h_2, and w from w_1 to w_2: then the increases in weight and height will be $\delta w = w_2 - w_1$ and $\delta h = h_2 - h_1$ respectively. We can call the quotient $\delta w/\delta h$ the "average rate of increase of w with respect to h". If $h_1 = 1{\cdot}00$, $h_2 = 1{\cdot}20$, then $\delta h = {\cdot}20$; $w_1 = 13{\cdot}00$, $w_2 = 22{\cdot}46$, and $\delta w = 9{\cdot}46$, so that $\delta w/\delta h = 47{\cdot}3$ kilograms per metre. Now if δh becomes very small the average rate of change $\delta w/\delta h$ will approximate to a number we call the "actual rate of change" or "derivative" of w with respect to h, and denoted by the symbols $D_h w$ or dw/dh. The derivative is calculated as before.

$$\delta w = w_2 - w_1 = 13h_2{}^3 - 13h_1{}^3$$
$$= 13(h_2 - h_1)(h_2{}^2 + h_1 h_2 + h_1{}^2)$$
$$\delta w/\delta h = 13(h_2{}^2 + h_1 h_2 + h_1{}^2)$$

and in the limit when h_2 tends to $h_1 = h$ we have $D_h w = 39h^2$.

By drawing the graph of w against h, choosing equal units of graduation for w and h, the derivative $D_h w$ can be interpreted as the slope $\tan\psi$ of the tangent. If different units are used for h and w, so that a unit difference in h corresponds to a length U_h on the h-axis, and a unit difference in w has length U_w, then the formula is

$$D_h w = (U_h \tan\psi)/U_w \quad . \quad . \quad . \quad (8.5)$$

PROBLEM

(1) Draw the graph $w = 13h^3$. By drawing tangents at the points at which $h = {\cdot}8$, $1{\cdot}0$, and $1{\cdot}2$ respectively estimate the derivatives at these points. Compare with the theoretically calculated values.

8.6 Limits

We have already met several situations of the following type: x and y are two variables connected by a functional relation $y = F(x)$. As x approaches a certain value, say X, y approaches more and more nearly to another value, say Y. For example the nearer x is to zero, the nearer $(\sin x)/x$ is to the constant H (Section 6.11). In fact, we rather implied by our argument that we could make the difference between

$(\sin x)/x$ and H as small as we please, by taking x sufficiently small, although that was not formally proved. In general, wherever we can make y approach as near as we please to Y by taking x as near as we like to X, we shall say that "Y is the limit of y as x tends to X", or in symbols

$$Y = \lim_{x \to X} y.$$

Alternatively we say that "y tends to Y as x tends to X", or

$$y \to Y \text{ as } x \to X.$$

These alternative phrases are identical in meaning. For a formal definition see Section 8.15. Thus

$$(\sin x)/x \to H \quad \text{as} \quad x \to 0$$

or

$$\lim_{x \to 0} [(\sin x)/x] = H;$$

and

$$\delta y/\delta t \to V \quad \text{as} \quad \delta t \to 0$$

or

$$\lim_{\delta t \to 0} (\delta y/\delta t) = V.$$

We also have a very similar limiting process in the case of an unending decimal. The equation $\frac{1}{3} = \cdot 3333 \ldots$ means that the decimals of the sequence $\cdot 3$, $\cdot 33$, $\cdot 333$, $\cdot 3333 \ldots$ approach more and more nearly to $\frac{1}{3}$. If we write $u_1 = \cdot 3$, $u_2 = \cdot 33$, $u_3 = \cdot 333$, etc., then we say that u_r "tends to $\frac{1}{3}$ as r tends to infinity" or in symbols

$$u_r \to \tfrac{1}{3} \quad \text{as} \quad r \to \infty$$

or

$$\lim_{r \to \infty} u_r = \tfrac{1}{3}.$$

The phrase "tends to infinity" is simply a shorthand way of saying "becomes larger and larger without bound", and has no philosophical implications about the nature of infinity.

Now we have assumed in our calculations certain reasonable properties of limits: that if t_2 approaches t_1 in value, then $t_1 + t_2$ approaches $2t_1$, and that $t_1 t_2$ approaches $t_1{}^2$. It is not difficult to justify some of these properties in a general sort of way. For example, the difference between $(t_1 + t_2)$ and $2t_1$ is $(t_1 + t_2) - 2t_1 = t_2 - t_1$, so that it is immediately evident that the smaller the difference $t_2 - t_1$ is the smaller is $(t_1 + t_2) - 2t_1$. Again the difference between $t_1 t_2$ and $t_1{}^2$ is $t_1 t_2 - t_1{}^2 = t_1(t_2 - t_1)$; so that if we keep t_1 fixed the smaller $(t_2 - t_1)$ is the smaller is $(t_1 t_2 - t_1{}^2)$. However, formal proofs of this are complicated; so we content ourselves here with a brief statement of what is true, without going into details (Section 8.15) which only confirm the commonsense view.

If x tends to X then x^2 tends to X^2, $x + C$ tends to $X + C$ (where C is a constant), and Cx tends to CX. Also $1/x$ tends to $1/X$ provided that X is not zero—a reasonable condition, for $1/X$ is not defined when

$X = 0$. It is always true that $\text{antilog}\, x \to \text{antilog}\, X$, $\sin x \to \sin X$, $e^x \to e^X$; and provided that X is positive it is true that $\sqrt{x} \to \sqrt{X}$ and $\log x \to \log X$. This again is reasonable since $\sqrt{X}$ and $\log X$ are only defined for positive X. It is reasonable and true that if y and z are functions of x, and $y \to Y$ and $z \to Z$ as $x \to X$, then $y + z \to Y + Z$ and $yz \to YZ$. Provided that $Z \neq 0$ it is also true that $y/z \to Y/Z$.

If y is a function of x and w a function of y, and if $y \to Y$ as $x \to X$ and $w \to W$ as $y \to Y$ then it is true that $w \to W$ as $x \to X$. For example, x might stand for the age of a child, y for its height (which can be considered as a function of x) and w for its weight (which can be considered as a function of y, and so as a function of x).

There is one small caution. It is tempting to suppose that if y is given as a function of x, say $y = F(x)$, and if $y \to Y$ as $x \to X$, then Y must be the value of y corresponding to $x = X$, i.e. $Y = F(X)$, provided that $F(X)$ is defined. We have seen that this is true if the relation is $y = x^2$, or $y = \sin x$, or $y = \text{antilog}\, x$. But occasionally it is untrue. Suppose that x is the weight of a letter, in ounces, and y the postage payable in pence. Then if $x = 3$, $y = 6$; if $x = 2{\cdot}1$, $y = 6$; if $x = 2{\cdot}001$, $y = 6$; for any weight exceeding 2 ounces by however little, y is 6. It follows that if $x > 2$ but $x \to 2$, $y \to 6$ as a limit. But if the weight is exactly 2 ounces the correct postage is 4d. and not 6d.

The reason for the discrepancy is a sudden jump in the postage rate at 2 ounces. (In case the reader feels inclined to argue that this is an artificial example we would remind him that such discontinuities do occur in nature: at the surface of a solid body the density seems to change suddenly, at least on an everyday scale of measurement. Seen from a molecular point of view the situation is no doubt different, but it is then even more complicated.)

Fortunately such discontinuities rarely interfere with calculations and, in brief, "limits" (in the mathematical sense of the word) usually behave exactly as one would like them to.

8.7 Differentiation of logarithmic and trigonometric functions

We can now find the rate of change of logarithms, antilogarithms, cosines and sines.

Consider the relation $y = \log t$ (using any system of logarithms). As usual let $y_1 = \log t_1$, $y_2 = \log t_2$, $\delta y = y_2 - y_1$, $\delta t = t_2 - t_1$. Then

$$\begin{aligned} \delta y &= \log t_2 - \log t_1 \\ &= \log \frac{t_2}{t_1} \\ &= \log \frac{t_1 + \delta t}{t_1} \\ &= \log(1 + \delta t/t_1) \end{aligned}$$

whence the average rate of change is

$$v = \frac{\delta y}{\delta t} = \frac{\log(1 + \delta t/t_1)}{\delta t}$$

$$= \frac{1}{t_1} \frac{\log(1 + \delta t/t_1)}{\delta t/t_1}$$

Now we know that when x is small $\log(1 + x)$ is approximately Mx, where M is the modulus of the system of logarithms. In fact we know that $\log(1 + x)/x \to M$ as $x \to 0$ (equation 6.29). Let $x = \delta t/t_1$; then as $\delta t \to 0$, so must $x \to 0$ (keeping t_1 fixed). Thus $\dfrac{\log(1 + \delta t/t_1)}{\delta t/t_1} \to M$, and so, as $\delta t \to 0$, the average rate of change $v \to M/t_1 = V$, the instantaneous rate of change. As usual we can now drop the suffix $_1$ and put

$$d(\log t)/dt = D_t \log t = M/t \quad . \quad . \quad . \quad (8.6)$$

In the case of common logs $M = \cdot 4343$. For natural logarithms $M = 1$, and the formula takes the simple form

$$d(\ln t)/dt = D_t \ln t = 1/t \quad . \quad . \quad . \quad (8.7)$$

For the antilogarithm function $y = \text{antilog}\, t$ we have, using the same notation

$$\begin{aligned} \delta y &= y_2 - y_1 \\ &= \text{antilog}\, t_2 - \text{antilog}\, t_1 \\ &= \text{antilog}\,(t_1 + \delta t) - \text{antilog}\, t_1 \\ &= \text{antilog}\, t_1 . \text{antilog}\, \delta t - \text{antilog}\, t_1 \\ &= \text{antilog}\, t_1 . (\text{antilog}\, \delta t - 1) \\ v = \delta y/\delta t &= \text{antilog}\, t_1 . (\text{antilog}\, \delta t - 1)/\delta t. \end{aligned}$$

Now as $\delta t \to 0$ the expression $(\text{antilog}\, \delta t - 1)/\delta t$ tends to M^{-1} (from equation 6.28), so that v tends to $M^{-1}\, \text{antilog}\, t_1$. Dropping the suffix we obtain

$$D_t \text{antilog}\, t = M^{-1} \text{antilog}\, t \quad . \quad . \quad . \quad (8.8)$$

In particular for the natural antilogarithm or exponential function, $\exp t = e^t$,

$$de^t/dt = D_t e^t = e^t . \quad . \quad . \quad . \quad (8.9)$$

This function is its own derivative.

These results can be explained geometrically. On the equiangular spiral defining the system of logarithms, with angle ϕ, let O be the centre, and I, P_1, and P_2 the points with polar co-ordinates $\{1, 0\}$, $\{r_1, \theta_1\}$, and $\{r_2, \theta_2\}$ respectively. Draw the arc P_1Z of a circle with centre O to meet OP_2 in Z (Fig. 8.4 overleaf). Then

$$\delta r = r_2 - r_1 = OP_2 - OP_1 = OP_2 - OZ = ZP_2,$$

and $$\delta\theta = \theta_2 - \theta_1 = \angle IOP_2 - \angle IOP_1 = \angle P_1OP_2.$$

The triangle P_1ZP_2 has angles of $90° - \phi$, $90°$, and ϕ at P_1, Z, and P_2 respectively, and the length of the arc P_1Z is $Hr_1\delta\theta$. Now when $\delta\theta$ becomes very small (and therefore so also does δr) the sides of the triangle P_1ZP_2 become very nearly straight, and so the ratio P_1Z/ZP_2 tends to $\tan \phi$ in the limit. That is, $Hr_1\delta\theta/\delta r \to \tan \phi$. On dividing both sides of this equation by Hr_1, and dropping the suffix from r_1 as usual, we see that $\delta\theta/\delta r \to (\tan \phi)/Hr = M/r$, and so also $\delta r/\delta\theta \to r/M$. Since $\theta = \log r$, and $r = \text{antilog } \theta$, this means that $D_r \log r = M/r$, $D_\theta \text{ antilog } \theta = (\text{antilog } \theta)/M$.

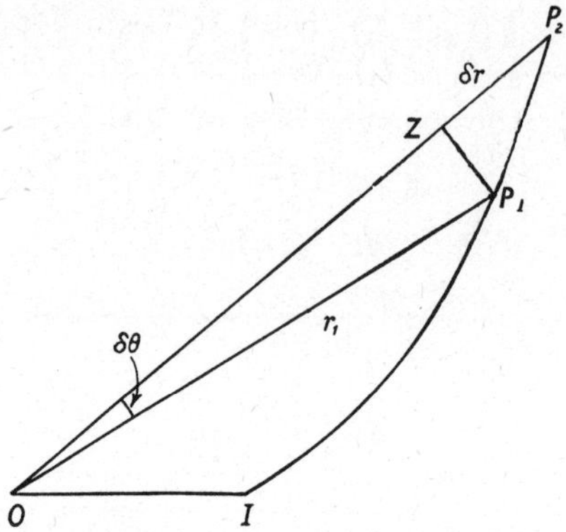

Fig. 8.4—Differentiation of logarithms and antilogarithms

To differentiate the relation $y = \cos t$ we can proceed as follows. $y_2 = \cos t_2 = \cos (t_1 + \delta t) = \cos t_1 . \cos \delta t - \sin t_1 . \sin \delta t$ by the addition law for cosines (5.11). Now when δt is small (Section 6.11) $\cos \delta t$ is very nearly equal to 1, and $\sin \delta t$ to $H . \delta t$, so that this reduces to $y_2 \simeq \cos t_1 - H \sin t_1 . \delta t$. By definition $y_1 = \cos t_1$, and so

$$\delta y = y_2 - y_1 \simeq - H \sin t_1 . \delta t$$

and

$$\delta y/\delta t \simeq - H \sin t_1.$$

In the limit as $\delta t \to 0$ we have

$$d(\cos t)/dt = D_t \cos t = - H \sin t \quad . \quad . \quad (8.10)$$

This holds whatever units t is measured in, provided that we insert the corresponding value of H. But in the calculus it is customary to measure angles in radians, and then $H = 1$, so that

$$D_t \cos t = - \sin t \text{ (for radian measure)} \quad . \quad . \quad (8.11)$$

Proceeding in the same way for sines we have, given that $y = \sin t$,

$$y_2 = \sin (t_1 + \delta t) = \sin t_1 . \cos \delta t + \cos t_1 . \sin \delta t$$
$$\simeq \sin t_1 + H \cos t_1 . \delta t \quad \text{when } \delta t \text{ is small;}$$

$y_1 = \sin t_1$, and so $\delta y = y_2 - y_1 \simeq H \cos t_1 \,.\, \delta t$
$\delta y/\delta t \simeq H \cos t_1$; i.e. on dropping the suffix from t_1,

$$\begin{aligned} D_t \sin t &= H \cos t \\ &= \cos t \text{ (for radian measure)} \end{aligned} \qquad . \quad (8.12)$$

An alternative proof of the formula $D_t \cos t = -H \sin t$ is as follows. We have $t_1 = \frac{1}{2}(t_2 + t_1) - \frac{1}{2}(t_2 - t_1) = \frac{1}{2}(t_2 + t_1) - \frac{1}{2}\delta t$, and $t_2 = \frac{1}{2}(t_2 + t_1) + \frac{1}{2}(t_2 - t_1) = \frac{1}{2}(t_2 + t_1) + \frac{1}{2}\delta t$. Therefore by the addition law, if $y = \cos t$ (Sections 5.6, 5.9),

$$\begin{aligned} y_2 &= \cos t_2 = \cos \tfrac{1}{2}(t_2 + t_1) \,.\, \cos \tfrac{1}{2}\delta t - \sin \tfrac{1}{2}(t_2 + t_1) \,.\, \sin \tfrac{1}{2}\delta t \\ y_1 &= \cos t_1 = \cos \tfrac{1}{2}(t_2 + t_1) \,.\, \cos \tfrac{1}{2}\delta t + \sin \tfrac{1}{2}(t_2 + t_1) \,.\, \sin \tfrac{1}{2}\delta t \end{aligned}$$

and on subtraction

$$\begin{aligned} \delta y &= y_2 - y_1 = -2 \sin \tfrac{1}{2}(t_2 + t_1) \,.\, \sin \tfrac{1}{2}\delta t. \\ \delta y/\delta t &= -\sin \tfrac{1}{2}(t_2 + t_1) \,.\, \sin(\tfrac{1}{2}\delta t)/(\tfrac{1}{2}\delta t) \end{aligned}$$

Now as $\delta t \to 0$, $t_2 \to t_1 = t$, $\sin \frac{1}{2}(t_2 + t_1) \to \sin t$ and $\sin(\frac{1}{2}\delta t)/(\frac{1}{2}\delta t) \to H$, so that $\delta y/\delta t \to -H \sin t = D_t y$.

PROBLEMS

(1) By a similar method prove that $D_t \sin t = H \cos t$. Can you see any advantage of this second method of proof as compared with the first?

(2) Using the fact that $\cos \theta$ and $\sin \theta$ are the x and y co-ordinates of the point P with polar co-ordinates $\{1, \theta\}$ on the circle $r = 1$, show graphically that $D_\theta \cos \theta = -H \sin \theta$ and $D_\theta \sin \theta = H \cos \theta$.

8.8 Derivative of a sum

Quite complicated expressions, even those such as $[t^t + \sqrt{t}]^2$ or $[\frac{1}{2}\{\log(1 + t^2)\}]^{-e}$, can be differentiated by means of the direct approach of finding the average velocity $\delta y/\delta t$ and taking the limit as $\delta t \to 0$. However, there is a simpler approach. We can think of such a complicated expression as built up by a series of steps, those of addition, multiplication, taking logarithms, and so on. An elaborate piece of machinery may be difficult to understand at first. But by decomposing it into its constituent rods, wheels, levers, etc., the whole will become clear. In the same way, by taking any expression step by step we can differentiate it, provided that we know how to deal with each step.

The first combination we consider is the sum. Let y and z be two variable quantities, each of which is a function of t. Let w be $y + z$. What is the rate of change of w?

Suppose for example that two rivers meet. Let y be the total quantity of water which has flowed down the first river to the junction, since, say, noon, and z the quantity which has flowed down the second river, then w will be the flow down the combined river. If t denotes time, then $D_t y$ is the rate of flow down the first river, say 10 cubic metres per second, $D_t z$ that down the second river, say 20 m³/sec, and $D_t w$ the combined flow, which clearly enough is $30 = 10 + 20$ m³/sec. Thus

$$D_t w = D_t(y + z) = D_t y + D_t z \qquad . \quad . \quad (8.13)$$

i.e. *the derivative of a sum is the sum of the derivatives.*

Formally, let y_1, z_1, and w_1 be the values of y, z, and w respectively at time t_1, and y_2, z_2, and w_2 the corresponding values at time t_2. Then $w_1 = y_1 + z_1$, $w_2 = y_2 + z_2$, and so by subtraction

$$\delta w = w_2 - w_1 = (y_2 - y_1) + (z_2 - z_1) = \delta y + \delta z.$$

On dividing by δt we have $\delta w/\delta t = \delta y/\delta t + \delta z/\delta t$. On letting $\delta t \to 0$,

$$D_t w = D_t y + D_t z.$$

Similarly for a difference we have:

$$D_t(y - z) = D_t y - D_t z.$$

The formula clearly generalizes to any sum or difference, e.g.:

$$D_t(y - z + u - \ldots) = D_t y - D_t z + D_t u - \ldots$$

EXAMPLES

(1) The rate of change of t is 1, that of t^2 is $2t$. Therefore

$$D_t(t + t^2) = 1 + 2t.$$

(2) $D_t(t^2 - \sin t + \cos t) = 2t - H\cos t - H\sin t.$

(3) Since any constant K has zero rate of change, $D_t(y + K) = D_t y$, i.e. $(y + K)$ and y always have the same derivative.

8.9 Derivative of a product

Consider now the product $w = yz$ where y and z are variable quantities. To make the matter more specific, suppose that $OYPZ$ is a rectangle of sides $OY = y$ and $OZ = z$. Then its area is $w = yz$. If the sides y and z are changing, what is the rate of change of w?

Let $OY_1P_1Z_1$ be the rectangle at time t_1, and $OY_2P_2Z_2$ its new shape at time t_2. Then $OY_1 = y_1$, $OY_2 = y_2$, and so $\delta y = y_2 - y_1 = Y_1Y_2$ (Fig. 8.5). Similarly $\delta z = Z_1Z_2$. Now $\delta w = w_2 - w_1 =$ the area $Y_1Y_2P_2Z_2Z_1P_1$. This area can be split into two parts by producing Z_1P_1 to meet Y_2P_2 in Q. Then the rectangle $Y_1Y_2QP_1$ has area $z_1\delta y$ and $Z_1QP_2Z_2$ has area $y_2\delta z$: i.e. $\delta w = z_1\delta y + y_2\delta z$. On dividing by δt we obtain

$$\delta w/\delta t = z_1 . \delta y/\delta t + y_2 . \delta z/\delta t$$

which is the relation between average rates of change.

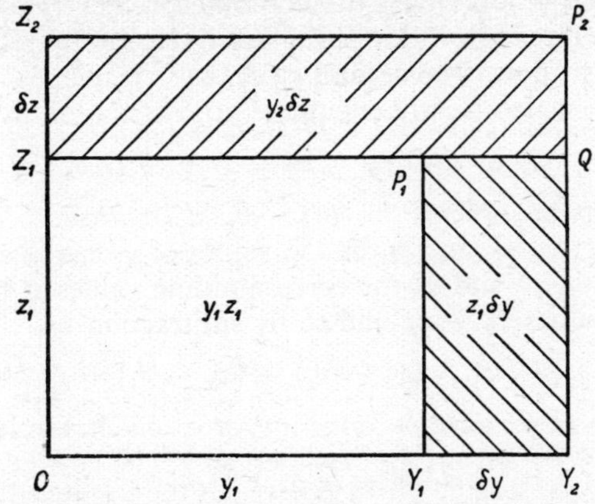

Fig. 8.5—The change in the area of a rectangle due to changes in its sides

We can also derive this algebraically:

$$\begin{aligned} \delta w &= w_2 - w_1 \\ &= y_2 z_2 - y_1 z_1 \\ &= (y_2 z_2 - y_2 z_1) + (y_2 z_1 - y_1 z_1) \\ &= y_2 (z_2 - z_1) + z_1 (y_2 - y_1) \\ &= y_2\, \delta z + z_1\, \delta y \end{aligned}$$

and $\delta w/\delta t = z_1\, \delta y/\delta t + y_2\, \delta z/\delta t.$

Now as $\delta t \to 0$, $y_2 \to y_1 = y$, dropping the suffix, and $z_1 = z$, so that

$$D_t w = D_t(yz) = z\, D_t y + y\, D_t z \qquad (8.14)$$

In words, *the derivative of the product of two quantities is the sum of the second times the derivative of the first plus the first times the derivative of the second.*

EXAMPLES

(1) Differentiate $t \ln t$

$$\begin{aligned} D_t\,(t \ln t) &= \ln t \,.\, D_t t + t \,.\, D_t \ln t \\ &= \ln t + 1. \end{aligned}$$

(2) $$\begin{aligned} D_t[(1 + t)\sin t] &= \sin t \,.\, D_t(1 + t) + (1 + t) \,.\, D_t \sin t \\ &= \sin t + H(1 + t)\cos t. \end{aligned}$$

(3) Prove, using the product rule, that $D_t\, t^2 = 2t$, $D_t\, t^3 = 3t^2$.

$$\begin{aligned} D_t\, t^2 &= D_t\,(tt) = tD_t t + tD_t t = t + t = 2t \\ D_t t^3 &= D_t(tt^2) = t^2 D_t t + tD_t t^2 = t^2 + t \,.\, 2t = 3t^2. \end{aligned}$$

(4) What is the derivative of Ky, where K is a constant?
We know that $D_t K = 0$; so putting K in place of z in (8.14).

$$D_t(Ky) = KD_t y \quad . \quad . \quad . \quad . \quad (8.15)$$

i.e. *the derivative of a constant multiple of a quantity is the constant times its derivative.*

(5) Differentiate $5t^3$.

$$D_t(5t^3) = 5D_t t^3 = 5 \cdot 3t^2 = 15t^2.$$

(6) Differentiate $3 + 4t + t^2$.

$$\begin{aligned} D_t(3 + 4t + t^2) &= D_t 3 + D_t(4t) + D_t t^2 \\ &= 0 + 4\,D_t t + D_t t^2. \\ &= 4 + 2t. \end{aligned}$$

It is a little more difficult to generalize the product rule to the product of 3 or more variables. The simplest way is perhaps to divide (8.14) through by $w = yz$; we obtain

$$\frac{D_t w}{w} = \frac{D_t y}{y} + \frac{D_t z}{z}$$

Now suppose that z in turn is the product of two variables u and x, so that $z = ux$, $w = yz = yux$. Then we must have

$$\frac{D_t z}{z} = \frac{D_t u}{u} + \frac{D_t x}{x},$$

or on substitution in the former equation

$$\frac{D_t w}{w} = \frac{D_t(yux)}{yux} = \frac{D_t y}{y} + \frac{D_t u}{u} + \frac{D_t x}{x} \quad . \quad . \quad . \quad (8.16)$$

This can be similarly extended to the product of any number of variables, e.g.

$$\frac{D_t(yuxs)}{yuxs} = \frac{D_t y}{y} + \frac{D_t u}{u} + \frac{D_t x}{x} + \frac{D_t s}{s}$$

For 3 variables we find, on multiplying (8.16) throughout by yux,

$$D_t(yux) = uxD_t y + yxD_t u + uyD_t x.$$

In the same way

$$D_t(yuxs) = uxsD_t y + yxsD_t u + yusD_t x + yuxD_t s$$

FURTHER EXAMPLE

(7) Find the rate of change of $w = t \sin t \,.\, \cos t$.

$$\begin{aligned} D_t w &= \sin t \,.\, \cos t \,.\, D_t t + t \cos t \,.\, D_t \sin t + t \sin t \,.\, D_t \cos t \\ &= \sin t \,.\, \cos t + H\, t\,(\cos t)^2 - H\, t\,(\sin t)^2 \end{aligned}$$

PROBLEMS

Find the rates of change (with respect to t) of the following expressions:

(1) $2 + 6t^2 + 7t^3$

(2) $(1 + 2t)\sqrt{t}$

(3) $t^2 \sin t$

(4) $2t \,.\, \ln t \,.\, \sin t$

(5) $\sqrt{t} \,.\, \ln t$.

8.10 Derivative of a quotient

If y, z are functions of t, what is the rate of change of $w = y/z$? At time t_1 we have $w_1 = y_1/z_1$, and at time t_2, $w_2 = y_2/z_2$. So

$$\begin{aligned}\delta w &= w_2 - w_1 = y_2/z_2 - y_1/z_1 \\ &= \frac{y_2 z_1 - y_1 z_2}{z_1 z_2} \text{ (by reduction to the common denominator } z_1 z_2) \\ &= \frac{y_2 z_1 - y_1 z_1 + y_1 z_1 - y_1 z_2}{z_1 z_2} \\ &= \frac{(y_2 - y_1) z_1 - y_1 (z_2 - z_1)}{z_1 z_2} \\ &= \frac{z_1 \delta y - y_1 \delta z}{z_1 z_2}\end{aligned}$$

And so on division by δt we find the formula for the average rate of change

$$\delta w/\delta t = \frac{z_1(\delta y/\delta t) - y_1(\delta z/\delta t)}{z_1 z_2}$$

On letting $\delta t \to 0$, $z_2 \to z_1 = z$, $y_1 = y$ we obtain the instantaneous rate of change

$$D_t w = D_t\left(\frac{y}{z}\right) = \frac{zD_t y - yD_t z}{z^2} \quad . \quad . \quad (8.17)$$

Given the derivatives of $\sin t$ and $\cos t$ this enables us to find those of other trigonometric functions. For example $\tan t = (\sin t)/\cos t$, so that

$$\begin{aligned}D_t \tan t &= \frac{\cos t \,.\, D_t \sin t - \sin t \,.\, D_t \cos t}{(\cos t)^2} \\ &= \frac{H \cos t \,.\, \cos t + H \sin t \,.\, \sin t}{(\cos t)^2} \\ &= H/(\cos t)^2 \\ &\quad [\text{since } (\cos t)^2 + (\sin t)^2 = 1] \\ &= H (\sec t)^2\end{aligned}$$

EXAMPLES

(1) Find the rate of change of $w = t^2 \ln t/(1 + t)$.

We first use the rule for a quotient, treating w as y/z where $y = t^2 \ln t$ and $z = 1 + t$. Thus

$$D_t w = \frac{z D_t y - y D_t z}{z^2}$$

$$= \frac{(1 + t) D_t (t^2 \ln t) - t^2 \ln t \,.\, D_t (1 + t)}{(1 + t)^2}$$

Now to find $D_t (t^2 \ln t)$ we use the product rule

$$\begin{aligned} D_t (t^2 \ln t) &= \ln t \,.\, D_t\, t^2 + t^2 D_t \ln t \\ &= \ln t \,.\, 2t + t^2/t \\ &= 2t \ln t + t. \end{aligned}$$

To find $D_t (1 + t)$ we use the sum rule,

$$D_t (1 + t) = D_t\, 1 + D_t\, t = 0 + 1 = 1$$

Thus finally

$$D_t\, w = \frac{(1 + t)(2t \ln t + t) - t^2 \ln t}{(1 + t)^2}$$

$$= \frac{2t \ln t + 2t^2 \ln t + t + t^2 - t^2 \ln t}{(1 + t)^2}$$

$$= \frac{2t \ln t + t^2 \ln t + t + t^2}{(1 + t)^2}$$

$$= \frac{t (2 + t) \ln t + t (1 + t)}{(1 + t)^2}.$$

(2) Find the rate of change of $z^{-1} = 1/z$, given the rate of change of z.

Put $y = 1$ in the quotient formula, then $D_t y = 0$

$$D_t (1/z) = -(D_t z)/z^2 \quad . \quad . \quad . \quad (8.18)$$

(3) Differentiate sec t.

We know that $\sec t = 1/\cos t$; so putting $z = \cos t$ in (8.18) we have

$$D_t \sec t = -H (\cos t)^{-2} (-\sin t) = H \sec t \,.\, \tan t.$$

(4) Differentiate e^{-t}.

$$\begin{aligned} e^{-t} &= 1/e^t, \text{ so putting } z = e^t \text{ in (8.18)} \\ D_t e^{-t} &= -(D_t e^t)/(e^t)^2 \\ &= -e^t/(e^t)^2 = -1/e^t = -e^{-t}. \end{aligned}$$

(5) Differentiate sinh t and cosh t.

By definition

$$\begin{aligned}\sinh t &= \tfrac{1}{2}(e^t - e^{-t})\\ D_t \sinh t &= \tfrac{1}{2} D_t (e^t - e^{-t})\\ &= \tfrac{1}{2}(D_t e^t - D_t e^{-t})\\ &= \tfrac{1}{2}(e^t + e^{-t}) = \cosh t.\\ \cosh t &= \tfrac{1}{2}(e^t + e^{-t})\\ D_t \cosh t &= \tfrac{1}{2} D_t (e^t + e^{-t})\\ &= \tfrac{1}{2}(e^t - e^{-t}) = \sinh t.\end{aligned}$$

(Compare with the formulas $D_t \sin t = \cos t$, $D_t \cos t = -\sin t$, where t is measured in radians.)

PROBLEMS

Prove that

(1) $D_t \operatorname{cosec} t = -H \cot t \,.\, \operatorname{cosec} t$

(2) $D_t \operatorname{cosech} t = -\coth t \,.\, \operatorname{cosech} t$

(3) $D_t \cot t = -H (\operatorname{cosec} t)^2$

(4) $D_t \coth t = -(\operatorname{cosech} t)^2$

(5) $D_t \tanh t = (\operatorname{sech} t)^2$.

(6) By using the fact that $t^{-n} = 1/t^n$, prove that

$$D_t t^{-n} = -nt^{-n-1}$$

(7) Differentiate with respect to t the expressions $1/(1 + t^2)$, $t/(1 + t)$, $1/\ln t$, $t \ln t - t$, $(t \ln t - t)^{-1}$.

8.11 Functions of functions

Suppose we have the following situation. A growing child has height x cm at age t years. x is then a function of t, and can be represented graphically. Suppose further that we plot its weight w kilograms against its height x. Then at a given age these graphs may show that the height x is increasing at the rate of 5 cm per year, and further that w is increasing at the rate of ·7 kilogram per centimetre increase in height. The weight w can also be considered as a function of the time t: can we say how rapidly it is increasing?

The obvious argument is to say that since the height increases by 5 cm in a year, the weight (counted at ·7 kg for each centimetre increase in height) must increase by $5 \times \cdot 7$ kg $= 3{\cdot}5$ kg (per year). In other words, since the rate of increase in height can be written $D_t x$, and the rate of increase in weight with respect to height is $D_x w$, we should expect that

$$D_t w = D_t x \,.\, D_x w \qquad (8.19)$$

But our argument only proves this for the average rates over a year.

However, we can readily show that the formula holds for momentary rates of increase. If in the interval of time δt the height x increases by δx and the weight w by δw, then

$$\frac{\delta w}{\delta t} = \frac{\delta x}{\delta t} \cdot \frac{\delta w}{\delta x}$$

identically, as follows on cancelling δx. Now as $\delta t \to 0$, $\delta w/\delta t$ and $\delta x/\delta t$ tend to their respective limits $D_t w$ and $D_t x$. Also as $\delta t \to 0$, it is also true that $\delta x \to 0$ and so $\delta w/\delta x \to D_x w$. Accordingly on taking limits as $\delta t \to 0$ we obtain equation (8.19).

This method of argument is perfectly general, and applies to any function x of t, and any function w of x, whatever the letters t, x, and w may represent. It can be stated in words as: *for 3 variables, of which the first is a function of the second and the second a function of the third, the derivative of the first with respect to the third is the product of the derivative of the first with respect to the second times the derivative of the second with respect to the third.*

In the alternative notation which is in common use, in which $D_t x$ is written as dx/dt, as if it was a quotient, the formula becomes

$$\frac{dw}{dt} = \frac{dx}{dt}\frac{dw}{dx}$$

This form of notation has accordingly the merit of suggesting the correct answer—although strictly speaking dw/dt is not itself a quotient but only the limit of a quotient. But we shall see that this simple form is no longer valid when we come to consider partial differentiation.

EXAMPLES OF THE USE OF EQUATION (8.19)

(1) To differentiate $w = \sin\sqrt{t}$.

The device used here is to put $x = \sqrt{t}$, so that $w = \sin x$. Since $x = \sqrt{t}$, we have $D_t x = 1/2\sqrt{t}$. Since $w = \sin x$, we have $D_x w = H \cos x = H \cos\sqrt{t}$. Thus

$$\begin{aligned} D_t w &= D_t x \,.\, D_x w \\ &= (1/2\sqrt{t}) . H \cos\sqrt{t} \\ &= H(\cos\sqrt{t})/2\sqrt{t}. \end{aligned}$$

(2) Differentiate $w = \ln(1 + t)$ with respect to t. The necessary substitution here is to put $1 + t = x$, so that $w = \ln x$. Then

$$\begin{aligned} D_t x &= D_t(1 + t) = 1 \\ D_x w &= D_x \ln x = 1/x = 1/(1 + t) \end{aligned}$$

Sometimes we may need two intermediate steps to perform the differentiation. Suppose that x is a function of t, y a function of x, and w a function of y. Then

$$\frac{\delta w}{\delta t} = \frac{\delta x}{\delta t} \cdot \frac{\delta y}{\delta x} \cdot \frac{\delta w}{\delta y}$$

identically. Therefore on taking the limit as $\delta t \to 0$,

$$D_t w = D_t x \,.\, D_x y \,.\, D_y w$$

or

$$\frac{dw}{dt} = \frac{dx}{dt} \cdot \frac{dy}{dx} \cdot \frac{dw}{dy}$$

as it may alternatively be written. For example, suppose we wish to differentiate $[\ln(1 + t)]^2$ with respect to t. The necessary substitutions are

$$\begin{aligned} x &= 1 + t; & D_t x &= 1 \\ y &= \ln(1 + t) = \ln x; & D_x y &= x^{-1} \\ w &= [\ln(1 + t)]^2 = y^2; & D_y w &= 2y \end{aligned}$$

so that

$$\begin{aligned} D_t w &= D_t x \,.\, D_x y \,.\, D_y w \\ &= 2x^{-1}y = 2(1 + t)^{-1} \ln(1 + t). \end{aligned}$$

This method can be applied to cases where there is a relation between variables which is not expressed explicitly in the form "y is a given function of t", i.e. for "implicit functions" (Section 5.14).

FURTHER EXAMPLES

(3) A ladder of length L is leaning against a wall, the foot of the ladder being at distance x from the foot of the wall, and the top of the ladder being at height y (Fig. 8.6). If the foot of the ladder is pulled away from the wall with velocity v, i.e. $D_t x = v$, at what rate $D_t y$ does the top of the ladder descend?

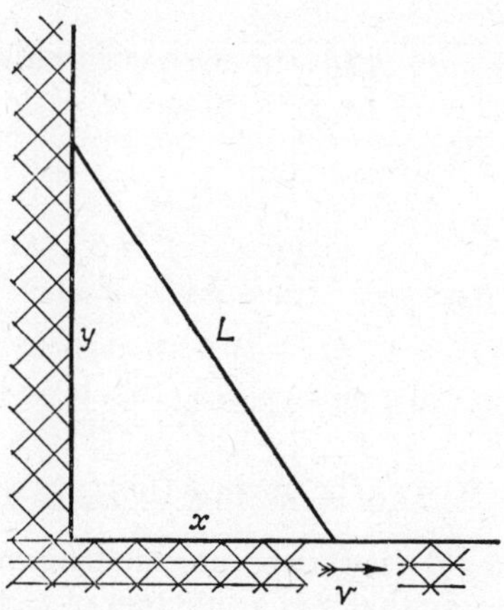

Fig. 8.6—Moving ladder problem

By Pythagoras's theorem,

$$x^2 + y^2 = L^2.$$

Differentiate this equation on both sides with respect to t: the new equation we obtain must be true, since the equation $x^2 + y^2 = L^2$ remains true as x and y vary. Now $D_t x^2 = D_t x . D_x x^2 = 2xv$, and $D_t y^2 = D_t y . D_y y^2 = 2y D_t y$. Also $D_t L^2 = 0$, since L^2 is constant. We have therefore

$$2xv + 2yD_t y = 0$$

i.e. $D_t y = -xv/y$.

The minus sign shows that (as expected) y is decreasing when x is increasing, and when v is positive. Since $y = \sqrt{(L^2 - x^2)}$ this can also be written as

$$D_t y = -xv/\sqrt{(L^2 - x^2)}$$

If we found that $L = 2{\cdot}6$ metres, and $v = 1$ metre per second, then at the moment when $x = 1$ metre, $y = 2{\cdot}4$ metres, and $D_t y = 1/2{\cdot}4 = {\cdot}42$ metres per second.

(4) Differentiate $y = \sin^{-1} t$ with respect to t.

The equation $y = \sin^{-1} t$ means by definition the same as $t = \sin y$. On differentiating both sides of this equation with respect to t we obtain

$$\begin{aligned} D_t t &= D_t \sin y \\ &= D_t y \,.\, D_y \sin y \\ &= H \cos y \,.\, D_t y. \end{aligned}$$

But $D_t t = 1$, and therefore $D_t y = 1/H \cos y$. And since $\sin y = t$, it follows that $\cos y = \pm\sqrt{(1 - t^2)}$. Thus

$$\begin{aligned} D_t \sin^{-1} t &= 1/H \cos \sin^{-1} t \\ &= \frac{\pm 1}{H\sqrt{(1 - t^2)}} \qquad . \quad . \quad . \quad . \quad (8.20) \end{aligned}$$

(5) Differentiate $y = \tan^{-1} t$. Since $\tan y = t$, and $D_y \tan y = (\sec y)^2$, we obtain by differentiation

$$H (\sec y)^2 D_t y = D_t t = 1,$$

so that $D_t y = 1/H (\sec y)^2$. Now $(\sec y)^2 = 1 + (\tan y)^2 = 1 + t^2$, and therefore

$$D_t \tan^{-1} t = \frac{1}{H(1 + t^2)}$$

(6) Differentiate $y = t^n$ for general (not necessarily integral) values of n.

By definition $y = e^{n \ln t}$ or

$$\ln y = n \,.\, \ln t.$$

On differentiating with respect to t,

$$y^{-1}D_t y = nt^{-1}$$

and on multiplying both sides by y,

$$D_t y = nyt^{-1} = nt^n t^{-1} = nt^{n-1}.$$

(7) Differentiate y^z with respect to t, where y and z are given functions of t. Let $w = y^z$, then $\ln w = z \ln y$. Hence by the product rule

$$D_t \ln w = \ln y . D_t z + z D_t \ln y$$
$$w^{-1} D_t w = \ln y . D_t z + zy^{-1} D_t y.$$

Multiplying through by $w = y^z$,

$$D_t y^z = y^z (\ln y . D_t z + zy^{-1} D_t y).$$

As a particular case let $y = K$, a constant, and $z = t$. Then

$$D_t y = 0,\ D_t z = 1 \text{ and}$$
$$D_t K^t = K^t \ln K.$$

Table 8.1—Standard forms

Function w	Derivative $D_t w = dw/dt$	Function w	Derivative $D_t w = dw/dt$
$y + z$	$D_t y + D_t z$	$\cosh t$	$\sinh t$
Ky	$KD_t y$	$\sinh t$	$\cosh t$
yz	$z D_t y + y D_t z$	$\tanh t$	$(\operatorname{sech} t)^2$
y/z	$z^{-2}[zD_t y - yD_t z]$	$\coth t$	$-(\operatorname{cosech} t)^2$
y/z^n	$z^{-n-1}[zD_t y - nyD_t z]$	$\operatorname{sech} t$	$-\operatorname{sech} t . \tanh t$
		$\operatorname{cosech} t$	$-\operatorname{cosech} t . \coth t$
$w = y^m z^n$	$w\left[\frac{m}{y} D_t y + \frac{n}{z} D_t z\right]$	$\cos^{-1} t$	$\pm[1 - t^2]^{-1/2}$
		$\sin^{-1} t$	$\pm[1 - t^2]^{-1/2}$
y^z	$y^z[zy^{-1} D_t y + \ln y . D_t z]$	$\cosh^{-1} t$	$\pm[t^2 - 1]^{-1/2}$
$w = F(y)$	$D_t y . D_y w$	$\sinh^{-1} t$	$[t^2 + 1]^{-1/2}$
		$\tan^{-1} t$	$1/[1 + t^2]$
K	0	$\cot^{-1} t$	$-1/[1 + t^2]$
t^n	nt^{n-1}	$\tanh^{-1} t$	$1/[1 - t^2]$
$\log Kt$	M/t	$\coth^{-1} t$	$1/[1 - t^2]$
$\ln Kt$	$1/t$		
$\operatorname{antilog} t$	$M^{-1} \operatorname{antilog} t$		
e^t	e^t		
$\cos t$	$-\sin t$		
$\sin t$	$\cos t$		
$\tan t$	$(\sec t)^2$		
$\cot t$	$-(\operatorname{cosec} t)^2$		
$\sec t$	$\sec t . \tan t$		
$\operatorname{cosec} t$	$-\operatorname{cosec} t . \cot t$		

8.12 Standard forms

We are now in a position to find the rate of change of any function which can be expressed explicitly or implicitly in terms of the operations of addition, subtraction, multiplication, division, the taking of powers and roots, logarithms, antilogarithms, trigonometric or hyperbolic functions, or any combination whatever of these operations and functions. It is useful to have a table of "standard forms" summarizing the rates of change of all the common functions. In Table 8.1 (p. 203) all differentiations are with respect to t. The letters K and n denote constants. $M = \log e$ is the modulus of common logarithms $= \cdot 4343 \ldots$ For all trigonometric functions, $\sin t$, $\cos t$, etc., t is measured in radians: this eliminates the constant H (which now becomes 1). y and z represent arbitrary functions of t.

It is scarcely necessary to remind the reader that skill in differentiation, as in any other activity, comes only with practice. In particular the derivatives of t^n, e^t, $\log t$, $\ln t$, and the trigonometric functions should be known by heart.

PROBLEMS

(1) Differentiate $w = (2 + t)^2$ with respect to t in the following ways:

(i) multiply the expression out in full and differentiate term by term according to the sum rule;

(ii) write $w = (2 + t)(2 + t)$ and differentiate by the product rule;

(iii) write $x = 2 + t$ and differentiate by the function-of-a-function rule;

(iv) use the inverse relation $t = \pm\sqrt{w} - 2$.

(2) A man is lifting a kilogram weight by raising the forearm only, keeping the elbow fixed. If the angle θ between the forearm and the vertical is decreasing at the rate of 1 radian per second, and the length of the forearm is 30 cm, at what rate is he doing work? (in kg weight . metre/sec).

(3) The velocity of sound in air is $v = 66{\cdot}3\sqrt{(T + 273)}$ feet per second, where T is the Centigrade temperature. What is the rate of rise of velocity per degree Centigrade at 10° C?

(4) If a hemispherical vessel of radius r cm contains water to a depth of x cm, then the volume of water contained is $V = \pi(rx^2 - \frac{1}{3}x^3)$ cc. Water enters such a vessel of radius 3 cm at the rate of 2 cc/sec. Find the rate at which the level is rising, expressing it in terms of the depth x. Draw a graph of this rate against x, and show that the rise is very rapid at first and afterwards slows down.

(5) The Schütz-Borisoff law with regard to the action of enzymes such as pepsin and rennin is expressed by the formula $x = K\sqrt{(Fat)}$, where F is the concentration of enzyme, a the initial concentration of the substrate (e.g. albumen or milk), t the time elapsed, x the amount transformed, and K a constant. What is the rate at which the transformation is taking place?

(6) A ball thrown upwards with initial velocity u metres per second has height $y = ut - 4{\cdot}905t^2$ metres after t seconds. What is its velocity at that time?

(7) A man is walking at the rate of 2 metres per second towards a camera of focal length 16 centimetres. When he is 10 metres distant from the camera at what rate must the lens (or ground-glass screen) be racked out to keep him in proper focus?

(8) A galvanometer mirror is 1 metre distant from the scale, and the spot of light is moving along it with velocity 15 cm per second. When it is deflected 17 cm what is the angular velocity of the beam of light, and what is the angular velocity of rotation of the mirror?

8.13 Warning—how not to differentiate!

It is sometimes easy to make an error like the following, if one is not aware of the danger. A stone thrown in the air has height

$$y = F(t) = 8t - 4{\cdot}9t^2 \text{ metres at time } t \text{ seconds.}$$

Find its velocity at time $t = 1$. We put $t = 1$ in the expression for y, and find $y = F(1) = 3{\cdot}1$. But the velocity v is the derivative of the height, and is therefore $v = D_t y = D_t(3{\cdot}1) = 0$. Similar calculations at other values of t would show that the velocity is always zero, which is clearly absurd.

The correct method is to differentiate first, and substitute particular numerical values afterwards. Thus

$$v = D_t y = D_t(8t - 4{\cdot}9t^2) = 8 - 9{\cdot}8t.$$

This gives the value of the velocity v at any time t. That is, it expresses the velocity as a function of the time t. If $y = F(t)$ is the relation connecting height y and time t, then we write the velocity as $v = F'(t)$, and call $F'(t)$ (here $= 8 - 9{\cdot}8t$) the "derived function". To find the velocity at time $t = 1$, put $t = 1$, getting $v = F'(1) = 8 - 9{\cdot}8 \times 1 = -1{\cdot}8$, which is correct.

But why is it possible to go wrong at all? Prof. K. Menger has shown that this, like many other difficulties, arises from ambiguity in the notation. Sometimes in the relation $y = F(t) = 8t - 4{\cdot}9t^2$ the symbols t, y are "variables". That is, the relation says that for *each* value t there is a corresponding value $y = F(t)$. The formula is equivalent to a graph of y against t, or to a moving picture of the stone's movement. It concerns not one value t and one value y, but an infinite number of such pairs of values. From such a formula we can certainly find the velocity v by differentiation. But when we put $t = 1$, we restrict attention to a single pair of values t, y, i.e. to a single point on the graph, or to an instantaneous snapshot. Such a single point or snapshot tells nothing about the velocity. Prof. Menger has also shown that by suitable small modifications the traditional notation can be made safer, more logical, and simpler at the same time, permitting one to write, e.g.

"cosec = 1/sin" instead of "cosec x = 1/sin x".

See his book, *Calculus, a Modern Approach* (Illinois Institute of Technology, 1953).

8.14 Motion in a curved path

So far we have considered motion along a straight line. Now let us imagine a point P to be moving in a plane along a curve C. We can specify its position at any time t by its cartesian co-ordinates x and y which will be functions of t. Now let $P_1 = (x_1, y_1)$ be its position at time t_1, and $P_2 = (x_2, y_2)$ at time t_2 (Fig. 8.7). Let L represent the

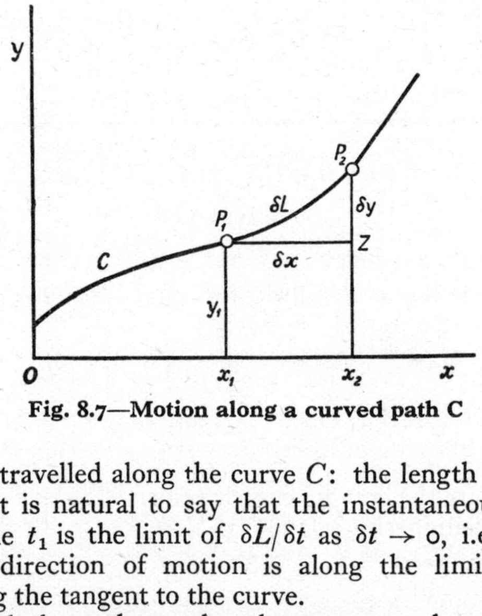

Fig. 8.7—Motion along a curved path C

distance P has travelled along the curve C: the length of the arc P_1P_2 is δL. Then it is natural to say that the instantaneous speed of the point P at time t_1 is the limit of $\delta L/\delta t$ as $\delta t \to 0$, i.e. D_tL, and the instantaneous direction of motion is along the limit of the chord P_1P_2, i.e. along the tangent to the curve.

It has already been shown that the tangent makes an angle ψ with the x-axis, where $\tan \psi = D_xy$. Since $D_ty = D_tx \,.\, D_xy$ this may alternatively be written as

$$\tan \psi = D_ty/D_tx \qquad . \quad . \quad . \quad (8.21)$$

provided that $D_tx \neq 0$.

To find D_tL we proceed as follows. Draw a line through P_1 parallel to the x-axis meeting one through P_2 parallel to the y-axis at Z (Fig. 8.7). Then

$$P_1Z = x_2 - x_1 = \delta x \quad \text{and} \quad ZP_2 = y_2 - y_1 = \delta y.$$

Now when δt is small the distances δx and δy will become small: and provided that C is a reasonably smooth curve the arc P_1P_2 will become very nearly straight. Thus by Pythagoras's theorem

$$(\delta L)^2 \simeq (\delta x)^2 + (\delta y)^2 \quad . \quad . \quad . \; (8.22)$$

The smaller δt is, the smaller the percentage error in this equation. Dividing through by $(\delta t)^2$,

$$(\delta L/\delta t)^2 \simeq (\delta x/\delta t)^2 + (\delta y/\delta t)^2$$

Take the limit as $\delta t \to 0$

$$(D_tL)^2 = (D_tx)^2 + (D_ty)^2$$

or

$$\text{instantaneous speed } D_tL = \sqrt{(D_tx)^2 + (D_ty)^2} \quad . \; (8.23)$$

NOTE 1. The square of δx is often written conventionally as δx^2. We could distinguish between $(\delta x)^2$ and $\delta(x^2)$ simply and consistently by using the notation proposed in Section 3.7, writing the first as $\delta x)^2$ and the second as $\delta x^2)$.

NOTE 2. If we divide equation (8.22) through by $(\delta x)^2$ and take the limit as $\delta x \to 0$ we obtain the important equation

$$(D_xL)^2 = 1 + (D_xy)^2$$

or

$$D_xL = \sqrt{1 + (D_xy)^2} \quad . \quad . \; (8.24)$$

EXAMPLES

(1) A projectile has height $y = 10t - 4{\cdot}9t^2$ metres and has covered a horizontal distance $x = 10t$ metres at time t seconds. Find its speed at time t.

By (8.23)
$$(D_tL)^2 = 10^2 + (10 - 9{\cdot}8t)^2$$
$$= 200 - 19{\cdot}6t + 96{\cdot}04t^2$$
$$\text{speed} = D_tL = \sqrt{(200 - 19{\cdot}6t + 96{\cdot}04t^2)} \text{ m/sec.}$$

(2) A point is moving along the circle $x^2 + y^2 = 1$ with uniform speed v. What are the components of velocity D_tx and D_ty?

On differentiating the relation $x^2 + y^2 = 1$ with respect to t we obtain $2xD_tx + 2yD_ty = 0$, i.e. either $y = 0$ and $D_tx = 0$ or else $D_ty = -xD_tx/y$. But

$$v^2 = (D_tx)^2 + (D_ty)^2$$
$$= (D_tx)^2 + x^2(D_tx)^2/y^2 \quad (\text{if } y \neq 0)$$
$$= (D_tx)^2(y^2 + x^2)/y^2$$
$$= (D_tx)^2/y^2 \quad \text{since } x^2 + y^2 = 1.$$

Therefore $(D_tx)^2 = y^2v^2$. If $y = 0$ this still holds, since then $D_tx = 0$. Thus in any case $D_tx = \pm yv$. Similarly $D_ty = \pm xv$.

PROBLEMS

(1) A body is moving according to the law $x = 2t$, $y = \frac{1}{2}t^2 - \ln t$. What is its velocity at any given time?

(2) A body is moving with constant speed v along the parabola $y = x^2$. Find $D_t x$ and $D_t y$.

(3) Show that if the position of a point P is given in polar co-ordinates as $\{r, \theta\}$, θ being measured in radians, then its speed $D_t L$ at any time is given by

$$(D_t L)^2 = (D_t r)^2 + (r D_t \theta)^2$$

(Hint: draw a diagram showing small changes in r and θ.)

(4) A point P in space can be specified by 3 cartesian co-ordinates (x, y, z). These co-ordinates are the distances of P from 3 fixed mutually perpendicular planes, such as two walls of a room and the floor. Show by repeated application of Pythagoras's theorem that the speed of motion $D_t L$ of P along any path C is given by

$$(D_t L)^2 = (D_t x)^2 + (D_t y)^2 + (D_t z)^2,$$

and that L is the distance gone along this path.

8.15 The formal definition of a limit

For the sake of completeness, and absolute accuracy, we now give the strict definition of a limit. This is mainly of interest from the theoretical and logical point of view, and the reader who is so inclined can omit these rather complicated definitions; they are not essential to the general understanding of the later chapters.

Suppose that x is a variable, and y a function of x. Let X be a particular value of x. Suppose we can find a number Y and a function $\phi(\epsilon)$ with the following properties:

(i) $\phi(\epsilon)$ is defined for all $\epsilon > 0$, and, for those values of ϵ, $\phi(\epsilon) > 0$. (The values of $\phi(\epsilon)$ for $\epsilon < 0$ are irrelevant to the definition, and $\phi(\epsilon)$ does not need to be defined for such values of ϵ.)

(ii) If $0 < |x - X| < \phi(\epsilon)$ then $|y - Y| < \epsilon$.

Then we say that by definition $y \to Y$ as $x \to X$, or $\lim_{x \to X} y = Y$.

In ordinary language this definition says that we can make y as near to Y as we wish, in fact we can make y differ from Y by less than ϵ, by making x differ from X by less than $\phi(\epsilon)$. The reader may think the definition rather complicated—which is why we have not stated it before—but it is in fact the simplest possible definition which agrees with one's commonsense idea of a limit. Any other definition either includes cases which one would not ordinarily consider as limits, or excludes some which are eligible.

EXAMPLES

(1) Suppose y is a constant C, show formally that $y \to C$ as $x \to X$.

Put $Y = C$. Choose any function $\phi(\epsilon)$ satisfying (i): the simplest choice is $\phi(\epsilon) = 1$ for all ϵ. Then (ii) is automatically satisfied since $y - Y = 0$ identically. Thus the limit is proved: we have found the required number Y and the required function $\phi(\epsilon)$ satisfying conditions (i) and (ii).

(2) Suppose $y = 3x$: show that $y \to Y = 3X$ as $x \to X$.

Choose as function $\phi(\epsilon) = \frac{1}{3}\epsilon$. Then condition (i) is evidently satisfied. To demonstrate (ii) we have

$$\begin{aligned} |y - Y| &= |3x - 3X| \\ &= |3(x - X)| \\ &= 3|x - X| \qquad \text{[by (4.1)]} \end{aligned}$$

so that if $|x - X| < \phi(\epsilon) = \frac{1}{3}\epsilon$ we must have $|y - Y| < \epsilon$.

(3) If $\lim_{x \to X} y = Y$, and $\lim_{x \to X} z = Z$ where y and z are two functions of x, show that $\lim_{x \to X} (y + z) = Y + Z$.

By our definition of a limit of y there must be a function $\phi_1(\epsilon)$ which is positive for all positive ϵ and which is such that if $0 < |x - X| < \phi_1(\epsilon)$ then $|y - Y| < \epsilon$. Similarly since $\lim z = Z$ there must be a second function $\phi_2(\epsilon) > 0$ such that if $0 < |x - X| < \phi_2(\epsilon)$ then $|z - Z| < \epsilon$. We wish to show that $(y + z) \to (Y + Z)$, which when written out in full means that there is a function $\phi_3(\epsilon)$ such that (i) $\phi_3(\epsilon) > 0$ if $\epsilon > 0$ and (ii) If $0 < |x - X| < \phi_3(\epsilon)$ it follows that $|(y + z) - (Y + Z)| < \epsilon$. Now using the theorems of Chapter 4 we have in any case

$$\begin{aligned} |(y + z) - (Y + Z)| &= |(y - Y) + (z - Z)| \\ &\leqslant |y - Y| + |z - Z| \quad . \quad (8.25) \end{aligned}$$

If we then take $\phi_3(\epsilon)$, given the value of ϵ to be the least of the two numbers $\phi_1(\frac{1}{2}\epsilon)$, $\phi_2(\frac{1}{2}\epsilon)$, we see that $\phi_3(\epsilon)$ is positive since by hypothesis $\phi_1(\frac{1}{2}\epsilon)$ and $\phi_2(\frac{1}{2}\epsilon)$ are positive. Also if $|x - X|$ is less than $\phi_3(\epsilon)$, it is less than $\phi_1(\frac{1}{2}\epsilon)$ and $\phi_2(\frac{1}{2}\epsilon)$, and so, by hypothesis, $|y - Y| < \frac{1}{2}\epsilon$ and $|z - Z| < \frac{1}{2}\epsilon$. By (8.25) it follows that $|(y + z) - (Y + Z)| < \epsilon$, and the result is established.

The proofs of further results, such as the limit of a product or a quotient, involve more complicated reasoning. For example, if we wish to show under the conditions we have given above ($y \to Y$ and $z \to Z$ as $x \to X$) that the product $yz \to YZ$ we use the identity

$$\begin{aligned} |yz - YZ| &= |(y - Y)(z - Z) + Y(z - Z) + Z(y - Y)| \\ &\leqslant |y - Y|\,|z - Z| + |Y|\,|z - Z| + |Z|\,|y - Y| \end{aligned}$$

Now let $\phi_4(\epsilon)$ be the least of the 4 positive numbers $\phi_1(\frac{1}{2}\sqrt{\epsilon})$, $\phi_2(\frac{1}{2}\sqrt{\epsilon})$, $\phi_2(\epsilon/4 \mid Y \mid)$, $\phi_1(\epsilon/4 \mid Z \mid)$ provided that Y and Z are both different from zero. (We leave the reader to work out the modifications needed if $Y = 0$ or $Z = 0$.) Then if $\mid x - X \mid < \phi_4(\epsilon)$ we must have $\mid y - Y \mid < \frac{1}{2}\sqrt{\epsilon}$ and $\mid z - Z \mid < \frac{1}{2}\sqrt{\epsilon}$, so that $\mid y - Y \mid \mid z - Z \mid < \frac{1}{4}\epsilon$, also $\mid z - Z \mid < \epsilon/4 \mid Y \mid$, so that $\mid Y \mid \mid z - Z \mid < \frac{1}{4}\epsilon$, and $\mid y - Y \mid < \epsilon/4 \mid Z \mid$, so that $\mid Z \mid \mid y - Y \mid < \frac{1}{4}\epsilon$. From this it follows that $\mid yz - YZ \mid < \frac{3}{4}\epsilon < \epsilon$, so the result is established. To establish formally our theorems that $\lim\limits_{x \to 0} \sin x = H$ and $\lim\limits_{x \to 0} \log(1 + x)/x = M$ involves much more difficult juggling, and we shall not go further into the matter.

PROBLEMS

(1) Show that if $y = f(x)$ can be written in the form $y = Y + \beta(x - X)$, where β is bounded (Section 4.4), i.e. $\mid \beta \mid$ is less than some fixed number B, then $y \to Y$ as $x \to X$.

(2) If $y = f(x)$, $Y = f(X)$, and if the ratio

$$\delta y/\delta x = (y - Y)/(x - X)$$

is bounded for all values of x in some range $X - k < x < X + k$, then $y \to Y$ as $x \to X$. (Here k stands for a given fixed positive number.)

(3) If $y \to Y$ as $x \to X$, and if k is some value of $\phi(\epsilon)$, then y is bounded for all values of x between $X - k$ and $X + k$.

(4) If dy/dx exists when $x = X$, then $y \to Y = f(X)$ as $x \to X$.

We can readily modify the above definition to cover the case of a limit of y as $x \to \infty$. We say that if there is a function $\phi(\epsilon)$ defined for all positive ϵ such that when $x > \phi(\epsilon)$ then $\mid y - Y \mid < \epsilon$, then by definition $y \to Y$ as $x \to \infty$. The proofs of other properties follow for this case by suitable modification.

8.16 Historical note

The idea of the calculation of rates of change, or "differential calculus" is due independently to Newton (1642–1727) and Leibnitz (1646–1716): it was also discovered about the same time by the Chinese. The exact formulation in terms of limits is a nineteenth-century development. Leibnitz seems to have thought of the derivative dy/dx as the ratio of two infinitely small changes dy and dx: indeed we still often use this idea of $D_x y$ or dy/dx being the value of $\delta y/\delta x$ when δx is small as a very helpful rough guide, although when we wish to be precise, we nowadays say that it is not the actual value but its limit as $\delta x \to 0$,

CHAPTER NINE

THE CALCULATION OF SMALL CHANGES

9.1 Small changes and errors

The derivative $D_x w = dw/dx$ is defined as the limit of $\delta w/\delta x$ as $\delta x \to 0$, where δw and δx are changes in the variables w and x respectively. It follows that when δx is very small the ratio $\delta w/\delta x$ is very nearly equal to $D_x w$. If we are working to a certain degree of approximation, as in practice we must, then the difference between $D_x w$ and $\delta w/\delta x$ may be quite inappreciable — or even if it is not absolutely inappreciable it may still be too small to be of any importance. In such a case

$$\delta w/\delta x \simeq D_x w$$

or on multiplying through by δx

$$\delta w \simeq D_x w . \delta x = (dw/dx) . \delta x \quad . \quad . \quad . \quad (9.1)$$

If there is a formula expressing w in terms of x it is easy to calculate $D_x w$. Equation (9.1) then shows how we can find the effect on w of a small change δx in x.

EXAMPLES

(1) A cube of side 1 centimetre is being used as a unit of volume. On a hot day it expands to 1·002 centimetres. What is its change in volume?

Let the side of the cube be x cm, and its volume w cc. Then $w = x^3$, and therefore $D_x w = 3x^2$. Thus $\delta w \simeq 3x^2 \delta x$ when δx is small. If $x = 1$ and $\delta x = \cdot 002$, then δw, the increase in volume, is $3 \times 1^2 \times \cdot 002 = \cdot 006$ cc.

(2) An electrical resistance of R ohms is being used for heating. A potential difference of V volts is maintained across it by batteries. If as the batteries run down the potential difference falls by δv volts, what is the fall in the rate of production of heat? Assume that by Ohm's law the current I flowing through the resistance is $I = V/R$ amps, and the rate W of production of heat is the product of the potential difference and current; $W = VI = V^2/R$ watts (or $V^2/4{\cdot}2R$ calories per second).

From this formula $D_V W = 2V/R$, and so $\delta W \simeq 2V . \delta V/R$ watts. If for example a voltage drop $V = 10$ volts is maintained across a resistance R of 20 ohms, and V falls by $\delta V = -\cdot 2$ volts (using a negative sign to indicate a fall), then the power falls by $\delta W = 2 \times 10 \times (-\cdot 2)/20 = -\cdot 2$ watts.

This argument can be applied to small experimental errors. Suppose we have a metal cube whose side is measured as x centimetres. Then its volume will be calculated to be $w = x^3$ cc. If the measurement of the side is in error by an amount δx then the volume will be wrong by approximately $D_x w \,.\, \delta x = 3x^2 \delta x$ cc. If we measure the side δx with an ordinary ruler with an error of not more than ·02 cm, then $| \delta x | < \cdot 02$, and $| \delta w | \simeq 3x^2 \, | \delta x | < \cdot 06x^2$. This shows us the maximum error we can expect in the calculated value of the volume.

It is important to be quite clear what sign to give to an error. It seems logical to call the difference (observed value − true value) the "error" in the measurement, and the opposite quantity (true value − observed value) the "correction for error"; i.e. the correction is what must be added to the observed value to give the true value. In the equation

$$\delta w \simeq D_x w \,.\, \delta x = (dw/dx) \,.\, \delta x$$

we can interpret δw and δx as errors in w and x respectively. Alternatively we can interpret them as corrections for error. It does not matter which convention we use, provided that we keep to the same convention throughout the calculation.

Often we are interested not in the actual error but in the proportional or percentage error. The error δx expressed as a fraction of x will be $\delta x/x$; as a percentage it will be $100 \, \delta x/x$ per cent. Now $D_x \ln x = 1/x$, so that if δx is small, $\delta \ln x \simeq D_x \ln x \,.\, \delta x = \delta x/x$. In other words a small proportional error in a measured quantity x is equal to the actual error in its natural logarithm $\ln x$. The percentage error is obtained by multiplying the proportional error by 100.

Considering again the example of a cube of side x and volume $w = x^3$, we have $\ln w = 3 \ln x$, so that $\delta \ln w = 3 \delta \ln x$. The proportional error in the volume is 3 times the proportional error in the side.

PROBLEMS

(1) An angle θ is measured in degrees correct to 1 decimal place, i.e. with a maximum error of about ·05°. Given the value of θ what is the maximum error in $\sin \theta$ and $\tan \theta$? What is the maximum proportional error?

(2) A circle has its radius R measured to the nearest tenth millimetre. How accurate is the calculated area πR^2?

(3) A pole measured as 12 metres high is found to throw a shadow 16 metres long. If the measurement of the height of the pole is too small by 1 per cent, what is the error in the calculated angle of elevation of the sun?

(4) An error of $\cdot 1\mu$ is made in measuring the diameter of a red blood cell as $7\cdot 5\mu$. Assuming that the cell is spherical, what is the percentage error in estimating its volume (as $\frac{4}{3}\pi R^3$) and surface area (as $4\pi R^2$)?

(5) The current I amps passing through a tangent galvanometer is estimated as $I = K \tan \theta$, where θ is the angle of deflection and K is a constant of the particular galvanometer used. Assuming that K is accurately known, and that the accuracy (expressed as maximum actual error) in the measurement of θ is the same for all values of θ, what angle of deflection will give the minimum percentage error in the calculated value of the current?

9.2 Partial differentiation

In all the cases discussed so far each variable quantity is a function of any other relevant variable. Thus in proving the formula $D_t w = D_t y \,.\, D_y w$ we considered the example of a growing child, whose height y was a function of the time t, and whose weight w was a function of y. It follows that w can also be considered as a function of t; and this is the most natural way to look at the example. We can think of the change in t, i.e. the child's growing older, as the cause of the change in y, the height, and w, the weight: we can say that y and w are "dependent" on t, i.e. functionally related to t, or determined in value by the value of t, while t itself is the "independent variable".

This is the commonsense approach. But considered from a purely mathematical point of view, in which we are interested only in the relationships between the values of t, y, and w, and not in stating which is the cause and which the effect, then we can equally well express t and w as functions of y, or t and y as functions of w. We do not think of the height y as "causing" the time t or the weight w; but to each value of y there will correspond a value of t and a value of w. (Conceivably there may be more than one value of t for a single value of y. But this does not really introduce any difficulty: we know that many functions considered in mathematics, such as $\sin^{-1}x$, for example, are not single-valued.) The important point is that any one variable can be called the "independent variable", and all the others will then be "dependent variables", i.e. determined in value when the value of the independent variable is known.

However, when we consider the effects of small changes it is very artificial to restrict ourselves to systems in which only one quantity can vary independently. The result of an experiment may be influenced by a considerable number of factors. The course of a chemical reaction will certainly depend on the concentrations of the reacting substances and on the temperature. The growth of an animal will depend on the amounts of the various proteins, vitamins, carbohydrates and fats it receives in its diet, and all these are capable of independent variation.

In calculating the results of an experiment we generally have to take into account not one but many measurements. Each measurement may be subject to an error which will affect the final result, and there is no way of inferring the error in one measurement from the error in another. Thus it is often necessary to consider a quantity w as a function of several others; $w = f(x, y)$, or $w = f(x, y, z)$. If x, y, and z are subject to small changes δx, δy, and δz respectively, what is the effect on w?

In experimentation there is a rule "study the effect of altering one variable at a time". In any particular experiment this may or may not be a sound principle, but it is certainly useful for our problem. Suppose w is a function of 3 variables x, y, and z, all of which can vary independently of one another. We begin by keeping y and z fixed and allowing only x to vary. w then becomes a function of one variable only, that is, a function of x, and so we can then differentiate w to get its derivative $D_x w$ and apply all the results previously obtained—for example that $\delta w \simeq D_x w \,.\, \delta x$.

In order to emphasize that y and z are being kept fixed we shall denote this derivative by the symbol $D_{x|y,z}\, w$ or $\left(\dfrac{\partial w}{\partial x}\right)_{y,z}$ and call it the "partial derivative of w with respect to x keeping y and z constant". The phrases "partial derivate" or "partial differential coefficient" mean the same as "partial derivative". The derivative is, of course, "partial" because when y and z are kept fixed the possibilities of variation are restricted to only a part of what they might otherwise be. It is to make this clear that we use the special letter ∂. This is simply the Russian form of the small letter d (the name of which, in Russian, is practically identical in sound with the standard English pronunciation of the word "dare"). But as a rule the symbol $\partial w/\partial x$ is spoken as "partial *dw* by *dx*" or "curly *dw* by *dx*".

EXAMPLES

(1) If $w = 2x + y^2$, then $D_{x|y}w = 2$, $D_{y|x}w = 2y$. For when we differentiate with respect to x, keeping y constant, the term y^2 will have a zero derivative. Conversely when x is held fixed the term $2x$ will contribute nothing to the derivative.

(2) If $w = x^2 + 2xy + yz^3$, $D_{x|y,z}w = 2x + 2y$, $D_{y|x,z}w = 2x + z^3$, $D_{z|x,y}w = 3yz^2$.

PROBLEMS

Find $D_{x|y}w = (\partial w/\partial x)_y$ and $D_{y|x}w = (\partial w/\partial y)_x$ when (1) $w = x \ln y$; (2) $w = x^2/y$; (3) $w = \sqrt{(x + y)}$.

We have shown above that if x alone varies by an amount δx, w will change by an amount δw given by the equation

$$\delta w \simeq (D_{x|y,z}\, w)\ \delta x = \left(\frac{\partial w}{\partial x}\right)_{y,z} \delta x.$$

In the same way it is possible to investigate the effect of a change in y alone, keeping x and z fixed. If we then differentiate w with respect to y we obtain the partial derivative $D_{y|x,z}w = \left(\frac{\partial w}{\partial y}\right)_{x,z}$ and when δy is small

$$\delta w \simeq (D_{y|x,z} w)\ \delta y = \left(\frac{\partial w}{\partial y}\right)_{x,z} \delta y$$

Similarly if only z varies

$$\delta w \simeq (D_{z|x,y} w)\ \delta z = \left(\frac{\partial w}{\partial z}\right)_{x,y} \delta z$$

These formulas show the effect of small changes in any one of the three variables separately. Suppose now we want to change all three; let us say that x changes from x_1 to x_2, y from y_1 to y_2, and z from z_1 to z_2, where all three changes $\delta x = x_2 - x_1$, $\delta y = y_2 - y_1$ and $\delta z = z_2 - z_1$ are small. Let us write $w = f(x, y, z)$, $w_1 = f(x_1, y_1, z_1)$ and $w_2 = f(x_2, y_2, z_2)$. Then as a result of the changes w is altered in value from w_1 to w_2. We wish to compute $\delta w = w_2 - w_1$.

The natural method is to change x first, then y, and finally z. If x is changed from x_1 to x_2 w will become $f(x_2, y_1, z_1)$ and the change in w is approximately given by the formula

$$f(x_2, y_1, z_1) - f(x_1, y_1, z_1) \simeq (D_{x|y,z} w)\ \delta x \quad . \quad . \quad (9.2)$$

where we take the value of the partial derivative $D_{x|y,z}w$ when $x = x_1$, $y = y_1$, $z = z_1$.

This first step is from (x_1, y_1, z_1) to (x_2, y_1, z_1). The second step is from (x_2, y_1, z_1) to (x_2, y_2, z_1): that is, we change y from y_1 to y_2, keeping x fixed at the value x_2 and z fixed at the value z_1. We then have the change in w

$$f(x_2, y_2, z_1) - f(x_2, y_1, z_1) \simeq (D_{y|x,z} w)\ \delta y \quad . \quad . \quad (9.3)$$

where the derivative $D_{y|x,z}w$ is now to be calculated for the values $x = x_2, y = y_1, z = z_1$. The final step is from (x_2, y_2, z_1) to (x_2, y_2, z_2), i.e. a change δz in z from z_1 to z_2 keeping x fixed at x_2 and y fixed at y_2. Accordingly

$$f(x_2, y_2, z_2) - f(x_2, y_2, z_1) \simeq (D_{z|x,y} w)\ \delta z \quad . \quad . \quad (9.4)$$

where $(D_{z|x,y}w)$ is to be calculated for $x = x_2, y = y_2, z = z_1$.

The total change $\delta w = w_2 - w_1 = f(x_2, y_2, z_2) - f(x_1, y_1, z_1)$ will be the sum of these three separate successive changes, so on adding these three equations (9.2), (9.3) and (9.4) we obtain

$$\delta w \simeq (D_{x|y,z} w)\ \delta x + (D_{y|x,z} w)\ \delta y + (D_{z|x,y} w)\ \delta z \quad . \quad (9.5)$$

Now we have said that in this equation the derivative $(D_{x|y,z}w)$ is to be calculated for the values $x = x_1$, $y = y_1$, $z = z_1$, the derivative $(D_{y|x,z}w)$ for the values $x = x_2$, $y = y_1$, $z = z_1$ and the derivative $(D_{z|x,y}w)$ for the values $x = x_2$, $y = y_2$, $z = z_1$. But in any practical example since x_1 and x_2 are very nearly equal it will not make any appreciable difference to the value of the partial derivatives whether we take the value when $x = x_1$ or when $x = x_2$, or whether $y = y_1$ or y_2 or $z = z_1$ or z_2. To be more precise, $(D_{x|y,z}w)$ occurs in the formula multiplied by δx, which is a small quantity, so that any small change in $(D_{x|y,z}w)$ when multiplied by δx will be quite negligible—the formula is only an approximation in any case, though the smaller δx, δy and δz become the closer is the approximation. The same remarks apply to the other two derivatives appearing in the formula—it does not matter whether we take x to be x_1 or x_2, y to be y_1 or y_2, and z to be z_1 or z_2. The formula is thus a general one. In the alternative notation it can be written

$$\delta w \simeq \left(\frac{\partial w}{\partial x}\right)_{y,z} \delta x + \left(\frac{\partial w}{\partial y}\right)_{x,z} \delta y + \left(\frac{\partial w}{\partial z}\right)_{y,z} \delta z \,. \quad (9.6)$$

For the purposes of illustration we have taken w to be a function of three independent variables x, y, and z. The same sort of argument will apply if we have more or less than three: if w was a function of two variables, x and y, the appropriate formula for the change in w produced by small changes δx and δy in x and y respectively would be

$$\begin{aligned}\delta w &\simeq (D_{x|y}w)\ \delta x + (D_{y|x}w)\ \delta y \\ &= \left(\frac{\partial w}{\partial x}\right)_{y} \delta x + \left(\frac{\partial w}{\partial y}\right)_{x} \delta y \quad . \quad . \quad (9.7)\end{aligned}$$

The formulas (9.5), (9.6) and (9.7) have a quite simple interpretation. In (9.5) we know that $(D_{x|y,z}w)\ \delta x$ is (approximately) the change in w produced by a small change δx in x alone, y and z being kept fixed. Similarly $(D_{y|x,z}w)\ \delta y$ is the change in w produced by the change δy acting on its own, and $(D_{z|x,y}w)\ \delta z$ the change produced by δz on its own. Thus the equation (9.6) asserts that the change in w produced by simultaneous *small* changes δx in x, δy in y, and δz in z is obtained to a sufficient accuracy by adding together the changes which would be produced by each of δx, δy, and δz acting separately. This is a very important result.

It is, of course, essential that the changes δx, δy, and δz should be small. It will not be true in general that the effect of a large change in x, y, and z simultaneously can be got by the simple addition of the effects of each of the changes acting on its own, without any change in the values of the other variables.

9.3 Calculation of the effects of several simultaneous changes

The formulas (9.5), (9.6) and (9.7) given in the previous paragraph have a somewhat formidable appearance owing to the large number of suffixes. This appearance is quite deceptive, for they are very easy to apply. It is only necessary to remember that the symbol $D_{x|y,z}$ means "differentiate with respect to x, treating y and z as constants". In doing this differentiation we can make use of any or all of the techniques we developed in Chapter 8.

For example, suppose that we are interested in the area of a rectangle of sides, say, y and z. What will be the effect of making small changes in the sides? We know that the area $w = yz$. Differentiation with respect to y only gives $D_{y|z}w = (\partial w/\partial y)_z = z$, differentiation with respect to z only, $D_{z|y}w = (\partial w/\partial z)_y = y$. Therefore

$$\begin{aligned}\delta w &\simeq (D_{y|z}w)\,\delta y + (D_{z|y}w)\,\delta z \\ &= z\,\delta y + y\,\delta z.\end{aligned}$$

The same principles apply to errors of measurement. If we have measured the side y with a small error δy and the side z with a small error δz, then the error in the measurement of the area is $z\,\delta y + y\,\delta z$.

The reader can compare this formula with Fig. 8.5 which shows that $\delta w = z_1\,\delta y + y_2\,\delta z$ exactly. If δy and δz are small then we can write y in place of y_2 and z in place of z_1; the error in replacing y_2 by y when multiplied by δz will give an exceedingly small quantity, quite negligible for practical purposes. The difference between the former application of this result and the present one is that in Chapter 8 we were concerned with the case in which the sides y and z were both functions of a single variable t, and so were connected by some relation, whereas here we suppose that y and z can vary independently.

EXAMPLES

(1) The volume V of a cylinder of radius R and length L is $\pi R^2 L$. If there is a maximum error of ·02 cm in the measurement of R and L (in cm) what is the maximum error in the estimate of V?

Differentiating with respect to R only we have

$$D_{R|L}V = 2\pi RL$$

Differentiating with respect to L only,

$$D_{L|R}V = \pi R^2.$$

Therefore

$$\begin{aligned}\delta V &= 2\pi RL\,\delta R + \pi R^2\,\delta L \\ &= \pi R(2L\,\delta R + R\,\delta L).\end{aligned}$$

If $|\,\delta R\,|$ and $|\,\delta L\,|$ are both less than ·02, $|\,\delta V\,| < \pi R(\cdot 04\,L + \cdot 02 R)$. For example, if $R = 1$, $L = 2$ $|\,\delta V\,| < \frac{1}{10}\pi = \cdot 31$ cc.

(2) The altitude of the sun is calculated by observing the length, l metres, of the shadow cast by a pole of height h metres. If the measurement h has an error δh, and l an error δl, what is the error in the measurement of the altitude α radians?

We have $h/l = \tan\,\alpha$, or $\alpha = \tan^{-1}(h/l)$. Therefore $D_{h|l}\alpha = D_{h|l}\tan^{-1}(h/l) = D_{(h/l)}\tan^{-1}(h/l)\,.\,D_{h|l}(h/l)$ (by the function of a function rule) $= [1 + (h/l)^2]^{-1}\,.\,(1/l)$. Similarly $D_{l|h}\alpha = [1 + (h/l)^2]^{-1}(-h/l^2)$. Thus

$$\begin{aligned}\delta\alpha &= [1 + (h/l)^2]^{-1}\,(\delta h/l) + [1 + (h/l)^2]^{-1}\,(-h\,\delta l/l^2)\\ &= \frac{1}{1 + (h/l)^2}\left(\frac{\delta h}{l} - \frac{h\,\delta l}{l^2}\right)\\ &= \frac{l\,\delta h - h\,\delta l}{l^2 + h^2}.\end{aligned}$$

An alternative and simpler solution is as follows. The equation $h/l = \tan\,\alpha$ can be written logarithmically

$$\ln \tan \alpha = \ln h - \ln l$$

$$\text{whence}\quad \delta \ln \tan \alpha = \delta \ln h - \delta \ln l$$

Now considering $\ln \tan \alpha$ as a function of the single variable α,

$$\begin{aligned}\delta \ln \tan \alpha &\simeq (D_\alpha \ln \tan \alpha)\,.\,\delta\alpha\\ &= (D_\alpha \tan \alpha)/\tan \alpha\,.\,\delta\alpha\\ &= (\sec \alpha)^2/\tan \alpha\,.\,\delta\alpha\\ &= \sec \alpha\,.\,\operatorname{cosec} \alpha\,.\,\delta\alpha.\end{aligned}$$

Similarly $\delta \ln h \simeq \delta h/h$, and $\delta \ln l \simeq \delta l/l$. Thus

$$\sec \alpha\,.\,\operatorname{cosec} \alpha\,.\,\delta\alpha = \delta h/h - \delta l/l.$$

On multiplying both sides of this equation by $1/(\sec \alpha\,.\,\operatorname{cosec} \alpha) = \sin \alpha\,.\,\cos \alpha$ we have

$$\delta\alpha \simeq (\delta h/h - \delta l/l) \sin \alpha\,.\,\cos \alpha.$$

$$\begin{aligned}\text{Now } \sin \alpha\,.\,\cos \alpha &= \tan \alpha\,(\cos \alpha)^2 = \tan \alpha/(\sec \alpha)^2\\ &= \tan \alpha/[1 + (\tan \alpha)^2]\\ &= (h/l)/[1 + (h/l)^2]\\ &= hl/(l^2 + h^2),\end{aligned}$$

so that $\delta\alpha = (l\,\delta h - h\,\delta l)/(l^2 + h^2)$ as before.

PROBLEMS

(1) A rectangle has width x and length y. What is the error in the estimation of the diagonal $\sqrt{(x^2 + y^2)}$ when x and y have small errors?

(2) The density of a cube of mass m and side s is m/s^3. What will be the effect of small errors in the measurement of m and s?

9.4 The meaning of partial differentiation

We can illustrate these results in the following way. Suppose that w is any quantity which varies from point to point over a plane. For example, w might be the degree of illumination, or the temperature, or the concentration of some substance diffusing through a thin film. At each point P, w will have a definite value. If we specify P by its cartesian co-ordinates (x, y), then w can be considered as a function of x and y.

Now when we differentiate w partially with respect to x, keeping y fixed, we are in effect limiting P to a horizontal line $y =$ constant (Fig. 9.1). The derivative $D_{x|y}w$ shows how rapidly w varies as P moves

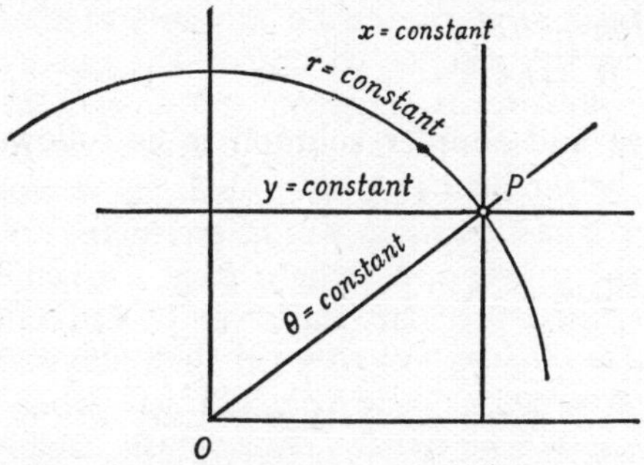

Fig. 9.1—Partial differentiation in a plane

along this line. In the same way the derivative $D_{y|x}w = (\partial w/\partial y)_x$ shows how w varies when P is restricted to move along a vertical line $x =$ constant (Fig. 9.1).

The point P can equally well be specified by its polar co-ordinates $\{r, \theta\}$ and therefore w can be considered as a function of r and θ. The derivative $D_{r|\theta}w$ then shows how w varies when P is restricted to the curve $\theta =$ constant, i.e. a straight line through P running out from the origin O, while $D_{\theta|r}w$ will give the rate of variation with respect to θ along $r =$ constant, a circle with O as centre.

These do not exhaust the possibilities. We could, for example, specify the point P by the co-ordinates x and r. This is an unusual combination, but it is certainly a possible one, and might indeed be useful in certain special problems. We can then calculate $D_{x|r}w$, meaning the rate of change of w (with respect to x) when P moves along a circle $r =$ constant. Comparing this with $D_{x|y}w$, which is the rate of change when P moves along the horizontal line $y =$ constant, we see that there is no reason in general why $D_{x|r}w$ and $D_{x|y}w$ should be the same: they are the answers to different questions, and the movement of the point P is restricted in different ways.

As an example we may take w to be equal to $r^2 = x^2 + y^2$. Expressing w in terms of x and r, $w = r^2$, $D_{x|r}w = 0$. This is natural, since when we keep r fixed the value of w is fixed too. On the other hand, since $w = x^2 + y^2$, $D_{x|y}w = 2x$, which is different from $D_{x|r}w$ except when $x = 0$. The moral is that in any partial differentiation it is essential to specify exactly what variables are being held constant.

An even clearer example occurs with the problem of the energy E contained in a mass of gas of volume V at temperature T and pressure P. We know that V, T, and P are connected by the gas equation $PV = RT$ where R is a constant. Thus only two of the three quantities P, V and T can be independent variables, since when two of them are given the value of the third can be deduced. The energy E can therefore be expressed either as a function of P and V, or else as a function of P and T, or finally as a function of T and V, and there are no obvious reasons for preferring any one of these to any other. Now $D_{V|P}E$ means the rate of change of the energy as the volume changes when the pressure is constant, and $D_{V|T}E$ the rate of change when the temperature is constant, and there is no reason why these two rates of change should be equal. Again it is essential to specify what is being kept constant when we differentiate.

Nevertheless there are cases where it is convenient to omit the suffix. If w is a function of two cartesian co-ordinates x and y then when differentiating with respect to x we shall almost always be interested in the derivative with y held constant, and when we differentiate with respect to y, x will be fixed. Thus it will usually be sufficiently clear to write $D_x w$ or $\partial w/\partial x$ for $D_{x|y}\, w$, and $D_y w$ or $\partial w/\partial y$ for $D_{y|x} w$. Again if we are using polar co-ordinates instead of cartesians, the expressions $\partial w/\partial r$ and $\partial w/\partial \theta$ will mean $(\partial w/\partial r)_\theta$ and $(\partial w/\partial \theta)_r$ as a rule. If we say explicitly "w is a function of three variables x, y, and z" (where x, y, and z denote *any* variables, not necessarily co-ordinates), or if we write $w = f(x, y, z)$, then $D_x w$ or $\partial w/\partial x$ will *always* be understood to mean $D_{x|y,z}\, w = (\partial w/\partial x)_{y,z}$, and similarly for $D_y w$ and $D_z w$. An alternative notation which is frequently used is to write w_x or $f_x(x, y, z)$ for $D_x w$, w_y or $f_y(x, y, z)$ for $D_y w$, and w_z or $f_z(x, y, z)$ for $D_z w$. Thus equation (9.5) can be written in any one of the forms

$$\begin{aligned}\delta w &\simeq D_x w \,.\, \delta x + D_y w \,.\, \delta y + D_z w \,.\, \delta z \\ &= \left(\frac{\partial w}{\partial x}\right) \delta x + \left(\frac{\partial w}{\partial y}\right) \delta y + \left(\frac{\partial w}{\partial z}\right) \delta z \\ &= w_x \,.\, \delta x + w_y \,.\, \delta y + w_z \,.\, \delta z \\ &= f_x(x, y, z)\, \delta x + f_y(x, y, z)\, \delta y + f_z(x, y, z)\, \delta z\end{aligned}$$

all of which have exactly the same meaning. It is important to note that $D_x w$ is, however, only a conventional shorthand form for $D_{x|y,z}\, w$, and the reader is advised to insert the subscripts indicating the quantities held constant whenever there is any possibility of doubt.

For example with this notation it will not be generally true that $\left(\frac{\partial r}{\partial x}\right) = 1 \Big/ \left(\frac{\partial x}{\partial r}\right)$, whereas for total differentiation, where w and z are any two quantities functionally related, we must have $\left(\frac{dw}{dz}\right) = 1 \Big/ \left(\frac{dz}{dw}\right)$.

If (x, y) are cartesian and $\{r, \theta\}$ polar co-ordinates we have, in fact,

$$r = \sqrt{(x^2 + y^2)}, \qquad \frac{\partial r}{\partial x} = \frac{2x}{2\sqrt{(x^2 + y^2)}} = \frac{x}{r}$$

$$x = r \cos \theta, \qquad \frac{\partial x}{\partial r} = \cos \theta = \frac{x}{r} = \frac{\partial r}{\partial x}$$

The reason for this apparently paradoxical result is simply that in calculating $\partial r/\partial x$ we keep y fixed, whereas in $\partial x/\partial r$ we keep θ fixed. Since we are keeping different quantities fixed we cannot expect the ordinary rule $\frac{\partial x}{\partial r} \cdot \frac{\partial r}{\partial x} = 1$ to hold. If we keep the *same* quantity fixed it will be true, so that

$$\left(\frac{\partial r}{\partial x}\right)_\theta = 1\Big/\left(\frac{\partial x}{\partial r}\right)_\theta \quad \text{and} \quad \left(\frac{\partial r}{\partial x}\right)_y = 1\Big/\left(\frac{\partial x}{\partial r}\right)_y$$

9.5 Change of variables in partial differentiation

From the above argument it appears that we need a rule which will tell us how to change the variables with respect to which we are differentiating. Such a rule will take the place of the "function of a function" rule in ordinary total differentiation.

Let us therefore consider a quantity w which is a function of three independent variables x, y, and z. The argument will readily generalize to cases in which we have less than or more than three independent variables. Suppose that we wish to replace x, y, and z by three new independent variables X, Y, and Z, which are connected with x, y, and z by known relations. (This includes the particular case in which we only replace one or more variables: we could change x to a new variable X, leaving y and z unaltered. This will be covered by the general formula by putting $Y = y$, $Z = z$). How can we now calculate the new derivatives

$$D_{X|Y,Z}\, w, \quad D_{Y|X,Z}\, w \quad \text{and} \quad D_{Z|X,Y}\, w?$$

To do this we take the general formula (9.5)

$$\delta w \simeq (D_{x|y,z} w)\, \delta x + (D_{y|x,z} w)\, \delta y + (D_{z|x,y}\, w)\, \delta z$$

and divide through by δX, obtaining

$$\frac{\delta w}{\delta X} = D_{x|y,z}\, w . \frac{\delta x}{\delta X} + D_{y|x,z}\, w . \frac{\delta y}{\delta X} + D_{z|x,y}\, w . \frac{\delta z}{\delta X}.$$

Now let us keep Y and Z fixed, and let δX tend to zero. Then by definition $\delta w/\delta X$ tends to the partial derivative $D_{X|Y,Z} w$, and similarly $\delta x/\delta X \to D_{X|Y,Z} x$, $\delta y/\delta X \to D_{X|Y,Z} y$ and $\delta z/\delta X \to D_{X|Y,Z} z$. Also the smaller δX is, the smaller δx, δy, and δz will become, and the nearer the approximate equality will be to exactness. Thus in the limit

$$D_{X|Y,Z}\,w = (D_{x|y,z}w)(D_{X|Y,Z}x) + (D_{y|x,z}w)(D_{X|Y,Z}y) + (D_{z|x,y}w)(D_{X|Y,Z}\,z) \quad . \quad . \quad . \quad (9.8)$$

This is the required formula. The formulas for the other two derivatives follow in exactly the same way:

$$D_{Y|X,Z}\,w = (D_{x|y,z}w)(D_{Y|X,Z}x) + (D_{y|x,z}w)(D_{Y|X,Z}y) + (D_{z|x,y}w)(D_{Y|X.Z}z)$$

$$D_{Z|X,Y}\,w = (D_{x|y,z}w)(D_{Z|X,Y}x) + (D_{y|x,z}w)(D_{Z|X,Y}y) + (D_{z|x,y}w)(D_{Z|X,Y}z).$$

The large number of suffixes makes these formulas look rather confusing. If we can safely adopt the convention that D_x stands for $D_{x|y,z}$ and D_X stands for $D_{X|Y,Z}$ (and similarly $D_y = D_{y|x,z}$, etc.) then they become rather simpler in appearance,

$$D_X w = (D_x w)(D_X x) + (D_y w)(D_X y) + (D_z w)(D_X z)$$

or

$$\frac{\partial w}{\partial X} = \frac{\partial w}{\partial x}\cdot\frac{\partial x}{\partial X} + \frac{\partial w}{\partial y}\cdot\frac{\partial y}{\partial X} + \frac{\partial w}{\partial z}\cdot\frac{\partial z}{\partial X}$$

or

$$w_X = w_x\,.\,x_X + w_y\,.\,y_X + w_z\,.\,z_X.$$

In words, "to find the partial derivative of w with respect to a new variable X we multiply each derivative of w with respect to an old variable x, y or z by the corresponding derivative of the old variable with respect to the new one X, and add all the products so formed". When expressed in this way the rule is not difficult to apply.

EXAMPLES

(1) w is a function of the position of a point in the plane, and is expressed in terms of the cartesian co-ordinates (x, y). Find its partial derivatives in terms of polars $\{r, \theta\}$.

For simplicity let us write w_x for $D_{x|y}w = (\partial w/\partial x)_y$, w_y for $D_{y|x}w$, and so on. Then

$$w_r = w_x x_r + w_y y_r$$

$$w_\theta = w_x x_\theta + w_y y_\theta$$

Now if θ is measured in radians, $x = r\cos\theta$, so that

$$x_r = \cos\theta, \quad x_\theta = -\,r\sin\theta.$$

Similarly $y = r\sin\theta$, and

$$y_r = \sin\theta, \quad y_\theta = r\cos\theta.$$

Substituting these values in the equations above we finally obtain

$$w_r = \cos\theta \,.\, w_x + \sin\theta \,.\, w_y$$
$$w_\theta = -\,r\sin\theta \,.\, w_x + r\cos\theta \,.\, w_y$$

For example, suppose that $w = xy$, then $w_x = y = r\sin\theta$, $w_y = x = r\cos\theta$, and therefore

$$\begin{aligned} w_r &= \cos\theta \,.\, r\sin\theta + \sin\theta \,.\, r\cos\theta \\ &= 2r\cos\theta \,.\, \sin\theta \\ &= r\sin 2\theta \\ w_\theta &= -\,r\sin\theta \,.\, r\sin\theta + r\cos\theta \,.\, r\cos\theta \\ &= r^2[-(\sin\theta)^2 + (\cos\theta)^2] \\ &= r^2\cos 2\theta. \end{aligned}$$

(2) *Metabolism in a bacterium*—The rate at which any chemical substance can be absorbed into a bacterium will be proportional to its surface area A. If we suppose that the substance has to be distributed throughout the whole volume V, the rate at which it can be delivered to any particular part will presumably be proportional to A/V. If we call this rate M, we shall therefore have $M = KA/V$ where K is a constant.

We may be interested in the question of how M is affected by changes in the shape and size of the bacterium. For simplicity suppose that it is of cylindrical form with length L and radius R with two hemispherical caps at the ends (Fig. 9.2). We shall require the formulas

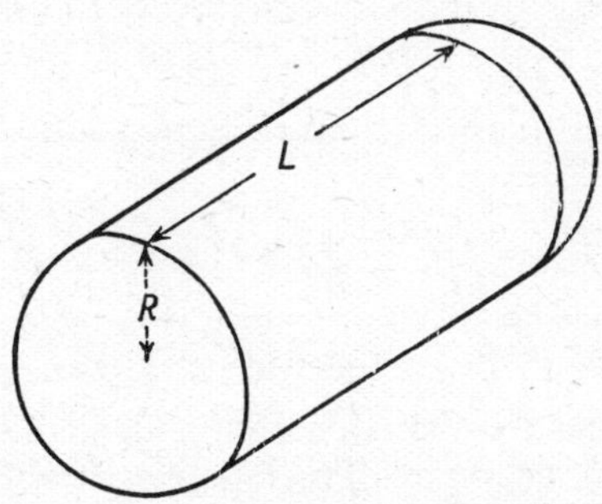

Fig. 9.2—Volume and area of a bacterium

for the areas and volumes of a cylinder and a hemisphere (see Sections 11.9 and 11.10). The volume of a cylinder is equal to the area of its cross-section times its length, i.e. in our case $\pi R^2 L$. The volume of a hemisphere is $\frac{2}{3}\pi R^3$. The total volume of the bacterium will therefore be $\frac{2}{3}\pi R^3 + \pi R^2 L + \frac{2}{3}\pi R^3 = \frac{4}{3}\pi R^3 + \pi R^2 L = V$. The area of a hemisphere is $2\pi R^2$. The area of the curved surface of a cylinder is $2\pi RL$. (For if it is slit along a line parallel to the axis it can then be opened out into a rectangle of length L and breadth equal to the circumference $2\pi R$ of the cylinder.) Thus the total area is $2\pi R^2 + 2\pi RL + 2\pi R^2 = 4\pi R^2 + 2\pi RL = A$. We have therefore

$$\begin{aligned} M = KA/V &= K(4\pi R^2 + 2\pi RL)/(\tfrac{4}{3}\pi R^3 + \pi R^2 L) \\ &= K(4R + 2L)/(\tfrac{4}{3}R^2 + RL) \end{aligned}$$

on cancelling out the factor πR from numerator and denominator.

To determine what effect changes of R and L have on this we must find the partial derivatives $D_{R|L}M$ and $D_{L|R}M$ or, as we shall denote them for conciseness, M_R and M_L. We can do this in several ways. One is to differentiate the expression directly as it stands. Another is to use a two-stage process, analogous to the "function-of-a-function" rule for the single independent variable. We write $u = 4R + 2L$, $v = \frac{4}{3}R^2 + RL$ so that $M = Ku/v$. We can first differentiate M partially with respect to u and v, and then use the formulas

$$\begin{aligned} M_R &= M_u u_R + M_v v_R \\ M_L &= M_u u_L + M_v v_L \end{aligned}$$

to find the derivatives with respect to R and L. (Here M_u is shorthand for $D_{u|v}M$ and M_v for $D_{v|u}M$.)

Since $M = Ku/v$, we have at once $M_u = K/v = K/(\frac{4}{3}R^2 + RL)$, and $M_v = - Ku/v^2 = - K(4R + 2L)/(\frac{4}{3}R^2 + RL)^2$.

Since $u = 4R + 2L$, $u_R = 4$ and $u_L = 2$. Since $v = \frac{4}{3}R^2 + RL$, $v_R = \frac{8}{3}R + L$ and $v_L = R$. Substituting these values in the equations we obtain

$$\begin{aligned} M_R &= \frac{4K}{\frac{4}{3}R^2 + RL} - \frac{K(\frac{8}{3}R + L)(4R + 2L)}{(\frac{4}{3}R^2 + RL)^2} \\ &= - 6K(8R^2 + 8RL + 3L^2)/(4R + 3L)^2 \end{aligned}$$

and

$$\begin{aligned} M_L &= \frac{2K}{\frac{4}{3}R^2 + RL} - \frac{K(4R + 2L)R}{(\frac{4}{3}R^2 + RL)^2} \\ &= - 12K/(4R + 3L)^2. \end{aligned}$$

If now R changes by a small amount δR and L by a small amount δL the expected change in rate of metabolism M will be $\delta M = M_R \delta R + M_L \delta L$. As both M_R and M_L are negative, this means that we can expect an increase in either the length or the radius to slow down the process. This is in accordance with common sense, since an increase in size will make the interior to some extent less accessible from the surface.

(3) Given the internal energy E of a gas as a function of its temperature and volume V, find the rate of change of E with respect to the temperature when the pressure P remains constant.

We shall use the notation $E_{T|V}$ for $D_{T|V}E = (\partial E/\partial T)_V$, and similarly for other derivatives. We know E as a function of T and V, and therefore we know $E_{T|V}$ and $E_{V|T}$. We wish to change the independent variables from T and V to T and P; the appropriate formula is

$$E_{T|P} = E_{T|V}\, T_{T|P} + E_{V|T}\, V_{T|P}.$$

Now if n is the mass of gas measured in molecular weight units (e.g. moles) then the gas law states that

$$PV = nRT$$

or, expressing T and V in terms of T and P,

$$T = T; \quad V = nRT/P$$

whence

$$T_{T|P} = 1; \quad V_{T|P} = nR/P$$

and, by substitution in the formula,

$$E_{T|P} = E_{T|V} + nRE_{V|T}/P$$

Observe that in this formula the suffixes indicating the quantities held constant cannot be omitted, for without them there would be no way of distinguishing between $E_{T|P}$ and $E_{T|V}$.

It is not possible to go any further by purely mathematical reasoning. However, there is a classical experiment which shows that if a gas enclosed in a thermally insulated container is allowed to expand into a vacuum, in the final state when it has again come to rest the temperature returns to its original value. Thus it has undergone a change in volume without change of temperature: $\delta T = 0$. Now since the container is insulated no energy is gained or lost by the gas in the form of heat. Furthermore it does no work on its surroundings, so that there is no change in internal energy. Thus at constant temperature $\delta E = 0$, so that $\delta E/\delta V = 0$. On taking the limit as $\delta V \to 0$ we have $E_{V|T} = 0$. On substituting this in the above equation we find that in fact

$$E_{T|V} = E_{T|P} \quad . \quad . \quad . \quad . \quad (9.9)$$

This is, however, a physical law and not a mathematical identity.

(4) *The specific heats of a gas*—Suppose that we take a unit mass of gas (in any units, not necessarily molecular weight units) and warm it, imparting to it an amount of heat δQ with the result that its temperature is raised by an amount δT. Then the ratio $\delta Q/\delta T$ may be called the "average specific heat of the gas" over this range of temperature, since it represents the average amount of heat required to raise unit mass through one unit of temperature measured over this range. If this average $\delta Q/\delta T$ tends to a limit c as $\delta T \to 0$ we shall naturally call this limit the actual specific heat at temperature T. It will, however, be important to state the conditions under which the heating takes

place; and in particular we can distinguish between the specific heat c_V when the volume is kept constant, and the specific heat c_P at constant pressure. (We can in theory make the same distinction for solids and liquids, but in those cases it is usually the specific heat at constant pressure that is meant: it would require very unusual conditions to keep the volume constant.)

Now when the volume is kept constant the gas can do no work on its surroundings, so that all the heat energy δQ received goes to increase the internal energy of the gas. Thus $\delta Q = \delta E$, $\delta Q/\delta T = \delta E/\delta T$ at constant volume, and accordingly in the limit as $\delta T \to 0$

$$c_V = E_{T|V} \quad . \quad . \quad . \quad . \quad (9.10)$$

On the other hand, if the gas is kept at constant pressure then as the temperature rises there will be an expansion by, say, δV. This will do an amount of work $P\delta V$ against the pressure exerted on the gas by its container. Thus the heat energy δQ supplied will go partly towards increasing the internal energy by an amount δE and partly to doing work $P\delta V$, or in other words

$$\delta Q = \delta E + P\delta V$$

at constant pressure. Now divide this equation through by δT and take the limit as $\delta T \to 0$; $\delta Q/\delta T$ will tend to the specific heat at constant pressure, and

$$c_P = E_{T|P} + PV_{T|P} \quad . \quad . \quad . \quad (9.11)$$

The equations (9.10) and (9.11) will hold for any substance whether gas, liquid, or solid: they amount in practice to definitions of the specific heats at constant volume and constant pressure in terms of partial derivatives. For a gas we have by (9.9) $E_{T|P} = E_{T|V} = c_V$. Furthermore if m is the molecular weight of the gas we are considering, then in unit mass there are $1/m$ molecular weight units, and therefore $PV = RT/m$, or $V = RT/mP$. By differentiation we find $V_{T|P} = R/mP$ or $PV_{T|P} = R/m$. Substituting these values in the equation (9.11) we obtain

$$c_P = c_V + R/m \quad . \quad . \quad . \quad . \quad (9.12)$$

the classical equation connecting the two specific heats.

Note. In this equation we suppose heat energy and mechanical energy to be measured in the same units. If heat energy is measured in calories and mechanical energy in joules, there being J joules in a calorie, then a specific heat c_P measured in calorie units will be equivalent to one of Jc_P measured in joule units, and similarly for c_V. Thus equation (9.12) must be modified to

$$Jc_P = Jc_V + R/m$$
$$\text{or} \quad c_P - c_V = R/mJ$$

(5) Let z be any function of any two independent variables, x, y, so that $z = f(x, y)$. Suppose that this relation can be rearranged to give x in terms of y and z, then provided that $D_{x|y}z \neq 0$, the formulas for the partial derivatives of x are

$$D_{y|z}x = -D_{y|x}z/D_{x|y}z$$
$$D_{z|y}x = 1/D_{x|y}z.$$

Proof. We shall write $z_{y|x}$ for $D_{y|x}z$, and similarly for other derivatives. Thus we have to prove that $x_{y|z} = -z_{y|x}/z_{x|y}$, $x_{z|y} = 1/z_{x|y}$, the quantities $z_{x|y}$ and $z_{y|x}$ being known quantities and $z_{x|y} \neq 0$. (The condition $z_{x|y} \neq 0$ has to be imposed to prevent the fractions having zero denominators. It can be shown that when $z_{x|y} \neq 0$ we can in fact always consider x to be a function of independent variables y and z; but the proof is somewhat complicated.)

Now the second relation $x_{z|y} = 1/z_{x|y}$ follows immediately from the ordinary rule for derivatives, $D_z x = 1/D_x z$, since we are keeping the same quantity y constant in both cases.

To prove the first relation $x_{y|z} = -z_{y|x}/z_{x|y}$ we consider the formula for change of variables from x, y to z, y. This is, for any quantity w,

$$\begin{aligned} w_{y|z} &= w_{x|y}x_{y|z} + w_{y|x}y_{y|z} \\ &= w_{x|y}x_{y|z} + w_{y|x} \end{aligned}$$

since $y_{y|z}$, the derivative of y with respect to y, is 1 (the suffix indicating that z is constant is irrelevant in $y_{y|z}$).

This holds for any function w. Take w to be z, then $w_{y|z} = z_{y|z} = 0$, since this means "the rate of change of z when z is constant". Therefore

$$0 = z_{x|y}x_{y|z} + z_{y|x}, \quad \text{i.e.}$$
$$z_{x|y}x_{y|z} = -z_{y|x}.$$

Since $z_{x|y}$ is not zero we can divide through by it obtaining the required formula $x_{y|z} = -z_{y|x}/z_{x|y}$.

If $z_{y|x} \neq 0$ this formula can be expressed in a very symmetrical form, for then $z_{y|x} = 1/y_{z|x}$ by the usual rule that $D_y z = 1/D_z y$, the same quantity x being kept constant in both cases. Thus we have $x_{y|z} = -1/y_{z|x}z_{x|y}$. On multiplication through by $y_{z|x}z_{x|y}$ this becomes an identity

$$x_{y|z}\, y_{z|x}\, z_{x|y} = -1$$

PROBLEMS

(1) If the bacterium of example (2) above is cylindrical, with flat ends instead of hemispherical ones, what is the effect on $M = KA/V$ of a small change in the length and radius?

(2) Differentiate $w = \tan(x + y) + \tan(x - y)$ partially with respect to x and y. [Change the variables to $(x + y)$ and $(x - y)$.]

(3) Differentiate $w = \ln(x^2 + y^2) - \ln(x^2 - y^2)$ partially with respect to x and y.

9.6 Formal definition of differentiability

In the discussion above we have used such arguments as that if w is a function of x and y, and if δx and δy are small then δw is very nearly equal to $w_{x|y}\,\delta x + w_{y|x}\,\delta y$, or as we can write it here, without ambiguity, $w_x\,\delta x + w_y\,\delta y$, where w_x and w_y are the partial derivatives of w with respect to x and y respectively.

We have also said that the smaller δx and δy are, the more accurately this formula holds. Now it is rather unsatisfactory to use such a vague term as "very nearly equal": it would be better if we could say more precisely what was meant. So, for the sake of logical accuracy, we shall examine more carefully the error

$$E = (w_x\delta x + w_y\delta y) - \delta w$$

in the formula for a small change. (The reader who is interested in the practical rather than the logical aspect can omit this discussion.)

E is the difference between the calculated change $(w_x\delta x + w_y\delta y)$ and the actual change δw in w. Roughly speaking what we want is this error to be very small compared with the small changes δx and δy. Unfortunately we cannot reasonably require δw to be always small in comparison both with δx and with δy. For δx and δy are independent quantities, and we can always if we wish take δx to be much smaller than δy, or even zero. But if we put $\delta x = 0$ the error E will be $w_y\,\delta y - \delta w$, which will not necessarily be zero, and it will be very large in comparison with the zero quantity δx. Thus this first and simplest attempt to clarify our ideas does not work.

We now make a second attempt, and say that we want E to be small compared with $|\delta x| + |\delta y|$. This does not run into the same objection, and we can put $\delta x = 0$ without encountering difficulties. For we then have

$$E = w_y\delta y - \delta w, \quad \text{i.e.} \quad E/\delta y = w_y - \delta w/\delta y.$$

Since w_y, the partial derivative, is by definition the limit of $\delta w/\delta y$ as $\delta y \to 0$ when x is constant $(\delta x = 0)$, it follows that the smaller δy becomes the smaller is the difference between w_y and $\delta w/\delta y$, so that the smaller is $E/\delta y$. In other words, the smaller we take δy to be the smaller the error E is in comparison with δy. So far we seem to be on the right track.

We can now go further and say that we want E to become small in comparison with $|\delta x| + |\delta y|$ when $|\delta x| + |\delta y|$ becomes very small. Put more formally, what we would like to be true would be that

$$\lim \frac{E}{|\delta x| + |\delta y|} = \lim \frac{(w_x\delta x + w_y\delta y) - \delta w}{|\delta x| + |\delta y|} = 0 \quad . \quad (9.13)$$

as $|\delta x| + |\delta y| \to 0$. For when (9.13) is true we shall be completely justified in using the approximate formula $\delta w \simeq w_x \delta x + w_y \delta y$.

We can show that the property (9.13) does hold for various simple functions. Let us take some examples.

Consider first the function $w = x + y$: then $w_x = 1$, $w_y = 1$, and therefore $E = \delta x + \delta y - \delta w$. Now with our usual notation

$$\begin{aligned} \delta x &= x_2 - x_1, \qquad \delta y = y_2 - y_1, \\ \delta w &= w_2 - w_1 = (x_2 + y_2) - (x_1 + y_1) \\ &= (x_2 - x_1) + (y_2 - y_1) = \delta x + \delta y. \end{aligned}$$

Thus $E = 0$ and so $\lim E/(|\delta x| + |\delta y|) = \lim 0 = 0$.

Secondly consider $w = x^2 + y^2$. Then $w_x = 2x$, $w_y = 2y$ and $E = 2x_1\, \delta x + 2y_1\, \delta y - \delta w$. Now

$$\begin{aligned} \delta w &= w_2 - w_1 = (x_2^2 + y_2^2) - (x_1^2 + y_1^2) = (x_2^2 - x_1^2) + (y_2^2 - y_1^2) \\ &= (x_2 + x_1)(x_2 - x_1) + (y_2 + y_1)(y_2 - y_1) = (x_2 + x_1)\delta x + (y_2 + y_1)\delta y. \end{aligned}$$

Therefore

$$E = (x_1 - x_2)\delta x + (y_1 - y_2)\delta y = -(\delta x)^2 - (\delta y)^2$$

and

$$|E| = |\delta x|^2 + |\delta y|^2 \leqslant |\delta x|^2 + 2|\delta x|\,|\delta y| + |\delta y|^2 = (|\delta x| + |\delta y|)^2.$$

Dividing by the positive number $|\delta x| + |\delta y|$ we have

$$\frac{|E|}{|\delta x| + |\delta y|} \leqslant |\delta x| + |\delta y|$$

and so as $|\delta x| + |\delta y| \to 0$ the left-hand side must also tend to 0.

Finally consider the function $w = xy$, for which $w_x = y$ and $w_y = x$. Here $E = y_1 \delta x + x_1 \delta y - \delta w$. But

$$\begin{aligned} \delta w &= w_2 - w_1 = x_2 y_2 - x_1 y_1 = (x_1 + \delta x)(y_1 + \delta y) - x_1 y_1 \\ &= y_1 \delta x + x_1 \delta y + \delta x\, \delta y, \end{aligned}$$

and therefore $E = \delta x\, .\, \delta y$. Now

$$(|\delta x| + |\delta y|)^2 = |\delta x|^2 + 2|\delta x|\,|\delta y| + |\delta y|^2,$$

and therefore

$$|E| = |\delta x|\,|\delta y| < (|\delta x| + |\delta y|)^2.$$

Thus

$$\frac{|E|}{|\delta x| + |\delta y|} \leqslant |\delta x| + |\delta y|$$

and must tend to zero as $|\delta x| + |\delta y|$ tends to zero.

In this way we can verify that equation (9.13) holds for all functions that one meets in practice. Accordingly we make the following definition.

Definition of differentiability of a function of two variables. If a function w of two variables x and y is such that equation (9.13) holds good as $|\delta x| + |\delta y| \to 0$ then w is said to be "differentiable with respect to the two variables".

In the same way if W is a function of three variables x, y and z, we shall say it is differentiable when

$$\frac{W_x\delta x + W_y\delta y + W_z\,\delta z - \delta W}{|\delta x| + |\delta y| + |\delta z|} \to 0 \text{ as } (|\delta x| + |\delta y| + |\delta z|) \to 0$$

and when that is so we can safely use the approximate formula $\delta W \simeq W_x\delta x + W_y\delta y + W_z\,\delta z$. This definition of differentiability allows us to give formal and logically accurate proofs of our formulas for differentiation. For example, if W is a differentiable function of x and y, and x and y are differentiable functions of X and Y, then the formulas for change of variable

$$W_X = W_x x_X + W_y y_X$$
$$W_Y = W_x x_Y + W_y y_Y$$

(where $W_X = D_{X|Y}W$, etc.) can be shown to be universally applicable. In principle all we have to do is to repeat the argument we have already given, but instead of saying "δW is approximately equal to $W_x\delta x + W_y\delta y$" we can now use a more precise definition in terms of limits. But we shall not pursue the matter further here. The important question for practical purposes is "when is a function differentiable, so that we can apply the required formulas?" The answer is that in practice all continuously variable functions can be considered to be differentiable. There are a few exceptions and qualifications, but they are almost always obvious. The function $1/x$ will not be differentiable when $x = 0$, since it then becomes infinite; the same applies to the function $\log x$. The function $(x^2 - y^2)/(x^2 + y^2)$ of the two variables x and y will not be differentiable when $x = y = 0$, since it then becomes $0/0$, which is indeterminate. A function which varies discontinuously, such as the population of a country, is not strictly speaking differentiable, although in practice it may be sufficiently nearly represented by a differentiable function. All these exceptions are fairly evident and will normally be avoided with sufficient care. In *theory* there are more subtle snags, but in practice these rarely if ever arise.

One further comment may be made. The definition of differentiability tells us that the error in the formula $\delta w = w_x\delta x + w_y\,\delta y$ can be made as small as we like in comparison with $|\delta x| + |\delta y|$ by making δx and δy sufficiently small. It does not tell us how great an error we can actually expect in any given example, at least not directly: but we shall later find formulas which answer this question.

9.7 Constrained variation

When a gas is free to vary in volume (V), pressure (P) and temperature (T) we know that any two of these can be chosen as independent variables, since they are connected by the gas law $PV = nRT$. Any property w of the gas, such as its internal energy, its specific heat, its thermal conductivity, or its entropy, can be expressed as a function of two of the variables, say the pressure P and volume V. We can therefore calculate the partial derivatives $D_{P|V}w$ and $D_{V|P}w$ of w.

Now imagine the gas contained in a rubber balloon. This imposes a further constraint on the gas, since the greater the volume the greater the pressure exerted by the balloon. Thus there will now be a definite relationship between pressure and volume, and they can now no longer vary independently. (Actually the situation will be complicated by the fact that the properties of the balloon are affected by temperature, but to avoid complicating the argument we shall ignore this effect. We shall also suppose that the external atmospheric pressure is effectively constant, so that all changes in pressure are due to the stretching of the balloon.) Then corresponding to each volume of the balloon there will now be a single value of the pressure, a single value of the temperature, and a single value of the property w; all quantities are now functionally connected. We can no longer speak of a *partial* derivative $D_{P|V}w$, the rate of change of w when V is kept constant: for if V is kept constant, P would also be constant, and $D_{P|V}w$ is a contradiction in terms. We can now only calculate the ordinary or total derivative $D_P w$.

Another case in which such a constraint would occur would be that of a mass of gas (not necessarily within a balloon) whose volume and pressure were changing with time. We could imagine recording instruments plotting graphs of the volume V and pressure P against the time t. P and V must now be considered as functions of the single variable t; and any other property of the gas, such as the temperature T, will also be a function of the single variable t, since it will be determined when t is known. Now suppose that we wish to estimate from these graphs how rapidly the temperature T is changing (with respect to the time). One way to do this would of course be to calculate from the gas law the values of T for each value of P and V and plot a graph of T against the time t; or if P and V could be expressed as mathematical functions of t, we could do this by substituting these functions in the equation $T = PV/nR$, and thereby express T directly in terms of t.

However, we can often tackle the problem more efficiently otherwise. We know that in general, whether there is an extra constraint or not, the gas law $T = PV/nR$ must apply. If the gas is unconstrained we have therefore that small changes δP and δV in the pressure and volume must correspond to a small change δT in temperature given by

$$\delta T \simeq (D_{P|V}T)\,\delta P + (D_{V|P}T)\,\delta V$$

But since $T = PV/nR$, $D_{P|V}T = V/nR$ and $D_{V|P}T = P/nR$, so that

$$\delta T \simeq (V/nR)\,\delta P + (P/nR)\,\delta V.$$

But written in this form the relation is simply a deduction from the gas law which must hold good however the pressure, volume, and temperature vary. It must therefore be equally true for the constrained system as well as the unconstrained system. Now divide through by δt; we obtain

$$(\delta T/\delta t) \simeq (V/nR)\,(\delta P/\delta t) + (P/nR)\,(\delta V/\delta t).$$

Take the limit as δt tends to zero.

$$D_t T = (V/nR)\,D_t P + (P/nR)\,D_t V.$$

This therefore is the equation we require which enables us to calculate the rate of change of T when we know the rates of change of P and V. But we calculated the coefficients (V/nR) and (P/nR) as the derivatives $D_{P|V}T$ and $D_{V|P}T$ for an unconstrained mass of gas: so we can write this relation in the form

$$D_t T = (D_{P|V}T)\,(D_t P) + (D_{V|P}T)\,(D_t V)$$

or alternatively in suffix notation

$$T_t = T_{P|V}P_t + T_{V|P}V_t$$

with the interpretation that $T_{P|V}$ and $T_{V|P}$ are partial derivatives calculated *before* the constraint is applied, and T_t, P_t and V_t are the derivatives for the constrained system in which P and V have become functions of the single variable t.

This is a general principle. Suppose that w is a function of several variables x, y, z. . . . Then so long as they are all allowed to vary independently,

$$\delta w \simeq w_{x|y,z\ldots}\,\delta x + w_{y|x,z\ldots}\,\delta y + w_{z|x,y\ldots}\,\delta z + \ldots$$

Now this equation must hold good however δx, δy, and δz vary. It must still be true if δx, δy and δz are limited by some relation, provided that we calculate the values of $w_{x|y,z\ldots}$, $w_{y|x,z\ldots}$, $w_{z|x,y\ldots}$, . . . for the free and unlimited system. Thus if x, y, z . . . are all constrained to be functions of a single variable t, then

$$\delta w/\delta t \simeq w_{x|y,z}\ldots\,\delta x/\delta t + w_{y|x,z}\ldots\,\delta y/\delta t + w_{z|x,y}\ldots\,\delta z/\delta t + \ldots$$

and in the limit as $\delta t \to 0$

$$w_t = w_{x|y,z\ldots}\,x_t + w_{y|x,z\ldots}\,y_t + w_{z|x,y\ldots}\,z_t + \ldots \qquad (9.14)$$

If we can safely omit the suffixes indicating the quantities held constant this equation takes the simple form

$$w_t = w_x x_t + w_y y_t + w_z z_t + \ldots$$

This equation is now exactly of the same form as that for a change of variable given in Section 9.5, which is a useful aid to the memory. But the interpretation is somewhat different. In the change-of-variable equation all symbols represent partial derivatives, whereas here w_x, w_y, w_z . . . represent the partial derivatives calculated for the unconstrained system, and w_t, x_t, y_t, z_t . . . are total derivatives for the constrained system. In the dw/dt notation we would write

$$\frac{\partial w}{\partial t} = \frac{\partial w}{\partial x}\frac{\partial x}{\partial t} + \frac{\partial w}{\partial y}\frac{\partial y}{\partial t} + \frac{\partial w}{\partial z}\frac{\partial z}{\partial t} + \dots$$

for the equation for change of variable, and

$$\frac{dw}{dt} = \frac{\partial w}{\partial x}\frac{dx}{dt} + \frac{\partial w}{\partial y}\frac{dy}{dt} + \frac{\partial w}{\partial z}\frac{dz}{dt} + \dots$$

for the constraint equation.

These equations can of course be proved more rigorously by using the formal definition of differentiability given in the previous section. The method of proof will be essentially the same, but it will be expressed more precisely in terms of limits. We shall not concern ourselves with these complications here.

These equations also enable us to solve the balloon problem. Suppose that w is any property of the gas which can be expressed in general in terms of its volume V and pressure P so that we know the partial derivatives $w_{P|V}$ and $w_{V|P}$ (for the unconstrained system). We now suppose that because of the constraint imposed by the balloon the volume V becomes a known function of the pressure P; we wish to calculate the derivative w_P of w considered now as a function of P only. Our formula gives

$$w_P = w_{P|V} P_P + w_{V|P} V_P.$$

But $P_P = D_P P = 1$, and V_P is supposed known. Thus the answer is

$$w_P = w_{P|V} + w_{V|P} V_P.$$

Suppose that the quantity we are interested in is the temperature T It has already been shown that $T_{P|V} = V/nR$, $T_{V|P} = P/nR$, so that $T_P = V/nR + PV_P/nR = (V + PV_P)/nR$. Suppose further that the properties of the balloon are such that its volume $V = \alpha/(\beta - P)^3$, where α and β are constants. Then by direct differentiation

$$V_P = V_{(\beta-P)}(\beta - P)_P = 3\alpha/(\beta - P)^4,$$

and therefore

$$V + PV_P = \alpha/(\beta - P)^3 + 3\alpha P/(\beta - P)^4 = \alpha\,(\beta + 2P)/(\beta - P)^4,$$

and

$$T_P = \alpha\,(\beta + 2P)/nR\,(\beta - P)^4.$$

This shows how rapidly the temperature changes in relation to the pressure: a small change δP in P corresponds to a small change $\delta T = T_P \delta P$ in T.

PROBLEMS

(1) Consider the bacterium of example (2), Section 9.5, consisting of a cylinder of length L and radius R with hemispherical caps on the ends. Supposing that the length L increases proportionately to the time, $L = \alpha t$, whereas the radius increases only as the cube root of the time, $R = \beta t^{\frac{1}{3}}$, where α and β are constants. How rapidly do the area and volume increase?

(2) A body thrown through space has cartesian co-ordinates x and y; the distance r from the origin is given by $r = \sqrt{(x^2 + y^2)}$. Find $r_{x|y}$ and $r_{y|x}$. If x and y vary according to the laws $x = 2t$, $y = 3t - 5t^2$, where t is the time, find the rate of change of r.

(3) The intensity of illumination I of the floor of a room due to a lamp at height h is $B/(x^2 + y^2 + h^2)$, where B is the brightness of the lamp and x and y are cartesian co-ordinates of the point P at which the illumination is measured, the co-ordinate axes being two perpendicular lines on the floor intersecting at O directly under the lamp. If $h = 2$ metres, x and y also being expressed in metres, find the partial derivatives with respect to x and y.

A man is walking across this room looking for a lost coin. If his position at time t seconds is given by $x = \frac{1}{2}t - 10$, $y = \frac{1}{3}t - 5$, find the rate at which the illumination of the floor under his feet is changing.

A constraint may not be so complete as those we have considered above, in which every quantity is reduced to a function of a single variable t. If a system has originally three independent variables x, y, and z, it is conceivable that a relation imposed on them might reduce them to two independent variables only. For example, consider gas contained in a vessel with an inlet tube. Normally we can vary the pressure, the volume and the mass of gas in the vessel independently. If, however, the vessel is a balloon then there will be a relation between the pressure and the volume, and only two variables such as the pressure and mass will be independently variable. The rule for this case is very similar. If w is a function of three variables x, y, and z, and if x, y, and z are constrained to be functions of two variables t and u, then when w is considered as a function of t and u its partial derivatives are

$$w_{t|u} = w_{x|y,z}x_{t|u} + w_{y|x,z}y_{t|u} + w_{z|x,y}z_{t|u}$$
$$w_{u|t} = w_{x|y,z}x_{u|t} + w_{y|x,z}y_{u|t} + w_{z|x,y}z_{u|t}$$

or if we can safely drop the extra suffixes,

$$w_t = w_x x_t + w_y y_t + w_z z_t$$
$$w_u = w_x x_u + w_y y_u + w_z z_u$$

Here w_x, w_y, and w_z are the three partial derivatives calculated before the constraint is applied. These formulas are of course of exactly the same form as those we have already had for constrained systems or change of variable without constraint, and the argument follows the same course. Either (or both) of the new variables t and u may be the same as one of the old variables x, y or z; the formula will still hold.

FURTHER PROBLEM

(4) If the mass of gas in the balloon is variable, and the pressure and volume are connected by the relation $V = \alpha/(\beta - P)^3$, what are the partial derivatives $T_{P|n}$ and $T_{n|P}$?

9.8 General form of constraint

Sometimes it may happen that two independent variables x and y have a constraint imposed on them which cannot be easily expressed in the form "x = a function of t, y = a function of t." For example, we might find that $x^3 + xy + y^3 = 0$. If w is any function of x and y we might want to know the derivative $w_x = D_x w$ after the constraint is applied. Now we know that

$$w_x = w_{x|y} x_x + w_{y|x} y_x$$
$$= w_{x|y} + w_{y|x} y_x$$

where $w_{x|y}$ and $w_{y|x}$ are the partial derivatives in the freely varying case. We can therefore solve the problem provided that we can find $y_x = D_x y$ for the constrained system.

The technique to be applied to this type of problem is as follows. Let the constraint be expressed by a relation which we can write $f(x, y) = 0$.

Now consider for the moment the situation where x and y can vary independently. $f(x, y)$ will then be a given function of x and y, and a small change δx in x and δy in y will produce a change δf in the value of f given by

$$\delta f \simeq f_{x|y} \cdot \delta x + f_{y|x} \cdot \delta y$$

where $f_{x|y}$ and $f_{y|x}$ are the partial derivatives calculated *with* x *and* y *independently variable.* We now reimpose the constraint $f = 0$. Since f is now constant, $\delta f = 0$ and therefore $f_{x|y}\,\delta x + f_{y|x}\delta y \simeq 0$, or on division by δx

$$f_{x|y} + f_{y|x}(\delta y/\delta x) \simeq 0.$$

Take the limit as $\delta x \to 0$

$$f_{x|y} + f_{y|x}\, y_x = 0.$$

If $f_{y|x} \neq 0$ we can divide through by it, obtaining finally

$$y_x = -f_{x|y}/f_{y|x} \quad . \quad . \quad . \quad (9.15)$$

(This can alternatively be deduced from the formula $x_{y|z} = - z_{y|x}/z_{x|y}$ of Example 5, Section 9.5, where z is any function of x and y. Write y, x, f in place of x, y, z respectively; the formula then becomes $y_{x|f} = -f_{x|y}/f_{y|x}$. But $y_{x|f}$ means the derivative of y with respect to x when f is constant, i.e. the derivative y_x under constraint.)

EXAMPLES

(1) If x and y are connected by the relation $x^2 + y^2 = K$ find $y_x = D_x y$.

Write $f(x,y) = x^2 + y^2 - K$: then this relation is $f(x, y) = 0$. Now when x and y vary independently $f_{x|y} = 2x$ and $f_{y|x} = 2y$, and therefore in the constrained case $y_x = - 2x/2y = - x/y$.

(2) If x and y are connected by the relation $e^x + e^y = K$ find y_x.

Write $f(x, y) = e^x + e^y - K$. Then the constraint is $f(x, y) = 0$. Now $f_{x|y} = e^x$, $f_{y|x} = e^y$, and therefore $y_x = - e^x/e^y = - e^{x-y}$.

The same technique can be applied to more complicated cases with more variables and/or more constraints. A curious feature of this formula is that the partial derivatives $f_{x|y}$ and $f_{y|x}$ need have no real significance in themselves, but may be only stepping stones in the process of finding the derivative y_x. If x and y are in reality connected by the relation $f(x, y) = 0$ (as, for example, $x^2 + y^2 - K = 0$) then in reality they cannot vary independently. But in imagination we can lift this restriction for a time and find what will happen to f when they do so. In the end the restriction is re-imposed and y_x calculated for the real and limited system. There is no harm in arguing in this way provided that what is being done is clearly understood. But it seems necessary to make these remarks since the process is sometimes used without adequate explanation and can at first appear not a little baffling.

FURTHER EXAMPLES

(3) The growth of a colony of bacteria obeys the law $n = N e^{kt}$, where n is the number of bacteria per cubic centimetre at time t, N the initial number per cubic cm when $t = 0$, and k is a constant. This law holds provided that the concentration n is sufficiently small for there to be no appreciable competition between bacteria or exhaustion of the food supply.

Now suppose that we make an error δn in measuring n, δN in measuring N, and δt in measuring t. What will be the effect of these errors on the value of k calculated from the observed values of n, N, and t?

There are several ways of solving this problem. One would be to solve the given equation for k, obtaining $k = (\ln n - \ln N)/t$, and then to apply the formula for a small error.

We can also derive the formula directly from the relation $n - Ne^{kt} = 0$. Let us put $n - Ne^{kt} = f(n, N, k, t)$. Then if for the moment n, N, k and t are imagined as independently variable we shall have

$$\delta f = f_n \delta n + f_N \delta N + f_k \delta k + f_t \delta t$$

where the partial derivatives are

$$f_n = 1, \quad f_N = -e^{kt}, \quad f_k = -Nte^{kt}, \quad f_t = -Nke^{kt}.$$

Now when we apply the constraint $f = 0$ we shall ensure that $\delta f = 0$, that is

$$\delta n - e^{kt}\,\delta N - Nte^{kt}\,\delta k - Nke^{kt}\,\delta t = 0$$

or

$$Nte^{kt}\,\delta k = \delta n - e^{kt}\,\delta N - Nke^{kt}\,\delta t$$

or

$$\delta k = \frac{\delta n}{Nte^{kt}} - \frac{\delta N}{Nt} - \frac{k\delta t}{t}$$

which is the formula we require to calculate δk from δn, δN, and δt.

(4) Given that x, y, and z are three variables connected by the relation $x^2 + y^2 + z^2 = r^2 =$ constant, so that only two of the variables are independent, find $z_{x|y}$ and $z_{y|x}$.

Write $f(x, y, z) = x^2 + y^2 + z^2 - r^2$. Now imagine for the moment that x, y, and z are allowed to vary independently: then

$$\begin{aligned} \delta f &\simeq f_{x|y,z}\,\delta x + f_{y|x,z}\,\delta y + f_{z|y,x}\,\delta z \\ &\simeq 2x\,\delta x + 2y\,\delta y + 2z\,\delta z. \end{aligned}$$

Re-impose the restriction $f=0$, then $\delta f=0$, i.e. $x\delta x + y\delta y + z\delta z \simeq 0$. Solving this equation for δz we have

$$\delta z \simeq -x/z \,.\, \delta x - y/z \,.\, \delta y.$$

Now $z_{x|y}$ is by definition the limit of $\delta z/\delta x$ as $\delta x \to 0$ with y constant, i.e. $\delta y = 0$. But on putting $\delta y = 0$ the above equation gives $\delta z \simeq -x/z \,.\, \delta x$, i.e. $\delta z/\delta x \simeq -x/z$. By taking the limit, we obtain $z_{x|y} = -x/z$. Similarly $z_{y|x} = -y/z$.

Alternative method. Use the formula for a constrained system to calculate $f_{x|y}$.

$$f_{x|y} = f_{x|y,z}x_{x|y} + f_{y|x,z}y_{x|y} + f_{z|x,y}z_{x|y}$$

where $f_{x|y,z} = 2x$, $f_{y|x,z} = 2y$, $f_{z|x,y} = 2z$ are the derivatives calculated for the freely varying system. But with the constraint $f = 0$ we must have $f_{x|y} = 0$. Also $x_{x|y} = 1$, and $y_{x|y} = 0$, since it is the rate of change of y when y is kept constant. Thus the equation for $f_{x|y}$ reduces to

$$2x + 0 + 2z \,.\, z_{x|y} = 0,$$

that is, $z_{x|y} = -x/z$.

CHAPTER TEN

RELATIONS INVOLVING RATES OF CHANGE

10.1 The relation between position and velocity

We have seen above that we can always find the instantaneous velocity v of a moving body when we know an equation connecting its position (measured by a co-ordinate y) and the time t. We write $v = D_t y$, or y_t, or dy/dt. The same applies if y is any variable quantity which is a function of t; we have a set of rules by which we can find the instantaneous rate of change of y_t. This we have called the derivative of y with respect to the time t.

Can we reverse this process? If we are given an expression for a velocity, can we deduce one for the position? If we know how rapidly a quantity is changing, can we find out what the quantity itself is? This is an important question, since it is, for example, fairly easy to measure the position of a point with reasonable accuracy, but very difficult to measure its instantaneous velocity. If we are able to deduce a value of its velocity from theoretical considerations, it is much easier to check whether the theory agrees with observation if we can convert it into one giving the position of the point as a function of the time. Again, in a chemical reaction it may be easy to deduce how rapidly the concentration of one of the reacting substances is increasing: but we cannot very well check this experimentally until it has been expressed as a relation between the concentration itself and the time.

Suppose that the quantity y is given by the law $y = t^2$; then its instantaneous rate of change is $v = y_t = D_t y = 2t$. If therefore we are given instead the law for the rate of change $v = 2t$, we can deduce that one possible expression for y would be t^2. But is this the only possible value of y? The answer is no: the law $y = t^2 + 1$ would also give the rate of change $v = 2t$, since the derivative of the constant 1 is zero. The law $y = t^2 + 15$ would also give the same rate of change $v = y_t = 2t$: and in general the law $y = t^2 + C$ where C is any constant will give $v = 2t$. Thus t^2, $t^2 + 1$, $t^2 + 15$ and in general $t^2 + C$ are all possible values of y which satisfy the law that the rate of change of y is $2t$. This is perhaps not surprising, for if we are given the rate of change of y at each time t we can only expect to be able to infer the amounts by which y changes from one time to another; we cannot expect to infer the absolute value of y. Thus the equation $y_t = 2t$ has the solution $y = t^2 + C$, where C is a constant. We can find the value of C when we are given as additional information the co-ordinate y of the point at any one instant. For example, if we know that $y = 0$ when $t = 0$, on substitution in the equation $y = t^2 + C$ we see that $0 = 0 + C$, i.e. $C = 0$. In fact in this particular case C is simply the value of y when $t = 0$.

We can perhaps illustrate this from one of the less well-known adventures of the famous detective Herlock Soames. "Follow that car," said Herlock, jumping into a taxi, "it is driven by Professor Moronami, the notorious diamond smuggler. But be careful! He sometimes carries clubs too. I advise you to go just as fast as he does: no quicker, no slower." So off started Moronami, and off started Soames, 100 yards behind. And when Moronami accelerated up to 50 miles per hour so did the taxi, and when he slowed down to 20 miles per hour, so did the taxi. And their speeds were always exactly the same. And what happened? Well, as the reader will have guessed, they always remained exactly 100 yards apart. Now Moronami was very

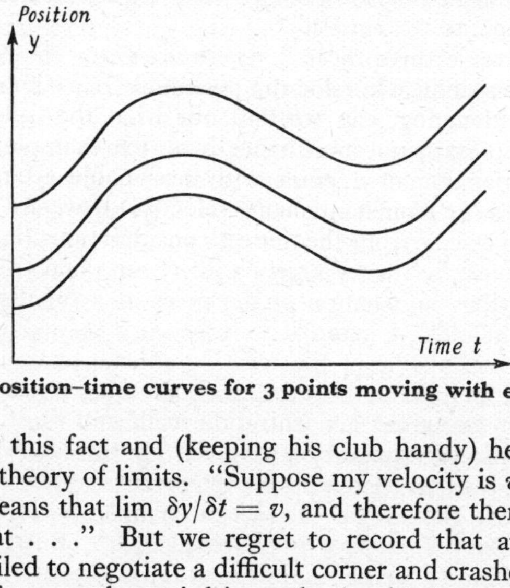

Fig. 10.1—Position–time curves for 3 points moving with equal velocity

fascinated by this fact and (keeping his club handy) he tried to work it out by the theory of limits. "Suppose my velocity is v," thought he, "then that means that $\lim \delta y/\delta t = v$, and therefore there is a function $f(\epsilon)$ such that . . ." But we regret to record that at this moment Moronami failed to negotiate a difficult corner and crashed into a lamp-post. And of course the taxi driver, obeying his instructions, followed suit and ran straight into another lamp-post, exactly 100 yards behind. And so they all were taken to hospital, and when they were recovering they passed the time by playing bridge with the Matron. And Soames still wondered how it was that Moronami and his partner always made five diamonds or five clubs, if not a slam. However, that is not our business: the moral as far as it concerns us is that two points whose velocities are always equal will move so as to remain a constant distance apart. Unfortunately, as no doubt Moronami discovered, this is by no means easy to prove formally. The reason is that if we deal with instantaneous velocities we have to cope with limits, and the definition of a limit is quite a complicated matter when written out in full. So we shall dodge the difficulty and take the result as proved. The importance of this result is that if we are told, for example, that $v = 2t$, and find that $y = t^2$ is a possible solution, then we know for certain that any solution differs from this by an arbitrary constant, and is therefore of the form $t^2 + C$. This includes every possible value of y, and nothing is overlooked. The same applies to any other formula for v. This is seen from Fig. 10.1, which shows several graphs for y with equal values of the velocity v. Any two of these graphs are separated by a constant distance in the vertical direction.

10.2 Integration

The process of finding a quantity y when its rate of change $v = y_t$ is known is called "integration", and y is known as the "indefinite integral of v". It is the opposite of differentiation, by which we find v from y: and y could also be called the "anti-derivative" of v. The reason for the name "integration" is this. In differentiating a variable y we take a small change δy in y corresponding to a small change δt in the time t, and find the velocity v by the formula $v = \delta y/\delta t$. More exactly the instantaneous velocity is the limit of this when $\delta t \to 0$; but this doesn't matter very much, since when δt is small enough $\delta y/\delta t$ will be inappreciably different from v. Conversely if v is known then we can find how much y changes during any given interval of time in the following way. We divide this interval into very small intervals δt; in each of these we find the change δy in y from the formula $\delta y \simeq v\,\delta t$. An addition of the small changes gives the whole or "integral" change in the complete interval (Latin *integer* = whole). This method is discussed further in Sections 11.3 and 11.4.

A much easier method of integrating a velocity or rate of change v is to look for an expression whose derivative is equal to v. This is a matter of inspired guessing: but there are certain rules of integration which are very helpful. Thus if $v = 3t^2$, since we know that $D_t t^3 = 3t^2$, we deduce that $y = t^3 + C$. If $v = \cos t$, since $D_t \sin t = \cos t$, we know that $y = \sin t + C$. If $v = 1/t$, since $D_t \ln t = 1/t$, $y = \ln t + C$. The constant C must never be forgotten; it is because of its presence that we call y the *indefinite* integral.

Now we know that we can deduce the derivative of Kt^2 from that of t^2 by the rule that when a function is multiplied by K so also is its derivative. Thus since $D_t t^2 = 2t$ we see that $D_t Kt^2 = 2Kt$. Looked at the other way round this means that since t^2 is an integral of $2t$, $2t^2$ must be an integral of $4t$, and Kt^2 must be an integral of $2Kt$. To get the general integral we must add a constant C, giving $Kt^2 + C$ as the general form of the indefinite integral of $2Kt$. Similarly since $D_t \sin t = \cos t$ we see that $D_t (K \sin t) = K \cos t$. Interpreted in reverse this means that $\sin t$ is an integral of $\cos t$, and $K \sin t$ is an integral of $K \cos t$. The general integral of $K \cos t$ is therefore

$K \sin t + C$; e.g. the integral of $\frac{1}{3} \cos t$ is $\frac{1}{3} \sin t + C$. In general we can say that if y is an integral of v, then Ky is an integral of Kv, and the general form of the integral is $Ky + C$. This enables us to integrate many expressions, for example powers of t. Since $D_t t^3 = 3t^2$, $D_t t^4 = 4t^3$, and in general $D_t t^{n+1} = (n + 1)t^n$, we see that $D_t \frac{1}{3}t^3 = t^2$, $D_t \frac{1}{4}t^4 = t^3$, and in general $D_t(n + 1)^{-1}t^{n+1} = t^n$. The integral of t^2 is therefore $\frac{1}{3}t^3 + C$, the integral of t^3 is $\frac{1}{4}t^4 + C$, and the integral of t^n is $t^{n+1}/(n + 1) + C$. Notice that fortunately we do not need to worry about the constant of integration C before the final stage of the calculation. If we wish to argue very carefully, we might say that the integral of $3t^2$ is $t^3 + C_1$, where C_1 is some constant. Hence one-third of this, i.e. $\frac{1}{3}t^3 + \frac{1}{3}C_1$, is an integral of t^2. But the general integral of t^2 is got by adding an arbitrary constant C_2 to any particular integral, and is accordingly $\frac{1}{3}t^3 + \frac{1}{3}C_1 + C_2$. But $(\frac{1}{3}C_1 + C_2)$ is itself a constant, and can just as well be written C: so the general integral of t^2 is $\frac{1}{3}t^3 + C$. This is what we get if we insert the constant C at the end of the calculation and ignore it at the intermediate steps.

The formula for the differentiation of a sum also gives us a useful rule for integration. We know that if v is the derivative of y, and w is the derivative of z, then $v + w$ is the derivative of $y + z$. In reverse this means that if y is any integral of v and z is any integral of w, then $y + z$ is an integral of $v + w$, and the general integral is $y + z + C$. (Again we can conveniently ignore the constant C until we come to the last step.) *The integral of a sum is the sum of the integrals plus an arbitrary constant* C. Thus since $\frac{1}{2}t^2$ is an integral of t, and $\frac{1}{3}t^3$ is an integral of t^2, the integral of $t + t^2$ must be $\frac{1}{2}t^2 + \frac{1}{3}t^3 + C$. Some expressions can be integrated by the use of both rules: the integral of $4t + 6t^2$ will be $4 \times \frac{1}{2}t^2 + 6 \times \frac{1}{3}t^3 + C = 2t^2 + 2t^3 + C$. The integral of $2 + 3 \cos t$ will be $2t + 3 \sin t + C$.

It is clear that we need a distinctive notation for an integral. One fairly obvious method of writing "the indefinite integral of v with respect to t" would be $y = D_t^{-1}v$; for just as $\sin^{-1}x$ means "an angle whose sine is x" so $D_t^{-1}v$ can be read as meaning "a function whose derivative is v". This notation is indeed sometimes used, though as a rule the suffix t is dropped and the integral written as $D^{-1}v$. But the usual notation for the integral of v is the original one, due to Leibnitz,

$$y = \int v\, dt$$

This is read as "integral $v\ d\ t$" and means exactly the same as $D_t^{-1}v$. Thus

$$D_t^{-1}t = \int t dt = \tfrac{1}{2}t^2 + C$$

$$D_t^{-1}t^2 = \int t^2 dt = \tfrac{1}{3}t^3 + C$$

$$D_t^{-1} \cos t = \int \cos t \,.\, dt = \sin t + C$$

The rules we have given for multiplication by a constant K and for addition can be written

$$\int K v\, dt = K \int v\, dt + C$$

or

$$D_t^{-1}(Kv) = KD_t^{-1}v + C$$

and

$$\int (v + w)dt = \int vdt + \int wdt + C$$

or

$$D_t^{-1}(v + w) = D_t^{-1}v + D_t^{-1}w + C.$$

The first rule has the simple interpretation that if P and Q are two moving points, and if the velocity Kv of Q is always exactly K times the velocity v of P, then the distance covered by Q is always exactly K times the distance covered by P. We leave the reader to think of an interpretation of the addition rule.

PROBLEMS

(1) Find $\int (3 + 2t)dt$, $\int (t^2 + \sin t)\, dt$, $\int (2 + t^{-1} + \cos t)\, dt$.

(2) The rate of growth v mm/day of a fungus is found to obey the law $v = 5 + \frac{1}{2}t$, where t is the age in days. Find the length of the fungus at age t.

(3) The rate of growth v of a colony of bacteria is found to obey the formula $v = \frac{1}{3}e^t$ (approximately: this is strictly speaking an idealization, since the growth is discontinuous). Find a formula for the total number y at time t.

10.3 Standard integrals

In order to be able to integrate rapidly and efficiently it is useful to have a list of the integrals of the functions which occur most frequently, such as t^n, $\ln t$, $\sin t$, e^t, etc. We give a list of standard integrals in Tables 10.1 and 10.2. Some of the more recondite ones, such as the integrals of sec t and coth t, will mostly be useful for reference when needed. But the reader is strongly advised to learn the simpler ones off by heart.

Before we present the tables there are a few comments which can usefully be made. The correctness of all the integrals can be checked by differentiation. The integral of t^n is in general $t^{n+1}/(n + 1)$, and this formula holds whether n is integral or fractional. But when $n = -1$ the formula fails, since $n + 1$ becomes zero and $1/(n + 1)$ infinite.

We know, however, that if t is positive $D_t \ln t = t^{-1}$, and therefore $\int t^{-1}dt = \ln t + C$. If t is negative this fails since a negative number has no logarithm. But $D_t \ln(-t) = t^{-1}$, and therefore if t is negative $\int t^{-1}dt = \ln(-t) + C$.

In practice too when we have to integrate a trigonometric function it will more often be something like sin $2t$ or cos $3t$ than the simple functions sin t and cos t. Accordingly the general integrals of sin at and cos at for any constant a are included in Table 10.2, although these can be quite easily deduced from the integrals for sin t and cos t by the method of change of variable explained in the next section. Many of the other integrals can also be obtained either by this method or by "integration by parts" (Section 10.5).

Table 10.1—Standard indefinite integrals

(I) FUNDAMENTAL INTEGRALS

Function v	Indefinite integral $y = D_t^{-1}v = \int v\,dt$	
0	C	
a	$at + C$	
$t^n \ (n \neq -1)$	$t^{n+1}/(n+1) + C$	
$1/t \ (t > 0)$	$\ln t + C$	
$1/t \ (t < 0)$	$\ln(-t) + C$	
$1/(1 + t^2)$	$\tan^{-1} t + C$	
$1/\sqrt{(1 + t^2)}$	$\sinh^{-1} t + C$	
$1/\sqrt{(t^2 - 1)}$	$\cosh^{-1} t + C$	$(t \geqslant 1)$
	or $\cosh^{-1}(-t) + C$	$(t \leqslant -1)$
$1/\sqrt{(1 - t^2)}$	$\sin^{-1} t + C$	
e^t	$e^t + C$	
$\sin t$	$-\cos t + C$	
$\cos t$	$\sin t + C$	
$\tan t$	$-\ln\cos t + C$	$(\cos t > 0)$
	or $-\ln(-\cos t) + C$	$(\cos t < 0)$
$\cot t$	$\ln\sin t + C$	$(\sin t > 0)$
	or $\ln(-\sin t) + C$	$(\sin t < 0)$

Table 10.2—Standard indefinite integrals

(II) OTHER USEFUL INTEGRALS

$$\int e^{at}\,dt = a^{-1}e^{at} + C$$

$$\int \ln t\,.\,dt = t\ln t - t + C$$

$$\int \log t\,.\,dt = t\log t - Mt + C$$

$$\int \cos at\,.\,dt = a^{-1}\sin at + C$$

$$\int \cosh at\,.\,dt = a^{-1}\sinh at + C$$

$$\int \sin at\,.\,dt = -a^{-1}\cos at + C$$

$$\int \sinh at\,.\,dt = a^{-1}\cosh at + C$$

$$\int \tan at\,.\,dt = -a^{-1}\ln\cos at + C \qquad (\cos at > 0)$$
$$\text{or } -a^{-1}\ln(-\cos at) + C \qquad (\cos at < 0)$$

$$\int \tanh at\,.\,dt = a^{-1}\ln\cosh at + C$$

$$\int \cot at\,.\,dt = a^{-1}\ln\sin at + C \qquad (\sin at > 0)$$
$$\text{or } a^{-1}\ln(-\sin at) + C \qquad (\sin at < 0)$$

$$\int \coth at\,.\,dt = a^{-1}\ln\sinh at + C \qquad (\sinh at > 0)$$
$$\text{or } a^{-1}\ln(-\sinh at) + C \qquad (\sinh at < 0)$$

$$\int (\sec at)^2\,dt = a^{-1}\tan at + C$$

$$\int (\text{sech}\, at)^2\,dt = a^{-1}\tanh at + C$$

$$\int (\text{cosec}\, at)^2\,dt = -a^{-1}\cot at + C$$

$$\int (\text{cosech}\, at)^2\,dt = -a^{-1}\coth at + C$$

$$\int \sec at\,.\,dt = a^{-1}\tanh^{-1}\sin at + C$$
$$\left.\begin{array}{l}\text{or } a^{-1}\ln(\sec at + \tan at) + C\\ = a^{-1}\ln\tan(\tfrac{1}{2}at + \tfrac{1}{4}\pi) + C\end{array}\right\}(\cos at > 0)$$
$$\left.\begin{array}{l}\text{or } a^{-1}\ln(-\sec at - \tan at) + C\\ = a^{-1}\ln\tan(-\tfrac{1}{2}at - \tfrac{1}{4}\pi) + C\end{array}\right\}(\cos at < 0)$$

$$\int \text{sech}\, at\,.\,dt = a^{-1}\tan^{-1}\sinh at + C$$

$$\int \text{cosec}\, at\,.\,dt = -a^{-1}\tanh^{-1}\cos at + C$$
$$\left.\begin{array}{l}\text{or } a^{-1}\ln(\text{cosec}\, at - \cot at) + C\\ = a^{-1}\ln\tan\tfrac{1}{2}at + C\end{array}\right\}(\sin at > 0)$$
$$\left.\begin{array}{l}\text{or } a^{-1}\ln(\cot at - \text{cosec}\, at) + C\\ = a^{-1}\ln\tan(-\tfrac{1}{2}at) + C\end{array}\right\}(\sin at < 0)$$

$$\int \text{cosech}\, at\,.\,dt = a^{-1}\ln\tanh\tfrac{1}{2}at + C \qquad (at > 0)$$
$$\text{or } a^{-1}\ln\tanh(-\tfrac{1}{2}at) + C \qquad (at < 0)$$

Table 10.2—Standard indefinite integrals *(contd.)*

$$\int \frac{dt}{a + b\cos t} = \frac{2}{\sqrt{a^2 - b^2}} \tan^{-1}\left(\sqrt{\frac{a-b}{a+b}} \tan\frac{t}{2}\right) + C \quad (a^2 > b^2)$$

$$\text{or} \quad \frac{2}{\sqrt{b^2 - a^2}} \tanh^{-1}\left(\sqrt{\frac{b-a}{b+a}} \tan\frac{t}{2}\right) + C$$

$$(b^2 > a^2 \quad \text{and} \quad \cos t > -a/b)$$

$$\text{or} \quad \frac{2}{\sqrt{b^2 - a^2}} \coth^{-1}\left(\sqrt{\frac{b-a}{b+a}} \tan\frac{t}{2}\right) + C$$

$$(b^2 > a^2 \quad \text{and} \quad \cos t < -a/b)$$

$$\int \sin^{-1} at . dt = t \sin^{-1} at + a^{-1} \cos(\sin^{-1} at) + C$$

$$\int \sinh^{-1} at . dt = t \sinh^{-1} at - a^{-1}\sqrt{(1 + a^2t^2)} + C$$

$$\int \cos^{-1} at . dt = t \cos^{-1} at - a^{-1} \sin(\cos^{-1} at) + C$$

$$\int \cosh^{-1} at . dt = t \cosh^{-1} at - a^{-1} \sinh(\cosh^{-1} at) + C$$

$$\int \tan^{-1} at . dt = t \tan^{-1} at - \tfrac{1}{2}a^{-1} \ln(1 + a^2t^2) + C$$

$$\int \tanh^{-1} at . dt = t \tanh^{-1} at + \tfrac{1}{2}a^{-1} \ln(1 - a^2t^2) + C$$

$$\int a^{bt}\, dt = a^{bt}/b \ln a + C$$

$$\int \frac{dt}{t^2 + a^2} = a^{-1} \tan^{-1}(t/a) + C$$

$$\int \frac{dt}{t^2 - a^2} = -a^{-1} \coth^{-1}(t/a) + C \qquad (t^2 > a^2)$$

$$\int \frac{dt}{a^2 - t^2} = a^{-1} \tanh^{-1}(t/a) + C \qquad (t^2 < a^2)$$

$$\int \frac{dt}{\sqrt{t^2 + a^2}} = \sinh^{-1}(t/a) + C \qquad (a > 0)$$

$$\int \frac{dt}{\sqrt{t^2 - a^2}} = \cosh^{-1}(t/a) + C \qquad (a > 0)$$

$$\int \frac{dt}{\sqrt{a^2 - t^2}} = \sin^{-1}(t/a) + C \qquad (a > 0)$$

$$\int \sqrt{t^2 + a^2}\, dt = \tfrac{1}{2}a^2 \sinh^{-1}(t/a) + \tfrac{1}{2}t\sqrt{t^2 + a^2} \qquad (a > 0)$$

$$\int \sqrt{t^2 - a^2}\, dt = -\tfrac{1}{2}a^2 \cosh^{-1}(t/a) + \tfrac{1}{2}t\sqrt{t^2 - a^2} \qquad (a > 0)$$

$$\int \sqrt{a^2 - t^2}\, dt = \tfrac{1}{2}a^2 \sin^{-1}(t/a) + \tfrac{1}{2}t\sqrt{a^2 - t^2} \qquad (a > 0)$$

These standard integrals can be used in conjunction with the rules for addition and multiplication by a constant to integrate a great variety of expressions.

EXAMPLES

(1) The velocity of a moving body is given by $v = 2 \sin 2t + 3 \cos 2t$. Find the relation between its position y and the time t.

$$y = \int v\,dt = 2 \int \sin 2t\,.\,dt + 3 \int \cos 2t\,.\,dt$$
$$= -2\left(\tfrac{1}{2}\cos 2t\right) + 3\left(\tfrac{1}{2}\sin 2t\right) + C$$
$$= -\cos 2t + \tfrac{3}{2}\sin 2t + C$$

(2) The rate of growth of a chicken embryo over a certain range is given by the formula $w_t = Kt^{2\cdot 6}$. Find the weight w as a function of the time t.

$$w = \int w_t\,dt = K \int t^{2\cdot 6}\,dt + C$$
$$= (K/3{\cdot}6)\,t^{3\cdot 6} + C$$

(H. A. Murray, *Jour. Gen. Physiol.*, **9** (1925), 48, gives the formula $w = {\cdot}668t^{3\cdot 6}$, corresponding to $K = 2{\cdot}405$, $C = 0$.)

(3) Integrate $v = (1 + t)^2/t$ with respect to t.

By expansion $v = t^{-1} + 2 + t$, whence

$$\int v\,dt = \ln t + 2t + \tfrac{1}{2}t^2 + C.$$

PROBLEMS

(1) Verify the integrals of Table 10.1 by showing that v is the derivative of y in each case.

Integrate the following expressions with respect to t:

(2) $1 + t + e^t$.
(3) $2t + e^{2t} + e^{-2t}$.
(4) $\sin t + \sin 2t + \sin 3t$.
(5) $(1 + t)^3/t^2$.
(6) $(1 - t^2)/(1 + t^2)$.
(7) $9/\sqrt{(t^2 + 4)} - 4/\sqrt{(t^2 - 9)}$.
(8) $\sin t + \cos t + \tan t$.
(9) $\sinh 3t + \cosh 3t$.
(10) $\sinh 3t - \cosh 3t$.
(11) $1 + t + t^2 + t^3$.
(12) $(1 + t^5)/(1 + t)$.

10.4 Change of variable

In Chapter 8 we found five rules which help us in differentiation. These rules were those for differentiating a sum, $y + z$, a product by a constant, Ky, a general product yz, a quotient y/z, and a function of a function. Any expression can be differentiated by splitting it up into simpler expressions connected by additions, multiplications, etc., and by using the appropriate rules. It is reasonable to suppose that a similar process can be applied to integration, and indeed we have already shown how to use the rules for a sum and for a product by a constant. The other rules can also be applied; in this section we shall study the function-of-a-function formula and in the next section the product formula. (The quotient y/z can be looked upon as the product of y and $1/z$, and does not give an essentially new result.) Unfortunately we shall see that although these formulas can be extremely helpful they no longer give an infallible and universally applicable method of obtaining the answer we require.

Suppose then that we are given the rate of change $v = y_t$ of a variable y with respect to a variable t, and that we wish to find y itself. Then $y = \int y_t \, dt = \int v \, dt$; that is the definition of an integral. Now in this definition there is no need for t to represent the time, although it will often do so. We can take any other suitable variable x, and if we know the derivative y_x we can find y by integration; $y = \int y_x dx$. This second integral might be easier to work out than the first one: the problem remains to find y_x. But this is given by the usual function-of-a-function rule (8.19); $y_x = y_t t_x = v t_x$, so that

$$y = \int y_x \, dx = \int v t_x \, dx \quad . \quad . \quad . \quad (10.1)$$

If instead of the notation t_x we write the derivative in the classical form dt/dx then equation (10.1) can be written

$$y = \int v \, dt = \int v \frac{dt}{dx} \, dx \quad . \quad . \quad . \quad (10.2)$$

This is the formula for the change of variable from t to x in an integral: writing it in this way has the advantage of suggesting the correct answer, since if we could treat the expression dx as if it was a number, and cancel it by the ordinary rule for fractions, we would have $\frac{dt}{dx} \, dx = dt$. Strictly speaking that is not logical, since "dx" has no independent existence and only forms part of expressions like $\frac{dt}{dx}$ and $\int u \, dx$; but still it is a useful aid to memory.

Naturally the constant C of integration must not be forgotten; but it can be left to the last step, as with the rules for addition and multiplication by a constant.

As a practical example consider $y = \int e^{t^2} t \, dt$: $v = e^{t^2} t$. Now we already know how to integrate an expression like e^t, but not one like

e^{t^2}. It is therefore natural to change the variable from t to $x = t^2$, so that e^{t^2} becomes e^x, and so try to simplify the integral. We have now to express it in terms of x. An inversion of the equation $x = t^2$ gives $t = \pm\sqrt{x}$, and $e^{t^2} = e^x$, so that v the integrand (quantity to be integrated) becomes $\pm e^x\sqrt{x}$. Also $\frac{dt}{dx} = \pm 1/2\sqrt{x}$, with the $+$ sign when $t = +\sqrt{x}$, and the $-$ when $t = -\sqrt{x}$. By our rule

$$\begin{aligned}\int v\,dt &= \int v\frac{dt}{dx}\,dx\\ &= \int(\pm e^x\sqrt{x})(\pm 1/2\sqrt{x})\,dx\\ &= \int \tfrac{1}{2}e^x\,dx\\ &= \tfrac{1}{2}e^x + C\\ &= \tfrac{1}{2}e^{t^2} + C\end{aligned}$$

since $x = t^2$. Thus the integration is completed.

Again suppose that a quantity y is known to have the rate of change $v = 1/(t + 1)$. Now our table of standard integrals allows us to integrate $1/t$, but not $1/(t + 1)$ with respect to t. Thus in order to find y it is natural to write $x = t + 1$, so that $v = 1/x$. Also since $t = x - 1$, we see that $dt/dx = t_x = 1$. Therefore by (10.2),

$$\begin{aligned}y &= \int v\,dt\\ &= \int v\frac{dt}{dx}\,dx\\ &= \int x^{-1}\,dx\\ &= \ln(\pm x) + C\\ &= \ln(\pm t \pm 1) + C\end{aligned}$$

where we take the $+$ signs if $t + 1 > 0$, and the $-$ signs if $t + 1 < 0$. In the same way $\int (t + a)^{-1}\,dt = \ln(\pm t \pm a) + C$.

Sometimes it may be necessary to make two or more substitutions to reduce the integral to a standard form. Consider $\int e^t(1 + e^t)^{-2}\,dt = \int v\,dt$ (say). The first substitution it is natural to make is to put $e^t = x$, or $t = \ln x$, so that $dt/dx = 1/x$. Now $v = e^t(1 + e^t)^{-2} = x(1 + x)^{-2}$ and so

$$\begin{aligned}y = \int v dt &= \int v\frac{dt}{dx}\,dx\\ &= \int x(1 + x)^{-2}x^{-1}\,dx\\ &= \int (1 + x)^{-2}\,dx\end{aligned}$$

It is now natural to try $1 + x = z$, giving $x = z - 1$, $dx/dz = 1$. Then

$$
\begin{aligned}
y &= \int (1 + x)^{-2}\,(dx/dz)\,dz \\
&= \int z^{-2} dz \\
&= -\,z^{-1} + C
\end{aligned}
$$

This completes the integration process, but the answer has to be translated back into a function of t. Since $z = (1 + x)$ and $x = e^t$ we have

$$
\begin{aligned}
y &= -\,z^{-1} + C \\
&= -\,(1 + x)^{-1} + C \\
&= -\,(1 + e^t)^{-1} + C
\end{aligned}
$$

It is clear that this method of change of variable is a very powerful one: many of the integrals of Table 10.2 are obtained in that way. Thus to get $\int \cos at\,.\,dt$ put $at = x$, $t = x/a$, $dt/dx = 1/a$. We have then

$$
\begin{aligned}
\int \cos at\,.\,dt &= \int \cos at\,.\,(dt/dx)\,.\,dx \\
&= \int \cos x\,.\,a^{-1}\,.\,dx \\
&= a^{-1} \sin x + C \\
&= a^{-1} \sin at + C.
\end{aligned}
$$

There is a particular form of integral, which is of frequent occurrence, in which some of the work can be short-circuited. This is when the integrand v is an obvious product of one variable (say u) by the derivative of another (say dx/dt),

$$v = u\,.\,dx/dt$$

in which case it is usually helpful to change the variable from t to x, for

$$
\begin{aligned}
\int v\,dt &= \int u\,.\,(dx/dt)\,dt \\
&= \int u\,dx \text{ [by (10.2)]} \qquad (10.3)
\end{aligned}
$$

Thus consider $y = \int \tan t\,.\,dt$. We know that $\tan t = \sin t/\cos t$, and $\sin t$ is the derivative of $x = -\cos t$. Thus

$$
\begin{aligned}
y &= - \quad (\cos t)^{-1} (dx/dt)\,.\,dt \\
&= -\int (\cos t)^{-1} dx \\
&= -\int x^{-1} dx \\
&= -\ln(\pm x) + C \\
&= -\ln(\pm \cos t) + C.
\end{aligned}
$$

Consider also the case $v = e^t \sin e^t$. Here we know that e^t is the derivative of $x = e^t$, so that $v = \sin e^t \,.\, dx/dt$, and

$$\begin{aligned}\int v\,dt &= \int \sin e^t\,(dx/dt)\,dt \\ &= \int \sin e^t\,dx \\ &= \int \sin x\,dx \\ &= -\cos x + C \\ &= -\cos e^t + C.\end{aligned}$$

In the same way if we want to integrate $v = e^{e^t}e^t$ it is natural to notice that e^t is the derivative of $x = e^t$, so that

$$\begin{aligned}\int v\,dt &= \int e^{e^t}\,(dx/dt)\,dt \\ &= \int e^x\,dx = e^x + C = e^{e^t} + C.\end{aligned}$$

The reader is recommended to practise the method of change of variable, as it is a most useful method of integration.

PROBLEMS

Integrate the following expressions with respect to t:

(1) $e^{\sqrt{t}}/\sqrt{t}$
(2) $\cos(t + 3)$
(3) $t/\sqrt{(1 - t^2)}$
(4) $t \sin(2t^2 + 3)$
(5) $\tan t \,.\, \ln \cos t$
(6) $\sin t \,.\, (\cos t)^3$
(7) $e^t \,.\, \ln(1 + e^t)$
(8) $\sqrt{(2t + 3)}$
(9) $\cos t/(1 + \sin t)^3$
(10) $\log(4t + 2)$
(11) $\sec t$, using $x = \sin t$
(12) $t\sqrt{(1 + t^2)}$
(13) $e^t/(1 + e^t)$
(14) $t^{-1}\sqrt{(1 + \ln t)}$
(15) $\sin t \,.\, \cos t$
(16) $t^2(1 + t^3)^{-1}\sqrt{\ln(1 + t^3)}$

10.5 Integration of a product (integration by parts)

We have seen above that if the quantity v to be integrated can be expressed as a product $u \,.\, dx/dt$ then

$$\int v\,dt = \int u\,dx$$

and this second form of the integral may be easier to deal with than the first.

Also in this case $v = u \,.\, dx/dt$ there is an alternative formula which may lead to a simpler integral. This depends on the formula for differentiation of a product

$$\frac{d(ux)}{dt} = u\frac{dx}{dt} + \frac{du}{dt}x$$

from which we have

$$v = u\frac{dx}{dt} = \frac{d(ux)}{dt} - \frac{du}{dt}x$$

Now the integral we want is accordingly

$$\begin{aligned} y &= \int v \, dt \\ &= \int \frac{d(ux)}{dt} \, dt - \int \frac{du}{dt} x \,.\, dt \end{aligned}$$

But integration is the reverse of differentiation, so that

$$\int \frac{d(ux)}{dt} \, dt = ux + C, \text{ and}$$

$$y = ux + C - \int \frac{du}{dt} x \,.\, dt$$

But the constant C is not really needed in this formula, since the integral $\int \frac{du}{dt} x \,.\, dt$ will have a constant of integration to be added on, and the sum of two arbitrary constants is equivalent to one. Thus finally we can state

$$y = \int u \frac{dx}{dt} \, dt = ux - \int \frac{du}{dt} x \, dt \qquad . \qquad . \quad (10.4)$$

As an example, consider the integration of $t \cos t$. We know that $\cos t$ is the derivative of $\sin t$, so v can be written as $u \,.\, dx/dt$, where $u = t$ and $x = \sin t$. Therefore by (10.4),

$$\begin{aligned} \int v \, dt &= \int t \cos t \,.\, dt \\ &= t \sin t - \int \frac{dt}{dt} \sin t \, dt \\ &= t \sin t - \int \sin t \,.\, dt \\ &= t \sin t + \cos t + C. \end{aligned}$$

Similarly when integrating te^t we observe that e^t is the derivative of $x = e^t$, so that

$$\begin{aligned}\int te^t\,dt &= tx - \int \frac{dt}{dt}\,x\,dt \\ &= te^t - \int e^t\,dt \\ &= te^t - e^t + C.\end{aligned}$$

For the ready use of formula (10.4) it is helpful to cast it into slightly different form. Let us put $dx/dt = w$, so that $v = uw$. Then x is the integral of w, and the equation can be rewritten as

$$y = \int uw\,dt = u\int w\,dt - \int \frac{du}{dt}\left(\int w\,dt\right)dt \quad . \quad (10.5)$$

or in words *the integral of a product* (uw) *equals the first* (u) *times the integral of the second* (w) *minus the integral of* [*the derivative of the first times the integral of the second*]. Here we can take for w any expression or function whose integral is known, and then $u = v/w$, to make $v = uw$.

EXAMPLES

(1) Integrate $v = (t + 1)\sin t$. Put $u = t + 1$, $w = \sin t$, so that $x = \int w\,dt = -\cos t$. (The constant C to be added is left until the final step.)

$$\begin{aligned}\int v\,dt &= ux - \int \frac{du}{dt}\,x\,dt \\ &= -(t + 1)\cos t + \int \cos t\,dt \\ &= -(t + 1)\cos t + \sin t + C\end{aligned}$$

(2) Sometimes it may be necessary to repeat the operation. Consider the case $v = t^2e^t$; put $u = t^2$, $w = e^t$, then $x = \int w\,dt = e^t$. Thus

$$\begin{aligned}\int v\,dt &= ux - \int \frac{du}{dt}\,x\,dt \\ &= t^2e^t - 2\int te^t\,dt\end{aligned}$$

But we have already shown by "integration by parts" that $\int te^t\,dt = te^t - e^t + \text{const.}$, so that

$$\int t^2e^t\,dt = t^2e^t - 2te^t + 2e^t + C$$

(3) There is no reason why we should not write $v = v \cdot 1$, so that $u = v$, $w = 1$, and $x = \int w\,dt = t$. In this case formula (10.4) gives

$$y = \int v\,dt = tv - \int \frac{dv}{dt}\,t\,dt \quad . \quad . \quad (10.6)$$

For example let $v = \ln t$, so that $dv/dt = 1/t$, then

$$\int \ln t \, dt = t \ln t - \int 1 \, dt$$
$$= t \ln t - t + C$$

Alternatively put $v = \sin^{-1} t$, then

$$\int \sin^{-1} t \, dt = t \sin^{-1} t \pm \int \frac{t \, dt}{\sqrt{1 - t^2}}$$

This second integral is dealt with by the method of change of variable. Let $1 - t^2 = z$, then $\frac{dz}{dt} = -2t$, so that

$$\int \frac{t \, dt}{\sqrt{1 - t^2}} = \int \frac{1}{\sqrt{z}} \cdot \left(-\tfrac{1}{2}\frac{dz}{dt}\right) dt$$
$$= -\tfrac{1}{2} \int \frac{1}{\sqrt{z}} \, dz$$
$$= -\tfrac{1}{2} \int z^{-\frac{1}{2}} \, dz$$
$$= -\tfrac{1}{2}/\tfrac{1}{2} \, . \, z^{\frac{1}{2}} + C$$
$$= -(1 - t^2)^{\frac{1}{2}} + C$$

whence

$$\int \sin^{-1} t \, . \, dt = t \sin^{-1} t \pm (1 - t^2)^{\frac{1}{2}} + C.$$

PROBLEMS

Find the following integrals $y = \int v \, dt$ where

(1) $v = t \cos t$
(2) $v = t \ln t$
(3) $v = t^2 e^t$
(4) $v = t^2 \sin t$
(5) $v = \sqrt{(1 + t^2)}$
(6) In the integral $\int \sin^{-1} t \, dt = t \sin^{-1} t \pm (1 - t^2)^{\frac{1}{2}} + C$ when must we take the plus sign, and when the minus?

10.6 Failure of methods of integration

Certain integrals cannot be coaxed into a standard form by any of the rules we have given. No amount of change of variable or integration by parts will give the values of $\int e^{-t^2} dt$, $\int \ln \sin t \, . \, dt$ or $\int \sqrt{(1 + t^3)} \, dt$. Nevertheless we can imagine a point moving in such a way that its velocity v at any instant is given by the formula $v = e^{-t^2}$, or by $v = \ln \sin t$, or $v = \sqrt{(1 + t^3)}$. The failure does not mean that there is no integral like $y = \int e^{-t^2}\, dt$: it merely means that we cannot express this in terms of the familiar functions formed by powers, logs, exponentials, trigonometric and hyperbolic functions. We have to regard $\int e^{-t^2}\, dt$ as an entirely new function of t, one which we have not met before; and by choosing suitable recalcitrant integrals we can get a wide variety of such functions, some of them of considerable practical importance. Later on we shall find ways of calculating their values numerically, in series, or in other suitable forms. The values of those which occur most frequently have been worked out and are given in tables [see A. Fletcher, J. C. P. Miller and L. Rosenhead, *An index of mathematical tables*, 1946 (London: Scientific Computing Service Ltd.)].

10.7 Differential equations

Integration is the process of finding the value of a variable quantity y when its rate of change $dy/dt = y_t$ is a known function of the time t. But it often happens that y_t is expressed as a function of y rather than of t. The rate of cooling of a hot body will depend in the first place on its own temperature y and on the temperature Y (say) of its surroundings. In fact, if the difference of temperature $y - Y$ is not too great, the rate of cooling will be proportional to this difference, so that we can put

$$y_t = -K(y - Y) \qquad . \quad . \quad . \quad (10.7)$$

where K is a positive constant.

Again if we have a population of bacteria we shall expect its rate of increase to be proportional to the number y of bacteria already present

$$y_t = Ky \qquad . \quad . \quad . \quad . \quad (10.8)$$

Of course, strictly speaking we cannot define an instantaneous rate of increase of a population, since it grows discontinuously. But if the numbers are large no harm will be done by treating it as if the number y in the population could change continuously.

Equation (10.8) will hold for an isolated population of bacteria where the density is so small that effects of overcrowding, exhaustion of food supply, and poisoning by metabolic products can be neglected. We might take a more complicated situation in which the population grows and is also increased by an immigration from outside. The rate of growth by cell division will be Ky. We suppose, for illustration, that the rate of immigration is a linear function $A + Bt$ of the time t. (This will include a constant rate of immigration if we put $B = 0$.) The total rate of growth will therefore be

$$y_t = Ky + A + Bt \qquad . \quad . \quad . \quad (10.9)$$

The equations (10.7), (10.8), and (10.9) express relations between the rate of change $y_t = dy/dt$ of a variable y, the variable y itself, and the time t. They are known as *first order differential equations*. From them we would like to be able to find what y itself is, expressed as a function of the time: this would be a solution or "integral" of the differential equation.

If the differential equation is of the form "$y_t =$ a function v of t" then we can find y by direct integration. If it is of any other form then some manipulation will be needed to obtain the solution. We shall not discuss here all the possible devices, which can be found in text-books of differential equations (e.g. H. T. H. Piaggio, *An elementary treatise on differential equations and their application*, 1946, G. Bell; or E. L. Ince, *Integration of ordinary differential equations*, 1949, Oliver & Boyd); but there are some simple and commonly occurring types of differential equation which deserve mention.

10.8 Some equations of growth of populations and similar processes

The next simplest type after the straightforward integral is the differential equation of the form

$$y_t = \text{a function of } y \text{ only.}$$

Examples of this are equation (10.7), $y_t = -K(y - Y)$, and equation (10.8), $y_t = Ky$. The correct procedure here is to reverse the natural line of attack, and instead of trying to find y in terms of t, to find t in terms of y. By the definition of an integral, $t = \int t_y dy$; but $t_y = 1/y_t$, and by hypothesis we already know this as a function of y. Thus considering first equation (10.8), which is simpler than (10.7), we have $y_t = Ky$, so that $t_y = 1/Ky$. Thus $t = \int t_y dy = (\ln y)/K + C$. [$y$ is by supposition positive, so we need not consider the integral $t = (\ln[-y])/K + C$.] We can now solve this equation for y in terms of t:

$$t - C = (\ln y)/K$$
$$K(t - C) = \ln y$$
$$y = e^{K(t-C)}$$

If we write $e^{-KC} = c$ this can alternatively be written

$$y = ce^{Kt} \quad . \quad . \quad . \quad . \quad (10.10)$$

which is therefore the law of growth of a population in the absence of competition or other limiting factor and with abundant food. The solution (10.10) is shown graphically in Fig. 10.2 overleaf.

We can solve (10.7), $y_t = -K(y - Y)$, in a similar way. We have $t_y = -1/K(y - Y)$ and therefore

$$t = \int t_y\, dy = -\int \frac{dy}{K(y - Y)}$$

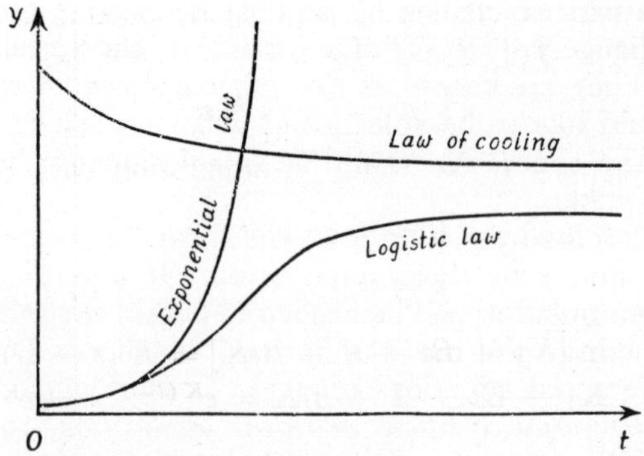

Fig. 10.2—Curves illustrating the solutions of differential equations

(i) exponential law of free growth, $y = ce^{Kt}$
(ii) logistic law of limited growth $y = \frac{1}{2}L\,(1 + \tanh[\frac{1}{2}L^2\,(t-C)/K])$
(iii) Newton's law of cooling, $y = Y + ce^{-Kt}$

To find this integral put $y - Y = z$, so that $y = z + Y$, and $dy/dz = 1$.

$$\begin{aligned} t &= -\int \frac{1}{K(y-Y)} \frac{dy}{dz}\, dz \\ &= -\int \frac{dz}{Kz} \\ &= -(\ln z)/K + C \\ &= -K^{-1} \ln (y - Y) + C \end{aligned}$$

This is the solution: to get it into the most useful form y must be expressed in terms of t. We have

$$\begin{aligned} t - C &= K^{-1} \ln (y - Y) \\ K(C - t) &= \ln (y - Y) \\ y - Y &= e^{K(C-t)} \\ &= ce^{-Kt} \quad \text{where } c = e^{KC} \end{aligned}$$

$$\therefore y = Y + ce^{-Kt} \quad . \quad . \quad . \quad . \quad (10.11)$$

This is therefore the law of cooling of a body at temperature y when the surroundings are at a fixed temperature Y.

Equation (10.9) can also be solved in this way provided that we make a simple change of variable. Let us put $Ky + Bt = z$, then

$$\begin{aligned} z_t &= Ky_t + B \\ &= K(Ky + A + Bt) + B \\ &= K(z + A) + B \\ &= Kz + (KA + B) \\ &= K[z + A + B/K] \end{aligned}$$

Now make a further change of variable by putting $z + A + B/K = w = Ky + Bt + A + B/K$. Then

$$w_t = z_t = Kw$$

so that $t_w = 1/w_t = K^{-1}w^{-1}$ and on integration with respect to w

$$t = \int t_w \, dw = K^{-1} \ln w + C$$

$$= K^{-1} \ln [Ky + Bt + A + B/K] + C$$

i.e.

$$\ln [Ky + Bt + A + B/K] = K(t - C)$$

$$Ky + Bt + A + B/K = e^{K(t-C)} = ce^{Kt}$$

where $c = e^{-KC}$,

$$Ky = ce^{Kt} - Bt - A - B/K$$

$$y = [ce^{Kt} - Bt - A - B/K]/K \quad . \quad . \quad (10.12)$$

This accordingly gives the law of increase of population with a steadily rising rate of immigration.

After solving a differential equation it is useful to check the correctness of the solution. From the mathematical standpoint this can be done by differentiation and substitution in the original equation. Thus from (10.10), $y = ce^{Kt}$, we have $y_t = cKe^{Kt} = Ky$, which agrees with the differential equation (10.8). We can also often see whether the answer agrees with common sense. Thus we know that e^{Kt}, with K positive, increases slowly at first, but rapidly gains speed and finally increases very rapidly indeed. This is just what we would expect of a population of bacteria. But we also see that when the period of extremely rapid multiplication has set in the population must soon approach a condition in which limiting factors become important, and the original equation will fail to hold. The equation of cooling $y = Y + ce^{-Kt}$ is also reasonable. e^{-Kt} approaches zero very rapidly at first, and soon becomes inappreciably different from zero. After that it decreases extremely slowly. Thus a cooling body will fairly soon attain the same temperature as its surroundings for all practical purposes—how soon depends on the constant K: the smaller K is, the slower the cooling. But in theory according to this equation absolutely exact equality will never occur in any finite time, since e^{-Kt} is never exactly zero. Equation (10.12) has a more surprising form, since one would not expect negative terms $-Bt - A - B/K$ to occur. But when t is large these terms will be small compared with ce^{Kt}, so that after a time the population will increase exponentially, with the natural growth overwhelming any effect due to immigration, until limiting factors intervene.

We notice too that in each of the solutions there is an arbitrary constant C. Thus for a growing population of bacteria $y = ce^{Kt}$ where $c = e^{-KC}$ and C is arbitrary; or, if we prefer, we can say that c is an arbitrary positive number. We can fix the value of this constant by specifying the value of y for one particular value of t; e.g. in the equation $y = ce^{Kt}$, when $t = 0$, $y = c$, so that c can be interpreted here as the number of bacteria at the start of the experiment. In general the solution of every differential equation has an arbitrary constant contained in it. We shall not give a formal proof of this, or state the exact conditions under which it is true: but it agrees in a general way with common sense, since the differential equation only gives the rate of change, and one would expect different initial conditions to give different solutions.

Another important equation soluble by the method given above is $y_t = Ky(1 - y/L)$. This represents the free growth of a population when limiting factors are taken into account. For when y is small this is approximately the same as the equation $y_t = Ky$ which we have already considered, as $1 - y/L \simeq 1$; but when the population y approaches L the growth slows down again as $(1 - y/L)$ approaches zero. From this $t_y = 1/Ky(1 - y/L)$ and

$$t = \int t_y dy = \int \frac{dy}{Ky(1 - y/L)}$$

One way of integrating this is to "complete the square" as for the solution of a quadratic equation:

$$\begin{aligned} Ky(1 - y/L) &= -Ky^2/L + Ky \\ &= -KL^{-1}(y^2 - Ly) \\ &= -KL^{-1}[(y - \tfrac{1}{2}L)^2 - \tfrac{1}{4}L^2] \end{aligned}$$

$$t = -K^{-1}L \int \frac{dy}{(y - \frac{1}{2}L)^2 - \frac{1}{4}L^2}.$$

Now change the variable of integration to $z = (y - \frac{1}{2}L)$; since $y = z + \frac{1}{2}L$, $y_z = 1$ and so by use of Table 10.2

$$\begin{aligned} t &= -K^{-1}L \int \frac{dz}{z^2 - (\frac{1}{2}L)^2} \\ &= K^{-1}L(\tfrac{1}{2}L)^{-1}\tanh^{-1}[(\tfrac{1}{2}L)^{-1}z] + C \\ &= 2K^{-1}\tanh^{-1}[2L^{-1}(y - \tfrac{1}{2}L)] + C \\ &= 2K^{-1}\tanh^{-1}[2y/L - 1] + C \\ t - C &= 2K^{-1}\tanh^{-1}[2y/L - 1] \\ \tfrac{1}{2}K(t - C) &= \tanh^{-1}[2y/L - 1] \\ 2y/L - 1 &= \tanh[\tfrac{1}{2}K(t - C)] \\ y &= \tfrac{1}{2}L(1 + \tanh[\tfrac{1}{2}K(t - C)]) \quad . \quad . \quad (10.13) \end{aligned}$$

This is the "Verhulst–Pearl logistic law of growth" and under good experimental conditions it is fairly closely obeyed. Since $\tanh u$ approaches -1 as $u \to -\infty$ and 1 as $u \to +\infty$ this means that the population increases very slowly at first, then has a period of rapid growth, and finally slows down again, tending asymptotically to the limiting population L (see Section 6.14 and Fig. 10.2).

10.9 Methods of solving differential equations

As with the problem of integration there is no golden rule which will solve every differential equation without fail. But there are a number of standard types which can be solved, and a certain amount of manipulation, coupled with judiciously chosen changes of variable, will often reduce an equation to one of these types. Some of these types are rather rare, and are given here mainly for reference if required.

(A) *Variables separable*

If the equation can be written in the form $y_t =$ a function of t divided by a function of $y = f(t)/g(y)$ (say), then it is solved thus. Multiply through by $g(y)$ giving $g(y)y_t = f(t)$, and integrate both sides with respect to t: $\int g(y)y_t dt = \int f(t)\, dt$. From our rule for change of variable in an integral we know that $\int g(y)y_t\, dt = \int g(y)\, dy$, so that the equation becomes

$$\int g(y)\, dy = \int f(t)\, dt$$

When we have (if possible) performed these integrations we have a relation between y and t, which is the required solution.

EXAMPLE

(1) $y_t = y/t = t^{-1}/y^{-1}$. Multiply both sides by y^{-1}, and integrate

$$\int y^{-1} y_t\, dt = \int y^{-1}\, dy = \int t^{-1}\, dt$$

That is,

$$\ln(\pm y) + C_1 = \ln(\pm t) + C_2$$

where C_1 and C_2 are constants of integration. This means that

$$\begin{aligned} \ln(\pm y) - \ln(\pm t) &= C_2 - C_1 = C \text{ (say)} \\ \ln(\pm y/t) &= C \\ y/t &= \pm e^C = c \text{ (say)} \\ y &= ct \end{aligned}$$

where c is an arbitrary constant.

(B) *Homogeneous equation*

A first-order differential equation is said to be "homogeneous" if it can be expressed in the form $y_t =$ a function of $(y/t) = f(y/t)$ say. In this case the change of variable from y to $z = y/t$ will give an equation in which the variables can be separated. For $z_t = (ty_t - y)/t^2 = (ty_t - tz)/t^2 = (y_t - z)/t$, and y_t is a function of z only.

EXAMPLES

(2) The equation $y_t = y/t$ which we have already considered above is also a homogeneous equation. Putting $y/t = z$ it becomes $y_t = z$, $z_t = (y_t - z)/t = 0$, so that $y/t = z =$ constant.

(3) The equation $t^2 + 2yt\, y_t - y^2 = 0$ is homogeneous, for on dividing by t^2 it becomes $1 + 2(y/t)y_t - (y/t)^2 = 0$, or $y_t = [(y/t)^2 - 1]/2(y/t) = (z^2 - 1)/2z$. Therefore $z_t = (y_t - z)/t = -(z^2 + 1)/2zt$. On bringing all terms containing z to the left, and t to the right, we have

$$-2zz_t/(z^2 + 1) = 1/t$$

Integrate both sides with respect to t:

$$-\int \frac{2zz_t\, dt}{z^2 + 1} = -\int \frac{2z\, dz}{z^2 + 1} = \int \frac{dt}{t}$$

The right-hand side is $\ln(\pm t) + C_1$, where C_1 is the constant of integration. The middle integral is evaluated by the substitution $z^2 + 1 = w$, since $w_z = 2z$, and so

$$-\int \frac{2z\, dz}{z^2 + 1} = -\int \frac{w_z\, dz}{w}$$

$$= -\int \frac{dw}{w} = -\ln(\pm w) + C_2.$$

Thus the equation becomes

$$-\ln(\pm w) + C_2 = \ln(\pm t) + C_1$$

This is effectively the solution of the equation. It can be reduced to a form giving y explicitly in terms of t in the following way. We first rearrange the solution to give

$$\ln(\pm t) + \ln(\pm w) = C_2 - C_1$$

i.e. $$\ln(\pm tw) = C_2 - C_1$$

i.e. $$tw = \pm e^{C_2 - C_1} = c \text{ (say)}.$$

But $w = z^2 + 1 = (y/t)^2 + 1 = y^2/t^2 + 1$, so that

$$t\,(y^2/t^2 + 1) = c$$

or on multiplying through by t,

$$y^2 + t^2 = ct$$

or $$y = \pm\sqrt{(ct - t^2)},$$

where c is an arbitrary constant.

(*C*) *Equation reducible to homogeneous form*

If the equation is expressible in the form $y_t =$ a function of $\dfrac{Hy + Jt + K}{hy + jt + k}$ where H, J, K, h, j, k are constants, then the change of variables to $Y = Hy + Jt + K$ and $T = hy + jt + k$ will generally reduce it to homogeneous form. For $Y_t = Hy_t + J$, $T_t = hy_t + j$, so that $Y_T = Y_t/T_t = (Hy_t + J)/(hy_t + j) =$ a known function of $y_t =$ a known function of Y/T. But if $Hj = hJ$ this method fails: the change of variable from y to $T = hy + jt + k$ will then be found to solve the equation. (The methods of integration described in Chapter 15 may sometimes be required to obtain the complete solution.)

(*D*) *Equation solvable for y*

Sometimes it may not be possible to express y_t in any convenient way in terms of y and t, but it may be possible to express y in terms of y_t and y. In this case write $y_t = v$, so that we can write $y = f(v, t)$. On differentiating this with respect to t we obtain a differential equation for v (since $y_t = v$) which may be soluble: having obtained v in terms of t we can find y from the original equation.

EXAMPLE

(4) Consider the equation $y = \frac{1}{3}y_t^3 - \frac{1}{2}y_t^2 = \frac{1}{3}v^3 - \frac{1}{2}v^2$.

On differentiating with respect to t we obtain

$$y_t = v = v^2 v_t - v v_t,$$
$$\text{i.e.} \quad 1 = (v - 1)\, v_t$$

The variables v and t are now separated; and on integrating with respect to t we have

$$t \pm C_1 = \int (v - 1)\, v_t\, dt$$
$$= \int (v - 1)\, dv = \tfrac{1}{2}v^2 - v + C_2$$

This gives us a quadratic equation for v

$$v^2 - 2v + (C_2 - C_1 - t) = 0$$

or writing $C_2 - C_1 - 1 = c$, a constant,

$$(v - 1)^2 + (c - t) = 0$$
$$v = 1 \pm \sqrt{(c - t)}$$

But we know that $y = \frac{1}{3}v^3 - \frac{1}{2}v^2$, so that we obtain y in terms of t.

(*E*) *Equation solvable for t*

Sometimes t can be expressed in terms of y_t and y, but y_t not in any convenient way in terms of y and t. But $y_t = 1/t_y$, so that then t can be expressed as a function of t_y and y; and this is simply the previous case (*D*) with the variables y and t interchanged.

(*F*) *The linear equation*

An equation of the form

$$y_t = y \,.\, P(t) + Q(t) \qquad (10.14)$$

where $P(t)$ and $Q(t)$ are functions of t only, is said to be "linear". It could arise biologically in the following way. We have seen that a population provided with ample food supplies will increase according to the law $y_t = Ky$, where K is the "growth constant". Now suppose that some external factor influencing the rate of growth, such as the temperature, is not constant but is a function of the time. Then the equation will become $y_t = y \,.\, P(t)$ instead of $y_t = Ky$. Suppose further that there is a rate of immigration $Q(t)$ which is also a known function of the time; then this must be added to the rate of increase of the population, which will become $y_t = yP(t) + Q(t)$. A linear equation is solved by the following special trick. Integrate $P(t)$; let w be any one indefinite integral. Then by definition $w_t = P(t)$. Now consider

$$\begin{aligned} z &= ye^{-w} \\ z_t &= y_t e^{-w} - ye^{-w}w_t \\ &= [yP(t) + Q(t)]e^{-w} - ye^{-w}P(t) \\ &= Q(t)e^{-w} \end{aligned}$$

But e^{-w} is a known function of t, so that we can integrate this to find z:

$$z = \int Q(t)e^{-w}\,dt$$

$$y = e^{w}z = e^{w}\int Q(t)e^{-w}\,dt.$$

EXAMPLE

(5) $y_t = y/t + qt$ (q constant). This corresponds to a population in which the natural rate of reproduction is being slowed down as $1/t$, but in which the rate of immigration is steadily increasing according to the law qt. Here $P(t) = 1/t$ and $Q(t) = qt$. Now one integral of $P(t)$ is $w = \ln t$, and $e^{w} = t$, $e^{-w} = 1/t$. Therefore

$$\begin{aligned} y &= e^{w}\int Q(t)e^{-w}\,dt \\ &= t\int qt/t \,.\, dt \\ &= t\int q\,dt = t\,(qt + C) \end{aligned}$$

Thus if t is large the population will increase roughly as qt^2.

(*G*) *The Bernoulli equation* $y_t = y.P(t) + y^n.Q(t)$.

This becomes linear on changing the variable to $z = y^{1-n}$.

(*H*) *The Riccati equation* $y_t = y^2 . P(t) + y . Q(t) + R(t)$.

This can be solved completely if one solution can be found. Let Y be this solution, and write $z = 1/(y - Y)$: then it will be found that the equation for z is linear.

EXAMPLE

(6) $y_t = ty^2 + (t^{-1} - 2t)y + (t - t^{-1})$.

One solution of this is $y = 1$, as is immediately verified by substitution. Set therefore $z = 1/(y - 1)$, then

$$\begin{aligned} z_t &= -y_t/(y-1)^2 \\ &= -z^2 y_t \\ &= -z^2\,[ty^2 + (t^{-1} - 2t)y + (t - t^{-1})] \end{aligned}$$

But $y = 1 + z^{-1}$: on substituting this value we find

$$z_t = -t^{-1}z - t$$

This is a linear differential equation for z. If $w = \int(-t^{-1})dt = -\ln t$ then the integral is, as shown in paragraph (*F*) above,

$$\begin{aligned} z &= e^w \int (-t)\,e^{-w}\,dt \\ &= t^{-1} \int (-t^2)\,dt \\ &= t^{-1}(-\tfrac{1}{3}t^3 + C) \\ &= -\tfrac{1}{3}t^2 + C/t \end{aligned}$$

whence $y = 1 + z^{-1} = 1 + t/(C - \tfrac{1}{3}t^3)$.

(*J*) *The "exact" equation*

If we take any relation between y and t, and write it in the form $f(y, t) = 0$, then on differentiation we shall obtain an equation connecting y, y_t and t. Let $f_{y|t}(y, t)$ and $f_{t|y}(y, t)$ be the partial derivatives of $f(y, t)$ when y and t are allowed to vary independently; then when y and t are subject to the constraint $f(y, t) = 0$ we shall have $f_t(y, t) = 0$, i.e. by the rules for constrained variables

$$f_{y|t}(y, t) . y_t + f_{t|y}(y, t) = 0 \qquad . \quad . \quad (10.15)$$

We shall get the same relation if we start with $f(y, t) = C$, since then $f_t(y, t) = C_t = 0$. Thus if we have an equation of the form (10.15) its general solution will be $f(y, t) =$ constant. Such an equation is said to be "exact". That is to say that if our equation is of the form $M(y, t)y_t + N(y, t) = 0$, where $M(y, t)$ and $N(y, t)$ are functions of y and t, then this equation will be exact if we can show that $M(y, t)$ is the partial derivative of a function $f(y, t)$ with respect to y, and $N(y, t)$ the partial derivative of the same function with respect to t. It may not be obvious at sight whether this is so or not, and fortunately there is a test (discussed in Section 12.12) which enables us to find out. An equation $M(y, t)y_t + N(y, t) = 0$ is exact if and only if

$$M_{t|y}(y, t) = N_{y|t}(y, t) \quad . \quad . \quad . \quad (10.16)$$

If this is so we can reconstruct the function $f(y, t)$ by methods to be given in the next section. If it is not so it is sometimes possible to multiply the original equation through by a function, known as an "integrating factor", which will make it exact. Unfortunately there is no rule which enables us to pick out such integrating factors in general: but sometimes they may be guessed.

EXAMPLE

(7) $y_t + (3t + y/t) = 0$.

If we multiply through by t we obtain $ty_t + 3t^2 + y = 0$, $M(y, t) = t$, $N(y, t) = 3t^2 + y$. These evidently satisfy $M_{t|y} = 1 = N_{y|t}$, and the equation is now exact.

In fact $N(y, t) = D_{t|y}(t^3 + ty)$ and $M(y, t) = D_{y|t}(t^3 + ty)$, so that the solution must be $t^3 + ty = C$, or $y = C/t - t^2$.

Note. In our illustrations of the methods of solution of differential equations, the equations have been divided by various expressions whenever it seemed convenient without considering whether these expressions were zero or not. Strictly speaking we should have been more careful: but such details would have interrupted the argument. As a result of neglecting this precaution several special solutions of the equations have been overlooked. Thus in solving equation (10.8), $y_t = Ky$, we divided by y. This will not be possible if $y = 0$: and in fact $y = 0$ is a solution, representing a population which is always zero and therefore can never grow. In solving (10.7), $y_t = -K(y - Y)$, we divided by $y - Y$; but $y - Y = 0$, i.e. $y = Y$ is a possible solution of the equation of cooling, representing a state in which the temperature of the body is always equal to the temperature of its surroundings, so that there is no exchange of heat. In the same way the logistic equation $y_t = Ky(1 - y/L)$ has solutions $y = 0$, a zero population, and $1 - y/L = 0$, $y = L$, i.e. a population which has attained the limiting value and stays there without further change. The reader can, if he wishes, investigate the other equations in a similar way.

PROBLEMS

Solve the following differential equations:

(1) $y_t = y^3$
(2) $y_t = \sec y$
(3) $y(t^2 - 1)y_t = t(y^2 - 1)$
(4) $2yt\, y_t = y^2 - t^2$
(5) $y_t = t^3 y^2$
(6) $y_t = \cot y \,.\, \cot t$
(7) $y_t = (y - t)/(y + t + 2)$
(8) $y_t = (t + 1)y/t + t - t^2$
(9) $y_t = e^{-t^2} y^2 + 2(1 - e^{-t^2})ty + y + (1 - t) + t^2(e^{-t^2} - 2)$
($y = t$ is one solution)
(10) $\sin y \,.\, \cos t + \cos y \,.\, \sin t \,.\, y_t = 0$

10.10 Partial differential equations

So far we have considered *ordinary* differential equations in which all variables are expressible as functions of a single one. What happens when we have more than one independent variable, so that we have to deal with partial, not ordinary, derivatives?

Suppose that we have a quantity y which is a function of two variables, say t and u,

$$y = f(t, u)$$

Then y will have two partial derivatives, one, y_t, with respect to t keeping u fixed, and one, y_u, with respect to u keeping t fixed. These derivatives are more usually written as $\partial y/\partial t$ and $\partial y/\partial u$, or $f_t(t, u)$ and $f_u(t, u)$ respectively. Let us first suppose that we know the value of y_t only: what then can we deduce about y?

By definition y_t is the derivative of y with u fixed. If then we take any particular value of u, say $u = u_1$, and, keeping u fixed, integrate with respect to t, we shall obtain the values of y for this value of u.

$$y = f(t, u_1) = \int y_t\, dt + C_1$$

where C_1 is a constant. Here t is allowed to vary, but u is not.

On taking another value of u, say $u = u_2$, we shall have in the same way

$$y = f(t, u_2) = \int y_t\, dt + C_2,$$

where C_2 is a constant. There is no reason why the two constants C_1 and C_2 should be equal: each of them can have any value we please. Our information only tells us how rapidly y changes with t when u is fixed: it tells us nothing about what happens when u varies.

Thus for each value of u we can say that $y = \int y_t\, dt + C$, where the value of C may be different for different values of u. In other words

C is a function of u, and only of u; we might write it as $\phi(u)$, and so obtain finally

$$y = \int y_t \, dt + \phi(u) \quad . \quad . \quad . \quad (10.17)$$

In this the integration is understood as performed while u is kept fixed, and $\phi(u)$ is now an *arbitrary function of u* instead of a constant of integration.

Looking at the matter graphically, suppose we plot y against u and t (Fig. 10.3), choosing t and u as horizontal co-ordinates in a plane, and

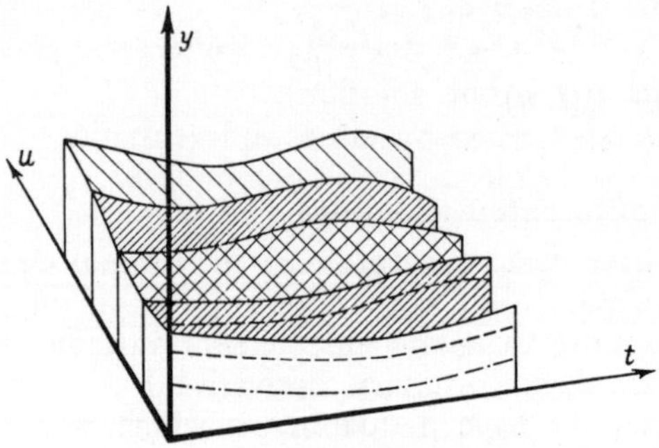

Fig. 10.3—The meaning of a partial differential equation "y_t = a known function"

Each section u = constant can be imagined raised or lowered by a constant amount, as indicated by the dotted lines. If, however, one section t = constant is also given the whole surface is determined

representing y by the vertical distance above the plane. The points corresponding to the values of y as t and u vary lie on a surface. Now a knowledge of y_t is equivalent to a knowledge of the slope in any section of this surface by a plane u = constant. In other words, we know the shapes of these sections, but we do not know how high each of them is above the (t, u) plane. If we are given one point in each of these sections then we can find the whole surface. For example, it is sufficient to know a single section t = constant, i.e. to know all values of y for a given value of t and varying values of u.

EXAMPLE

(1) Consider the equation $\partial y/\partial t = y_t = u + t$. This gives $y = \int (u + t)\, dt + \phi(u) = ut + \frac{1}{2}t^2 + \phi(u)$. Here we can take any function we like for $\phi(u)$; $ut + \frac{1}{2}t^2$, $ut + \frac{1}{2}t^2 + 2u$, $ut + \frac{1}{2}t^2 + 27{\cdot}5$, and $ut + \frac{1}{2}t^2 + \sqrt{[e^u - \frac{1}{2} \sin \tanh u^3]}$ are all solutions. This can be seen on differentiating with respect to t to verify the original equation, for the partial derivative of $ut + \frac{1}{2}t^2$ is $(u + t)$, and the derivative

of the function $\phi(u)$ is zero when u is held fixed, since $\phi(u)$ is then also fixed.

It will be seen from the above argument that if we are given both partial derivatives y_t and y_u we can reconstruct the original function y except for a constant C of integration to be added on. In fact we have more information than we need. From the value of y_t we can deduce that $y = \int y_t \, dt + \phi(u) = I(t, u) + \phi(u)$, where $I(t, u)$, the integral of the given function y_t, is known. Differentiation with respect to u shows that

$$y_u = I_u(t, u) + \phi_u(u),$$

or $\phi_u(u) = y_u - I_u(t, u)$.

Now $I_u(t, u)$ is obtained by direct differentiation, and is therefore known. It follows that if we know the value of y_u *for any one given value of* t we can determine the derivative $\phi_u(u)$ of the unknown function $\phi(u)$. We have then only to integrate to find ϕ, and hence y:

$$\phi(u) = \int \phi_u(u) \, du + C$$

Graphically this means that if we know y_t for all values of t and u, and also the shape (but not the height) of any one section $t =$ constant of the y surface, then we know the shape of the whole surface: but it can still be moved up or down as a whole according to the value of the constant C.

EXAMPLE

(2) Consider again the equation $y_t = (u + t)$; suppose we also know that $y_u = 0$ when $t = 0$ for all values of u. It has already been shown in example (1) that

$$y = ut + \tfrac{1}{2}t^2 + \phi(u)$$

whence $y_u = t + \phi_u(u)$.

Putting $t = 0$, $y_u = 0$, so that $\phi_u(u) = 0$ for all values of u. This shows that $\phi(u)$ is a constant, C, say, and the complete solution is

$$y = ut + \tfrac{1}{2}t^2 + C.$$

Any equation connecting the partial derivatives y_t and y_u with the variables t, u and y is known as a "(first-order) partial differential equation". The solution of such an equation will in general contain an arbitrary function ϕ in some way, just as the solution of an ordinary differential equation contains an arbitrary constant C. Many natural laws are expressible by such equations—e.g. the laws of flow of heat, or of electricity and magnetism. Unfortunately the methods of solution of such equations form a large and complicated subject, and we shall merely give a few simple and important examples here.

EXAMPLES

(3) Consider the equation $\partial y/\partial t = 0$, where y is a function of two variables t and u.

From (10.17) we have at once $y = \phi(u)$, since $y_t = 0$; that is, y is a function of u only. Conversely if $y = \phi(u)$, $y_t = \partial y/\partial t = 0$.

This is a commonsense result; the equation $y_t = 0$ means that when u is fixed the rate of change of y is zero, i.e. changing the value of t without changing u has no effect on y.

(4) Consider the equation $\partial y/\partial t = \partial y/\partial u$ or $y_{t|u} = y_{u|t}$, where y is again a function of t and u.

Here the device we use to solve the equation is to change our variables from t and u to t and $s = t + u$. By the equation for change of variables $y_{t|s} = y_{t|u}t_{t|s} + y_{u|t}u_{t|s}$.

But $t_{t|s} = 1$, and since $u = s - t$, $u_{t|s} = -1$, so that

$$y_{t|s} = y_{t|u} - y_{u|t} = 0 \text{ (by hypothesis).}$$

By our previous example this means that y is a function of $s = t + u$ only; or $y = \phi(t + u)$.

(5) Find the value of $y = \sin t . \cos u + \cos t . \sin u$.

By direct differentiation $\partial y/\partial t = \cos t . \cos u - \sin t . \sin u$. Similarly $\partial y/\partial u = \cos t . \cos u - \sin t . \sin u = \partial y/\partial t$. Therefore by the previous example y is a function of $t + u$, say

$$y = \phi(t + u) = \sin t . \cos u + \cos t . \sin u$$

To find what function ϕ is, put $u = 0$; this gives

$$y = \phi(t) = \sin t + 0.$$

Thus ϕ is the sine function and

$$\sin(t + u) = \sin t . \cos u + \cos t . \sin u.$$

The reader should compare this method of proving the addition theorems with the direct method (Section 5.6).

PROBLEMS

(1) Solve the equation $t . \partial y/\partial t = u . \partial y/\partial u$ by changing the variables to t and $s = tu$.

(2) Given that $d \ln t/dt = 1/t$, and $\ln 1 = 0$, show that $\ln t + \ln u = \ln(tu)$.

(3) Using example (4) show also that $\cos t . \cos u - \sin t . \sin u = \cos(t + u)$.

(4) Solve $t^{-1} \partial y/\partial t + u^{-1} \partial y/\partial u = 0$.

CHAPTER ELEVEN

LENGTHS, AREAS, AND VOLUMES

11.1 Definite integrals

Consider a point moving with a velocity v, which we take to be possibly varying, but to be a known function of the time. Then its position y at any time is given by $y = \int v\,dt$: that is the definition of the "indefinite integral" $\int v\,dt$. But we know that this integral is not uniquely determined by the velocity v; if $Y(t)$ is one integral, then $Y(t) + C$ is also a possible integral:

$$y = \int v\,dt = Y(t) + C \qquad (11.1)$$

We have to be told exactly where the point started from, as well as its velocity, in order to fix its position for certain.

Suppose, however, we are interested not in the absolute position y of the point, but simply in the distance it has moved in a given time, say from time t_1 to time t_2. If y_1 is its co-ordinate at time t_1, and y_2 at time t_2, then the distance travelled will be $y_2 - y_1$. We should then expect this distance to be completely determined by the velocity. For when the velocity of a point is known, we expect to be able to deduce how far it has gone. In fact, since y_1 is by definition the value of y when $t = t_1$, we see from (11.1) that $y_1 = Y(t_1) + C$, and in the same way $y_2 = Y(t_2) + C$. By subtraction we obtain

$$\begin{aligned} \text{distance gone} &= y_2 - y_1 \\ &= [Y(t_2) + C] - [Y(t_1) + C] \\ &= Y(t_2) - Y(t_1) \qquad (11.2) \end{aligned}$$

The unknown constant C has cancelled out. This formula can be expressed in words: *to find the distance travelled between times t_1 and t_2, take any one indefinite integral $Y(t)$ of the velocity v, find the values* $Y(t_1)$ and $Y(t_2)$ *of this integral at times t_1 and t_2 respectively, and finally subtract $Y(t_1)$ from $Y(t_2)$.*

EXAMPLE

(1) A point is moving with velocity $v = 2t$. What is the distance it travels between times $t_1 = 1$ and $t_2 = 2$?

One possible integral of v is $Y(t) = \int 2t \, . \, dt = t^2$. Thus $Y(1) = 1^2 = 1$, $Y(2) = 2^2 = 4$, and the distance travelled is $Y(2) - Y(1) = 3$ units.

It should perhaps be made clear that this "distance" $y_2 - y_1$ means simply the distance between the initial position y_1 and the final position y_2. It is positive if y_2 is on the positive side of y_1, i.e. if $y_2 > y_1$, and negative if $y_2 < y_1$. If the point returns to the place it started from then the total distance covered is counted as zero in this formula. To evaluate the distance covered by the point in all its wanderings, not counting the signs positive or negative, we must take the integral $\int |v| dt$ instead of $\int v \, dt$. This, however, is rarely needed in practice.

EXAMPLE

(2) Suppose $v = \sin t$. What is the change in position between times $t_1 = 0$ and $t_2 = 2\pi$?

$$Y(t) = \int \sin t \, . \, dt = - \cos t.$$

The change in position in therefore

$$y_2 - y_1 = Y(2\pi) - Y(0) = (-1) - (-1) = 0;$$

the point returns to its original position.

This change in position $y_2 - y_1$ is called the "definite integral of v from time t_1 to time t_2" and is denoted by the symbol $\int_{t_1}^{t_2} v \, dt$. [This is read "integral $v \, dt$ from t_1 to t_2"]. Thus a definite integral means simply a *change in position* or *distance travelled*: it is "definite" because it does not contain any unknown constant C, and it is an "integral" because it is intimately connected with the indefinite integral $Y(t) = \int v \, dt$. In fact

$$\int_{t_1}^{t_2} v \, dt = Y(t_2) - Y(t_1) \quad . \quad . \quad (11.3)$$

Here we must also introduce a second notation which is both very convenient and in common use. This is to write $Y(t_2) - Y(t_1)$ as $[Y(t)]_{t_1}^{t_2}$. Here the bracket symbol $[\quad]_{t_1}^{t_2}$ surrounding the function $Y(t)$ means "subtract the value of this function when $t = t_1$ from its value when $t = t_2$". Thus $[t^2]_{t_1}^{t_2}$ means $t_2^2 - t_1^2$, and $[3 \sin t]_{t_1}^{t_2} = 3 \sin t_2 - 3 \sin t_1$. In this notation equation (11.3) is written in the compact form

$$\int_{t_1}^{t_2} v \, dt = [Y(t)]_{t_1}^{t_2} \quad . \quad . \quad . \quad (11.4)$$

The working in examples (1) and (2) above would be set out briefly as

$$\int_1^2 2t\,dt = [t^2]_1^2 = 2^2 - 1^2 = 3$$

$$\int_0^{2\pi} \sin t\,dt = [-\cos t]_0^{2\pi} = (-1) - (-1) = 0$$

The values t_1 and t_2 of t between which the integration is performed are sometimes called the "limits" or "termini" of integration. This is of course a sense of the word "limit" quite different from that special technical sense which is used in phrases such as "dy/dt is the limit of $\delta y/\delta t$ as $\delta t \to 0$."

The definite integral can be applied to other rates of change besides a velocity, or rate of change of position. If v stands for the rate of increase of weight of an animal, then $\int_{t_1}^{t_2} v\,dt$ will be the gain in weight between times t_1 and t_2. If v is the rate at which some metabolic product is being excreted by an organism, then $\int_{t_1}^{t_2} v\,dt$ will be the total amount excreted in the specified interval of time. If I is the electric current in a wire at time t, then $\int_{t_1}^{t_2} I\,dt$ is the total quantity of electricity which has passed down the wire during the given interval.

PROBLEMS

Find the following integrals:

(1) $\int_0^1 3t\,dt$; (2) $\int_1^2 (t^3 + 1)\,dt$; (3) $\int_a^\beta e^t\,dt$; (4) $\int_0^1 te^t\,dt$; (5) $\int_{-\pi}^{\pi} \cos t\,dt$.

11.2 Areas as integrals

A more unexpected use of integration is that of calculating lengths, areas, and volumes.

We are all used to the ideas of length, area, and volume in everyday life; we are accustomed to measuring the lengths of strings and wires, the volumes of liquids, or the area of a wall when it has to be painted. So we rarely think seriously about exactly what we mean by a length or a volume, and still less do we subject our ideas to careful mathematical analysis. In fact such analysis is normally scarcely ever required. If we have to find the length of a complicated path, such as a route traced out on a map, we can do it by running a thread along the path, straightening the thread out, and measuring it along a ruler. If we want to know the volume of an oddly shaped vessel, we can fill it with water, and then pour the water into a measuring glass. Such empirical methods are very useful for particular problems. But they do not give us a general line of attack from which we can obtain a formula for the volume of a vessel of given shape, such as a cone, cylinder, or sphere, or the length of a given type of curve. It is the integral calculus which provides just such generality.

First let us consider an area of the following shape. Suppose we have a plane in which there are two co-ordinate axes OX and OY at right angles to each other (Fig. 11.1). We shall take OX to be the

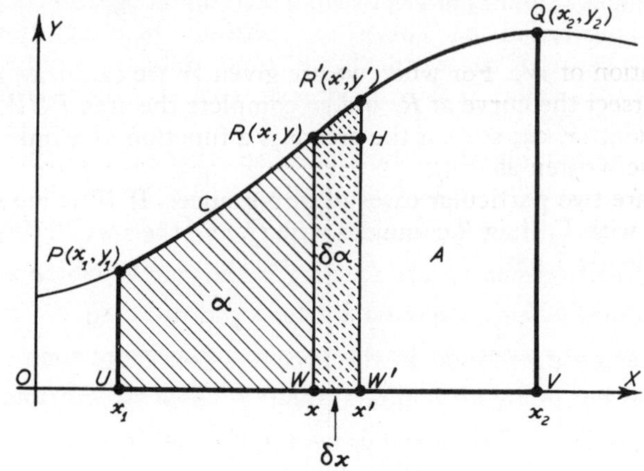

Fig. 11.1—An area considered as an integral

axis of x, and OY the axis of y, in the usual way. Suppose further that we have a curve C in this plane which is the graph of some function $y = f(x)$: for the present we shall suppose that C lies entirely *above* the x-axis. Let $P = (x_1, y_1)$ and $Q = (x_2, y_2)$ be two points on this curve. Draw PU and QV perpendiculars from P and Q respectively onto the x-axis; suppose further that U is to the left of V, i.e. that $x_1 < x_2$. Then we shall prove that the area A contained between the straight lines PU, UV, VQ and the curve QP is simply the definite integral $\int_{x_1}^{x_2} f(x)\,dx$. Thus, for example, suppose that the curve C was simply the straight line $y = x$, so that $f(x) = x$, $UP = y_1 = x_1$, $VQ = y_2 = x_2$, and $A = \int_{x_1}^{x_2} x\,dx = [\frac{1}{2}x^2]_{x_1}^{x_2} = \frac{1}{2}x_2^2 - \frac{1}{2}x_1^2$, if the above formula is correct. This agrees with the usual formula for the area of a quadrilateral with two parallel sides, which is usually stated in textbooks as $A = \frac{1}{2}$ (distance between the 2 parallel sides) $\times$ (sum of the lengths of the parallel sides)

$$= \tfrac{1}{2}\,UV \times (UP + VQ) = \tfrac{1}{2}(x_2 - x_1)(x_2 + x_1)$$
$$= \tfrac{1}{2}(x_2^2 - x_1^2).$$

To see why this formula holds we first notice that the value of the area A depends on the points P and Q we choose on the curve. If P

and Q are imagined as moving along C, then the perpendicular PU or QV will also move, and the area A will alter. Suppose that P is kept fixed, but the right-hand boundary of the area is allowed to move. Instead of taking QV as this boundary, let us take RW, where R is a point (x, y) on the curve, and W the point below it on the x-axis: we shall suppose that R is free to move up and down the curve as we choose, but keeping always at or to the right of P. We shall now consider the area $PUWR$, which will be called α. This area will depend on the position of W. For when we are given W we can draw a vertical line to intersect the curve at R, and so complete the area $PUWR$. Now W is the point $(x, 0)$; so that the area α is a function of x only, and can therefore be written as $\alpha(x)$.

There are two particular cases of importance. If W is moved along to coincide with V, then R coincides with Q, and the area $PUWR = \alpha(x_2)$ becomes identical with $PUVQ = A$. So, since now $x = x_2$,

$$\alpha(x_2) = A \quad . \quad . \quad . \quad . \quad (11.5)$$

On the other hand, if W is moved down towards U the area α shrinks, and finally as W coincides with U the area becomes that of $PUUP$, or zero. This is the case when $x = x_1$, so that

$$\alpha(x_1) = 0$$

Combining this equation with (11.5) we can evidently write

$$A = \alpha(x_2) - \alpha(x_1) \quad . \quad . \quad . \quad (11.6)$$

Now consider what happens to α as x varies. Suppose that W is moved to the right to a new point $W' = (x', 0)$, so that R moves to $R' = (x', y')$, and the area α changes from $\alpha(x) =$ area of $PUWR$ to $\alpha(x') = PUW'R'$. Then (Fig. 11.1) the change in x is

$$\delta x = x' - x = OW' - OW = WW'$$

and the change in the area α is

$$\delta\alpha = \alpha(x') - \alpha(x) = \text{area } PUW'R' - \text{area } PUWR = \text{area } RWW'R'.$$

Now draw RH perpendicularly onto $W'R'$. Then

$$\text{area of the rectangle } RWW'H = WR \times WW' = y\ \delta x$$

In the figure the curve C lies above RH, and we see that the difference between the areas $RWW'R' = \delta\alpha$ and $RWW'H = y\delta x$ is the area of the triangle RHR'. Now when δx is small, so that W' is near W, this triangle will have a very small area—small even in comparison with the rectangle $RWW'H$. That is to say that the areas $RWW'R'$ and $RWW'H$ are very nearly equal when δx is small:

$$\delta\alpha \simeq y\ \delta x$$

and the smaller δx is the more nearly this will be true. A little thought will show that this will be true for any reasonable shape of the curve C: the areas $RWW'R'$ and $RWW'H$ will always be nearly equal whether C lies entirely above RH, entirely below it, or partly above and partly below. In all cases $\delta a \simeq y\,\delta x$, and the smaller δx the smaller the error will be in comparison with δx. But we can write this equation as $\delta a/\delta x \simeq y$; and if we take the limit as $\delta x \to 0$ we find that $da/dx = y = f(x)$.

Expressed in another way this means that $a = a(x)$ is an indefinite integral of $y = f(x)$. Equation (11.6) shows that A is the difference between the values of this indefinite integral when $x = x_1$ and $x = x_2$; that is to say that A is by definition the definite integral

$$A = \int_{x_1}^{x_2} y\,dx = \int_{x_1}^{x_2} f(x)\,dx \quad . \quad . \quad (11.7)$$

The area A under a curve C is the integral of the ordinate y.

The importance of this result is two-fold. Firstly it shows how to calculate an area. Secondly it gives us a method of picturing an integral. Suppose, for example, that we know the velocity v of a moving body at time t. Then the distance moved between times t_1 and t_2 will be $\int_{t_1}^{t_2} v\,dt$. Now draw the graph of v against t, v being plotted vertically and t horizontally (Fig. 11.2). Then the distance moved will

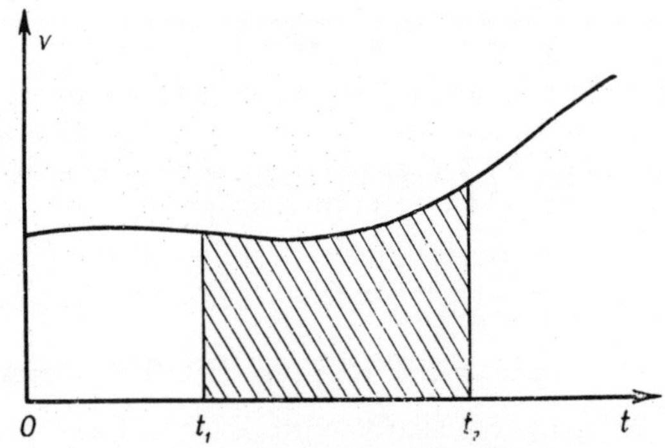

Fig. 11.2—Distance travelled expressed as an area under the velocity–time graph

be equal to the vertical area (shaded in the figure) under the curve between the points t_1 and t_2, bounded above by the curve, below by the t-axis, and on the left and the right by vertical lines through the points $(t_1, 0)$ and $(t_2, 0)$. The only question which arises here is that of the unit of area. If the scale divisions along the t and v axes are each equal to the ordinary unit of length, then the integral $\int_{t_1}^{t_2} v\,dt$ will be equal to the area in ordinary units. If, however, we use for convenience some other scales, then, in order to agree with the integral, the area will have to be measured in units equal to the area of a rectangle whose width is equal to one t-scale division and whose height is one v-scale division; for such a rectangle represents the distance travelled in unit time with unit velocity.

EXAMPLES

(1) Find the area under the parabola $y = x^2$ between the vertical lines $x = x_1$ and $x = x_2$ (Fig. 11.3).

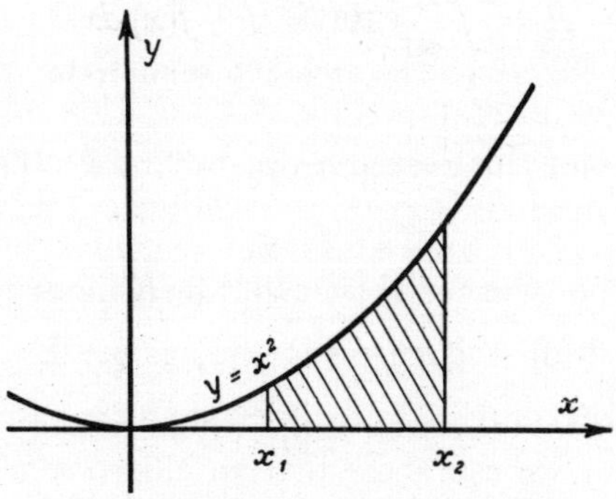

Fig. 11.3—The area under a parabola

The area

$$A = \int_{x_1}^{x_2} y\,dx = \int_{x_1}^{x_2} x^2\,dx$$
$$= [\tfrac{1}{3}x^3]_{x_1}^{x_2}$$
$$= \tfrac{1}{3}(x_2^3 - x_1^3).$$

(2) Find the area under the rectangular hyperbola $y = 1/x$ between the vertical lines $x = 1$ and $x = x_2$.

The area $A = \int_1^{x_2} \frac{dx}{x} = [\ln x]_1^{x_2} = \ln x_2$.

This property is sometimes taken as the *definition* of a natural logarithm; it is fairly easy to deduce from it all the other properties of the logarithm, including our definition by means of a spiral.

(3) Find the area of a semicircle.

A semicircle can be regarded as the area above the x-axis contained within the curve $x^2 + y^2 = R^2$, which is the equation of a circle centre O and radius R. The value of y for points above the x-axis is therefore $\sqrt{(R^2 - x^2)}$, with the positive value of the square root (Fig. 11.4).

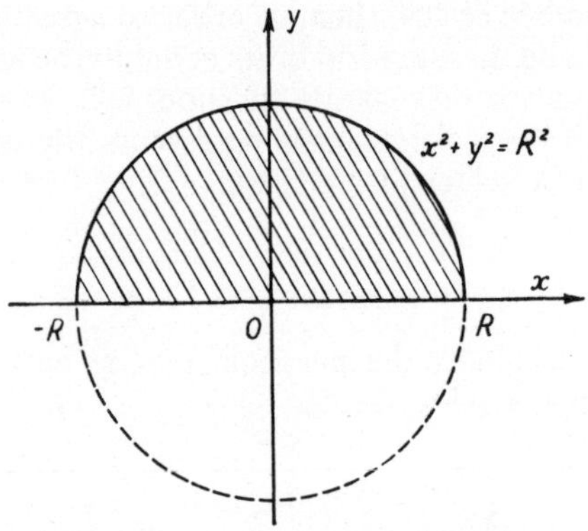

Fig. 11.4—The area of a semicircle

Furthermore the values of x range from $-R$ to R. Thus the required area is (see Table 10.2)

$$A = \int_{-R}^{R} \sqrt{R^2 - x^2}\, dx = \tfrac{1}{2}\left[R^2 \sin^{-1}(x/R) + x\sqrt{R^2 - x^2}\right]_{-R}^{R}$$
$$= \tfrac{1}{2}[R^2 \,.\, \tfrac{1}{2}\pi + 0] - \tfrac{1}{2}[R^2(-\tfrac{1}{2}\pi) + 0] = \tfrac{1}{2}\pi R^2$$

By doubling this we find the area of the whole circle to be πR^2.

In the same way we can show that an ellipse $x^2/a^2 + y^2/b^2 = 1$ of major axis of length $2a$ and minor axis of length $2b$ has area πab. This is almost evident from the fact that the ellipse is a compressed circle. The original circle has radius a and area πa^2; it is compressed vertically in the ratio b/a, so that the resulting ellipse should have area $\pi a^2 \,.\, b/a = \pi ab$.

PROBLEMS

(1) Find the area under the curve $y = \sqrt{x}$ between $x = 0$ and $x = 2$.

(2) Find the area under the curve $y = e^x$ between $x = 0$ and $x = 1$.

11.3 Calculation of areas by the summation of thin vertical strips

There is a second and alternative way of thinking of the calculation of an area or of a definite integral.

Suppose again that we wish to find the area A under a curve $y = f(x)$ between the vertical lines $x = x_1$ and $x = x_2$ and above the x-axis (Fig. 11.5). We have already shown that this area is the definite integral $\int_{x_1}^{x_2} y\, dx$. But, instead of performing the calculation in that way, let us suppose that the area is divided up into small strips by a number of vertical lines. If, for example, we divide it by lines through the points $X_1 = x_1, X_2, X_3, \ldots X_{n+1} = x_2$, then it will be divided into n strips: Fig. 11.5 illustrates the division into six strips by seven vertical lines (including the left-hand and right-hand edges and five intermediate lines). Let us suppose that the corresponding values of y on the curve are $Y_1 = y_1, Y_2, Y_3, \ldots Y_{n+1} = y_2$.

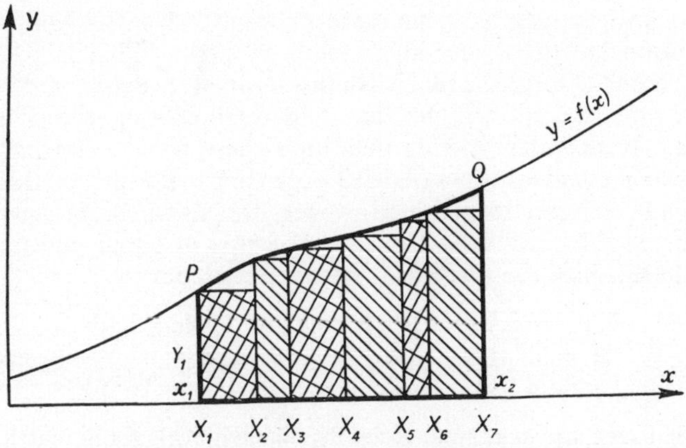

Fig. 11.5—An area as the limit of a sum

Now if the first strip is sufficiently narrow we can approximate to it by a rectangle of height Y_1, shown shaded in Fig. 11.5, by drawing a horizontal line across the strip through the first point P on the curve. There will be a small difference between the area of the strip and the rectangle: but if the strip is sufficiently narrow the error will be inappreciable. Similarly by drawing a horizontal line at height Y_2 through the second point (X_2, Y_2) on the curve, we can find a rectangle in the second strip approximating to its area. By continuing in this way we can place in each strip a rectangle approximating to its area.

Let us add the areas of all these thin rectangles together, and call the sum S. Then S will be an approximation to the area A; and the more strips we have, and the thinner they are, the nearer S will be to A.

We can write down an expression for S. The first rectangle has height Y_1 and width $X_2 - X_1$; its area is therefore $Y_1(X_2 - X_1)$. The second rectangle has height Y_2, width $X_3 - X_2$, and area $Y_2(X_3 - X_2)$. The further rectangles follow in a similar way, so that

$$S = Y_1(X_2 - X_1) + Y_2(X_3 - X_2) + \ldots + Y_n(X_{n+1} - X_n) \qquad (11.8)$$

If we agree to write $X_2 - X_1$, the width of the first strip, as simply δX_1, and $X_3 - X_2$ as δX_2, and so on, then (11.8) can be written as

$$S = Y_1\delta X_1 + Y_2\delta X_2 + Y_3\delta X_3 + \ldots + Y_n\delta X_n$$

or in a very convenient notation

$$S = \Sigma(Y\delta X) \quad . \quad . \quad . \quad . \quad (11.9)$$

where the capital sigma is simply an abbreviation for "the sum of", so that $\Sigma(Y\delta X)$ means "the sum of all quantities of the form $Y\delta X$", i.e. $Y_1\delta X_1 + \ldots + Y_n\delta X_n$.

If now we take more and more strips, and make them thinner, then the sum S will approach A; or more precisely, A is the limit of S as the maximum width of any strip tends to zero. This is practically common sense; a formal proof that the limit of S exists and is equal to the definite integral will be found in textbooks on the Theory of Functions. Notice that in this definition there is no need for all the strips to be of equal width, provided only that the width of the widest strip is made to tend to zero. However, for all practical purposes it is sufficient to consider only the case of strips of equal width, and in that way to simplify the formula. We shall then have

$$\begin{aligned} \delta X_1 &= \delta X_2 = \ldots = \delta X_n = \delta \text{ (say), and} \\ S &= Y_1\delta + Y_2\delta + \ldots + Y_n\delta \\ &= \delta.(Y_1 + Y_2 + \ldots + Y_n) = \delta.\Sigma Y. \end{aligned}$$

Also since there are n strips of width δ each, the total width of the area must be $n\delta$ and this must be equal to $(x_2 - x_1)$. Thus $\delta = (x_2 - x_1)/n$ and

$$A = \lim_{n\to\infty} S = \lim_{n\to\infty} (x_2 - x_1)(Y_1 + Y_2 + \ldots + Y_n)/n \qquad (11.10)$$

since as $n \to \infty$ the width $\delta = (x_2 - x_1)/n$ of each strip tends to zero. This provides an alternative method of calculating an area.

11.4 Arithmetic series

The basic formula (11.10) can only be applied in practice if we have a method of evaluating sums of the form $Y_1 + Y_2 + \ldots + Y_n$. Suppose, for example, that it is to be applied to find the area A under the straight line $y = \alpha + \beta x$, where α and β are constants. Then $Y_1 = \alpha + \beta X_1$, $Y_2 = \alpha + \beta X_2$, $Y_3 = \alpha + \beta X_3$, and so on; and

$$\begin{aligned} Y_1 + Y_2 + \ldots + Y_n &= (\alpha + \beta X_1) + (\alpha + \beta X_2) + \ldots + (\alpha + \beta X_n) \\ &= n\alpha + \beta(X_1 + X_2 + \ldots + X_n) \end{aligned}$$

Now $X_1, X_2 \ldots X_n$ are equally spaced points: $X_2 - X_1 = \delta = X_3 - X_2 = X_4 - X_3 = \ldots$ The numbers $X_1, X_2 \ldots X_n$ are said to form an "arithmetic series" or "arithmetic progression". Since such arithmetic series are of fairly common occurrence, their properties are worth mentioning for their own sake.

Consider therefore a series of equally spaced numbers X_1, X_2, $X_3, \ldots$ such that $X_2 - X_1 = X_3 - X_2 = \ldots = \delta$. (Examples are 1, 2, 3, 4, . . . , with $\delta = 1$, and 1, 4, 7, 10, 13, . . . , with $\delta = 3$.)

This spacing δ is known as the "common difference" of the series. We then have

$$\begin{aligned} X_2 &= X_1 + \delta \\ X_3 &= X_2 + \delta = (X_1 + \delta) \ \ + \delta = X_1 + 2\delta \\ X_4 &= X_3 + \delta = (X_1 + 2\delta) + \delta = X_1 + 3\delta \\ X_5 &= X_4 + \delta = (X_1 + 3\delta) + \delta = X_1 + 4\delta \end{aligned}$$

The general formula for the rth term X_r is evidently

$$X_r = X_1 + (r - 1)\,\delta \quad . \quad . \quad . \quad (11.11)$$

PROBLEMS

(1) An arithmetic series has first term 5 and common difference 3. What is the rth term X_r? If there are 7 terms in all, what is the last term?

(2) An arithmetic series of 10 terms begins with 5 and ends with 104. What is the common difference, and what are the terms of the series?

Suppose now that we wish to find the sum $s = X_1 + X_2 + \ldots + X_n$ of such an arithmetic series. We shall illustrate the general process on the particular series 3, 5, 7, 9. This is shown in Fig. 11.6: the first

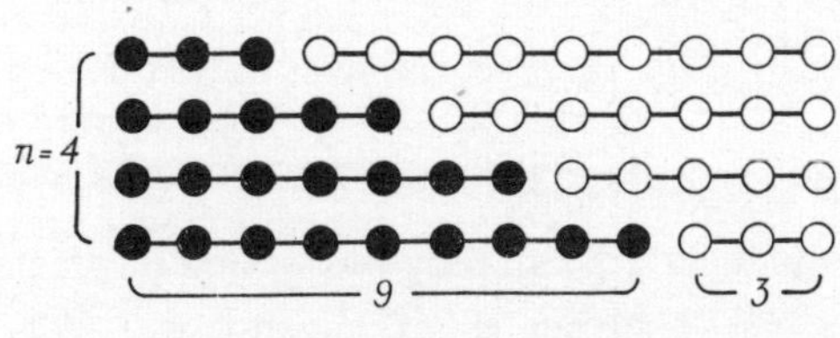

Fig. 11.6—The sum of an arithmetic series

term, 3, is represented by the line of 3 black circles joined together; the second term, 5, by a line of 5 black circles, and the remaining 2 terms by lines of 7 and 9 black circles. The sum s will then be equal to the total number of black circles.

Now add to this series an equal series written in reverse order: $9 + 7 + 5 + 3$. This is represented by the white circles in Fig. 11.6. The total number of circles, black and white together, will now be $2s$. But we now find that every row is of equal length. The first row is of length $3 + 9 = 12$. In the second row the number of blacks is increased by the common difference 2, while the number of whites is diminished by 2. Thus the total is unaltered. The same applies to the third row, and so on throughout. Thus the circles are arranged in a rectangular array, and the total number $2s$ can be calculated by multiplying the length of a row by the number of rows, i.e. $12 \times 4 = 48$. Dividing by 2 we find $s = 24$.

If we repeat this process with the general series $X_1 + X_2 + \ldots + X_n$ we shall again obtain a rectangular array. The length of the first row will be $X_1 + X_n$, the sum of the first and last terms (represented by the black and white circles respectively). The number of rows will be the number of terms, n. We have therefore

$$2s = \text{number of circles in rectangle} = n(X_1 + X_n)$$

whence

$$s = n \cdot \tfrac{1}{2}(X_1 + X_n) \quad . \quad . \quad . \quad (11.12)$$

or in words *the sum of an arithmetic series is the product of the number of terms into the average of the first and last terms.* By using the fact that $X_n = X_1 + (n - 1)\delta$ this can also be written

$$s = \tfrac{1}{2}n\,[2X_1 + (n - 1)\delta] \quad . \quad . \quad . \quad (11.13)$$

These formulas are very useful. For example, let us take the simplest possible series $1 + 2 + 3 + 4 + \ldots + n$. Applying (11.12) we find its sum to be $n \times$ average of first and last terms $= \tfrac{1}{2}n(n + 1)$.

Another interesting series is that of the odd numbers $1 + 3 + 5 + \ldots$ with common difference 2. The successive sums are then simply the squares in order: $1 = 1^2$, $1 + 3 = 4 = 2^2$, $1 + 3 + 5 = 9 = 3^2$. To show this in general for n terms we apply formula (11.13) with first term $X_1 = 1$ and common difference $\delta = 2$. We get

$$s = \tfrac{1}{2}n\,[2 + 2\,(n - 1)] = \tfrac{1}{2}n \cdot 2n = n^2$$

This is illustrated in Fig. 11.7, where we see that $1 + 3$ fills a square of side 2, $1 + 3 + 5$ a square of side 3, and $1 + 3 + 5 + 7$ a square of side 4.

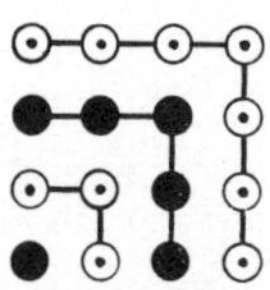

Fig. 11.7—The sum of successive odd numbers is a square

Now let us apply this formula to find the area under a straight line $y = \alpha + \beta x$ between ordinates at x_1 and x_2. We have already shown that if we divide this area into n strips of equal width by ordinates at $X_1 = x_1$, X_2, $X_3, \ldots, X_{n+1} = x_2$ then the area is approximated to by the sum (formula 11.10)

$$S = (x_2 - x_1)\,(Y_1 + Y_2 + \ldots + Y_n)/n$$

and

$$Y_1 + Y_2 + \ldots + Y_n = n\alpha + \beta\,(X_1 + X_2 + \ldots + X_n)$$

But by (11.13),

$$X_1 + X_2 + \ldots + X_n = \tfrac{1}{2}n\,[2X_1 + (n - 1)\delta]$$

Furthermore since there are n strips making a total width of $x_2 - x_1$ the width of each strip, or common difference δ, is $(x_2 - x_1)/n$. Thus, combining these results,

$$\begin{aligned} X_1 + X_2 + \ldots + X_n &= \tfrac{1}{2}n\,[2X_1 + (n-1)(x_2 - x_1)/n] \\ &= \tfrac{1}{2}n\,[2x_1 + (x_2 - x_1)(1 - 1/n)] \\ &= \tfrac{1}{2}n\,[x_1 + x_2 - (x_2 - x_1)/n] \\ Y_1 + Y_2 + \ldots + Y_n &= n\,\{\alpha + \tfrac{1}{2}\beta\,[x_1 + x_2 - (x_2 - x_1)/n]\} \end{aligned}$$

$$\begin{aligned} S &= (x_2 - x_1)\,\{\alpha + \tfrac{1}{2}\beta\,[x_1 + x_2 - (x_2 - x_1)/n]\} \\ &= (x_2 - x_1)\alpha + \tfrac{1}{2}(x_2{}^2 - x_1{}^2)\beta - \tfrac{1}{2}(x_2 - x_1)^2\beta/n \end{aligned}$$

Now the area A is actually the limit of this sum as n, the number of strips, tends to infinity. The first two terms in the expression for S do not contain n, and will not be affected. But as $n \to \infty$, $1/n \to 0$, and therefore the third term $-\tfrac{1}{2}\beta(x_2 - x_1)^2/n$ will also tend to 0, so

$$A = \lim S = (x_2 - x_1)\alpha + \tfrac{1}{2}(x_2{}^2 - x_1{}^2)\beta.$$

The reader can compare this method of finding the area with that of direct integration

$$\begin{aligned} A = \int_{x_1}^{x_2} y\,dx &= \int_{x_1}^{x_2} (\alpha + \beta x)\,dx \\ &= [\alpha x + \tfrac{1}{2}\beta x^2]_{x_1}^{x_2} \\ &= [\alpha x_2 + \tfrac{1}{2}\beta x_2{}^2] - [\alpha x_1 + \tfrac{1}{2}\beta x_1{}^2] \\ &= \alpha(x_2 - x_1) + \tfrac{1}{2}\beta(x_2{}^2 - x_1{}^2). \end{aligned}$$

This gives the same answer, and is evidently the simpler and quicker way of performing the calculation. But, as will be seen later, the *principle* of the strip method is of great importance.

PROBLEM

(3) Sum the series $1 + 5 + 9 + 13 + \ldots$ to n terms.

11.5 The sigma notation

We have already noted the use of the Greek capital sigma Σ as a sign of summation: $\Sigma\, x$ means "the sum of all quantities x".

If we wish to be more precise we denote a sum like $x_1 + x_2 + x_3 + \ldots + x_n$ by the symbol $\sum\limits_{r=1}^{n} x_r$, read as "sigma from 1 to n (of) x_r". That is, $\sum\limits_{r=1}^{n}$ means "sum for all values of r beginning with 1 and ending with n". This could also be denoted by $\sum\limits_{s=1}^{n} x_s$, or $\sum\limits_{a=1}^{n} x_a$. All these mean exactly the quantity $x_1 + x_2 + \ldots + x_n$: and usually it will be sufficiently clear to write simply $\Sigma_1^n\, x_r$, or $\Sigma_1^n\, x_a$, or even frequently just $\Sigma\, x_a$. (Sometimes the letter S is used instead of Σ.)

In this book we shall almost invariably follow a useful convention, suggested by Prof. C. A. Rogers, that any suffix which is summed over in this way will be denoted by a Greek letter. Thus $\sum_1^n x_\alpha$ will mean $x_1 + x_2 + \ldots + x_n$; $\sum_1^n x_\alpha^2$ will mean $x_1^2 + x_2^2 + \ldots + x_n^2$ (and must be carefully distinguished from $[\sum_1^n x_\alpha]^2$, the square of $\sum_1^n x_\alpha$, which is read as "sigma from 1 to n of x_α, all squared"), and $\sum_1^n \alpha x_\alpha$ will mean $1 x_1 + 2x_2 + 3x_3 + \ldots + nx_n$. The effect of this convention is that complicated expressions can be much simplified: thus $\Sigma x_{\alpha r}$, where there are two suffixes, is to be summed over the Greek α, but not over the Latin r, i.e. it means $x_{1r} + x_{2r} + x_{3r} + \ldots$ (In fact one can with sufficient care even leave out most of the sigma signs, writing x_α or x_α^2 and leaving the Greek suffix to indicate that these are summed as Σx_α or Σx_α^2. But we shall not do that here.)

This use of sigma explains the notation for an integral. We have seen that an area A can be considered as the limit of a sum $\Sigma Y \delta X$, where Y denotes the ordinate in a strip and δX the width of the strip. When we take the limit we obtain the integral denoted by $\int y\, dx$. Here the sigma is replaced by a $\int$, which is simply a form of the letter S standing for Sum, and δX is replaced in the limit by dx, as in differential calculus. Although the symbols $\int$ and dx have strictly speaking no meaning in isolation they serve to remind us that the integral can be approximated to by a ∫um of small *d*ifferences or small strips.

The great advantage of the sigma notation is its compactness. But this compactness is bought at the price of making the symbolism a little harder to comprehend. It is easy to take in at a glance the meaning of the series $(1 + 2 + 3 + \ldots + n)$; it is less easy when it is written as $\sum_{\alpha=1}^{n} \alpha$, and some practice is needed in getting used to such expressions. We therefore give some formal rules of operation which may be helpful.

The expression $\sum_1^n (x_\alpha + y_\alpha)$ means $(x_1 + y_1) + (x_2 + y_2) + \ldots + (x_n + y_n)$ when written out in full. But this can be rearranged as $(x_1 + x_2 + \ldots + x_n) + (y_1 + y_2 + \ldots + y_n)$, or $\sum_1^n x_\alpha + \sum_1^n y_\alpha$. Thus

$$\sum_1^n (x_\alpha + y_\alpha) = \sum_1^n x_\alpha + \sum_1^n y_\alpha \quad . \quad . \quad (11.14)$$

In the same way

$$\sum_{1}^{n}(x_a - y_a) = \sum_{1}^{n} x_a - \sum_{1}^{n} y_a$$

$$\sum_{1}^{n}(x_a + y_a + z_a) = \sum_{1}^{n} x_a + \sum_{1}^{n} y_a + \sum_{1}^{n} z_a$$

$$\sum_{1}^{n}(x_a + x_a^2 + x_a^3) = \sum_{1}^{n} x_a + \sum_{1}^{n} x_a^2 + \sum_{1}^{n} x_a^3.$$

If k is a constant (independent of a) then $\sum_{1}^{n}(kx_a)$ means $kx_1 + kx_2 + \ldots + kx_n$, or $k(x_1 + x_2 + \ldots + x_n)$, or $k\sum_{1}^{n} x_a$. Thus we have the rule for multiplication by a constant:

$$\sum_{1}^{n} kx_a = k\sum_{1}^{n} x_a \qquad . \quad . \quad . \quad (11.15)$$

If instead of multiplying by a constant we add it, we see that $\sum_{1}^{n}(k + x_a)$ means $(k + x_1) + (k + x_2) + \ldots + (k + x_n)$ which equals $nk + (x_1 + x_2 + \ldots + x_n) = nk + \sum_{1}^{n} x_a$. Thus

$$\sum_{1}^{n}(k + x_a) = kn + \sum_{1}^{n} x_a \qquad . \quad . \quad . \quad (11.16)$$

These relations will be sufficient for our present purposes. Certain other important ones are given later (Section 13.14). From our relations we can infer such equations as

$$\sum_{1}^{n}(2 + 3x_a + x_a^2) = 2n + 3\sum_{1}^{n} x_a + \sum_{1}^{n} x_a^2.$$

11.6 Generalized arithmetic series

It is sometimes necessary to find the sums of more complicated series, such as $1^2 + 2^2 + 3^2 + \ldots + n^2$, where the rth term is r^2, or $1 + 3 + 6 + 10 + \ldots + \frac{1}{2}n(n + 1)$, where the rth term is $\frac{1}{2}r(r + 1)$. We may call a series one of "generalized arithmetic type" if the rth term x_r can be expressed as a polynomial in r:

$$x_r = A + Br + Cr^2 + \ldots + Hr^h.$$

In an ordinary arithmetic series the rth term is $X_r = X_1 + (r - 1)\delta = (X_1 - \delta) + \delta r$, so that this is included in the form $A + Br$ with $A = X_1 - \delta$ and $B = \delta$. These series do not occur so often in general biological work, but are sometimes needed in statistical calculations, and it seems convenient to consider them here.

The key to the problem of summing such series is contained in the following propositions:

$$1 + 2 + 3 + \ldots + n = \sum_{1}^{n} \alpha = \tfrac{1}{2}n(n+1)$$

$$1 \cdot 2 + 2 \cdot 3 + 3 \cdot 4 + \ldots + n(n+1) = \sum_{1}^{n} \alpha(\alpha+1) = \tfrac{1}{3}n(n+1)(n+2)$$

$$1 \cdot 2 \cdot 3 + 2 \cdot 3 \cdot 4 + 3 \cdot 4 \cdot 5 + \ldots + n(n+1)(n+2)$$
$$= \sum_{1}^{n} \alpha(\alpha+1)(\alpha+2) = \tfrac{1}{4}n(n+1)(n+2)(n+3) \quad . \quad . \quad (11.17)$$

and so on.

To see how these formulas arise let us prove the second one: this will illustrate the general method of proof, which can equally well be applied to the others. We want therefore to find an expression for the sum

$$s = 1 \cdot 2 + 2 \cdot 3 + 3 \cdot 4 + \ldots + (n-1)n + n(n+1)$$

First multiply by 3, obtaining

$$3s = 3(1 \cdot 2) + 3(2 \cdot 3) + 3(3 \cdot 4) + \ldots + 3(n-1)n + 3n(n+1).$$

Now $3 = -1 + 4 = -2 + 5 = \ldots = -(n-2) + (n+1) = -(n-1) + (n+2)$, so the expression can be rewritten as

$$\begin{aligned} 3s &= 1 \cdot 2 \cdot 3 + (-1 + 4) \cdot 2 \cdot 3 + (-2 + 5) \cdot 3 \cdot 4 + \ldots \\ &\quad + [-(n-2) + (n+1)](n-1)n + [-(n-1) + (n+2)]n(n+1) \\ &= 1 \cdot 2 \cdot 3 - 1 \cdot 2 \cdot 3 + 2 \cdot 3 \cdot 4 - 2 \cdot 3 \cdot 4 + 3 \cdot 4 \cdot 5 - \ldots \\ &\quad -(n-2)(n-1)n + (n-1)n(n+1) - (n-1)n(n+1) + n(n+1)(n+2) \\ &= n(n+1)(n+2) \end{aligned}$$

since the first two terms cancel out, so do the next two, and so on in pairs, and only the last term is left. Dividing through by 3 we finally obtain the required answer

$$s = \tfrac{1}{3}n(n+1)(n+2).$$

PROBLEM

(1) Repeat the procedure with the series
$1 + 2 + \ldots + n$ and $1 \cdot 2 \cdot 3 + 2 \cdot 3 \cdot 4 + \ldots + n(n+1)(n+2)$.

A sum such as $1^2 + 2^2 + 3^2 + \ldots + n^2$ can now be found by the following device.

$$1^2 = 1 \cdot 1 = 1(2 - 1) = 1 \cdot 2 - 1$$
$$2^2 = 2 \cdot 2 = 2(3 - 1) = 2 \cdot 3 - 2$$
$$3^2 = 3 \cdot 3 = 3(4 - 1) = 3 \cdot 4 - 3$$

and in general

$$a^2 = a \cdot a = a[(a + 1) - 1] = a(a + 1) - a.$$

This rearrangement splits the original series into two series whose sums have already been found.

Thus, with all summations running from 1 to n,

$$\begin{aligned}\Sigma a^2 &= \Sigma a(a + 1) - \Sigma a \\ &= \tfrac{1}{3}n(n + 1)(n + 2) - \tfrac{1}{2}n(n + 1) \\ &= n(n + 1)[\tfrac{1}{3}(n + 2) - \tfrac{1}{2}] \\ &= n(n + 1)[\tfrac{1}{6} \cdot 2(n + 2) - \tfrac{1}{6} \cdot 3] \\ &= \tfrac{1}{6}n(n + 1)[2(n + 2) - 3] \\ &= \tfrac{1}{6}n(n + 1)(2n + 1)\end{aligned}$$

i.e. $1^2 + 2^2 + \ldots + n^2 = \tfrac{1}{6}n(n + 1)(2n + 1)$.

The same device can be used to sum $1^3 + 2^3 + 3^3 + \ldots + n^3$. Here the general term a^3 must be expressed in terms of $a(a + 1)(a + 2)$, $a(a + 1)$, and a, which are series we already know how to sum. Now

$$a(a + 1) = a^2 + a$$

$$a(a + 1)(a + 2) = a^3 + 3a^2 + 2a.$$

Since $a(a + 1)$ and a do not give terms in a^3, we must take the expression $a(a + 1)(a + 2)$ to give an a^3, and we can write

$$a^3 = a(a + 1)(a + 2) - 3a^2 - 2a.$$

But we have already seen that $a^2 = a(a + 1) - a$, so that substituting this in the above expression we find that

$$a^3 = a(a + 1)(a + 2) - 3a(a + 1) + a$$

and therefore

$$\begin{aligned}\Sigma a^3 &= \Sigma a(a + 1)(a + 2) - 3\,\Sigma a(a + 1) + \Sigma a \\ &= \tfrac{1}{4}n(n + 1)(n + 2)(n + 3) - \tfrac{3}{3}n(n + 1)(n + 2) + \tfrac{1}{2}n(n + 1) \\ &= \tfrac{1}{4}n^2(n + 1)^2\end{aligned}$$

We can continue in this way to find the sums of higher powers. The calculations are straightforward but laborious, and lead to the following results.

Table 11.1—Sums of Powers

$$
\begin{aligned}
1 + 2 + \dots + n &= \tfrac{1}{2}n(1 + n) \\
&= \tfrac{1}{2}(n + n^2) \\
1^2 + 2^2 + \dots + n^2 &= \tfrac{1}{6}n(1 + n)(1 + 2n) \\
&= \tfrac{1}{6}(n + 3n^2 + 2n^3) \\
1^3 + 2^3 + \dots + n^3 &= \tfrac{1}{4}n^2(1 + n)^2 \\
&= \tfrac{1}{4}(n^2 + 2n^3 + n^4) \\
1^4 + 2^4 + \dots + n^4 &= \tfrac{1}{30}n(1 + n)(1 + 2n)(-1 + 3n + 3n^2) \\
&= \tfrac{1}{30}(-n + 10n^3 + 15n^4 + 6n^5) \\
1^5 + 2^5 + \dots + n^5 &= \tfrac{1}{12}n^2(1 + n)^2(-1 + 2n + 2n^2) \\
&= \tfrac{1}{12}(-n^2 + 5n^4 + 6n^5 + 2n^6) \\
1^6 + 2^6 + \dots + n^6 &= \tfrac{1}{42}n(1 + n)(1 + 2n)(1 - 3n + 6n^3 + 3n^4) \\
&= \tfrac{1}{42}(n - 7n^3 + 21n^5 + 21n^6 + 6n^7) \\
1^7 + 2^7 + \dots + n^7 &= \tfrac{1}{24}n^2(1 + n)^2(2 - 4n - n^2 + 6n^3 + 3n^4) \\
&= \tfrac{1}{24}(2n^2 - 7n^4 + 14n^6 + 12n^7 + 3n^8) \\
1^8 + 2^8 + \dots + n^8 &= \tfrac{1}{90}n(1+n)(1+2n)(-3+9n-n^2-15n^3+ \\
&\qquad 5n^4+15n^5+5n^6) \\
&= \tfrac{1}{90}(-3n+20n^3-42n^5+60n^7+45n^8+10n^9)
\end{aligned}
$$

For further information about the properties of these sums, and for a simpler method of calculation, see T. J. I'A. Bromwich, *An Introduction to the Theory of Infinite Series* (2nd edn., 1931, Macmillan), Chapter XI.

One curious identity emerges from these formulas:

$$(1^3 + 2^3 + 3^3 + \dots + n^3) = (1 + 2 + 3 + \dots n)^2.$$

This is illustrated in Fig. 11.8, where we see that a cubical arrangement of objects of side 2 can be cut into 3 rectangular slices, 1×2, 2×2, 1×2. By rearranging these slices as in the lower half of the diagram we can make $1^3 + 2^3$ objects fill up a square of side $1 + 2$. Similarly a cube of side 3 can be cut into slices 1×3, 2×3, 3×3, 2×3, 1×3, and by adding these on to the square of side $1 + 2$ we get one of side $1 + 2 + 3$, and similarly for larger cubes.

These generalized arithmetic series can be used to find the area under the parabola $y = x^2$ between the ordinates $x_1 = 0$ and $x_2 = 1$. Let this area be divided into n strips of equal width. Then by (11.10) the area A is the limit of the sum

$$S = (x_2 - x_1)(Y_1 + Y_2 + \dots + Y_n)/n.$$

Now $x_2 - x_1 = 1$, and $Y_1 = X_1^2$, $Y_2 = X_2^2$, ... $Y_n = X_n^2$. Thus

$$S = (X_1^2 + X_2^2 + \dots + X_n^2)/n.$$

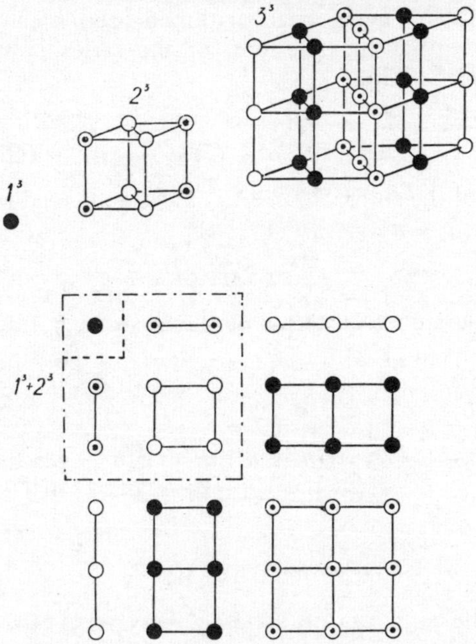

Fig. 11.8—The sum of successive cubes

Now if δ is the width of each strip, $X_2 = X_1 + \delta = \delta$ since $X_1 = 0$. Similarly $X_3 = 2\delta$, $X_4 = 3\delta$, and $X_n = (n-1)\delta$. So

$$S = (\delta^2/n) \,.\, [1^2 + 2^2 + \ldots + (n-1)^2].$$

But by replacing n by $(n-1)$ in the formula for $1^2 + 2^2 + \ldots + n^2$ we see that

$$1^2 + 2^2 + \ldots + (n-1)^2 = \tfrac{1}{6}(n-1)n(2n-1).$$

Furthermore since the total width of the area is 1, and it is divided into n strips, the width of each strip must be $\delta = 1/n$. Making these substitutions we find finally

$$S = (1/n^3) \,.\, \tfrac{1}{6}(n-1)n(2n-1)$$
$$= \frac{1}{6}\left(1 - \frac{1}{n}\right)\left(2 - \frac{1}{n}\right).$$

Now, as $n \to \infty$, $1/n \to 0$, and therefore $S \to A = \frac{1}{6} \,.\, 1 \,.\, 2 = \frac{1}{3}$. This is the required area.

Alternatively by direct integration we have

$$A = \int_0^1 y\,dx = \int_0^1 x^2\,dx = [\tfrac{1}{3}x^3]_0^1 = \tfrac{1}{3} - 0 = \tfrac{1}{3}$$

(Again the integration method turns out to be simpler.)

The sum of any generalized arithmetic series can now be easily found. Suppose that the rth term of the series is $x_r = A + Br + Cr^2 + \ldots + Hr^h$. Then

$$\begin{aligned} \Sigma x\alpha &= \Sigma(A + B\alpha + C\alpha^2 + \ldots H\alpha^h) \\ &= nA + B\Sigma\alpha + C\Sigma\alpha^2 + \ldots + H\Sigma\alpha^h \end{aligned}$$

and the values of the sums $\Sigma\alpha$, $\Sigma\alpha^2$, ... $\Sigma\alpha^h$ can be read from Table 11.1.

EXAMPLE

(1) Find the sum to n terms of the series 1, 7, 19, ... whose rth term is $x_r = 1 - 3r + 3r^2$.

We have

$$\begin{aligned} \Sigma x_r &= n - 3\,\Sigma\alpha + 3\,\Sigma\alpha^2 \\ &= n - 3 \cdot \tfrac{1}{2}(n + n^2) + 3 \cdot \tfrac{1}{6}(n + 3n^2 + 2n^3) \\ &= n - \tfrac{3}{2}(n + n^2) + \tfrac{1}{2}(n + 3n^2 + 2n^3) \\ &= n^3 \end{aligned}$$

PROBLEMS

(2) Find the sum of the series $5 + 23 + 53 + \ldots$ to n terms, given that the rth term is $6n^2 - 1$.

(3) Find a formula for the rth term of the series $2 + 10 + 24 + 44 + 70 + 102 + 140 + \ldots$, and for the sum to n terms.

11.7 Integrals as the limit of sums

We have seen that an area can be considered either as an integral $\int_{x_1}^{x_2} y\,dx$ or as the limit of a sum $\Sigma Y\delta X$. The importance of this result does not lie in providing alternative methods of calculating an area, for, as we have seen, the definite integral is almost always much easier to calculate than the limit of the sum. It lies rather in showing that the two expressions are equal, so that whenever a quantity can be expressed as the limit of a sum like $\Sigma Y\delta X$ we know that it must be the integral $\int y\,dx$.

Consider the problem of finding the work done in stretching an elastic cord or spring from length x_1 to length x_2. If f is the force of tension in the spring at length x, then f will be a function of x. (In general the greater the length x of the cord the greater the tension f.) Now imagine the extension done in a series of small steps; first from length $X_1 = x_1$ to length X_2, secondly from length X_2 to X_3, thirdly from X_3 to X_4, and so on (ending at $X_{n+1} = x_2$). Let the tension at the beginning of each of these steps be F_1, F_2, F_3 ... respectively. Then if the first step is sufficiently small the tension will be approximately F_1 throughout, and the work done in extending the length by

$\delta X_1 = X_2 - X_1$ will be very nearly $F_1 \delta X_1$. In the second step the work done will be nearly equal to $F_2 \delta X_2$, where $\delta X_2 = X_3 - X_2$; and in the whole extension the total work will be $\Sigma F_a \delta X_a$ nearly. The smaller the steps the better the approximation will be. In the limit as the greatest step δX tends to zero the work must be exactly expressed by the integral

$$W = \int_{x_1}^{x_2} f\,dx.$$

If the cord obeys Hooke's Law, $f = A + Bx$ where A and B are constants, then

$$\begin{aligned} W &= \int_{x_1}^{x_2} (A + Bx)\,dx \\ &= [Ax + \tfrac{1}{2}Bx^2]_{x_1}^{x_2} \\ &= [Ax_2 + \tfrac{1}{2}Bx_2^2] - [Ax_1 + \tfrac{1}{2}Bx_1^2] \end{aligned}$$

From now on we shall abandon the use of capital letters in the sum $\Sigma Y \delta X$; we only needed them to distinguish between the limits of integration x_1 and x_2 and the boundary points of the strips X_1, $X_2, \ldots, X_{n+1}$. We can put our rule in the form

If a small change δQ in any quantity Q can be expressed approximately in the form $\delta Q \simeq y\,\delta x$, where y and x are other quantities functionally related to Q, then, when x changes from x_1 to x_2, Q changes by the integral $\int_{x_1}^{x_2} y\,dx$.

For instance, any force F acting along a line will do work $W = \int_{x_1}^{x_2} F\,dx$ in moving a body from position x_1 to x_2 in this line: for the work done in a small change of position δx is $\delta W \simeq F\,\delta x$. If the force itself does not act in the line in question the above formula will still hold, provided that we take F to be the component of the force along the line. If we plot the graph of F against x, then the work done will be represented by the area under the curve.

Consider a gas at pressure P and volume V. The work done by the gas in expanding by a small amount δV will be approximately $P\delta V$. Thus the work done in expanding from a volume V_1 to a volume V_2 is approximated to by the sum $\Sigma P \delta V$, or in the limit it is exactly expressed by the integral $\int_{V_1}^{V_2} P\,dV$. If the gas obeys the law $PV = nRT$ and the expansion takes place at constant temperature T, then $P = nRT/V$, and the work done is

$$\begin{aligned} W &= \int_{V_1}^{V_2} nRT/V \,.\, dV = [nRT \ln V]_{V}^{V_2} \\ &= nRT\,(\ln V_2 - \ln V_1). \end{aligned}$$

If a body is moving with velocity v at time t, then the distance δy travelled during a short interval of time δt will be approximately the product $v\ \delta t$ of the velocity and the length of time, δt. In a finite interval, from t_1 to t_2, the distance moved will therefore be $\int_{t_1}^{t_2} v\,dt$.

11.8 Areas of loops

The areas considered in previous sections have been of a very special form, with three straight sides and one curved one. However, it is an easy matter to obtain the area of any curved figure. Take first a simple loop *ABCDA* (Fig. 11.9) with left-hand end *A* and right-hand end *C*.

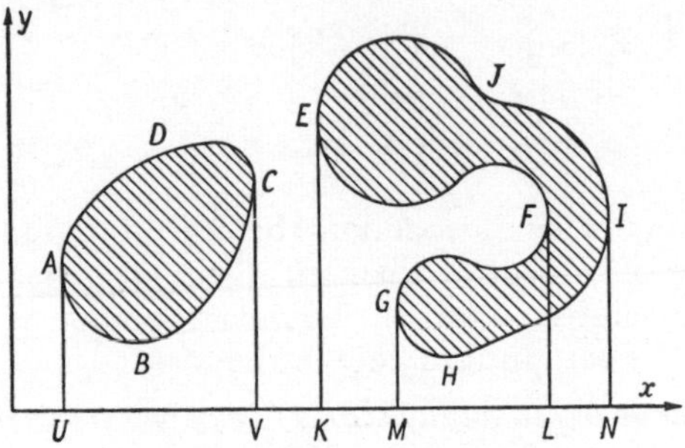

Fig. 11.9—The area of a loop

Draw perpendiculars *AU*, *CV* on to the *x*-axis. The area *AUVCDA* can then be evaluated by integrating *y* with respect to *x* along the upper part of the loop

$$\text{area } AUVCDA = \int_{ADC} y\,dx.$$

The area *AUVCBA* can similarly be found by integrating *y* along the lower part of the loop. But the area of the loop is simply the difference, area *AUVCDA* − area *AUVCBA*.

A more complicated figure such as *EFGHIJE* in Fig. 11.9 can be dealt with by a similar procedure. Here the area of the loop = area *EKNIJE* − area *EKLFE* + area *GMLFG* − area *GMNIHG*, and each of these areas can be calculated by integration.

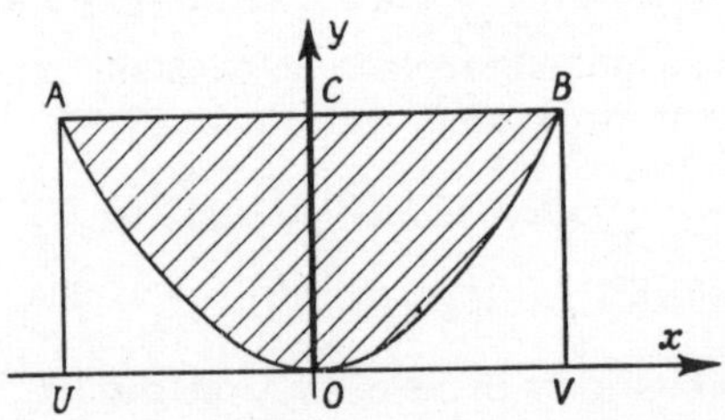

Fig. 11.10—The area of a segment of a parabola

EXAMPLE

(1) Find the area AOB between the parabola $y = x^2$ and the line $y = 1$ (Fig. 11.10).

A and B, the intersections of the parabola and the line, are the points $(-1, 1)$ and $(1, 1)$ respectively. Drawing perpendiculars AU, BV onto the x-axis we have therefore

$$\text{area } AUVBCA = \int_{-1}^{1} 1\,dx = 2$$

$$\text{area } AUVBOA = \int_{-1}^{1} x^2\,dx = \left[\tfrac{1}{3}x^3\right]_{-1}^{1} = \tfrac{2}{3}$$

$$\text{area of loop} = 2 - \tfrac{2}{3} = \tfrac{4}{3}.$$

11.9 Areas divided into curved strips

So far we have considered areas as divided into a large number of thin straight strips, and thereby in the limit expressed as an integral. The calculations can often be simplified by using curved rather than straight strips.

The basis of this method is that a strip of length L and uniform thickness w, where w is small, has an area very nearly equal to wL (Fig. 11.11). For imagine this strip cut into small segments of length

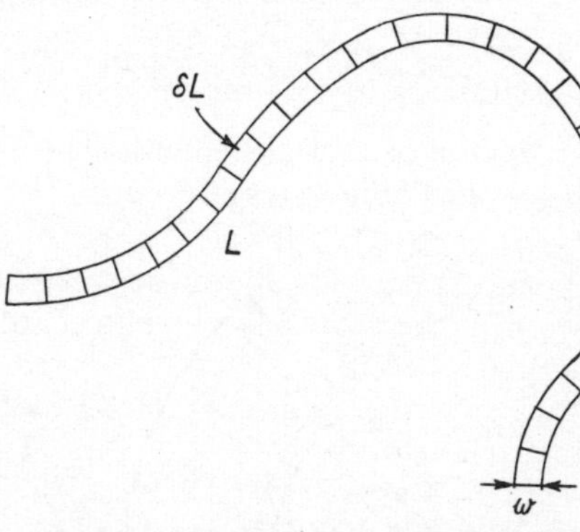

Fig. 11.11—The area of a curved strip

δL by lines perpendicular to the strip. Each of these segments will be nearly rectangular in form, and will therefore have area $\simeq$ length $\times$ breadth $= w\,\delta L$. By the addition of all these small segments we find the total area to be $\Sigma w\,\delta L = w\Sigma\,\delta L$ (since w is constant) $= wL$.

EXAMPLES

(1) The area of a circle of radius R (Fig. 11.12).

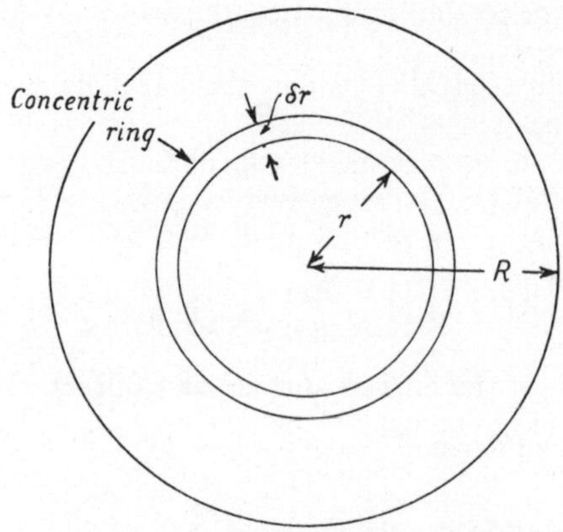

Fig. 11.12—The area of a circle

Let us imagine the circle divided up into a number of rings by concentric circles of radius r varying from 0 to R. Take any one such ring, say of radius r and width δr. Then its area δA will be its length $2\pi r$ (Section 6.10) times its width δr; $\delta A = 2\pi r\ \delta r$. The area of the whole circle will be approximately the sum of the areas of all these rings, or $\Sigma\delta A = \Sigma 2\pi r\ \delta r$; and the narrower the rings the more accurate this formula will be. Thus in the limit the area will be exactly

$$\int_0^R 2\pi r\, dr = \left[\pi r^2\right]_0^R = \pi R^2 \qquad (11.18)$$

(2) The area of a sector of a circle bounded by two radii making an angle θ with one another (Fig. 11.13).

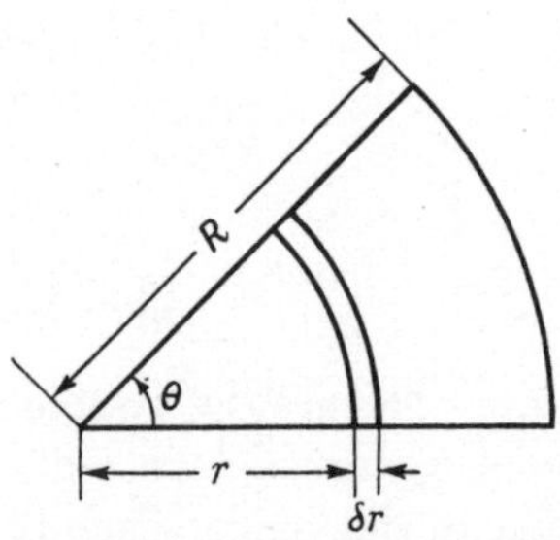

Fig. 11.13—The area of a sector of a circle

Again the sector is to be divided into strips of uniform width between the arcs of concentric circles. If θ is measured in radians then the length of any such strip of radius r will be $r\theta$, and if its width is δr its area will be approximately $\delta A = r\theta\ \delta r$. The total area of the sector will accordingly be obtained approximately by summing these strips to find $\Sigma\, r\theta\, \delta r$; or in the limit when the width δr tends to 0 it will be exactly expressed as

$$\int_0^R r\theta\, dr = \left[\tfrac{1}{2}r^2\theta\right]_0^R = \tfrac{1}{2}R^2\theta.$$

So far we have considered only plane areas. This method of cutting the area up into strips is equally applicable to curved surfaces.

FURTHER EXAMPLES

(3) The area of the curved surface of a (circular) cylinder of length L and radius R (Fig. 11.14).

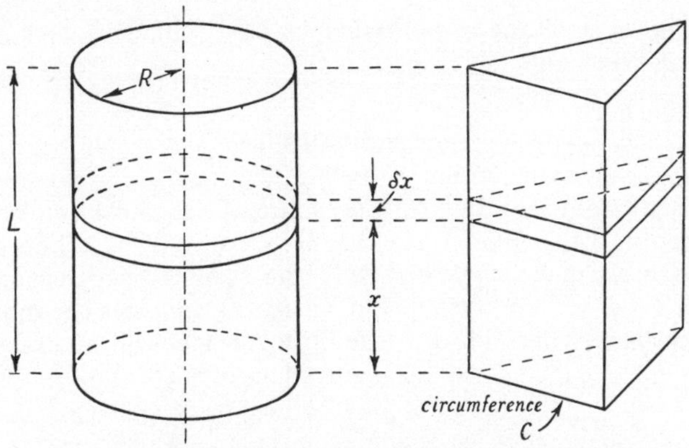

Fig. 11.14—The surface area of a cylinder or prism

We shall imagine the base of the cylinder to be horizontal, and the axis vertical, as in the figure. Let the surface be divided into thin circular strips by cutting it by a large number of horizontal planes. Call the height (above the base) of a typical cutting plane "x": then the width of such a strip is the change in x between the bottom and top of the strip; i.e. it is δx. The length of the strip is the circumference $2\pi R$ of the cylinder, and its area is therefore approximately $2\pi R\, \delta x$. The area of the whole curved surface is approximately $\Sigma\, 2\pi R\, \delta x$, or exactly $\int_0^L 2\pi R\, dx$, since x varies from 0 at the base to L at the top. But $\int_0^L 2\pi R\, dx = \left[2\pi Rx\right]_0^L = 2\pi RL$.

A similar formula holds for the general cylinder or prism, defined as the volume swept out by any plane figure moved a distance L perpendicular to itself. Such a triangular prism is shown in Fig. 11.14; again we imagine the base to be horizontal. If the circumference of the base is C, the area of the vertical surface is CL. For we again cut the surface into strips of width δx by planes parallel to the base. The area of such a strip is approximately $C\delta x$, and the area of the whole is $\Sigma C\ \delta x$, or exactly $\int_0^L C\,dx$ in the limiting case when the number of strips increases indefinitely, and the greatest width tends to zero. But $\int_0^L C\,dx = [Cx]_0^L = CL$. One can in fact take a rectangle of paper or similar material of sides C and L and wrap it exactly round the cylinder or prism, thus showing in another way that its area must be CL.

If it is the whole surface area of the cylinder which is required, then the areas of the two flat ends must be included. For a circular cylinder each end will have area πR^2 (Example 1), and the total area will be $2\pi RL + 2\pi R^2 = 2\pi R(R + L)$.

(4) The area of the curved surface of a (circular) cone with base of radius R and height H (Fig. 11.15).

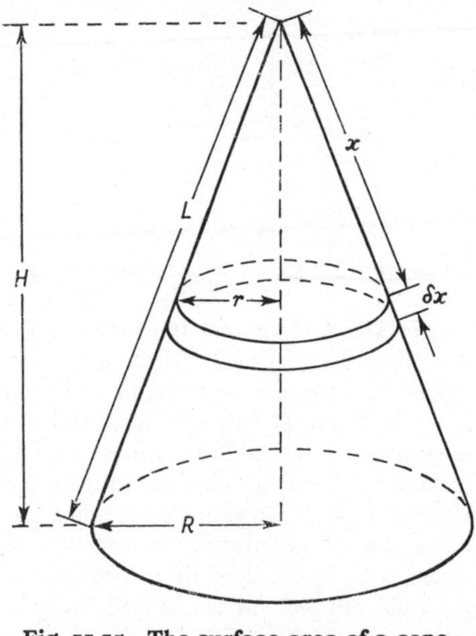

Fig. 11.15—The surface area of a cone

We again imagine the base of the cone to be horizontal, and the axis vertical, as in the figure. If L is the "slant height" of the cone, i.e. the distance from vertex to base measured on the surface, then by Pythagoras's Theorem

$$L = \sqrt{(R^2 + H^2)}$$

Again let the surface be divided into thin horizontal strips, and let x be the distance of a typical strip from the vertex, measured down the sloping surface of the cone. Then the width of the strip will be the change in the value of x between its two edges, i.e. δx. The length of the strip will be $2\pi r$, where r is its radius. But by simple proportion (see Fig. 11.15) $r/x = R/L$, i.e. $r = xR/L$, and the length of the strip is therefore $2\pi xR/L$, and its area (approximately) $2\pi xR\ \delta x/L$. Adding all these strips together we find the total curved surface area to be approximately $\Sigma\ 2\pi xR\delta x/L$, or, in the limit, exactly

$$\int_0^L 2\pi xR\ dx/L = \left[\pi x^2R/L\right]_0^L$$
$$= \pi RL$$
$$= \pi R\sqrt{(R^2 + H^2)}.$$

For the total surface area of the cone the base must also be included, giving the final formula $\pi R\,[R + \sqrt{(R^2 + H^2)}]$.

An alternative approach is to cut out in paper a sector of a circle of radius L and arc $2\pi R$. This sector can be bent to fit exactly over the slanting surface of the cone. Now if θ is the angle in this sector, measured in radians, then the arc is $L\theta$ and the area is $\frac{1}{2}L^2\theta$. Thus $L\theta = 2\pi R$, whence $\theta = 2\pi R/L$, and the slanting area $= \frac{1}{2}L^2(2\pi R/L) = \pi RL$.

(5) Area of a sphere of radius R (Fig. 11.16).

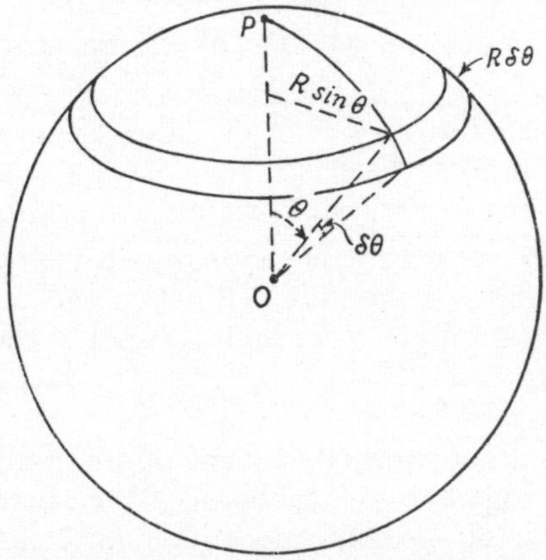

Fig. 11.16—The surface area of a sphere

Let O be the centre of the sphere. It is convenient to take a fixed point P on the sphere, which we may call the "north pole", and cut the sphere into thin strips by "circles of latitude" surrounding P. Let θ be the angle subtended at O (or "difference in latitude") between P and such a strip. Then if the strip corresponds to a difference $\delta\theta$ in θ, its width will be $R\delta\theta$ (see Fig. 11.16). Furthermore its radius will be $R \sin\ \theta$, and therefore the length of the strip will be $2\pi R \sin\ \theta$, and its area $(2\pi R \sin \theta)(R\delta\theta) = 2\pi R^2 \sin \theta\, \delta\theta$. Now θ varies from 0 at the north pole P to π (= 180°) at the south pole and so we see that the total area obtained by adding all strips is

$$\begin{aligned}\int_0^\pi 2\pi R^2 \sin\theta\, dv &= [-\,2\pi R^2 \cos\theta]_0^\pi \\ &= [-\,2\pi R^2(-\,1 - 1)] \\ &= 4\pi R^2.\end{aligned}$$

PROBLEMS

(1) Show that the area of a parallelogram is the product of the length of the base times the height (measured at right angles to the base).

(2) Show that the area of a triangle is half the product of the length of the base and the height.

(3) Find the area of the segment of a circle of radius R cut off by a straight line at distance a from the centre.

(4) A goat is tethered in a circular field of radius R by a chain of length L attached to a point on the circumference of the field. Find the area over which the goat can graze. Taking $R = 1$, plot a graph of this for different lengths L of the chain. Estimate the length L for which the goat can graze over half the area of the field.

(5) Find the area of a cylinder by taking strips parallel to the axis.

(6) A cylindrical birthday cake of radius R and height H is cut by two plane cuts through the axis at angle θ (radians) with one another. Find the surface area of the piece cut out.

(7) A "frustum" of a cone is the part cut out by two planes perpendicular to the axis. If the radii of the two ends are R_1 and R_2 and the thickness of the frustum is T, find the area of the sloping part.

11.10 Volumes

Most volumes can be found by cutting up into thin slices of uniform thickness, adding together the volumes of the slices, and taking the limit as the number of slices increases indefinitely and their thickness tends to 0. This gives a definite integral.

The volume of a thin slice based on a flat or curved area A and of thickness w is approximately wA. This is almost obvious: but it can be proved by cutting the slice into small nearly rectangular parts (Fig. 11.17). (The parts near the edge may give irregularly shaped parts, but these will have only a small total volume, as is practically obvious from the diagram, and could be rigorously proved under suitable conditions by a more complicated argument.) Now each of these parts, if we take the thickness of the slice into account, is a nearly

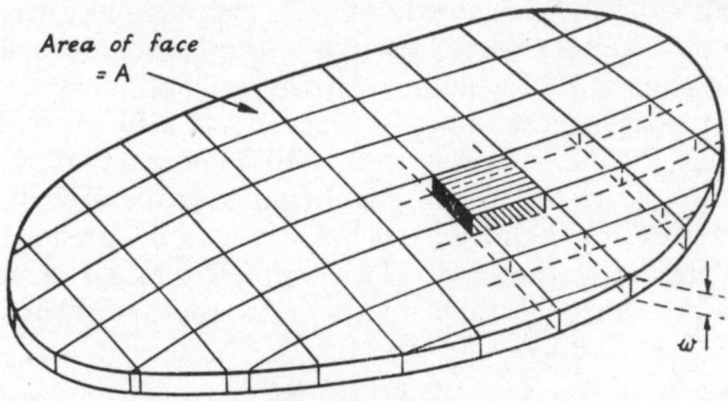

Fig. 11.17—The volume of a thin slice

rectangular block: its volume is therefore the product of its thickness, breadth, and length, i.e. the thickness w times the area of its upper face. If we add all these blocks together we get a total volume $\simeq w$ times the area A of the face. The smaller w is, the nearer this formula will be to exact truth.

EXAMPLES

(1) The volume of a cylinder of radius R and height H (Fig. 11.18).

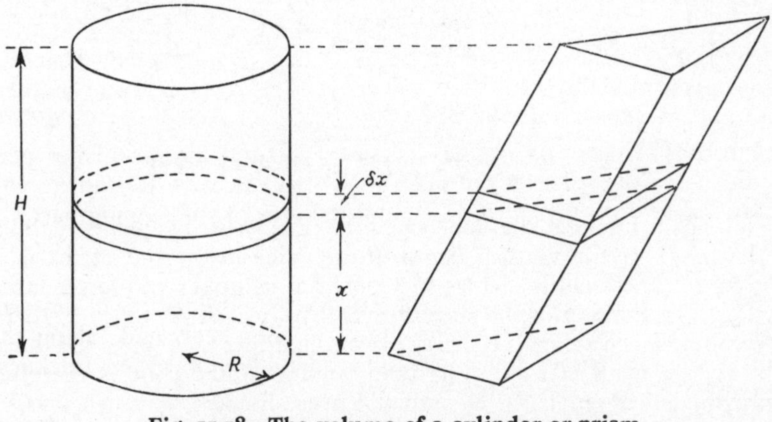

Fig. 11.18—The volume of a cylinder or prism

Again consider the cylinder cut into thin strips by planes parallel to the base. If a typical strip is at height x, and of thickness δx, its volume is the product of δx times the area of the base, $= \pi R^2\, \delta x$. The total volume is therefore approximately $\Sigma\, \pi R^2\, \delta x$, or exactly $\int_0^H \pi R^2\, dx = \pi R^2 H$.

A similar formula holds for the general cylinder or prism: its volume is equal to the product of the area A of the base times the height H. This is true even if the cylinder or prism is "oblique", i.e. if it is the volume swept out by the area A moved parallel to itself (Fig. 11.18), provided that the height H is measured between the two end faces and perpendicular to the base. The formula for the *area* of the sloping surface of an oblique cylinder can be shown to be the height H times the circumference of the base. The simplest method of showing this is to cut it into strips parallel to the axis: the details are left to the reader.

(2) The volume of a cone of radius R and height H (Fig. 11.19).

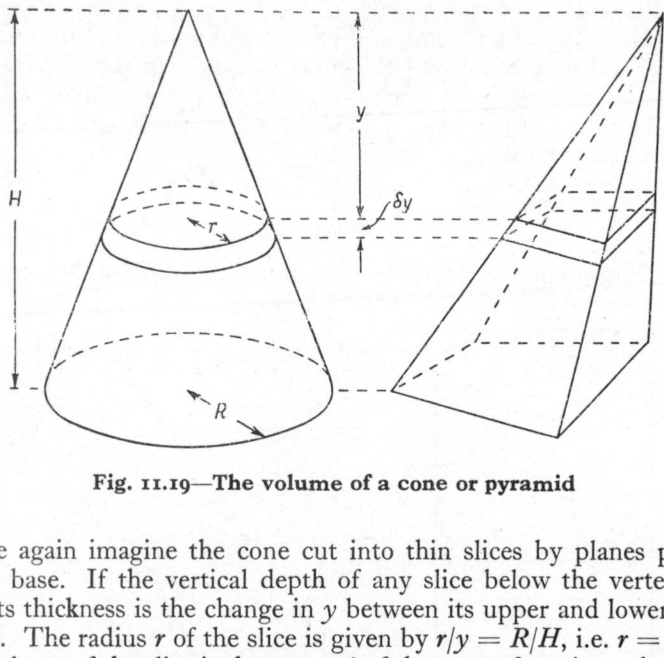

Fig. 11.19—The volume of a cone or pyramid

We again imagine the cone cut into thin slices by planes parallel to the base. If the vertical depth of any slice below the vertex is y, then its thickness is the change in y between its upper and lower faces, i.e. δy. The radius r of the slice is given by $r/y = R/H$, i.e. $r = yR/H$. The volume of the slice is the area πr^2 of the upper face times the thickness δy, i.e. $\pi(yR/H)^2\ \delta y$. The total volume of the cone is therefore approximately $\Sigma\ \pi(yR/H)^2\ \delta y$ summed over all slices, or since y varies from 0 to H, it is accurately

$$\begin{aligned} V &= \int_0^H \pi(yR/H)^2\, dy \\ &= \pi(R/H)^2 \int_0^H y^2\, dy \\ &= \pi(R/H)^2\, [\tfrac{1}{3}y^3]_0^H \\ &= \tfrac{1}{3}\pi(R/H)^2 H^3 = \tfrac{1}{3}\pi R^2 H. \end{aligned}$$

A similar argument applies to any cone or pyramid, including an oblique cone or pyramid (Fig. 11.19). Such a cone is defined as the volume enclosed between a plane base of area A and all the straight lines joining points on the circumference of A to a fixed point P called the "vertex" of the cone. If the perpendicular distance of P from the plane containing the base A is H, then the volume of the cone is $\frac{1}{3}AH$. (Its area is much more difficult to obtain.)

(3) The volume of a sphere of radius R (Fig. 11.20).

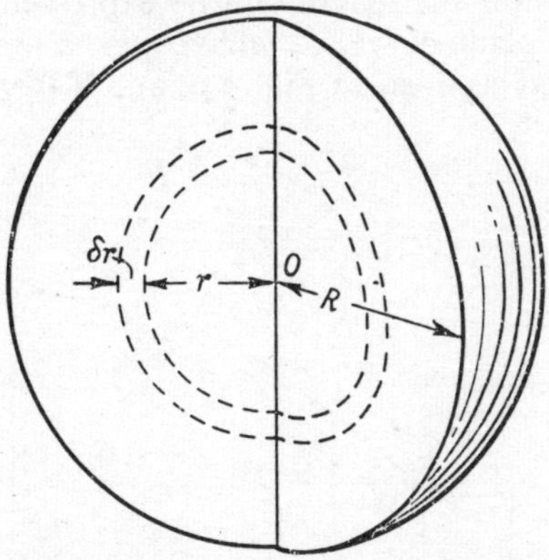

Fig. 11.20—Construction for determining the volume of a sphere

Imagine the sphere cut into concentric spherical slices. Let us take any one such slice, of inner radius r: then its thickness is the change in r between the inner and outer surfaces, i.e. δr. The area of the inner face is $4\pi r^2$, by the formula for the area of a sphere. The volume of the slice is accordingly approximately $4\pi r^2 \delta r$, and by summing and taking the limit in the usual way we find for the volume of the sphere the integral

$$\int_0^R 4\pi r^2\, dr = [4\pi r^3/3]_0^R = \tfrac{4}{3}\pi R^3.$$

PROBLEM

(1) Show that the volume of a "parallelopipedon"—i.e. a box with three pairs of opposite parallel faces, as in Fig. 11.21, but not necessarily

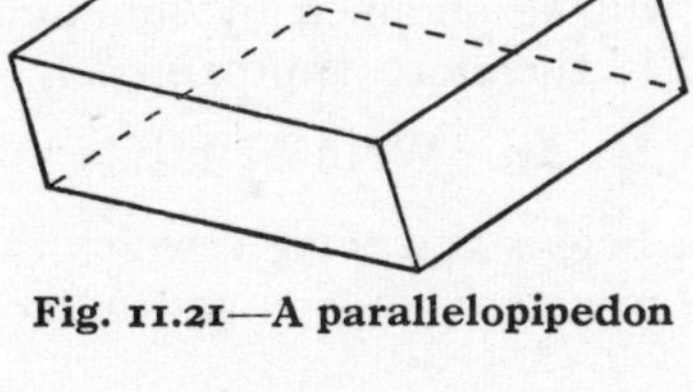

Fig. 11.21—A parallelopipedon

rectangular—is the product of the area of the base and the perpendicular distance between the base and top surface.

11.11 Length

The length of a curve is calculated by cutting it up into small portions, or arcs, finding the approximate length of each, and summing these. The smaller the arcs are into which the curve is cut the more nearly their approximate sum will approach the true length of the curve: and in the limit the length will be expressed as an integral.

Consider first a plane curve, *C*, which we shall take as the graph of a function y plotted against x (Fig. 11.22). Take two neighbouring

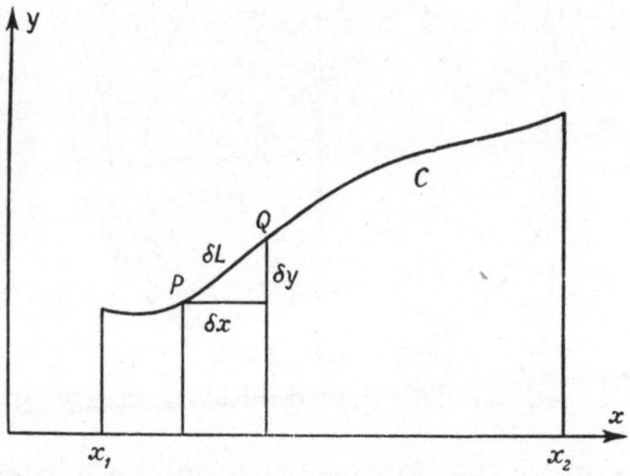

Fig. 11.22—Calculation of the length of a plane curve C

points P and Q on this curve: let P have co-ordinates (x, y) and Q $(x + \delta x, y + \delta y)$. If P and Q are sufficiently near together then the arc PQ, of length δL say, is very nearly a straight line. Its length is therefore given by Pythagoras's theorem (see Fig. 11.22).

$$(\delta L)^2 \simeq (\delta x)^2 + (\delta y)^2$$
$$= (\delta x)^2 \cdot [1 + (\delta y/\delta x)^2]$$
$$\text{i.e.} \quad \delta L \simeq \delta x \cdot \sqrt{[1 + (\delta y/\delta x)^2]}$$

Now when P and Q are near together the quotient $\delta y/\delta x$ is very nearly equal to the derivative $D_x y$, or y_x as we shall write it for the sake of brevity. Thus $\delta L \simeq \delta x \sqrt{[1 + y_x^2]}$.

The total length will be obtained by summing the lengths of all these small arcs. It is therefore approximately $\Sigma\, \delta x \sqrt{[1 + y_x^2]}$, or accurately $\int \sqrt{[1 + y_x^2]} \cdot dx$. We have finally to fill in the limits of integration. If the arc begins at a point at which the x-co-ordinate is a, and ends with x-co-ordinate b, then a and b are the required limits (since we are integrating with respect to x), and the length of the whole curve is

$$L = \int_a^b \sqrt{[1 + y_x^2]} \, . \, dx$$

EXAMPLES

(1) Find the length of the arc of the parabola $y = x^2$ from $x = 0$ to $x = 1$. Here $y_x = 2x$, so that

$$\begin{aligned} L &= \int_0^1 \sqrt{[1 + 4x^2]}\, dx \\ &= 2 \int_0^1 \sqrt{[\tfrac{1}{4} + x^2]}\, dx \\ &= 2 \, . \, (\tfrac{1}{4}/2)\, [\sinh^{-1} 2x + (x/\tfrac{1}{4}) \sqrt{(\tfrac{1}{4} + x^2)}]_0^1 \\ &= \tfrac{1}{4}\, [\sinh^{-1} 2 + \sqrt{20}] \end{aligned}$$

(2) Find the length of the catenary $y = \cosh x$ from $x = 0$ to $x = 1$.

Here $y_x = \sinh x$, so that

$$\begin{aligned} L &= \int_0^1 \sqrt{[1 + (\sinh x)^2]}\, dx \\ &= \int_0^1 \cosh x \, . \, dx \\ &= [\sinh x]_0^1 = \sinh 1. \end{aligned}$$

If we have a curve in 3-dimensional space, and if the points are specified by 3 cartesian co-ordinates (x, y, z), then the length can similarly be shown to be the integral $L = \int_a^b \sqrt{[1 + y_x^2 + z_x^2]}\, dx$, where $x = a$ at the beginning of the curve and $x = b$ at the end.

11.12 Surfaces of revolution

The sphere, circular (right) cylinder, and circular (right) cone are all examples of "surfaces of revolution", the sort of surfaces which can be produced by a lathe. We can look on a sphere as produced by a semicircle rotated round its diameter, which is held fixed. Similarly a cylinder can be produced by rotating a rectangle round one side, and a cone by rotating a right-angled triangle round one side (not the one opposite the right angle). The general surface of revolution will be obtained by rotating a plane curve C around the x-axis, which will become the central axis of the surface.

There are general formulas for the volume and area of such a surface. Let r be the radius of a section at a distance x along the axis (Fig. 11.23): the section, taken perpendicular to the axis, will be circular, and of area πr^2. If we take a small slice of thickness δx lying between two such sections, its volume will be approximately $\pi r^2 \delta x$.

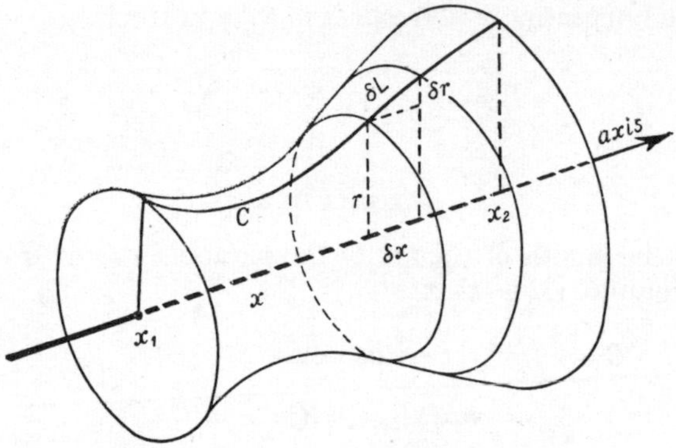

Fig. 11.23—A surface of revolution

It follows that the total volume contained within the surface between the planes $x = a$ and $x = b$ will be the limit of the sum

$$\Sigma\, \pi r^2\, \delta x, \quad \text{i.e. it will be } \int_a^b \pi r^2\, dx.$$

The two parallel-section planes at distance δx apart will also cut off a small strip of the surface. The length of this strip will be the circumference of the section, i.e. $2\pi r$; let us call its width δL. If we take a section of the surface by a plane through the axis, then we obtain a curve, C. It is in fact this curve which we imagine rotated round the axis to form the surface of revolution. The width δL of the strip is then equal to the length of the small arc of the curve cut off between x and $x + \delta x$; and, just as in the case of the method of finding the length of a curve (Section 11.11) we must have

$$(\delta L)^2 \simeq (\delta x)^2 + (\delta r)^2$$

where δr is the change in the radius r when x increases by δx. Written in another form,

$$\begin{aligned} \delta L &\simeq \sqrt{[(\delta x)^2 + (\delta r)^2]} \\ &= \sqrt{[(\delta x)^2 \{1 + (\delta r/\delta x)^2\}]} \\ &= \delta x \,.\, \sqrt{[1 + (\delta r/\delta x)^2]} \\ &\simeq \delta x \sqrt{[1 + r_x{}^2]} \end{aligned}$$

where r_x is the derivative $D_x r = dr/dx$. It follows that the area of the strip is approximately $2\pi r\, \delta L \simeq 2\pi r \sqrt{[1 + r_x{}^2]}\, \delta x$. By adding all these strips together we find the total area of the surface to be approximately $\Sigma\, 2\pi r \sqrt{[1 + r_x{}^2]} \,.\, \delta x$, or in the limit exactly

$$2\pi \int_a^b r \sqrt{[1 + r_x{}^2]} \,.\, dx$$

This formula gives the area of the curved surface between the planes $x = a$ and $x = b$, and does not include the area of the two circular ends.

EXAMPLES

(1) The area and volume of a cylinder calculated from the general formulas.

A cylinder has constant radius $r = R$ for all values of x: thus $r_x = 0$. Its length is $b - a = H$. By our formulas

$$\begin{aligned}\text{Volume } V &= \int_a^b \pi r^2 dx \\ &= \int_a^b \pi R^2 dx \\ &= [\pi R^2 x]_a^b = \pi R^2 b - \pi R^2 a \\ &= \pi R^2(b - a) = \pi R^2 H.\end{aligned}$$

Surface area A (excluding ends)

$$\begin{aligned} &= \int_a^b 2\pi r \sqrt{[1 + r_x^2]}\, dx \\ &= \int_a^b 2\pi R dx \\ &= [2\pi Rx]_a^b = 2\pi R\,(b - a) = 2\pi RH.\end{aligned}$$

(2) The area and volume of a spheroid.

A "spheroid" is the body formed by rotating an ellipse about one of its axes. If the ellipse is rotated about the minor (shorter) axis the figure obtained is an "oblate" spheroid, similar to the shape of the earth, which is flattened at the poles. If the axis of rotation is the major (longer) axis we obtain a "prolate" spheroid. Since an ellipse can be regarded as a circle squashed or pulled out in the direction of an axis, a spheroid will be the figure got by contracting or expanding a sphere in a given direction in a given ratio.

Suppose therefore that the axis about which the ellipse is rotated has length $2a$, and the perpendicular axis has length $2b$. The relation between x and r will then be given by the usual equation for an ellipse (5.28)

$$x^2/a^2 + r^2/b^2 = 1.$$

Solving this we have

$$\begin{aligned} r^2 &= b^2(1 - x^2/a^2) \\ \text{or} \quad r &= b\sqrt{(1 - x^2/a^2)}\end{aligned}$$

On differentiating this with respect to x we find

$$r_x = -ba^{-2}x/\sqrt{(1 - x^2/a^2)}$$

The values of x will range from $-a$ to a, and therefore the volume is

$$\begin{aligned} V &= \int_{-a}^{a} \pi r^2\, dx \\ &= \pi \int_{-a}^{a} b^2 (1 - x^2/a^2)\, dx \\ &= \pi b^2 [x - \tfrac{1}{3}x^3/a^2]_{-a}^{a} \\ &= \pi b^2 [(a - \tfrac{1}{3}a^3/a^2) - (-a + \tfrac{1}{3}a^3/a^2)] \\ &= \pi b^2 [\tfrac{2}{3}a - (-\tfrac{2}{3}a)] \\ &= \tfrac{4}{3}\pi a b^2 \end{aligned}$$

The area is

$$\begin{aligned} A &= 2\pi \int_{-a}^{a} r \sqrt{(1 + r_x{}^2)}\, dx \\ &= 2\pi \int_{-a}^{a} b\sqrt{1 - x^2a^{-2}} \sqrt{1 + \frac{b^2x^2a^{-4}}{1 - x^2a^{-2}}}\, dx \\ &= 2\pi b \int_{-a}^{a} \sqrt{(1 - x^2a^{-2} + b^2x^2a^{-4})}\, dx \\ &= 2\pi b \int_{-a}^{a} \sqrt{1 - x^2a^{-4}(a^2 - b^2)}\, dx \\ &= 2\pi b \int_{-a}^{a} \sqrt{a^{-4}(a^2 - b^2)\left[\frac{1}{a^{-4}(a^2 - b^2)} - x^2\right]}\, dx \end{aligned}$$

Now in this formula we have two possibilities, since we have not stated which axis is longer, the one about which the ellipse is being rotated (of length $2a$) or the perpendicular axis (of length $2b$). First suppose $a > b$, i.e. the spheroid is prolate (a sphere pulled out). Then $a^2 - b^2$ is positive. We can simplify the formula by writing $\sqrt{\frac{1}{a^{-4}(a^2 - b^2)}} = H$, so that the integral for the area becomes a standard form (p. 233)

$$\begin{aligned} A &= \frac{2\pi b}{H} \int_{-a}^{a} \sqrt{(H^2 - x^2)}\, dx \\ &= \frac{2\pi b}{H}\left[\frac{H^2}{2}\right]\left[\sin^{-1}\frac{x}{H} + \frac{x\sqrt{(H^2 - x^2)}}{H^2}\right]_{-a}^{a} \\ &= 2\pi b H \left[\sin^{-1}\frac{a}{H} + \frac{a\sqrt{(H^2 - a^2)}}{H^2}\right] \end{aligned}$$

On the other hand if $a < b$ we have an oblate or flattened spheroid. In that case $(b^2 - a^2)$ is positive, and we can write $\sqrt{\frac{1}{a^{-4}(b^2 - a^2)}} = K$.

The integral then simplifies to the standard form

$$A = \frac{2\pi b}{K}\int_{-a}^{a} \sqrt{(K^2 + x^2)}\, dx$$

$$= \frac{2\pi b}{K}\left[\frac{K^2}{2}\right]\left[\sinh^{-1}\frac{x}{K} + \frac{x\sqrt{(K^2 + x^2)}}{K^2}\right]_{-a}^{a}$$

$$= 2\pi bK\left[\sinh^{-1}\frac{a}{K} + \frac{a\sqrt{(K^2 + a^2)}}{K^2}\right]$$

PROBLEMS

(1) Find the area and volume of a cone, using the general formulas.

(2) A wooden ring is made by boring a cylindrical hole through a sphere, the axis of the cylinder passing through the centre of the sphere. If the length of the hole (measured parallel to the axis) is L, what is the volume of the wood in the ring?

11.13 Summary of principal results

(All integrals have limits of integration x_1 and x_2.)

LENGTHS

Circumference of circle = 2π (radius).
Arc of circle = (radius) (angle at centre in radians).
General plane curve = $\int \sqrt{(1 + y_x^2)}\, dx$.

General curve in space = $\int \sqrt{(1 + y_x^2 + z_x^2)}\, dx$.

AREAS

Rectangle or parallelogram = (base) (perpendicular height).
Triangle = $\frac{1}{2}$ (base) (perpendicular height).
Circle = π (radius)^2.
Sector of circle = $\frac{1}{2}$ (radius)^2 (angle at centre in radians).
Ellipse = π (semi-major axis) (semi-minor axis).
Circular cylinder (excluding end faces) = 2π (radius) (perpendicular height).
General cylinder or prism (excluding end faces) = (circumference of base) (perpendicular height).
Right circular cone (excluding base) = π (radius of base) (slant height).
Sphere = 4π (radius)^2.
Area under general plane curve = $\int y\, dx$.

Area of surface of revolution (excluding end faces)

$$= 2\pi \int r \sqrt{(1 + r_x^2)}\, dx.$$

VOLUMES

Box or parallelopipedon or cylinder or prism
$= $ (area of base) (perpendicular height).
Circular cylinder $= \pi$ (radius)^2(perpendicular height).
Cone or pyramid $= \frac{1}{3}$ (area of base) (perpendicular height).
Circular cone $= \frac{1}{3}\pi$ (radius)^2(perpendicular height).
Sphere $= \frac{4}{3}\pi$ (radius)^3.
Spheroid $= \frac{1}{6}\pi$ (axis length) (perpendicular diameter)^2.
Surface of revolution $= \pi \int r^2\,dx$.

EXAMPLES

(1) Assuming metabolism to be proportional to amount of surface area, compare the rate of metabolism in man, whose surface area is about 260 square centimetres per kilo of body weight, with that of a cylindrical *bacillus coli*, whose size is $\mu \times 2\mu$, assuming its specific gravity to be 1, and $\mu = 10^{-4}$ cm.

$$\text{Volume of bacillus} = \pi r^2 h = \pi\left(\frac{\mu}{2}\right)^2 2\mu = \frac{\pi \times 10^{-12}\ \text{cc}}{2}$$

$$\therefore \text{Weight of bacillus} = \frac{\pi \times 10^{-12}}{2}\ \text{gram}$$

$$\text{Surface of bacillus} = 2\pi r(r+h) = 2\pi\left(\frac{\mu}{2}\right)\left(\frac{\mu}{2} + 2\mu\right)$$

$$= \frac{5\pi \times 10^{-8}}{2}\ \text{sq cm}$$

$$\therefore \text{Surface per gram of bacillus} = \frac{5\pi \times 10^{-8}}{\pi \times 10^{-12}} = 5 \times 10^4\ \text{sq cm}$$

$$\text{and surface per kilogram of bacillus} = 5 \times 10^7\ \text{sq cm}$$

$$\therefore \frac{\text{Surface per kilo of bacillus}}{\text{Surface per kilo of man}} = \frac{5 \times 10^7}{260} = \text{about } 2 \times 10^5$$

Hence, metabolism in bacteria is about 200,000 times as quick as in man. It is on account of this enormous rate of metabolism and absorption of food material that bacterial growth is so rapid—division taking place at the rate of about once in half an hour.

(2) It has been found that the average diameter of an adult's pulmonary air-cell = 0·2 mm, whilst that of an infant's air-cell (at birth) = 0·07 mm. Assuming that these air-cells are spherical, and that the total volume of the lungs = 1617 cc in the adult, and 67·7 cc in the new-born infant, find the total number of air-cells and their total surface in the adult and in the new-born infant.

Volume of single air-cell in adult $= \frac{4}{3}\pi(0\cdot1)^3$ cu mm
$= 0\cdot004$ cu mm
Volume of single air-cell in new-born $= \frac{4}{3}\pi(0\cdot035)^3$ cu mm
$= 0\cdot00018$ cu mm

$\therefore$ total number of air-cells in adult $= \dfrac{1617 \times 10^3}{0\cdot004} = 404 \times 10^6$

And total number of air-cells in new-born $= \dfrac{67\cdot7 \times 10^3}{0\cdot00018} = 376 \times 10^6$

i.e. *the number is approximately the same at birth as in the full-grown adult, viz. about* 4×10^8.

Surface of single air-cell in adult $= 4\pi(0\cdot1)^2 = 0\cdot125$ sq mm
Surface of single air-cell in new-born $= 4\pi(0\cdot035)^2 = 0\cdot0154$ sq mm
$\therefore$ Total surface of air-cells in adult $= 4 \times 10^8 \times 0\cdot125$ sq mm
$= 50$ sq metres

And total surface of air-cells in new-born $= 0\cdot0154 \times 4 \times 10^8$ sq mm
$= 6$ sq metres.

$\therefore$ Total surface of air-cells in new-born is about $\frac{1}{8}$ that in the adult.

Hence we see that whilst the volume of the infant's lungs is only about $\frac{1}{24}$ that in the adult, the *total surface of the alveoli is as much as* $\frac{1}{8}$ *that in the adult*—showing that the gaseous exchange is more active in young infants, i.e. about three times as active as in the adult. Moreover, as the area of the infant's skin surface is $\frac{1}{8}$ that in the adult, it is seen that the amount of gaseous interchange per unit of body surface is the same in the infant as in the adult. (See W. M. Feldman, *Principles of Ante-Natal and Post-Natal Child Physiology*, Longmans, 1920, and *Principles of Ante-Natal and Post-Natal Child Hygiene*, John Bale, Sons & Danielsson, 1927.)

PROBLEMS

(1) The radius of each particle of cholesterol in a sol containing 0·0005 gram of that substance per cc is 10^{-6} cm. Find the total surface area of these particles. (Assume sp. gr. = 1.)

(2) The average diameter of a human capillary is $\frac{1}{100}$ mm; the linear velocity of blood in it is $\frac{1}{2}$ mm per second.

Find the volume of outflow from a capillary per second.

(3) Experiments on animals have shown that the circulation time is equal to 28 heart-beats. Assuming this to hold good for man, and also assuming the total volume of blood in the body to be 4000 cc, find the number of capillaries in the human body, using the data of the last example, and assuming the pulse rate to be 72 per minute.

(4) The following has been found to be the percentage composition of ordinary bacteria: water, 85 per cent; solids, 15 per cent, of which 1 part in a thousand consists of sulphur.

Assuming that the weight of an atom of any element $= A \times 8{\cdot}6 \times 10^{-22}$ mgm, where $A =$ atomic weight of the element, how many atoms of sulphur does a micrococcus of diameter $0{\cdot}15\mu$ ($\mu = \frac{1}{1000}$ mm) contain?

11.14 Areas with signs

In Section 11.3 it was shown that the integral $\int_{x_1}^{x_2} f(x)dx$ could be interpreted under certain conditions as the area under the curve $y = f(x)$ between the ordinates x_1 and x_2. These conditions were (i) $x_1 \leqslant x_2$, (ii) $f(x)$ is never negative for values of x between x_1 and x_2. We shall now investigate what happens when these conditions are relaxed.

First we shall keep the condition $x_1 \leqslant x_2$, but allow $f(x)$ to become negative.

If $f(x)$ is negative for all values of x between x_1 and x_2, the integral $\int_{x_1}^{x_2} f(x)\,dx$ must also be negative. This follows directly from the definition of the integral. Let $I(x)$ be the indefinite integral $\int f(x)\,dx = \int y\,dx$: then $\int_{x_1}^{x_2} f(x)\,dx$ means $I(x_2) - I(x_1)$. But $D_x\,I(x) = I_x(x) = f(x)$, by the definition of an indefinite integral, and so $I(x)$ has a *negative* rate of change $f(x)$. This means that $I(x)$ decreases as x increases, and in particular as x increases from x_1 to x_2 $I(x)$ must decrease from $I(x_1)$ to $I(x_2)$, and the difference $I(x_1) - I(x_2)$ must be negative.

This will equally apply to other interpretations of the integral. If t stands for the time, and v the velocity of a moving body, then $\int_{t_1}^{t_2} v\,dt$ means the distance gone, or change in position, between the times t_1 and t_2. If v is always negative the body will always be moving backwards instead of forwards. It will lose ground instead of gaining it, and the total distance covered will accordingly be counted as negative. Again if F is a force acting on a body moving along a line, and x is the co-ordinate of the body, then $\int_{x_1}^{x_2} F\,dx$ represents the work done by the force in moving from position x_1 to x_2. If F is negative the body is always moving against the force, and it has to do work instead of having work done on it, i.e. the work done will be counted as negative.

Finally, we can still regard the integral as the limit of a sum. If the interval from x_1 to x_2 is divided into n strips by numbers $X_1 = x_1$, $X_2, X_3, \ldots X_{n+1} = x_2$, with corresponding y values $Y_1, Y_2 \ldots Y_{n+1}$, then the integral will still be the limit of $S = \Sigma Y_a (X_{a+1} - X_a) = \Sigma Y \delta X$ as the number of strips tends to ∞ and their maximum width to zero. If y is always negative the sum S will be negative, since each term $Y\delta X$ is negative: and therefore the limit of S must be negative, or at any rate not positive.

Consider what this means in terms of areas. Draw the curve $y = f(x)$: this must now lie entirely *below* the x-axis. Let P be the first point (x_1, y_1) on the curve, and Q the last, (x_2, y_2) (Fig. 11.24).

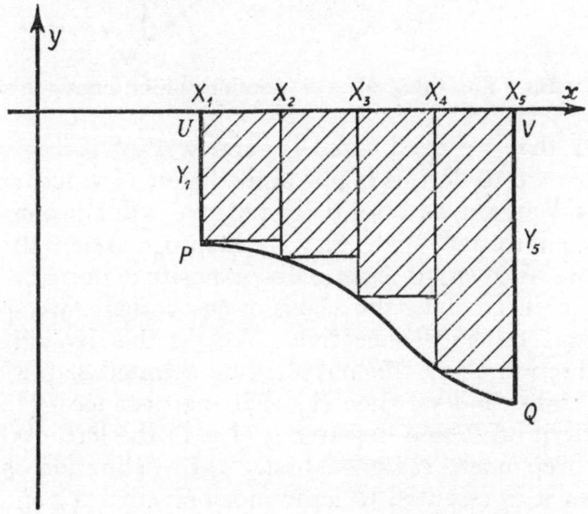

Fig. 11.24—A negative area above a curve

Draw perpendiculars PU, QV onto the x-axis. Divide the area $UPQV$ into n strips by ordinates at the points $X_1, X_2, X_3 \ldots X_{n+1}$. (In the figure $n = 4$.) Then $Y_1 \delta X_1 = Y_1(X_2 - X_1)$ is *minus* the area of a rectangle approximately equal to the first strip, since Y_1 is negative. Similarly $Y_2 \delta X_2$ is approximately minus the area of the second strip, and so on: and the sum $S = \Sigma Y\delta X$ approximates to minus the area of the whole figure $UPQV$. In the limit we see that $\int_{x_1}^{x_2} y\, dx = -$ area of $UPQV$ = the area between the curve and the x-axis, but with a negative sign.

From this we can readily see what the interpretation will be when y is sometimes positive and sometimes negative. Suppose, for example, that the curve $y = f(x)$ starts above the x-axis at P, crosses it at W, and ends below it at Q (Fig. 11.25). Then the loop PUW will be counted as positive and the loop WVQ as negative and $\int_{x_1}^{x_2} y\, dx =$ area $PUW -$ area WVQ.

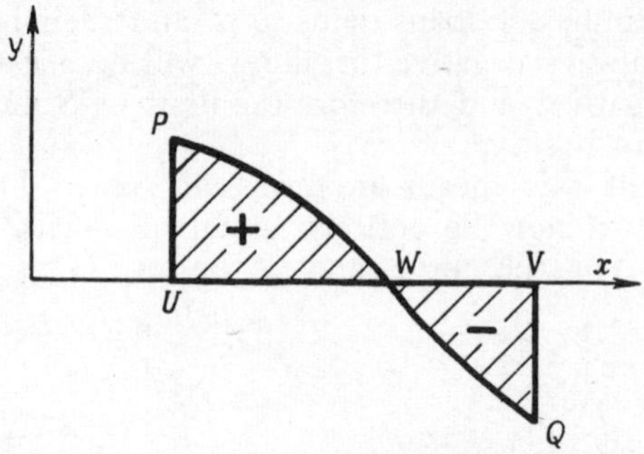

Fig. 11.25—The integral of a function which changes sign

It is clear that we need some general way of giving a sign to an area. How we are to do this is not immediately obvious, since an area, as usually understood, is essentially positive. But a consideration of the cases of length and angle may be helpful. A length or distance, considered on its own, is essentially a positive number. But when lengths are measured along the x-axis of co-ordinates we speak of some as positive and others as negative. We do this by agreeing that a particular direction along the axis shall be counted as positive, namely from left to right; and we then say that the distance PQ is positive if Q is to the right of P, and negative if Q is to the left. Why do we do this? For three main reasons: firstly the distinction is important. In most cases it is essential to know not only that a point has moved such a distance, say 10 cm, but also which direction it has moved in: the consequences of motion one way may be very different from those the other way. Secondly, the sign convention simplifies calculations. If P, Q, R are three points on a line, then the relation $PQ + QR = PR$ holds whenever Q lies between P and R. If we count all distances as positive it will no longer be true in other circumstances, but will be replaced by some other relation such as $PQ - QR = PR$. But if distances are given their proper signs then we can write $PQ + QR = PR$ without exception. The third advantage is that the convention of signs agrees with that needed for other branches of mathematics, such as trigonometry and differential calculus.

A similar situation holds for angles in a plane. We agree to consider $\angle AOB$ as positive if the rotation from OA to OB is anti-clockwise. Again it always holds that $\angle AOB + \angle BOC = \angle AOC$ when the proper signs are used.

Areas lying in a plane can also be provided with signs. Most areas are contained within some form of loop, and we can imagine this loop to be traced out by a moving point P (Fig. 11.26). If P describes the loop in an anti-clockwise direction we say that the area A contained within it is positive. If P goes round in a clockwise direction then the area

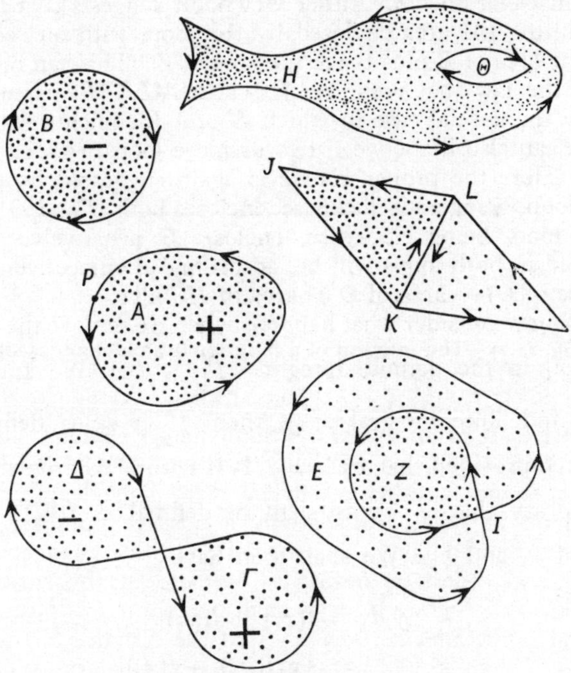

Fig. 11.26—The sign convention for areas

is negative, as with area *B*. If the loop crosses itself then those areas round which the loop runs anti-clockwise are positive (such as *Γ*) and those contained in a clockwise loop (such as *Δ*) are negative. This convention agrees with the correct sign for an integral, provided that we take the area under the arc *PQ* to be that contained in the loop *PUVQP*, described in that direction (Fig. 11.1, 11.24, 11.25). In Fig. 11.1 the loop is anti-clockwise and the integral is positive, in Fig. 11.24 it is clockwise and the integral negative, and in Fig. 11.25 the loop splits into two, with corresponding signs.

With this convention it also follows that if the loops containing two areas have a part *KL* in common, but described in opposite directions, then we can always write area *JKLJ* + area *MLKM* = area *JKMLJ*, using the appropriate signs (Fig. 11.26).

Sometimes an area may be contained more than once within a loop. For example, in the case of the areas *E* and *Z* in the figure we see that *E* is contained once within the loop, but *Z* is surrounded twice in consequence of the way the loop crosses itself. Or we may look at it another way. We can start from the point *I*, go once round the outer part of the loop returning to *I*, and enclosing both areas *E* and *Z* in an anti-clockwise direction: we can then go round the inner part of the loop, enclosing only Z. Either viewpoint suggests that the proper way of counting the area enclosed by the loop, with our convention, is to take the area of E plus twice the area of Z. This can be shown to agree with the law that area $JKLJ$ + area $MLKM$ = area $JKMLJ$: it is simply the special case in which K and L coincide with I.

An area can also lie between two or more loops, as area H does in the figure. Here the proper course is again suggested by the figure. The outer loop, being anti-clockwise, encloses both H and Θ positively. The inner loop, being clockwise, encloses Θ negatively. The total area enclosed in both loops will be, according to our convention, (area of H + area of Θ) − area of Θ = area of H only.

We shall now consider what happens when we remove the restriction that $x_2 \geqslant x_1$ in the definite integral $\int_{x_1}^{x_2} y\,dx$. We know that if $I(x)$ is the indefinite integral of y, then $\int_{x_1}^{x_2} y\,dx$ is defined to be $I(x_2) - I(x_1)$, provided that $x_2 \geqslant x_1$. It is natural to drop this restriction, and to say that $\int_{x_1}^{x_2} y\,dx$ shall be defined as $I(x_2) - I(x_1)$ for all values of x_1 and x_2. We shall then have

$$\begin{aligned}\int_{x_2}^{x_1} y\,dx &= I(x_1) - I(x_2)\\ &= -[I(x_2) - I(x_1)]\\ &= -\int_{x_1}^{x_2} y\,dx \qquad . \quad . \quad . \ (11.19)\end{aligned}$$

so that interchanging the limits of integration merely changes the sign of the integral. This agrees with our definition of the integral as the area $PUVQ$ (Fig. 11.1) taken with the proper sign, for the integral $\int_{x_2}^{x_1} y\,dx$ with limits x_1 and x_2 interchanged will mean the area $QVUP$, i.e. the same area except that the containing loop is described in the opposite direction. It also agrees with our interpretation of $\int_{x_1}^{x_2} F\,dx$ as the work done by a force F in moving from position x_1 to position x_2; for $\int_{x_2}^{x_1} F\,dx$ is then the work done in moving back from x_2 to x_1, and must be equal in magnitude but opposite in sign.

We can extend this sign convention to curved surfaces, by fixing an arbitrary sense of rotation on the surface, and to volumes, by giving an appropriate sign to the surfaces which contain them. But to consider these matters in detail would take us too far off our course. The important point is the connection between the sign of an area and the sign of an integral.

11.15 Area of a polygon

Using this idea of sign we can readily find a formula for the area of a polygon such as *PQRSTP* constructed by joining the points *P*, *Q*, *R*, *S*, *T* by straight lines (Fig. 11.27). Let *P* have co-ordinates

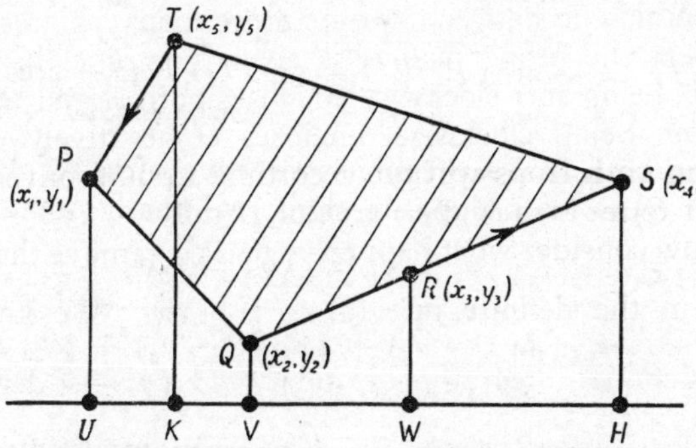

Fig. 11.27—The area of a polygon

(x_1, y_1), $Q\ (x_2, y_2)$, and so on up to $T = (x_5, y_5)$. Draw perpendiculars *PU*, *QV*, *RW*, *SH*, *TK* on to the *x*-axis. Consider first the area *PUVQ*. This is $\int_{x_1}^{x_2} y\, dx$, where y represents the height of the point on the line *PQ* with *x*-co-ordinate *x*. But since *PQ* is a straight line, it will have an equation of the form $y = A_1 + B_1x$, where A_1 and B_1 are certain constants. They can be determined from the fact that *P* and *Q* lie on this line, so that

$$\left.\begin{aligned} y_1 &= A_1 + B_1x_1 \\ y_2 &= A_1 + B_1x_2 \end{aligned}\right\} \quad . \quad . \quad . \quad . \quad (11.20)$$

Now the area of *PUVQ*, with the proper sign, is

$$\begin{aligned} \int_{x_1}^{x_2} y\, dx &= \int_{x_1}^{x_2} (A_1 + B_1x)\, dx \\ &= [A_1x + \tfrac{1}{2}B_1x^2]_{x_1}^{x_2} \\ &= A_1(x_2 - x_1) + \tfrac{1}{2}B_1(x_2^{\,2} - x_1^{\,2}) \\ &= \tfrac{1}{2}A_1(x_2 - x_1) + \tfrac{1}{2}x_2(A_1 + B_1x_2) - \tfrac{1}{2}x_1(A_1 + B_1x_1) \\ &= \tfrac{1}{2}(x_2y_1 - x_1y_2) + \tfrac{1}{2}x_2y_2 - \tfrac{1}{2}x_1y_1 \end{aligned}$$

using equations (11.20).

In the same way

$$\begin{aligned} \text{area } QVWR &= \tfrac{1}{2}(x_3y_2 - x_2y_3) + \tfrac{1}{2}x_3y_3 - \tfrac{1}{2}x_2y_2 \\ \text{area } RWHS &= \tfrac{1}{2}(x_4y_3 - x_3y_4) + \tfrac{1}{2}x_4y_4 - \tfrac{1}{2}x_3y_3 \end{aligned}$$

and so on.

Now in Fig. 11.27 the areas *PUVQ*, *QVWR*, *RWHS* are positive, and the areas *SHKT*, *TKUP* are negative. If therefore we add all these areas together we shall obtain the area of the loop *PQRSTP* with its sign changed: all the other parts cancel out. The reader can satisfy himself by drawing polygons of other shapes that this is a general result. Therefore

$$\text{area } PQRSTP = -\text{ area } PUVQ - \text{area } QVWR - \text{area } RWHS - \text{area } SHKT - \text{area } TKUP$$

and on substituting the expressions we have already obtained, and cancelling out terms with opposite signs, we finally obtain:

$$\begin{aligned}\text{area } PQRSTP &= \tfrac{1}{2}(x_1y_2 - x_2y_1 + x_2y_3 - x_3y_2 + x_3y_4 - x_4y_3 \\ &\quad + x_4y_5 - x_5y_4 + x_5y_1 - x_1y_5) \\ &= \tfrac{1}{2}x_1(y_2 - y_5) + \tfrac{1}{2}x_2(y_3 - y_1) + \tfrac{1}{2}x_3(y_4 - y_2) \\ &\quad + \tfrac{1}{2}x_4(y_5 - y_3) + \tfrac{1}{2}x_5(y_1 - y_4) \qquad (11.21)\end{aligned}$$

A similar expression holds for a polygon with any number of vertices. If the polygon crosses itself at any point the area given by this formula will agree with the convention of signs explained above.

EXAMPLE

(1) Find the area of the triangle with vertices (4, 2), (3, 3), and (2, 1).

$$\begin{aligned}\text{The area } &= \tfrac{1}{2}x_1(y_2 - y_3) + \tfrac{1}{2}x_2(y_3 - y_1) + \tfrac{1}{2}x_3(y_1 - y_2) \\ &= \tfrac{1}{2}\cdot 4 \cdot (2) + \tfrac{1}{2}\cdot 3 \cdot (-1) + \tfrac{1}{2}\cdot 2 \cdot (-1) \\ &= 4 - \tfrac{3}{2} - 1 = \tfrac{3}{2}.\end{aligned}$$

PROBLEMS

(1) Find by the formula the area of the square with vertices (1, 0), (0, 1), (− 1, 0), (0, − 1).

(2) Show that for the rectangle with vertices at (x_1, y_1), (x_2, y_1), (x_2, y_2), and (x_1, y_2), the formula for the area reduces to the product $(x_2 - x_1)(y_2 - y_1)$, i.e. length × breadth.

(3) Prove algebraically that if three points *P*, *Q*, *R* lie on a straight line the area of the triangle *PQR* is zero.

11.16 Numerical integration

Sometimes it happens that it is necessary to find the area of a figure—such as a section of a bone—which can be drawn, but not expressed by any simple algebraic formula. Sometimes although a formula may be available it may have no simple integral. In either case we can resort to "numerical integration", that is, the use of approximate methods to give a numerical value for the area or the integral.

We already know one approximation to the definite integral or area $\int_{x_1}^{x_2} y\,dx$, namely the sum $S = \Sigma\, y\,dx$, obtained by cutting the area into vertical strips. For if the strips are made sufficiently narrow S will become very close in value to the integral. However, although this method is theoretically perfect, it is of little practical use as a very large number of strips are needed to give a reasonably accurate value.

There are other formulas which give very much better results. The most suitable one for most purposes is "Simpson's rule", which combines simplicity with a high degree of accuracy.

Suppose that the required integral is $\int_{x_1}^{x_2} y\,dx$, and that we are given a graph of y against x (Fig. 11.28). The interval from x_1 to x_2

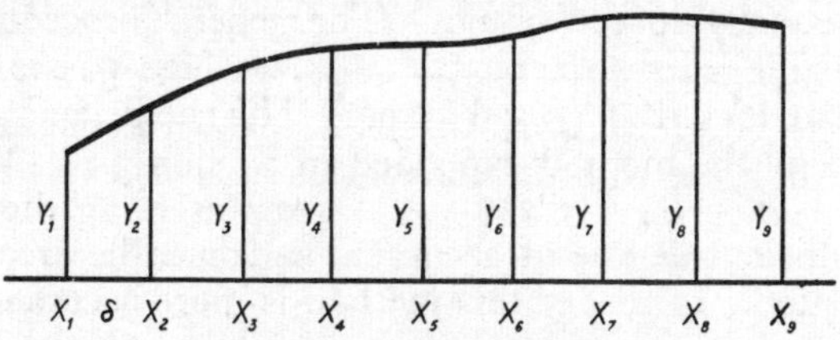

Fig. 11.28—Simpson's rule

must then be divided into an *even* number $2n$ of equal intervals of width δ, so that $2n\delta = x_2 - x_1$. (In the figure $2n = 8$.) Let the dividing points be $X_1 = x_1, X_2, X_3, \ldots X_{2n}, X_{2n+1} = x_2$, and let the corresponding values of y be $Y_1, Y_2, \ldots Y_{2n+1}$.

Simpson's rule (which we give without proof) states that

$$\int_{x_1}^{x_2} y\,dx = \text{area under the curve}$$
$$\simeq \tfrac{1}{3}\delta\,[Y_1 + 4Y_2 + 2Y_3 + 4Y_4 + 2Y_5 + \ldots + 4Y_{2n} + Y_{2n+1}] \quad . \quad . \quad . \quad . \quad (11.22)$$

or, in words, $\frac{1}{3}$ *the width of a strip times* [*first ordinate* + *last ordinate* + *twice the sum of the remaining odd ordinates* + 4 *times the sum of the even ordinates*].

EXAMPLE

(1) Find the value of $\ln 2 = \int_1^2 x^{-1}\,dx$. First let us divide the interval from 1 to 2 into 2 strips, so that the dividing points are $X_1 = 1$, $X_2 = \frac{3}{2}$, $X_3 = 2$. The corresponding values of $y = x^{-1}$ are $Y_1 = 1$, $Y_2 = \frac{2}{3}$, $Y_3 = \frac{1}{2}$, and the width of the strip is $\delta = \frac{1}{2}$. The area is therefore approximately

$$\tfrac{1}{3}\delta\,(Y_1 + 4Y_2 + Y_3) = \tfrac{1}{6}[1 + \tfrac{8}{3} + \tfrac{1}{2}] = \cdot 69444.$$

To obtain a more accurate answer we divide the area into 6 strips, with $X_1 = 1$, $X_2 = \frac{7}{6}$, $X_3 = \frac{8}{6}$, $X_4 = \frac{9}{6}$, up to $X_7 = 2$. The corresponding values of y are $Y_1 = 1$, $Y_2 = \frac{6}{7}$, $Y_3 = \frac{6}{8}$, $Y_4 = \frac{6}{9}$, $Y_5 = \frac{6}{10}$, $Y_6 = \frac{6}{11}$ and $Y_7 = \frac{6}{12}$, and the area is approximately $\frac{1}{3} \cdot \frac{1}{6} \cdot [Y_1 + 4Y_2 + 2Y_3 + 4Y_4 + 2Y_5 + 4Y_6 + Y_7] = \cdot 69350$. As the true value of ln 2 is ·69315 to 5 figures, it is evident that this method gives a remarkably good result with very little trouble.

Sometimes it is not possible to divide an area or integral conveniently into an even number of strips in this manner. If the area is divided into an odd number of equal strips, then the area of the first three of these will be approximately $\frac{3}{8}\delta\,(Y_1 + 3Y_2 + 3Y_3 + Y_4)$, and the remaining area can be calculated by Simpson's rule. If the values of y, say $Y_1, Y_2, Y_3, \ldots$ are known for values $X_1, X_2, X_3 \ldots$ of x, then we can always find an approximate value for the area, whether the X's are equally spaced or not. The general procedure is to find the polynomial $y = A + Bx + Cx^2 + \ldots$ which passes through the points (X_1, Y_1), (X_2, Y_2), (X_3, Y_3), etc. The coefficients $A, B, C, \ldots$ can be found by the method explained in Section 3.10. We then take the curve $y = A + Bx + Cx^2 + \ldots$ as representing the given curve for y to a sufficient degree of accuracy, and find the area under it by direct integration. In theory this method is more accurate than Simpson's rule, but the gain in accuracy is usually worth the extra trouble only if for some special reason it is inconvenient to divide the interval into an even number of equal strips. Some of these most accurate formulas are tabulated in a convenient form for strips of equal width by W. G. Bickley in *Mathematical Gazette*, **23** (1939), p. 352, and in Fisher & Yates, *Statistical Tables for Biological, Agricultural, and Medical Research* (4th edn., Oliver and Boyd, 1953).

Simpson's and similar rules apply to a smooth, continuous curve. They are not so accurate when the curve has sudden breaks, changes in direction, or a vertical tangent. When calculating the area of a smooth rounded loop we shall lose a certain accuracy on applying such a rule to the whole loop because there will be points on either side where the tangent is vertical (e.g. in Fig. 11.29A). This loss in accuracy can be overcome by cutting a vertical slice through the centre of the loop, dividing it into an even number of equal strips by ordinates at $X_1, X_2, X_3 \ldots X_{2n+1}$. If the vertical widths of the loop at these values of x are $W_1, W_2, \ldots W_{2n+1}$ respectively, then the area of the vertical slice is by Simpson's rule

$$\tfrac{1}{3}(\text{horizontal width of strip})\,(W_1 + 4W_2 + 2W_3 + 4W_4 + \ldots + W_{2n+1}) \qquad (11.23)$$

The areas of the segments at the ends not included in the vertical slice can be calculated by dividing them into horizontal strips and applying Simpson's rule.

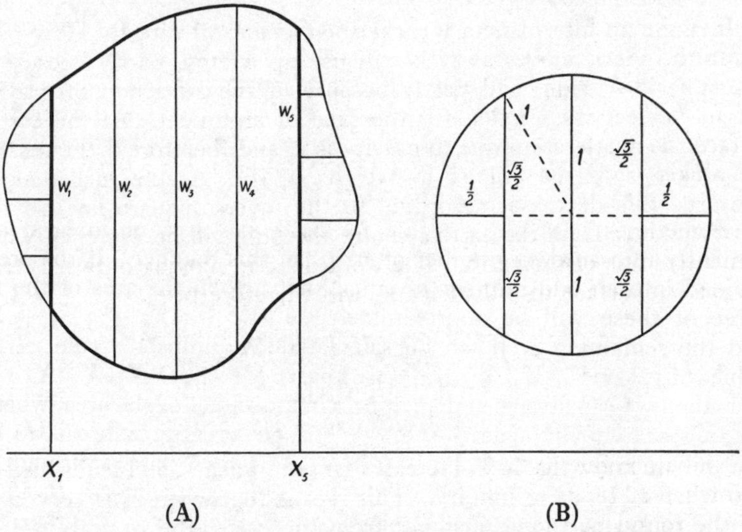

Fig. 11.29—Approximate estimation of the areas of a loop and of a circle of unit radius

FURTHER EXAMPLE

(2) To find approximately the area of a circle of unit radius using this method (Fig. 11.29 at B).

Take a vertical slice of width 1 centrally through the circle. Divide this slice into 2 strips by a vertical line through the centre. Then the vertical widths at the left-hand edge, centre and right-hand edge of the slice are $W_1 = \sqrt{3}$, $W_2 = 2$, $W_3 = \sqrt{3}$ respectively, and the horizontal width of a strip is $\frac{1}{2}$. The area is therefore approximately $\frac{1}{3} \cdot \frac{1}{2} \cdot (\sqrt{3} + 4 \times 2 + \sqrt{3}) = (4 + \sqrt{3})/3$. We also divide the left-hand segment into two horizontal strips by a horizontal line through the centre of the circle, of length $\frac{1}{2}$, and (in order to apply Simpson's rule) by two horizontal lines (which will have zero length) through the lowest and highest point of the segment. The vertical width of the strips is $\frac{1}{2}\sqrt{3}$. The area of this segment is therefore approximately $\frac{1}{3} \cdot \frac{1}{2}\sqrt{3} \cdot [0 + 4 \times \frac{1}{2} + 0] = \frac{1}{3}\sqrt{3}$. The area of the right-hand segment is equal to this. The total area of the circle including the vertical slice and the two ends is therefore $\frac{1}{3}(4 + \sqrt{3}) + \frac{1}{3}\sqrt{3} + \frac{1}{3}\sqrt{3} = 3{\cdot}065$. In fact the true area is $\pi = 3{\cdot}142$, so that this simple estimate has less than 3 per cent error. Naturally if we took a larger number of strips we should get a very much more accurate value.

Since volumes and lengths can usually be expressed as integrals this gives us a convenient method of calculating them too.

11.17 Infinite integrals

Imagine an interplanetary rocket being sent off directly away from the earth. As it moves away it will use up energy, partly because of atmospheric friction, and partly because of the attraction of gravity. Let us neglect the friction for the sake of argument. At the earth's surface the acceleration due to gravity is g, and therefore if the mass of the rocket is M the pull of its weight is Mg. As the rocket moves outwards this decreases according to the inverse square law, so that at a distance r from the earth's centre the pull will be MgR^2/r^2, where R is the radius of the earth. The work done in moving outwards from the earth's surface to a distance r_2 will therefore be

$$\begin{aligned} W &= \int_R^{r_2} (MgR^2/r^2)\,dr \\ &= [-MgR^2/r]_R^{r_2} \\ &= MgR^2(1/R - 1/r_2) \qquad (11.24) \end{aligned}$$

Now we know that as r_2 increases $1/r_2$ diminishes, and tends towards zero when r_2 tends to infinity. This means that when r_2 is very large, i.e. the rocket is a great distance from the earth, the work W will be very nearly equal to $MgR^2/R = MgR$, which is a finite quantity. The rocket only needs to use up an amount of energy MgR to get as far away from the earth as it chooses. [If we assume that $M = 10{,}000$ kg, $g = 9{\cdot}8$ metres/sec^2 and $R = 6{,}500{,}000$ metres, we see that such a rocket would need about $6{\cdot}4 \times 10^{11}$ joules to escape from the earth.]

Thus the integral $W = \int_R^{r_2} (MgR^2/r^2)\,dr$ tends to a finite limit MgR as r_2 tends to infinity. It is natural to express this as

$$\int_R^{\infty} (MgR^2/r^2)\,dr.$$

This expression is known as an "infinite integral", and is to be understood as the limit of the finite or ordinary integral $\int_R^{r_2}(MgR^2/r^2)\,dr$ in the above sense.

Another example of an infinite integral is the amount of work done by an electric charge e_1 in moving away from a fixed charge e_2. Here the law of force is again the inverse square law, $F = e_1e_2/\kappa_0 r^2$: as the distance increases from r_1 to r_2 the work done is $\int_{r_1}^{r_2} F\,dr = \int_{r_1}^{r_2} (e_1 e_2/\kappa_0 r^2)\,dr = [-e_1 e_2/\kappa_0 r^2]_{r_1}^{r_2} = e_1 e_2 [r_1^{-1} - r_2^{-1}]/\kappa_0$. As r_2 increases indefinitely this tends to the finite limit $\int_{r_1}^{\infty} F\,dr = e_1e_2/r_1\kappa_0$.

Another example would be the decay of a radioactive element. The radiation R from this will tend to decay according to an exponential law: $R = ce^{-Kt}$, where t is the time and c and K are constants. The total radiation given off between times t_1 and t_2 will therefore be

$\int_{t_1}^{t_2} ce^{-Kt}\,dt = [-cK^{-1}e^{-Kt}]_{t_1}^{t_2} = c(e^{-Kt_1} - e^{-Kt_2})/K$. As t_2 tends to infinity this remains finite, tending to the limit $\int_{t_1}^{\infty} R\,dt = ce^{-Kt_1}/K$ which accordingly is the total amount of radiation emitted in the whole of time after the instant t_1.

We can also have integrals such as $\int_{-\infty}^{x_2} y\,dx$, meaning

$$\lim_{x_1\to-\infty} \int_{x_1}^{x_2} y\,dx,$$

and $\int_{-\infty}^{\infty} y\,dx$, meaning $\lim\limits_{x_1\to-\infty}\ \lim\limits_{x_2\to\infty} \int_{x_1}^{x_2} y\,dx$.

EXAMPLES

(1) Find $\int_{-\infty}^{\infty} \frac{dx}{1+x^2}$

We have $\int_{x_1}^{x_2} \frac{dx}{1+x^2} = [\tan^{-1} x]_{x_1}^{x_2}$

$$= \tan^{-1}x_2 - \tan^{-1}x_1.$$

Now as $x_1 \to -\infty$, $\tan^{-1} x_1 \to -\frac{1}{2}\pi$ $(= -90°)$, and as $x_2 \to \infty$, $\tan^{-1} x_2 \to \frac{1}{2}\pi$. Thus

$$\int_{-\infty}^{\infty} \frac{dx}{1+x^2} = \tfrac{1}{2}\pi - (-\tfrac{1}{2}\pi) = \pi$$

(This example shows how the number π, defined in the first instance as the ratio of the circumference of a circle to its diameter, can enter into calculations which at first sight have no connection whatever with circles.)

(2) Find $\int_{-\infty}^{0} e^x\,dx$

$$\int_{x_1}^{0} e^x\,dx = [e^x]_{x_1}^{0} = e^0 - e^{x_1} = 1 - e^{x_1}$$

But as $x_1 \to -\infty$, $e^{x_1} \to 0$, so that in the limit $\int_{-\infty}^{0} e^x\,dx = 1$.

PROBLEM

Find the following integrals:

(1) $\int_1^{\infty} \frac{dx}{x^3}$, (2) $\int_2^{\infty} e^{-x}\,dx$, (3) $\int_0^{\infty} \operatorname{sech} x\,dx$.

11.18 Change of variable in a definite integral

In Sections 10.2, 10.4 and 10.5 we have found rules for the calculation of indefinite integrals. These rules apply equally well to definite integrals. Consider, for example, the rule for multiplication by a constant. Let $I(x) = \int y\,dx$ be the indefinite integral of y, and $J(x) = \int ky\,dx$ the indefinite integral of ky. Then we know that $J(x) = kI(x) + C$, where C is an arbitrary constant. It follows that

$$\begin{aligned}\int_{x_1}^{x_2} ky\,dx &= J(x_2) - J(x_1) \text{ (by definition)}\\ &= [kI(x_2) + C] - [kI(x_1) + C]\\ &= k\,[I(x_2) - I(x_1)]\\ &= k\int_{x_1}^{x_2} y\,dx.\end{aligned}$$

Multiplication of y by a constant also multiplies its definite integral by the same constant.

In the same way

$$\int_{x_1}^{x_2} (y + z)\,dx = \int_{x_1}^{x_2} y\,dx + \int_{x_1}^{x_2} z\,dx$$

(addition rule);

if Z is any indefinite integral of z,

$$\int_{x_1}^{x_2} yz\,dx = [yZ]_{x_1}^{x_2} - \int_{x_1}^{x_2} y_x Z\,dx$$

("integration by parts").

The rule for change of variable in an indefinite integral is

$$\int y\,dx = \int y\frac{dx}{dz}\,dz$$

When we come to put in the limits of integration we must remember that in the right-hand side we are integrating with respect to z, not x, so that the limits for z will be z_1 and z_2 (say), the corresponding values of z when $x = x_1$ and x_2 respectively.

$$\int_{x_1}^{x_2} y\,dx = \int_{z_1}^{z_2} y\frac{dx}{dz}\,dz \quad . \quad . \quad . \quad (11.25)$$

There is also a further rule which applies only to definite integrals:

$$\int_{x_1}^{x_2} y\,dx + \int_{x_2}^{x_3} y\,dx + \int_{x_3}^{x_4} y\,dx = \int_{x_1}^{x_4} y\,dx \qquad (11.26)$$

i.e. the sum of the integrals over the ranges from x_1 to x_2, from x_2 to x_3, from x_3 to x_4, is equal to the integral over the whole range from x_1 to x_4. Interpreted in terms of areas it means that the area under a curve between (say) P, (x_1, y_1) and Q, (x_2, y_2), plus the area between Q and R, (x_3, y_3), plus the area between R and S (x_4, y_4) sums to the total area between P and S. It also follows from the definition of a definite integral:

$$\begin{aligned}&\int_{x_1}^{x_2} y\,dx + \int_{x_2}^{x_3} y\,dx + \int_{x_3}^{x_4} y\,dx\\ &\quad = [I(x_2) - I(x_1)] + [I(x_3) - I(x_2)] + [I(x_4) - I(x_3)]\\ &\quad = I(x_4) - I(x_1) = \int_{x_1}^{x_4} y\,dx.\end{aligned}$$

(Although we have taken the case of three integrals for the sake of illustration, this is evidently a general rule.)

But there is one point where some caution is necessary in applying these rules, especially the one for change of variable, and that is when we have many-valued functions. We know, for instance, that $\int (1 + x^2)^{-1}\, dx = \tan^{-1} x$, and therefore $\int_{x_1}^{x_2} (1 + x^2)^{-1}\, dx$ is by definition $(\tan^{-1} x_2 - \tan^{-1} x_1)$. But $\tan^{-1} x$ is a many-valued function; $\tan^{-1} 0$ can be 0, or π, or 2π, or any multiple of π. On the other hand $\int_{x_1}^{x_2} (1 + x^2)^{-1}\, dx$ is a perfectly definite quantity, being the integral of a single-valued function, so that we are limited in the possible values we can take for $\tan^{-1} x_2$ and $\tan^{-1} x_1$ if we are to make their difference equal to the integral in question. The correct way to look at the problem is this. If $I(x)$ is any indefinite integral of a function y, the definite integral $\int_{x_1}^{x_2} y\, dx$ means the change in the value of $I(x)$ when x changes in value from x_1 to x_2. We must imagine x as moving continuously from x_1 to x_2, passing through all intermediate values on the way: if $I(x)$ starts with some definite value $I(x_1)$ it will then change continuously and end with some definite value $I(x_2)$. It is the difference between these two determinate values which equals the definite integral. For example, the integral $\int (1 + x^2)^{-1}\, dx = \tan^{-1} x$ can be illustrated geometrically as follows. Let X be the point on the x-axis with co-ordinate x, X_1 that with co-ordinate x_1, and X_2 that with co-ordinate x_2. Furthermore let O be the origin, and H the point (0, 1) at unit distance vertically above O (Fig. 11.30). Then by definition $\angle OHX = \theta$

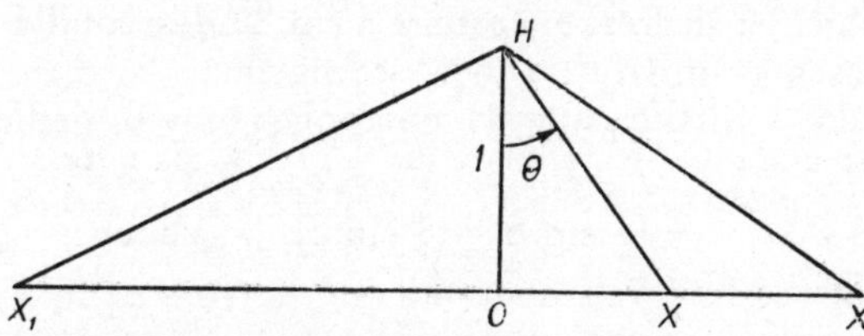

Fig. 11.30—Determination of the value of $\tan^{-1} x$ in an integral

(measured with its proper sign) is one value of $\tan^{-1} x$. Here we mean by θ the angle lying between $-\frac{1}{2}\pi = -90°$ and $\frac{1}{2}\pi = 90°$, as suggested by the diagram. Now so long as X moves continuously up and down the x-axis θ must remain within these limits. In particular, as X moves from X_1 to X_2, $\theta = \tan^{-1} x$ goes from $\tan^{-1} x_1 = \angle OHX_1$ to $\tan^{-1} x_2 = \angle OHX_2$, both angles lying between $-\frac{1}{2}\pi$ and $\frac{1}{2}\pi$, and the integral $= \tan^{-1} x_2 - \tan^{-1} x_1 = \angle OHX_2 - \angle OHX_1 = \angle X_1HX_2$. This fixes the integral completely.

In the same way if we change the variable from x to z, we must make sure that as x changes continuously from x_1 to x_2 so z changes continuously from z_1 to z_2; and it is as well to notice carefully *how* z changes.

EXAMPLES

(1) Evaluate $\int_1^2 x \cos x^2\,dx$.

Here the natural substitution is to put $z = x^2$, so that $dz/dx = 2x$, $dx/dz = 1/2x$. Then, applying this to the indefinite integral,

$$\begin{aligned}\int x \cos x^2\,dx &= \int x \cos x^2\,(dx/dz)\,dz \\ &= \tfrac{1}{2}\int \cos z\,dz \\ &= \tfrac{1}{2}\sin z\end{aligned}$$

Now as x increases continuously from 1 to 2, z increases from $1^2 = 1$ to $2^2 = 4$. This is accordingly quite straightforward, and we can insert the limits of integration 1 and 4 for z.

$$\begin{aligned}\int_1^2 x \cos x^2\,dx &= [\tfrac{1}{2}\sin z]_1^4 \\ &= \tfrac{1}{2}\sin 4 - \tfrac{1}{2}\sin 1.\end{aligned}$$

(2) Evaluate $\int_{-1}^1 x \cos x^2\,dx$.

As before, putting $x^2 = z$, we have $\int x \cos x^2\,dx = \tfrac{1}{2}\int \cos z\,dz$ $= \tfrac{1}{2}[\sin z]$. But now as x increases from -1 to 1, z at first decreases from 1 to 0 and then increases again from 0 to 1. If we want to play safe we can split the range of integration for z into two parts, firstly the part in which z decreases from 1 to 0, and secondly the part in which z increases again from 0 to 1, saying that

$$\begin{aligned}\int_{-1}^1 x \cos x^2\,dx &= \int_1^0 \tfrac{1}{2}\cos z\,dz + \int_0^1 \tfrac{1}{2}\cos z\,dz \\ &= [\tfrac{1}{2}\sin 0 - \tfrac{1}{2}\sin 1] + [\tfrac{1}{2}\sin 1 - \tfrac{1}{2}\sin 0] \\ &= 0.\end{aligned}$$

Actually in this example we are erring on the side of caution, since the integral $\tfrac{1}{2}\sin z$ is a single-valued function, and we shall be justified in a straightforward substitution

$$\begin{aligned}\int_{-1}^1 x \cos x^2\,dx &= [\tfrac{1}{2}\sin z]_1^1 \\ &= \tfrac{1}{2}\sin 1 - \tfrac{1}{2}\sin 1 = 0.\end{aligned}$$

(3) Evaluate $\int_{-2}^2 x\sqrt{x^2 - 1}\,dx$.

Here the obvious substitution is $z = x^2 - 1$, so that $dx/dz = 1/2x$. For the indefinite integral we have

$$\int x\sqrt{x^2 - 1}\,dx = \int x\sqrt{x^2 - 1}/2x\,.\,dz$$
$$= \tfrac{1}{2}\int \sqrt{z}\,.\,dz$$
$$= \tfrac{1}{2}\,.\,\tfrac{2}{3}z^{3/2} = \tfrac{1}{3}z^{3/2}.$$

Now when $x = -2$, $z = 3$, and when $x = 2$, $z = 3$, and by direct substitution we have

$$\int_{-2}^{2} x\sqrt{x^2 - 1}\,dx = [\tfrac{1}{3}z^{3/2}]_3^3 = 0,$$

and this is wrong. For if we look more closely at the integral we see that as x increases from -2 to 2 it passes through a range of values for which $x^2 - 1$ is negative, so that $\sqrt{x^2 - 1}$ does not exist. The "integral" of $x\sqrt{x^2 - 1}$ cannot therefore be evaluated between the limits -2 and 2, and our apparent answer 0 is purely a fallacy.

(4) Evaluate $\int_{-1}^{1} \sqrt{x^2 + x^4}\,dx$.

Here the obvious transformation is to take $z = x^2$.
Thus for the indefinite integral we have

$$\int \sqrt{x^2 + x^4}\,dx = \int \sqrt{x^2 + x^4}\,\frac{dx}{dz}\,dz$$
$$= \int \frac{\sqrt{x^2 + x^4}}{2x}\,dz.$$

Now since $\sqrt{(x^2 + x^4)}$ means by convention the *positive* square root we have

$$\text{If } x > 0, \quad \sqrt{(x^2 + x^4)}/x = \sqrt{(1 + x^2)}$$
$$\text{If } x < 0, \quad \sqrt{(x^2 + x^4)}/x = -\sqrt{(1 + x^2)}.$$

If $x = 0$ we have an indeterminate result: but this will not affect the integral (just as a possible indeterminacy of one single point of a curve will not affect the area under the curve by any finite amount). It therefore follows that we must consider negative and positive values of x separately and write

$$\int_{-1}^{1} \sqrt{x^2 + x^4}\,dx = \int_{-1}^{0} \sqrt{x^2 + x^4}\,dx + \int_{0}^{1} \sqrt{x^2 + x^4}\,dx$$

(by 11.26)

$$= \int_{x=-1}^{x=0} -\tfrac{1}{2}\sqrt{1 + x^2}\,dz + \int_{x=0}^{x=1} \tfrac{1}{2}\sqrt{1 + x^2}\,dz$$
$$= -\tfrac{1}{2}\int_{1}^{0} \sqrt{1 + z}\,dz + \tfrac{1}{2}\int_{0}^{1} \sqrt{1 + z}\,dz$$

since the values $-1, 0, 1$ of x correspond to the values $1, 0, 1$ of $z = x^2$.

It is now natural to make the further transformation $1 + z = w$. Here $dz/dw = 1$, $\sqrt{1 + z} = \sqrt{w}$, and the integral becomes

$$-\tfrac{1}{2}\int_2^1 \sqrt{w}\,dw + \tfrac{1}{2}\int_1^2 \sqrt{w}\,dw$$
$$= -\tfrac{1}{2}\cdot\tfrac{2}{3}\cdot[w^{3/2}]_2^1 + \tfrac{1}{2}\cdot\tfrac{2}{3}\cdot[w^{3/2}]_1^2$$
$$= -\tfrac{1}{3}[1 - \sqrt{8}] + \tfrac{1}{3}[\sqrt{8} - 1]$$
$$= \tfrac{2}{3}[\sqrt{8} - 1].$$

PROBLEMS

(1) Find $\int_{x_1}^{x_2} xe^{-\frac{1}{2}x^2}\,dx$

(2) Find $\int_{-\infty}^{\infty} xe^{-\frac{1}{2}x^2}\,dx$

(3) Find $\int_{-1}^{1} x\sqrt{1 + x^2}\,dx$. Compare the result with example (4) above.

(4) Find $\int_0^1 x \cos x^2\,dx$.

CHAPTER TWELVE

ACCELERATION: MAXIMUM AND MINIMUM VALUES

12.1 Acceleration

Imagine a point P moving up and down a fixed straight line. The distance OP of P from a fixed point O will be called y. This distance y we shall suppose measured with the appropriate sign, that is, positive if P is on one side of O, and negative on the other side. We shall call the time t, so that y will be a function of t and can be plotted graphically

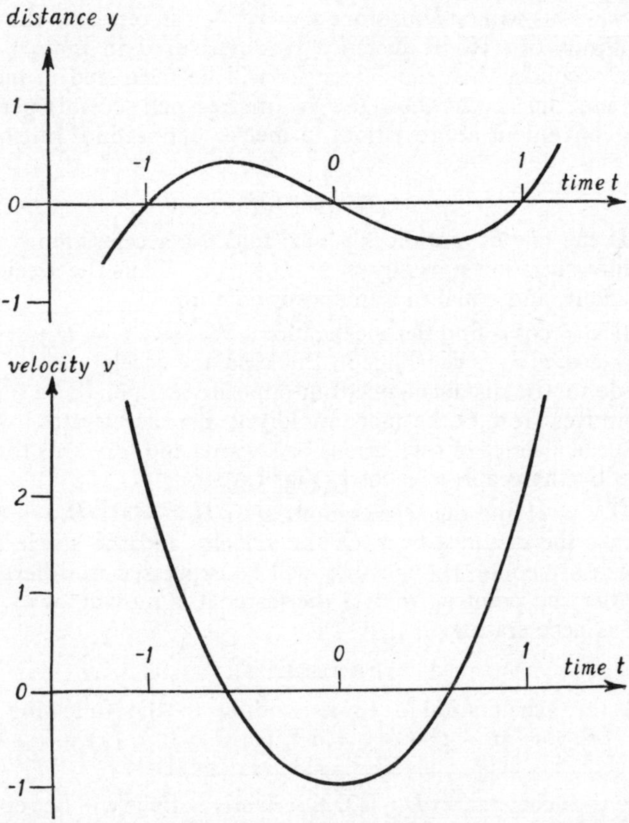

Fig. 12.1—Graphs of position y and velocity v for a moving point

against t. If the law of motion is $y = t^3 - t$ we obtain the graph shown in the top portion of Fig. 12.1. This shows that the point P moves forward, slows down, stops, goes backward through O again (at time $t = 0$), again stops, reverses and once again passes through O in the forward direction and gradually gains speed.

The velocity v of the moving point can be calculated by differentiation. $v = D_t y = 3t^2 - 1$ since $y = t^3 - t$. At each time t there will be a velocity v given by this formula, and we can plot a graph of v against t (Fig. 12.1, lower diagram). In interpreting this graph remember that a positive value of v represents a forward velocity, and corresponds to an upward slope of the graph of y against t. A negative velocity v means a backward motion, and corresponds to a downward slope of the y graph. With this in mind we see from the v graph that the velocity decreases continuously, through zero, to a negative value of -1 when $t = 0$, and then gradually increases again, becoming positive after a short time.

We can now speak of the "acceleration" or rate of change of the velocity v. This will be $D_t v$; since $v = 3t^2 - 1$ in our case, the acceleration is $D_t v = 6t$. If the distance y is measured in metres, and the time t in seconds, then the velocity v will be measured in metres per second, and the acceleration $D_t v$ in (metres per second) per second, or, by a convenient abbreviation, in metres per second² [m/sec²].

EXAMPLES

(1) If the law of motion is $y = t^2$ find the acceleration.

By differentiation $v = D_t y = 2t$, $D_t v = 2$. Thus the acceleration is now constant, and equal to 2 in appropriate units.

(2) If $y = \cos t$, find the acceleration. We have $v = D_t y = -\sin t$, $D_t v = -\cos t = -y$. Thus in this case the acceleration is equal in magnitude to the distance gone, but opposite in sign. The further the point P moves from O the more rapidly it then accelerates towards O. The result is a series of oscillations backwards and forwards through O, as shown by the graph $y = \cos t$ (Fig. 5.9).

(3) If $y = e^t$ find the acceleration. $v = D_t y = e^t$; $D_t v = e^t$. Thus in this case the distance from O, the velocity and the acceleration are all numerically equal, though they will be expressed in different units. The further the point is from O the faster it is moving away, and the faster it is accelerating.

PROBLEMS

Find the acceleration a corresponding to the following laws of motion: (1) $y = 2t + t^2$, (2) $y = 1/t$, (3) $y = \ln t$, (4) $y = 1/(1 + t^2)$.

Since the acceleration $D_t v$ is the derivative of v it will be represented in the v graph by the slope—upwards for a positive acceleration and downwards for a negative acceleration or retardation. In terms of the original graph of y the velocity v corresponds to the slope, and therefore the acceleration corresponds to the rate of change of slope. If the acceleration is positive the slope will become more and more upwards—or in other words the graph is curving upwards. On the other hand if the acceleration is negative the graph curves downwards. This can be seen in Fig. 12.1. The acceleration is $6t$ and is therefore negative if t is negative and positive if t is positive. So the y graph is concave downwards up to the time $t = 0$, and thereafter concave upwards.

Now the velocity v is the derivative of the distance y. The acceleration is therefore the derivative of the derivative of y, or as we shall say, the "second derivative" of y. Since $v = D_t y$, we can write the acceleration of $D_t v$ as $D_t(D_t y)$; but this is usually contracted in writing to $D_t^2 y$, just as the product yy is contracted to y^2. Each commonly recognized way of writing the first derivative $D_t y$ will give rise to a notation for the second derivative. Thus if we write v as Dy, we shall write the acceleration a as D^2y; if v is written y', a will be written as y'', and if v is written y_t, a will be written y_{tt} [not as y_{t^2}, as that would mean $D_{(t^2)}y$, i.e. $dy/d(t^2)$]. Finally if v is written as $\frac{dy}{dt}$ or $\left(\frac{d}{dt}\right)y$, the acceleration will be $\left(\frac{d}{dt}\right)^2 y$, the symbol $\left(\frac{d}{dt}\right)$ being used as an alternative to D_t. This last symbol is often written $\frac{d^2y}{dt^2}$, and this may be considered as the standard way of writing the second derivative ("second derivate", or "second differential coefficient" are alternative names). Unfortunately there seems nothing to be said in favour of the notation $\frac{d^2y}{dt^2}$ except that it is commonly accepted; it does not look very natural, and if taken at its face value can lead to errors.* We shall not use it in this book.

---

* As an example consider a point moving according to the law $y = t^2$. We have already shown that $\frac{dy}{dt} = D_t y = 2t$, $\frac{d^2y}{dt^2} = 2$. Also $\frac{dy}{dy} = D_y y = 1$, and $\frac{d^2y}{dy^2} = \frac{d}{dy}\left(\frac{dy}{dy}\right) = 0$. Now $\frac{d^2y}{dt^2} = \frac{d^2y}{dy^2} \cdot \frac{dy^2}{dt^2} = 0$; but $\frac{d^2y}{dt^2} = 2$, therefore $2 = 0$, which is absurd. The fallacy lies in the suggestion made by the notation that we can cancel out the dy^2 in $\frac{d^2y}{dy^2} \cdot \frac{dy^2}{dt^2}$ to get $\frac{d^2y}{dt^2}$. When we try to do such a cancellation we get an absurd answer. But if we write it properly as $D_y^2 y\,(D_t y)^2$ there is no longer any temptation to contract it fallaciously to $D_t^2 y$. But this condemnation does not apply to the ordinary way of writing the first derivative as $\frac{dy}{dt}$. This can be quite helpful. It suggests the correct formula $\frac{dw}{dt} = \frac{dw}{dy}\frac{dy}{dt}$ for change of variable, and also reminds us that $\frac{\delta y}{\delta t}$ is very nearly equal to $\frac{dy}{dt}$ when δy and δt are small.

From the acceleration we can go on to further derivatives. The rate of change of acceleration $D_t a$ will be the "third derivative" $D_t^3 y$, and this when differentiated again will give us the fourth derivative $D_t^4 y$, and so on. These can be easily calculated when we have a mathematical expression giving y in terms of t. But although it is easy to visualize the meaning of the first two derivatives, the velocity, or rate of change of position, and acceleration, or rate of change of velocity, it is much more difficult to form any clear mental picture of the rate of change of acceleration; while for still higher derivatives it is practically hopeless. They are quantities exactly defined in mathematical terms and useful in calculation but rather difficult to grasp imaginatively except in the vaguest manner.

FURTHER EXAMPLES

(4) If $y = t^4$ find the successive derivatives of y.

The velocity $v = D_t y = 4t^3$. The acceleration is the rate of change of velocity $= D_t^2 y = 4 \times 3t^2 = 12t^2$. The rate of change of acceleration $= D_t^3 y = 4 \times 3 \times 2t = 24t$. The fourth derivative $D_t^4 y = 4 \times 3 \times 2 \times 1 = 24$. Since this is a constant the fifth and all subsequent derivatives will be zero.

(5) If $y = e^t$ find the successive derivatives of y.

Since $D_t y = e^t = y$, we have $D_t^2 y = e^t$, and all succeeding derivatives are equal to e^t.

(6) If $y = e^{Kt}$ find the successive derivatives of y.

We have $D_t y = Ke^{Kt}$, $D_t^2 y = K \,.\, Ke^{Kt} = K^2 e^{Kt}$, $D_t^3 y = K \,.\, K^2 e^{Kt} = K^3 e^{Kt}$, and in general $D_t^n y = K^n e^{Kt}$.

(7) If $y = \cos t$ find the successive derivatives of y.

$D_t y = -\sin t$, $D_t^2 y = -\cos t$, $D_t^3 y = \sin t$, $D_t^4 y = \cos t$, and the process then repeats itself, $D_t^5 y = -\sin t$, etc. In general $D_t^{4n} y = \cos t$, $D_t^{4n+1} y = -\sin t$, $D_t^{4n+2} y = -\cos t$, $D_t^{4n+3} y = \sin t$.

PROBLEMS

(5) If $y = \cosh t$ find the successive derivatives of y.

(6) If $y = \cos Kt$ find the successive derivatives of y.

(7) If $y = A + Bt + Ct^2 + e^t$ find $D_t^n y$.

(8) If $y = \ln (1 + t)$ find $D_t^n y$.

12.2 Maxima and minima

We know that if a heavy body, such as a cricket ball, is thrown into the air, it will rise until it reaches a certain maximum height, and then fall again until it hits the ground, when its height will be a minimum. A piece of down, on the other hand, will be blown about by the wind, rising and falling for quite a time until it comes to rest. In general any function y of t considered for a certain range of values of t (say from $t = t_1$ to $t = t_2$) will have one or more maximum and minimum values in this range.

We have already introduced a scheme of classification for such maxima and minima. A value of y which is greater than all other values in its immediate neighbourhood is a *local* maximum. If it is greater than all other values without restriction, or at any rate not less than any other, it is an *absolute* maximum. Generally speaking therefore an absolute maximum is necessarily a local maximum, but the converse need not be true. (We could in theory have a value which would be an absolute maximum and *equal* to all the neighbouring values, so that it would strictly speaking not be a local maximum. But such a case is unlikely to occur in practice, and not difficult to deal with if it should occur.)

We shall concentrate our attention on the local maxima and minima, firstly because they are of interest on their own account, and secondly because the simplest method of finding the absolute maximum is to find all the local maxima and test their values individually to see which is the greatest. Accordingly we shall usually omit the qualifying word "local", speaking simply of "maxima" and "minima".

A further classification can be made. A maximum or minimum value which occurs at the beginning or end of the curve will be called "terminal", one occurring within the range of values will be "intermediate". Thus in the graph of y shown in Fig. 12.2, y has a terminal

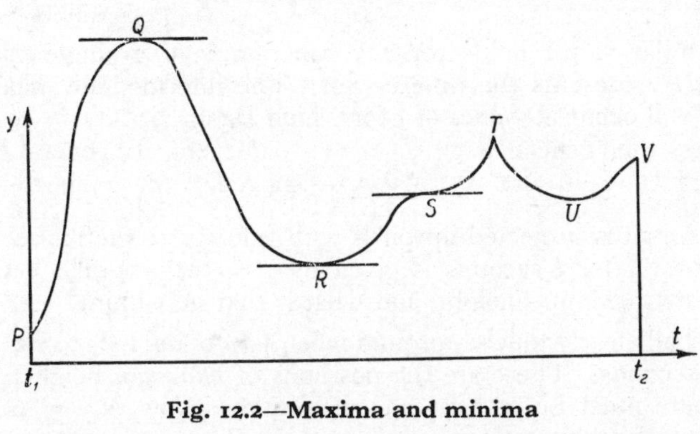

Fig. 12.2—Maxima and minima

absolute minimum at P, an intermediate absolute maximum at Q, another intermediate maximum at T, intermediate minima at R and U, and a terminal maximum at V.

The terminal maxima and minima are as a rule easy to deal with. If the function is increasing at the starting-point of the curve, P, as in Fig. 12.2, then P will be a minimum; if the function is decreasing then P will be a maximum. Thus we have the simple conditions for the initial point of the curve, $t = t_1$, that $D_t y > 0$ corresponds to a terminal minimum, and $D_t y < 0$ to a maximum. At the final point, $t = t_2$, the conditions are reversed, $D_t y > 0$ corresponding to a maximum and $D_t y < 0$ to a minimum.

Now consider the intermediate maxima and minima. Near a maximum point, such as Q, we see that the curve must be rising on the left-hand side of Q, and falling on the right-hand side. This can happen as at T in the figure by a sudden change in direction. But more often there will be a smooth maximum as at Q. Now at such a point $D_t y$ cannot be positive, i.e. the tangent cannot be sloping upwards, for then there would be points of the curve on the right of Q where y is greater that at Q, contrary to our supposition that it is a local maximum. Similarly the slope $D_t y$ cannot be negative at Q. Thus the only possibility remaining is that $D_t y = 0$, and the tangent at Q is horizontal. For similar reasons the tangent to the curve at a minimum point R must be horizontal: for if it sloped upwards there would be points just on the left of R where there would be still smaller values of y, while if it sloped downwards there would be such points on the right of R. Thus we have the following general test for an intermediate maximum or minimum that (subject to rare exceptions like T already noted) the slope $D_t y$ will be zero at such a point. Conversely, generally speaking a point where $D_t y = 0$ will be a maximum or minimum. Occasionally there will be a point such as S in Fig. 12.2 where the tangent is horizontal, but which is neither a maximum nor minimum, but merely a momentary halt in the general rise or fall of the curve.

A similar result holds for any function y of a single variable t, whether t represents the time or not. The intermediate maxima and minima will occur at values of t for which $D_t y = 0$.

EXAMPLES

(1) A ball is projected upwards with velocity 10 metres/sec, so that its height y after t seconds is given by $y = 10t - 4{\cdot}9t^2$. When does it reach its maximum height, and what is that maximum?

The ball clearly touches ground when $y = 0$, i.e. $t = 0$ or $t = 10/4{\cdot}9 = 2{\cdot}04$ seconds. These are the positions of *minimum* height: between them there must be at least one maximum. Now $D_t y = 10 - 9{\cdot}8t$; so that we have a stationary point when $D_t y = 0$, i.e. $t = 10/9{\cdot}8 = 1{\cdot}02$ seconds. Since this is the only stationary point, and since there are no sudden changes in the slope $D_t y$, it follows that this must be the intermediate maximum point. Substituting in the equation $y = 10t - 4{\cdot}9t^2$ we find the maximum height to be $5{\cdot}1$ metres.

This problem could also be solved by the method of "completing the square" of Section 3.4, since y is a quadratic function of t. The method of differentiation is, however, more general, as we shall see from the next example.

(2) A farmer wishes to construct a rectangular sheep-fold of area 25 square metres. What length must he make the sides so as to make the length of the boundary fence a minimum?

Suppose that the sides of the rectangle are x and y metres respectively, then $xy = 25$, so that $y = 25/x$. Now the length of the fence must be $L = 2x + 2y$ metres. In order to find when this is a maximum or minimum we must first express it in terms of a *single* independent variable, say x. Using the relation $y = 25/x$ we have $L = 2x + 50/x$. We must now solve the equation $D_xL = 0$ [here L takes the place of y and x the place of t in the formula $D_ty = 0$], i.e. $2 - 50/x^2 = 0$, i.e. $1/x^2 = 1/25$. So $x = 5$ metres, and therefore $y = 5$ metres and the enclosure is square. It remains to see whether this is a maximum or minimum. Now if we make x very small, $y = 25/x$ becomes very large, and the length $L = 2x + 2y$ becomes very large. If x is large it also follows that $L = 2x + 2y$ is large. Thus somewhere between these two extremes there must be at least one minimum; and since we have found only one stationary point, that must be the minimum in question, i.e. of all rectangles with given area a square has the smallest perimeter.

In these examples we have deduced from the general form of the curve that the point at which the derivative vanishes is a maximum in the first example, and a minimum in the second. But there is also a simple general method of distinguishing between intermediate maxima and minima. Suppose that y has a stationary point when $t = T$, so that then $D_ty = 0$. Calculate the value of the second derivative D_t^2y at this point. If this is positive, it means that the slope D_ty of the curve is increasing around that point. Since it is zero when $t = T$, it follows that it must be negative for values of t slightly smaller than T, and positive for t slightly greater than T. In other words the curve is sloping downwards as far as the stationary point $t = T$, and thereafter slopes upwards, and the stationary point must be a minimum. On the other hand if D_t^2y is negative when $t = T$ the point will be a maximum. This can also be shown geometrically. A positive value of D_t^2y means that the curve of y is concave upwards, as at the minimum point R in Fig. 12.2, whereas a negative value of D_t^2y means concavity downwards, as at the maximum point Q. If $D_t^2y = 0$ we can have either a maximum or a minimum or a "point of inflexion" such as S in Fig. 12.2: such a case will be discussed later.

Thus in example (1) above, $y = 10t - 4{\cdot}9t^2$, $D_t^2y = -9{\cdot}8 < 0$, so that any stationary point for y must be a maximum. In example (2), $L = 2x + 50/x$, $D_x^2L = 100/x^3$ and is positive, so that the stationary point is a minimum.

FURTHER EXAMPLES

(3) Robertson (*Child Physiology*, p. 249) has found on theoretical grounds that the growth in the weight of an infant up to nine months is an autocatalytic phenomenon, the relation between the weight w (ounces) and age t (months) being

$$\ln w - \ln(341{\cdot}5 - w) = K(t - 1{\cdot}66).$$

At what age will the infant grow most rapidly?

The rate of growth $v = D_t w$. This can be found by differentiating Robertson's equation with respect to t, obtaining

$$v/w + v/(341{\cdot}5 - w) = K$$

i.e. on reduction to a common denominator,

$$341{\cdot}5v/[w(341{\cdot}5 - w)] = K$$
$$v = Kw(341{\cdot}5 - w)/341{\cdot}5.$$

Now the maximum rate of growth will occur when $D_t v = 0$. But

$$D_t v = D_w v \,.\, D_t w = K(341{\cdot}5 - 2w)v/341{\cdot}5$$

This will be zero when $341{\cdot}5 - 2w = 0$, i.e. when $w = 170{\cdot}75$ ounces. It remains to test whether this is a maximum or a minimum point. On differentiating this expression with respect to t by the product rule we have

$$D_t^2 v = [K/341{\cdot}5]\{D_t(341{\cdot}5 - 2w) \,.\, v + (341{\cdot}5 - 2w)D_t v\}$$

But $D_t(341{\cdot}5 - 2w) = -2D_t w = -2v$, while at the stationary point $D_t v = 0$ by definition. Therefore $D_t^2 v = [K/341{\cdot}5][-2v^2]$, and is negative at this point (since v^2 is necessarily positive). Thus, in fact, we have the maximum rate of growth.

The age t at which this maximum occurs is found by substituting the value $w = 170{\cdot}75$ in the original equation connecting w and t. We find

$$\ln 170{\cdot}5 - \ln 170{\cdot}5 = K(t - 1{\cdot}66)$$

whence $t = 1{\cdot}66$ months $=$ about 7 weeks.

(4) A cylinder has to have a given volume V. How must the length L and radius R be chosen so as to make the surface area A a minimum?

The formulas for the area and volume of a cylinder are

$$A = 2\pi R(R + L)$$
$$V = \pi R^2 L$$

Since V is given there is a relation between L and R, and the first step in finding the minimum value of A is to express it as a function of a single variable, say of R. From the second relation we have

$L = V/\pi R^2$, and so $A = 2\pi R(R + V/\pi R^2) = 2\pi R^2 + 2V/R$. (Here V is a known constant, by hypothesis.) A maximum or minimum value of A will occur when $D_R A = 0$, that is

$$4\pi R - 2V/R^2 = 0$$
$$\text{i.e.}\quad R^3 = V/2\pi$$
$$R = [V/2\pi]^{1/3}$$

whence $L = V/\pi R^2 = 2[V/2\pi]^{1/3} = 2R$. Thus the length of the cylinder is equal to its diameter.

To test whether this is a maximum or minimum we must find the sign of $D_R{}^2 A = 4\pi + 4V/R^3$ at the stationary point. Since V and R are positive, $D_R{}^2 A$ must also be positive, and therefore this gives us the minimum area.

(5) John Hunter wrote as follows: "To keep up a circulation sufficient for the part and no more, Nature has varied the angles of origin of the arteries accordingly." Let us put the following interpretation on this statement. Let AB be a main trunk, of radius R, and C a point to be fed by a branch artery PC of radius r leaving AB at P (Fig. 12.3). We suppose that when R, r and the point C are given,

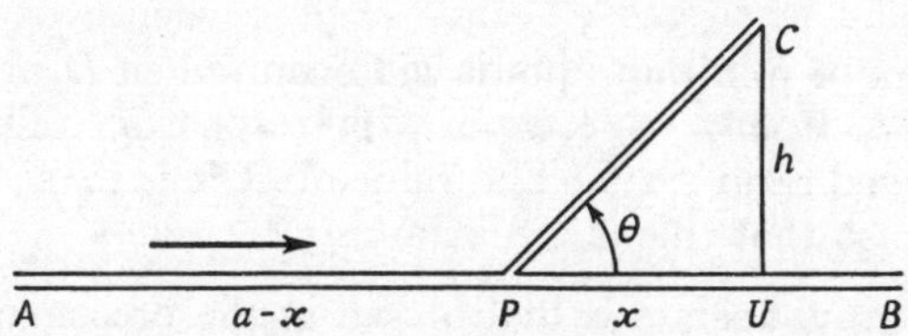

Fig. 12.3—Flow of blood in a branched artery

the point of origin P of the branch artery is to be chosen in such a way that the fall in pressure between A and C is a minimum.

Let p be this fall in pressure, supposed mainly due to friction against the arterial walls. Then by Hess's law p is proportional to $AP/R + PC/r = y$ (say). Now draw CU perpendicularly from C onto AB, and let the distances $CU = h$, $AU = a$, and the unknown $PU = x$, also $\angle UPC = \theta$. Then $AP = a - x$, $PC = \sqrt{(x^2 + h^2)}$, so that

$$y = (a - x)/R + (x^2 + h^2)^{\frac{1}{2}}/r$$
$$D_x y = -1/R + x(x^2 + h^2)^{-\frac{1}{2}}/r$$

This is zero when $r/R = x/\sqrt{(x^2 + h^2)} = PU/PC = \cos\theta$.

Also $$D_x{}^2 y = D_x(D_x y)$$
$$= (x^2 + h^2)^{-\frac{1}{2}}/r - x^2(x^2 + h^2)^{-\frac{3}{2}}/r$$
$$\text{(by the product rule)}$$
$$= (x^2 + h^2)^{-\frac{3}{2}}/r \,.\, [(x^2 + h^2) - x^2]$$
$$= h^2(x^2 + h^2)^{-\frac{3}{2}}/r > 0$$

so that this value of θ gives a minimum.

We see from this that the angle θ depends only on the ratio of the radii of the main and branch arteries. If this theory is correct, then when r is very small compared with R the branch should come off almost at right angles, while if r is nearly equal to R, $\cos\theta$ is nearly 1, and θ nearly 0. Thus a branch of large calibre can be expected to come off nearly parallel to the main trunk (as happens often though not invariably, e.g. the external and internal carotids).

(6) In a submarine cable with an insulating sheath the rate of signalling is proportional to $-x^2 \ln x$, where x is the ratio of the radius of the core to the outer radius of the sheath. What value of x gives the most rapid transmission? Assuming that a nerve acts in the same manner, does this agree well with the observed ratio 1 : 1·6 of the radii of the axon and myelin sheath? (W. M. Feldman, *Proc. Physiol. Soc.*, 1923.)

If $v = -Kx^2 \ln x$ is the velocity of the impulse, then

$$\begin{aligned} D_x v &= -K(2x)\ln x - Kx^2/x \\ &= -Kx(2\ln x + 1) \\ D_x^2 v &= -K(2\ln x + 1) - Kx(2/x) \\ &= -K(2\ln x + 3) \end{aligned}$$

A maximum or minimum point will occur when $D_x v = 0$, i.e. when $2\ln x + 1 = 0$, $\ln x = -\frac{1}{2}$, or $x = e^{-\frac{1}{2}} = 1/1{\cdot}65$. This agrees well with the observed ratio 1/1·6. The value of $D_x^2 v$ is then $-K(-1+3) = -2K < 0$, so that this is a maximum.

(7) When ethyl acetate is hydrolysed in the presence of acetic acid as a catalyst the following reaction occurs:

$$CH_3COOC_2H_5 + H_2O \rightarrow CH_3COOH + C_2H_5OH$$

The acetic acid formed thus gradually increases in amount and tends to cause an acceleration in the velocity of the reaction. But at the same time the active mass of the ester diminishes, thus causing a retardation in the reaction velocity. At what point will the velocity be a maximum?

Let the initial concentration of acetic acid be a moles/litre, and of ester b moles/litre. Suppose further that x moles are hydrolysed after a time t, producing x moles of acetic acid. Then there will be $(a + x)$ moles of acid and $(b - x)$ moles of ester remaining, and the reaction velocity is proportional to $v = (a + x)(b - x)$. A maximum or minimum will occur when $D_x v = 0$, i.e. $(b - x) - (a + x) = 0$, or $x = \frac{1}{2}(b - a)$.

PROBLEMS

(1) An open tank is constructed with a square base and vertical sides so as to contain a given volume V of water. What must be the

relation between depth and width so as to make the expense of lining it with lead a minimum?

(2) If $y = t^3 - 6t^2 + 11t - 6$, for what values of t will y be a maximum or a minimum?

(3) For what positive value of t is $t^{-1} \ln t$ a maximum? Sketch a graph of the function.

(4) What is the minimum value of $t^{1/t}$ for positive t, and what is its maximum value? Sketch the graph.

(5) What number exceeds its square by the greatest number possible, and by how much?

(6) Two light bulbs of 32- and 4-candle power respectively are placed 3 metres apart on a table. If the intensity of illumination due to a source of light is inversely proportional to the square of the distance, what point on the line joining the lamps is least illuminated?

(7) A window is in the shape of a rectangle surmounted by a semi-circle. If the perimeter is 6 metres, find the dimensions so that the greatest possible amount of light may be admitted.

(8) A photographic lens of 25 cm focal length forms the image of an object on a screen. What is the minimum possible distance between object and image?

(9) A piece of wire 2 metres long is cut into two parts. One part is bent to form a square, the other a circle. What are the lengths of the parts when the sum of the areas formed is a minimum?

NOTE.—It has been suggested that bees build honeycombs in such a way that they use the minimum amount of wax to contain the maximum honey. This leads to a problem in calculus (see d'Arcy W. Thompson, *Growth and Form*) and the solution does correspond reasonably well with the measured shape of the honeycomb. But probably this is little more than a coincidence, since it is doubtful whether this does save any appreciable amount of wax. There may be other factors determining the form of the cell.

12.3 Improved formulas for small changes

If y is a differentiable function of t then the ratio

$$\delta y/\delta t = (y_2 - y_1)/(t_2 - t_1)$$

tends to the limit $D_t y$ when $\delta t = t_2 - t_1$ tends to 0. We have already suggested that when δt is sufficiently small the difference between $\delta y/\delta t$ and $D_t y$ may be negligible for practical purposes, and it may be quite good enough to say that $\delta y/\delta t = D_t y$, or $\delta y = (D_t y)\delta t$. This gives an approximate formula for the calculation of δy. But it is not in general absolutely correct, and we shall now investigate how we can improve its accuracy.

The equation $\delta y/\delta t = D_t y$ will be exact if the velocity $v = D_t y$ is constant: for then the average velocity, as well as the instantaneous velocity, will always be equal to v. For the graph of y plotted against t will be a straight line of slope v, and the line joining any two points (t_1, y_1), (t_2, y_2) on the graph has a slope equal to $\delta y/\delta t$. But this line must coincide with the original graph, so that $\delta y/\delta t = v$ exactly when v is constant.

This suggests that the main cause for inaccuracy in the formula $\delta y = (D_t y)\delta t$ will be curvature of the graph. This curvature corresponds to a change in the slope of the graph, i.e. to a change in the velocity or rate of change $D_t y$. This change in the velocity corresponds to the existence of an acceleration $D_t v = D_t^2 y$ (or a second derivative, if t represents some variable other than the time).

Now when the velocity is changing the simple formula $\delta y = (D_t y)\delta t$ is ambiguous, since it does not state explicitly at what time t we are to calculate the velocity (or rate of change) $D_t y$. The simplest choice is to take the time t to be t_1. We shall denote this by a suffix 1, writing $v_1 = (D_t y)_1$ for the value when $t = t_1$. The first approximation to the value of $\delta y = y_2 - y_1$ is therefore

$$\delta y \simeq (D_t y)_1 \cdot \delta t \qquad . \quad . \quad . \quad (12.1)$$

and this formula is exactly true on the assumption that the velocity is constant and always equal to $(D_t y)_1$. It is sufficiently nearly true if the velocity does not change appreciably between times t_1 and t_2.

This suggests that for our next approximation we should take the acceleration $a = D_t^2 y$ to be sensibly constant, as follows.

12.4 Motion with constant acceleration

The equation $D_t^2 y = y_{tt} = a$ involving the second derivative is called an "ordinary differential equation of the second order". In general any equation connecting t, y, $v = D_t y = y_t$ and $a = D_t^2 y = y_{tt}$ is a second-order equation. Such equations are of many types and will not be discussed in detail here. But the special type $y_{tt} = a$ is easily solved, whether a is a constant or not. We can get from the acceleration a to the velocity v by integrating with respect to t, since a is the derivative of v; and we can then get from v to y by a further integration.

$$v = C_1 + \int a\,dt$$

where C_1 is a constant of integration;

$$y = C_2 + \int v\,dt = C_2 + C_1 t + \int\left(\int a\,dt\right)dt \qquad . \ (12.2)$$

where C_2 is a second constant of integration. Thus there are two integration constants in the solution of a second-order equation. This is a general property, true for all such equations. Such constants can be determined if we are given, in addition to the original equation, the position y at two different times, or the position and velocity at one time, or any two similar items of information.

If a is constant, then $\int a\,dt = at$, $\int(\int a\,dt)dt = \int at\,dt = \frac{1}{2}at^2$, so that the equation (12.2) becomes

$$y = C_2 + C_1t + \tfrac{1}{2}at^2 \quad . \quad . \quad . \quad (12.3)$$

We can interpret the numbers C_1 and C_2 as follows. When $t = 0$, $y = C_2$, i.e. C_2 is the *initial position* and $y - C_2$ is the distance travelled. Also on differentiating (12.3) we obtain $v = y_t = C_1 + at$, so that when $t = 0$, $v = C_1 =$ the initial velocity. Thus (12.3) can be written

$$y - C_2 = \text{distance gone} = (\text{initial velocity})t + \tfrac{1}{2}at^2$$

(a formula to be found in many text-books on mechanics).

EXAMPLES

(1) Suppose the acceleration $a = e^t$, what is the position y?

By (12.2), since $\int a\,dt = e^t$, $\int(\int a\,dt)dt = e^t$,

$$y = C_2 + C_1t + e^t$$

(2) Suppose the acceleration $a = 1/t^2$, what is the position y?

$$\int a\,dt = -1/t, \ \int(\int a\,dt)dt = -\ln t,$$

so that

$$y = C_2 + C_1t - \ln t.$$

PROBLEMS

(1) A body is moving with acceleration $a = \sin t$. Supposing that it starts at time $t = 0$ from position $y = 0$ with velocity $v = 1$, what is its position at time t?

(2) A body has acceleration $(1 + t)$; find a formula for its position.

Now from our point of view it is the time $t = t_1$ rather than the time $t = 0$ which is our starting-point. So it is convenient to change the variable from t to $T = t - t_1$. For when $t = t_1$, $T = 0$, and when $t = t_2$, $T = \delta t$.

Now if Q is any function whatever of t, we know that $Q_T = Q_t t_T$ (Section 8.11). But $t = t_1 + T$, so that $t_T = 1$, and $Q_T = Q_t$. That is to say, differentiation with respect to T is the same as differentiation with respect to t. It follows that, whatever Q may be, $Q_{TT} = (Q_T)_T = (Q_T)_t = Q_{tt}$, and similarly $Q_{TTT} = Q_{ttt}$, and so on for all higher derivatives. In particular, if y represents the position, then $y_t = y_T = v$ is the velocity, and $y_{tt} = y_{TT} = a$ is the acceleration. Now suppose that at time $t = t_1$ the body has position y_1, velocity $v_1 = (y_t)_1$, and acceleration $a_1 = y_{tt}$, which is a constant. Then we can write

$$y_{TT} = a_1$$

Integrating this with respect to T, we have

$$v = y_T = \int a_1 dT = C_1 + a_1 T.$$

But when $T = 0$ the body has velocity v_1, and therefore from the above equation $C_1 = v_1$, i.e.

$$v = y_T = v_1 + a_1 T.$$

Now integrate this again with respect to T:

$$y = \int v\, dT = C_2 + v_1 T + \tfrac{1}{2} a_1 T^2.$$

But substituting $T = 0$ we obtain $y_1 = C_2$, i.e. finally

$$\begin{aligned} y &= y_1 + v_1 T + \tfrac{1}{2} a_1 T^2 \\ &= y_1 + [y_t]_1 (t - t_1) + \tfrac{1}{2} y_{tt} (t - t_1)^2 \quad . \quad . \quad (12.4) \end{aligned}$$

This is therefore the general formula giving the position y of a body moving with constant acceleration a_1, when its position y_1 and velocity v_1 at time t_1 are known. At least, we have found it convenient to talk of "positions" y, "velocities" v and "accelerations" a, but the same formula will hold for any function y of any variable t, when the second derivative y_{tt} is constant and equal to a_1.

12.5 Second approximation to a small change

Formula (12.4) is exact when the second derivative $y_{tt} = a_1$ is constant. It will be a good approximation if the time interval is so short that the second derivative can be regarded as substantially constant. It can also be written

$$y - y_1 = v_1 T + \tfrac{1}{2} a_1 T^2.$$

If we put $t = t_2$, then $T = t_2 - t_1 = \delta t$, $y_2 - y_1 = \delta y$, so that this becomes

$$\delta y = y_2 - y_1 \simeq v_1 \delta t + \tfrac{1}{2} a_1 (\delta t)^2 \quad . \quad . \quad (12.5)$$

This is the second approximation to δy, and differs from the first approximation only in the inclusion of an extra term, namely $\tfrac{1}{2} a_1 (\delta t)^2 = \tfrac{1}{2} [y_{tt}]_1 (\delta t)^2$, representing the effect of the acceleration.

EXAMPLES

(1) A cubical box of side 1 cm expands under heat to a side 1·01 cm. What is its change in volume?

The relation between volume V and side x is $V = x^3$, whence $V_x = 3x^2$, $V_{xx} = 6x$. Therefore

$$\delta V \simeq (V_x)_1 \delta x + \tfrac{1}{2}(V_{xx})_1(\delta x)^2$$

where the suffix 1 indicates that we take the value when $x = 1$, i.e. $(V_x)_1 = 3$, $(V_{xx})_1 = 6$. But $\delta x = \cdot 01$, whence

$$\begin{aligned}\delta V &\simeq 3 \times \cdot 01 + \tfrac{1}{2} \times 6 \times (\cdot 01)^2 \\ &= \cdot 0303\end{aligned}$$

to the second approximation. The first approximation would be

$$\delta V \simeq (V_x)_1\, \delta x = 3 \times \cdot 01 = \cdot 03.$$

Actually here we can find the true value for δV, for the final volume is $1{\cdot}01^3 = 1{\cdot}030301$ and the initial volume $1^3 = 1$, whence $\delta V = 1{\cdot}030301 - 1 = \cdot 030301$ exactly. Thus the second approximation will suffice for most purposes whereas the first has about 1% error.

(2) Find ln 1·1.

Let $y = \ln t$, $t_1 = 1$, $T = t - 1$. Then by (12.4)

$$y = \ln(1 + T) \simeq y_1 + [y_t]_1 T + \tfrac{1}{2}[y_{tt}]_1 T^2.$$

But $y_t = 1/t$, $y_{tt} = -1/t^2$, so that putting $t = 1$

$$y_1 = 0, \quad [y_t]_1 = 1, \quad [y_{tt}]_1 = -1, \quad \text{and}$$
$$\ln(1 + T) \simeq T - \tfrac{1}{2}T^2.$$

If we put $T = \cdot 1$ we obtain

$$\ln 1{\cdot}1 \simeq \cdot 1 - \tfrac{1}{2}(\cdot 1)^2 = \cdot 095.$$

Actually ln 1·1 = ·09531, so that this second approximation is quite reasonable.

PROBLEMS

(1) Find the first and second approximations to $e^{0\cdot 1}$, taking $y = e^t$, $t_1 = 0$.

(2) Find the first and second approximations to $\sqrt{1{\cdot}1}$.

The equation (12.4) can also be applied to the problem of maxima and minima. Let t_1 be a value of t for which y is stationary, i.e. $y_t = v_1 = 0$ when $t = t_1$. Then any neighbouring value of y is approximately $y_1 + \tfrac{1}{2}a_1 T^2$, where a_1 is the value of the second derivative y_{tt}

when $t = t_1$. Now $\frac{1}{2}T^2$ is always positive: so if a_1 is positive we must have

$$y \simeq y_1 + \tfrac{1}{2}a_1T^2 > y_1$$

i.e. y_1 is less than every neighbouring value of y, and so is by definition a local minimum. On the other hand if a_1 is negative then $y < y_1$, and y_1 is a local maximum. This justifies the rule for distinguishing maxima and minima given in Section 2.2.

12.6 Orders of small quantities

If δt, say, is small, then $(\delta t)^2$ is much smaller, and $(\delta t)^3$ much smaller still. For example, if $\delta t = \cdot 001$, then $(\delta t)^2 = \cdot 000001$ and $(\delta t)^3 = \cdot 000000001$.

It is often convenient to classify the various small changes we have to deal with into " orders of smallness". We can select one of them, say δt, and call it a "small quantity of the first order". Any quantity which is of about the same size as δt, such as $2\delta t$, or $\pi^2\delta t$, or $\frac{1}{2}t\ \delta t$, will also be considered as of the first order. Any quantity which is comparable in size with $(\delta t)^2$, such as $3{\cdot}5(\delta t)^2$, is considered as the second order; one comparable with $(\delta t)^3$ is of the third order, and so on. Naturally this is rather a rough division: there is no exact line separating the first order from the second, or the second order from the third. But all the same it is very useful. For we may have a good idea of what accuracy we require in our calculations—possibly small quantities of the second order can be neglected for our purposes, possibly only third-order quantities. But if we know that, say, second-order quantities are negligible, then they can be omitted from our calculations without appreciable loss, and often with considerable saving of effort. Thus we know that $[1 + \delta t]^3 = 1 + 3\delta t + 3(\delta t)^2 + (\delta t)^3$ exactly: but if we are allowed to neglect $(\delta t)^2$ and $(\delta t)^3$ this simplifies to $[1 + \delta t]^3 \simeq 1 + 3\delta t$. If we can neglect $(\delta t)^3$ but not $(\delta t)^2$ we shall have $[1 + \delta t]^3 \simeq 1 + 3\delta t + 3(\delta t)^2$ sufficiently accurately. As another example we have

$$[1 + 2\delta t][2\delta t + 3(\delta t)^2] = 2\delta t + 7(\delta t)^2 + 6(\delta t)^3 \text{ exactly.}$$

But if $(\delta t)^3$ is negligible this simplifies to $2\delta t + 7(\delta t)^2$, and if $(\delta t)^2$ is negligible it is enough to write $2\delta t$ for the product.

There are certain rules which help us in manipulating small quantities:

(A) The sum of two quantities of the first order is also of the first order; and in general the sum of two nth-order quantities is of the nth order. For if x and y are both about the same size as $(\delta t)^n$ so also is $(x + y)$.

(B) The sum of a first-order and a second-order small quantity is of the first order, since the second quantity is small compared with the first. The same applies to the sum of an mth-order and an nth-order quantity; if $n > m$, the sum is of the mth (i.e. smaller) order.

(C) The product of two first-order quantities is of the second order: for $\delta t \times \delta t = (\delta t)^2$. Similarly the product of an mth-order and an nth-order quantity is of order $(m + n)$.

(D) If δt is of the first order and y is any function of t then the corresponding change δy in y is also of the first order. For $\delta y \simeq D_y t \,.\, \delta t$. But at points where the derivative $D_t y$ is zero or infinite, and near such points, this will no longer necessarily be true: δy may then be of a higher or lower order than δt.

12.7 The general approximation

The first-order approximation to y, $y = y_1 + [y_t]_1 T = y_1 + [y_t]_1(t - t_1)$ will be satisfactory if T^2 is so small that it can be neglected. If it cannot we must use (12.4),

$$y = y_1 + [y_t]_1 T + \tfrac{1}{2}[y_{tt}]_1 T^2$$

which takes the second derivative into consideration. We can obtain the nth-order approximation, counting all powers of T up to T^n, by taking all the first n derivatives of y into consideration.

Let us illustrate by considering the fourth-order approximation, taking into account the derivatives y_t, y_{tt}, y_{ttt}, y_{tttt}. If we write $T = t - t_1$, we know that $y_t = y_T$, $y_{tt} = y_{TT}$, $y_{ttt} = y_{TTT}$, $y_{tttt} = y_{TTTT}$, since differentiation with respect to T is equivalent to differentiation with respect to t. Let $[y_t]_1$, $[y_{tt}]_1$, etc., denote the values of these derivatives when $t = t_1$, i.e. when $T = 0$.

Now let us assume that we can take the fourth derivative y_{TTTT} to be constant to a sufficient order of accuracy. Then we can denote its value by $[y_{tttt}]_1$, i.e. $y_{TTTT} = [y_{tttt}]_1$. On integrating this with respect to T we get

$$y_{TTT} = C_1 + [y_{tttt}]_1 T.$$

But when $T = 0$, $C_1 = [y_{TTT}]_1 = [y_{ttt}]_1$, so that

$$y_{TTT} = [y_{ttt}]_1 + [y_{tttt}]_1 T.$$

An integration of this with respect to T gives us

$$y_{TT} = C_2 + [y_{ttt}]_1 T + \tfrac{1}{2}[y_{tttt}]_1 T^2.$$

Putting $T = 0$ we see that C_2 is the value of $y_{TT} = y_{tt}$ when $t = t_1$, i.e. $C_2 = [y_{tt}]_1$, and

$$y_{TT} = [y_{tt}]_1 + [y_{ttt}]_1 T + \tfrac{1}{2}[y_{tttt}]_1 T^2.$$

Again on integration

$$y_T = C_3 + [y_{tt}]_1 T + \tfrac{1}{2}[y_{ttt}]_1 T^2 + \frac{1}{2\,.\,3}\,[y_{tttt}]_1 T^3$$

$$= [y_t]_1 + [y_{tt}]_1 T + \tfrac{1}{2}[y_{ttt}]_1 T^2 + \frac{1}{2\,.\,3}\,[y_{tttt}]_1 T^3$$

(where the dot means multiplication; $2.3 = 2 \times 3$) since $C_3 = [y_t]_1$

follows on putting $T = 0$. Finally by a last integration

$$y = y_1 + [y_t]_1 T + \tfrac{1}{2}[y_{tt}]_1 T^2 + \frac{1}{2 \cdot 3}[y_{ttt}]_1 T^3 + \frac{1}{2 \cdot 3 \cdot 4}[y_{tttt}]_1 T^4$$

approximately. This gives the appropriate formula for y when T^5 is negligible.

The general pattern is now clear. To get the nth order approximation we continue the series

$$y_1 + \frac{1}{1}[y_t]_1 T + \frac{1}{1 \cdot 2}[y_{tt}]_1 T^2 + \frac{1}{1 \cdot 2 \cdot 3}[y_{ttt}]_1 T^3 + \frac{1}{1 \cdot 2 \cdot 3 \cdot 4}[y_{tttt}]_1 T^4 + \ldots$$

as far as the term in T^n.

It is useful to have a symbol for the product $1 \cdot 2 \cdot 3 \cdot 4 \ldots . (n-1)n$ of the first n integers. This is called "factorial n" and is written $\lfloor n$ or $n!$. Thus

$$\begin{aligned}
\lfloor 1 &= 1! = 1. \\
\lfloor 2 &= 2! = 1 \cdot 2 = 2. \\
\lfloor 3 &= 3! = 1 \cdot 2 \cdot 3 = 6. \\
\lfloor 4 &= 4! = 1 \cdot 2 \cdot 3 \cdot 4 = 24. \\
\lfloor 5 &= 5! = 1 \cdot 2 \cdot 3 \cdot 4 \cdot 5 = 120. \\
\lfloor 6 &= 6! = 1 \cdot 2 \cdot 3 \cdot 4 \cdot 5 \cdot 6 = 720. \\
\lfloor 7 &= 7! = 1 \cdot 2 \cdot 3 \cdot 4 \cdot 5 \cdot 6 \cdot 7 = 5040. \\
\lfloor 8 &= 8! = 1 \cdot 2 \cdot 3 \cdot 4 \cdot 5 \cdot 6 \cdot 7 \cdot 8 = 40320. \\
\lfloor 9 &= 9! = 1 \cdot 2 \cdot 3 \cdot 4 \cdot 5 \cdot 6 \cdot 7 \cdot 8 \cdot 9 = 362880. \\
\lfloor 10 &= 10! = 1 \cdot 2 \cdot 3 \cdot 4 \cdot 5 \cdot 6 \cdot 7 \cdot 8 \cdot 9 \cdot 10 = 3628800.
\end{aligned}$$

Notice that $\lfloor 3 = 1 \cdot 2 \cdot 3 = 3(1 \cdot 2) = 3 \lfloor 2$, $\lfloor 4 = 1 \cdot 2 \cdot 3 \cdot 4 = 4(1 \cdot 2 \cdot 3) = 4 \lfloor 3$, and so on, and in general

$$\lfloor n = n \lfloor n-1 \quad . \quad . \quad . \quad . \quad (12.6)$$

This relation enables us to attach a value to $\lfloor 0$. For if we suppose $\lfloor 0$ to be defined in such a way that (12.6) remains true when $n = 1$, we have $\lfloor 1 = 1 \lfloor 0$. But $\lfloor 1 = 1$, and therefore we must set $\lfloor 0 = 1$. This is a little unexpected, but we shall find that it fits in with a great number of formulas, where otherwise a special exception would have to be made.

We can therefore write in general

$$y \simeq y_1 + \frac{1}{\underline{|1}}[y_t]_1 T + \frac{1}{\underline{|2}}[y_{tt}]_1 T^2 + \frac{1}{\underline{|3}}[y_{ttt}]_1 T^3 + \ldots + \frac{1}{\underline{|n}}[D_t^n y]_1 T^n$$

$$= y_1 + \sum_{a=1}^{n} \frac{1}{\underline{|a}}[D_t^a y]_1 (t - t_1)^a \qquad . \quad . \quad (12.7)$$

(or $\sum_{a=0}^{n} \frac{1}{\underline{|a}}[D_t^a y]_1 (t - t_1)^a$ if we interpret $D_t^0 y$ to mean y).

EXAMPLES

(1) Find an approximate formula for $1/t$ near $t_1 = 1$.

Put $T = t - t_1 = t - 1$, then $t = 1 + T$. Also write $y = 1/t$, so that $y_t = (-1)t^{-2}$, $y_{tt} = (-1)(-2)t^{-3} = \underline{|2}\, t^{-3}$, $y_{ttt} = (-1)(-2)(-3)\, t^{-4} = -\underline{|3}\, t^{-4}$, and in general $D_t^r y = (-1)^r \underline{|r}\, t^{-r-1}$. So when $t = t_1 = 1$, $y_1 = 1$, $[y_t]_1 = -1$, $[y_{tt}]_1 = \underline{|2}$, $[y_{ttt}]_1 = -\underline{|3}$, and in general $[D_t^r y]_1 = (-1)^r \underline{|r}$. Substituting these values in (12.7),

$$y = \frac{1}{t} = \frac{1}{T+1} \simeq 1 - \frac{\underline{|1}}{\underline{|1}} T + \frac{\underline{|2}}{\underline{|2}} T^2 - \frac{\underline{|3}}{\underline{|3}} T^3 + \ldots$$

$$= 1 - T + T^2 - T^3 + \ldots + (-1)^n T^n$$

For example, the successive approximations to $1/1{\cdot}1$ are ($T = {\cdot}1$), 1, $1 - {\cdot}1 = {\cdot}9$, $1 - {\cdot}1 + ({\cdot}1)^2 = {\cdot}91$, $1 - {\cdot}1 + ({\cdot}1)^2 - ({\cdot}1)^3 = {\cdot}909$, $1 - {\cdot}1 + ({\cdot}1)^2 - ({\cdot}1)^3 + ({\cdot}1)^4 = {\cdot}9091$, and so on. As $1/1{\cdot}1 =$ the recurring decimal ·909090 . . . , we see that successive approximate values rapidly approach the true value as a limit.

(2) Find an approximation to e^t near $t = t_1 = 0$.

Put $y = e^t$, and $T = t - t_1 = t$.

Now by successive differentiation $y_t = e^t = y_{tt} = y_{ttt} = \ldots$, so that the values y_1, $[y_t]_1$, $[y_{tt}]_1$, . . which are obtained by setting $t = t_1 = 0$ are all equal to 1. Thus

$$e^t \simeq 1 + \frac{1}{\underline{|1}} T + \frac{1}{\underline{|2}} T^2 + \frac{1}{\underline{|3}} T^3 + \ldots + \frac{1}{\underline{|n}} T^n$$

$$= 1 + t/\underline{|1} + t^2/\underline{|2} + t^3/\underline{|3} + \ldots + t^n/\underline{|n}.$$

(3) Find an approximation to antilog t when t is small.

We know that antilog $t = e^{t/M}$, where $M = \log e = {\cdot}4343 \ldots$ Therefore we have merely to substitute t/M for t in the result of the last example, obtaining

$$\text{antilog}\; t \simeq 1 + t/M + t^2/M^2\,\underline{|2} + t^3/M^3\,\underline{|3} + \ldots + t^n/M^n\,\underline{|n}$$

By taking n, the order of the approximation, sufficiently large this can be used to calculate antilog t to any desired degree of accuracy.

PROBLEMS

Find approximate formulas for the following functions for small values of T:

(1) $\ln(1 + T)$.

(2) $\log(1 + T)$.

(3) $\sin T$.

(4) $\cos T$.

(5) $\sqrt{(1 + T)}$ as far as the term containing T^4. Use the approximation to calculate $\sqrt{1 \cdot 1}$.

12.8 The general rule for intermediate maxima and minima

We can apply the formula (12.7) to more difficult cases of maxima and minima. If t_1 is a value of t which makes $[y_t]_1$ zero then (12.7) can be written

$$y \simeq y_1 + [y_{tt}]_1(t - t_1)^2/\underline{|2} + [y_{ttt}]_1(t - t_1)^3/\underline{|3} + \ldots$$

Now if $[y_{tt}]_1$, the second derivative at t_1, is positive, we know that we have a minimum, and if $[y_{tt}]_1$ is negative, we have a maximum. We therefore consider only the remaining case when $[y_{tt}]_1 = 0$. In this case $y \simeq y_1 + [y_{ttt}]_1(t - t_1)^3/\underline{|3}$ as far as the third order of small quantities. Now $(t - t_1)^3$ can be made positive by choosing a value of $t > t_1$, and negative if we take $t < t_1$. Thus the term $[y_{ttt}]_1(t-t_1)^3/\underline{|3}$ can be made to have either sign by choosing t to lie on one side or the other of t_1, provided that $[y_{ttt}]_1$ is not zero, and so we can find values of y greater than y_1, and values of y less than y_1. y_1 is accordingly neither a maximum nor a minimum.

If the third derivative $[y_{ttt}]_1$ is zero at t_1, as well as the first and second derivatives, we must go on to consider the fourth-order approximation

$$y \simeq y_1 + [y_{tttt}]_1(t - t_1)^4/\underline{|4}.$$

Now $(t - t_1)^4$ is always positive: so if $[y_{tttt}]_1$ is positive we must have $y > y_1$ for any value of y near t_1. Thus y_1 is a minimum. If on the other hand $[y_{tttt}]_1$ is negative y_1 is a maximum. If $[y_{tttt}]_1$ is zero we go on to the fifth-order approximation, and so on until we come to a non-zero derivative.

The general rule is:

(i) If the first non-zero derivative when $t = t_1$ is of odd order we have neither a maximum nor minimum;

(ii) if the first non-zero derivative is of even order and positive then y_1 is a minimum;

(iii) if the first non-zero derivative is of even order and negative then y_1 is a maximum.

PROBLEMS

(1) Find the stationary points of $y = 2 + 7t + 9t^2 + 5t^3 + t^4$, and investigate whether they give maxima or minima.

(2) Find the stationary points of $y = t^4 - t^5$, and investigate whether they are maxima or minima.

(3) What is the general rule for a terminal maximum or minimum?

12.9 Limits of error in approximation

It is always helpful when using an approximate formula to know how large an error we can expect to find. We can do this fairly easily for formula (12.7) in the following way.

Imagine a hare, a tortoise and a snail having a race. Suppose they start off together, but that the hare always runs faster than the tortoise, and the tortoise faster than the snail. Then at all subsequent times we shall expect the hare to be in the lead, the tortoise second, and the snail third. Elementary, my dear Watson? The result is common-sense and true, but a formal proof is a little tricky, because of the rather complicated nature of the definition of a velocity as a limit. So we shall simply take it for granted.

In this analogy the tortoise corresponds to the function y we are studying. The hare and the snail correspond to specially constructed functions between which we can expect y to lie. There is one observation on this principle which may be made here for the sake of clarity. In saying that the hare is running faster than the tortoise we mean more precisely that the difference (velocity of hare minus velocity of tortoise) is positive, where the velocities are taken with their proper signs. It is not good enough to say that the absolute value of the velocity of the hare is greater than that of the tortoise irrespective of sign. For if the hare runs rapidly to and fro, but the tortoise creeps steadily forward, it is possible for the tortoise to outrun the hare.

Consider now any function y of a variable t. Let $v = y_t$ be its rate of change. Denote by the symbol v_{max} the greatest value of v in the range of values of t we are considering, and its least value by v_{min}, so that $v_{min} \leqslant v \leqslant v_{max}$ for all values of v.

We now define two functions, the "snail" $y_{min} = y_1 + v_{min}(t - t_1)$ and the "hare" $y_{max} = y_1 + v_{max}(t - t_1)$. Then when $t = t_1$ all three functions take the same value y_1. But $D_t y_{min} = v_{min}$, $D_t y_{max} = v_{max}$, $D_t y = v$, so that y_{min} has always a smaller rate of increase than y, and y always a smaller rate of increase than y_{max}. It follows by the

" hare, tortoise, and snail" principle that for all later values of t, y lies between y_{min} and y_{max}; that is

$$y_1 + v_{min}(t - t_1) \leqslant y \leqslant y_1 + v_{max}(t - t_1).$$

By a similar argument if $t < t_1$, working *backwards* from t_1,

$$y_1 + v_{min}(t - t_1) \geqslant y \geqslant y_1 + v_{max}(t - t_1).$$

In any case y always lies between the two values $y_1 + v_{min}(t - t_1)$ and $y_1 + v_{max}(t - t_1)$. This can be considered as a precise expression of the approximate formula, $y \simeq y_1 + v_1(t - t_1)$.

This argument can be extended to the nth-order formula to show that if $[D_t^n y]_{min}$ is the least value of $[D_t^n y]$ over the range of values of t we are considering, and $[D_t^n y]_{max}$ the greatest value, then y always lies between

$$(y_1 + [y_t]_1\, T + [y_{tt}]_1\, T^2/\underline{|2} + \ldots + [D_t^n y]_{min} T^n/\underline{|n})$$

and $$(y_1 + [y_t]_1\, T + [y_{tt}]_1\, T^2/\underline{|2} + \ldots + [D_t^n y]_{max}\, T^n/\underline{|n}).$$

EXAMPLE

(1) If $y = \cos t$, what is the greatest error that can occur in using the formula $\delta y \simeq -\sin t_1 \,.\, \delta t$ provided that $|\delta t|$ does not exceed ·01?

If $y = \cos t$ then $y_t = -\sin t$ and $y_{tt} = -\cos t$. We know therefore that y_2 must lie between

$$y_1 + [y_t]_1 (t_2 - t_1) + \tfrac{1}{2}[y_{tt}]_{min} (t_2 - t_1)^2 \quad \text{and}$$
$$y_1 + [y_t]_1 (t_2 - t_1) + \tfrac{1}{2}[y_{tt}]_{max} (t_2 - t_1)^2.$$

That is, $\delta y = y_2 - y_1$ must lie between

$$(-\sin t_1)\, \delta t + \tfrac{1}{2}[y_{tt}]_{min} (\delta t)^2 \quad \text{and}$$
$$(-\sin t_1)\, \delta t + \tfrac{1}{2}[y_{tt}]_{max} (\delta t)^2$$

so that the error in taking δy to be $(-\sin t_1)\, \delta t$ must lie between $-\tfrac{1}{2}[y_{tt}]_{min} (\delta t)^2$ and $-\tfrac{1}{2}[y_{tt}]_{max} (\delta t)^2$. But $y_{tt} = -\cos t$, and cannot be greater than 1 or less than -1, so that the error must lie between $-\tfrac{1}{2}(\delta t)^2$ and $\tfrac{1}{2}(\delta t)^2$. Since $|\delta t|$ does not exceed ·01, by hypothesis, the error of the formula cannot exceed $\tfrac{1}{2}(\cdot 01)^2 = \cdot 00005$.

12.10 Repeated partial differentiation

Let $y = f(t, u)$ be a function of two variables t and u. Then we can find by differentiation the derivatives $D_{t|u}y = D_t y = f_t(t, u) = y_t$ (with respect to t keeping u constant), and $D_{u|t}y = D_u y = f_u(t, u) = y_u$ (keeping t constant). In turn these derivatives themselves have derivatives, so that we can find four second derivatives $D_t(D_t y) = f_{tt}(t, u) = y_{tt}$, and $D_u(D_t y) = f_{tu}(t, u) = y_{tu}$, the derivatives of y_t, and y_{ut} and y_{uu} the derivatives of y_u.

For example if $y = t + u^2t^2$ then $y_t = 1 + 2u^2t$, $y_u = 2ut^2$, whence $y_{tt} = D_t y_t = 2u^2$, $y_{tu} = D_u y_t = 4ut$, $y_{ut} = D_t y_u = 4ut$, $y_{uu} = D_u y_u = 2t^2$.

The second derivative $D_u D_t y = y_{tu}$ is also written as

$$\left(\frac{\partial}{\partial u}\right)\left(\frac{\partial}{\partial t}\right)y, \text{ or } \frac{\partial^2 y}{\partial u\,\partial t}.$$

In the same way, if y is a function of three variables, say $y = f(t, u, v)$ then we can define higher-order derivatives such as

$$D_t D_u D_v y = y_{vut} = \left(\frac{\partial}{\partial t}\right)\left(\frac{\partial}{\partial u}\right)\left(\frac{\partial}{\partial v}\right)y.$$

12.11 Approximation formula for a function of several variables

We know that if $y = f(t, u)$ is a function of two variables, t and u, then a small change δt in t and a small change δu in u produce together a change δy in y given by the approximate formula

$$\delta y \simeq y_t \delta t + y_u \delta u.$$

As with the one-variable case more accurate expressions can be found by taking the higher derivatives into account. As usual we proceed by examining the effect of changing the variables one at a time.

Suppose we wish to know the value of $y = f(t, u)$ for values of t and u in the neighbourhood of t_1 and u_1. Let us put $t - t_1 = T$ and $u - u_1 = U$. Now if as a first step we keep t fixed we can consider y as a function of u only, and therefore the approximation formula for a single variable will be applicable:

$$y = f(t, u) \simeq f(t, u_1) + \frac{1}{\lfloor 1}f_u(t, u_1)\,U + \frac{1}{\lfloor 2}f_{uu}(t, u_1)\,U^2 + \dots \qquad (12.8)$$

This formula relates the value of the function at (t, u) to its value and the value of its derivatives at (t, u_1). The second step is to relate these values to those at (t_1, u_1), i.e. to keep u fixed at u_1, and to vary t. We can then consider $f(t, u_1)$ as a function of the variable t only and so the single variable formula shows that

$$f(t, u_1) \simeq f(t_1, u_1) + \frac{1}{\lfloor 1}f_t(t_1, u_1)T + \frac{1}{\lfloor 2}f_{tt}(t_1, u_1)T^2 + \dots .$$

$$= y_1 + \frac{1}{\lfloor 1}[y_t]_1 T + \frac{1}{\lfloor 2}[y_{tt}]_1 T^2 + \dots$$

where the suffix $(_1)$ here indicates that the values are to be taken for

$t = t_1$ and $u = u_1$ after the differentiation has been completed. In the same way $f_u(t, u_1)$, $f_{uu}(t, u_1)$, etc., can be considered as functions of t, giving

$$f_u(t, u_1) = f_u(t_1, u_1) + \frac{1}{\underline{|1}} f_{ut}(t_1, u_1)T + \frac{1}{\underline{|2}} f_{utt}(t_1, u_1)T^2 + \ldots$$

$$= [y_u]_1 + \frac{1}{\underline{|1}}[y_{ut}]_1 T + \frac{1}{\underline{|2}}[y_{utt}]_1 T^2 + \ldots$$

$$f_{uu}(t, u_1) = [y_{uu}]_1 + \frac{1}{\underline{|1}}[y_{uut}]_1 T + \frac{1}{\underline{|2}}[y_{uutt}]_1 T^2 + \ldots$$

A substitution of these values into (12.8) and a slight rearrangement of the sum provides the approximation formula

$$\begin{aligned} y = y_1 & \\ & + [y_t]_1 T + [y_u]_1 U \\ & + \frac{1}{\underline{|2}}[y_{tt}]_1 T^2 + \frac{1}{\underline{|1}\,\underline{|1}}[y_{ut}]_1 TU + \frac{1}{\underline{|2}}[y_{uu}]_1 U^2 \\ & + \frac{1}{\underline{|3}}[y_{ttt}]_1 T^3 + \frac{1}{\underline{|2}\,\underline{|1}}[y_{utt}]_1 T^2 U + \frac{1}{\underline{|1}\,\underline{|2}}[y_{uut}]_1 TU^2 + \frac{1}{\underline{|3}}[y_{uuu}]_1 U^3 \\ & + \text{etc.} \quad \ldots \quad (12.9) \end{aligned}$$

where $T = t - t_1$, $U = u - u_1$, and the series is to be continued up to the point at which further terms become negligible.

If T and U are small quantities of the first order, then the second row of the right-hand side of (12.9) is made up of first-order terms, the third row consists of second-order terms, the fourth row of third-order terms, and we can proceed in this way to any desired order of approximation. The general term can be written in the form $[D_t^h D_u^k y]_{(t_1, u_1)} T^h U^k / \underline{|h}\, \underline{|k}$, where h and k are zero or positive integers. If $k = 0$, $D_u^0 y$ (i.e. y differentiated zero times) is to be interpreted as y itself: similarly $D_t^h D_u^k y$ is to be read as $D_u^k y$ when $h = 0$.

We can, of course, state this equation more precisely using the hare, tortoise, and snail principle. But the result is not so simple as in the one variable case, and we shall not give it here. Such a precise formulation is necessary when it is essential to be quite certain that there are no loopholes or fallacies in the argument, but it is rather complicated. In what follows we shall try to put the argument in a reasonably convincing form, without troubling about the finer details.

EXAMPLES

(1) $y = e^{t+u}$. Here *all* partial derivatives y_t, y_u, y_{ut}, y_{tt}, etc., are equal to $y = e^{t+u}$. If, therefore, we take $t_1 = u_1 = 0$ we have $T = t$, $U = u$, and $[y]_1 = [y_t]_1 = [y_u]_1$ etc. $= 1$, so that

$$e^{t+u} \simeq 1 + t + u + t^2/\underline{|2} + tu/\underline{|1}\,\underline{|1} + u^2/\underline{|2} + t^3/\underline{|3} + \ldots$$

the approximation being carried on until the terms become negligible.

(2) $y = e^t + e^u$. Here all the "mixed" partial derivatives y_{ut}, y_{uut}, etc., are zero, $y_t = y_{tt} = y_{ttt}$ etc. $= e^t$, and $y_u = y_{uu} = y_{uuu}$ etc. $= e^u$, so that

$$e^t + e^u = 2 + t + u + t^2/\underline{|2} + u^2/\underline{|2} + t^3/\underline{|3} + u^3/\underline{|3} + \ldots.$$

For a function of three variables $y = f(t, u, v)$ there is a similar formula,

$$\begin{aligned} y \simeq y_1 &+ [y_t]_1T + [y_u]_1U + [y_v]_1V + \tfrac{1}{2}[y_{tt}]_1T^2 + \tfrac{1}{2}[y_{uu}]_1U^2 \\ &+ \tfrac{1}{2}[y_{vv}]_1V^2 + [y_{vt}]_1TV + [y_{vu}]_1UV + [y_{ut}]_1UT \\ &+ \tfrac{1}{6}[y_{ttt}]_1T^3 + \text{etc.} \qquad . \quad . \quad . \quad . \quad . \quad . \quad (12.10) \end{aligned}$$

where $T = t - t_1$, $U = u - u_1$, $V = v - v_1$, and the suffix $(_1)$ means the variables t, u, v must be given the values t_1, u_1, v_1 *after* the differentiations have been performed. The general term is

$$[D_t{}^hD_u{}^kD_v{}^ly]_1T^hU^kV^l/\underline{|h}\,\underline{|k}\,\underline{|l}.$$

PROBLEMS

(1) Find the approximation to $\sin(t + u)$ as far as terms of the third order.

(2) Find the approximation to $\sqrt{(1 + t + 2tu)}$ as far as terms of the second order.

12.12 Change in the order of differentiation

Consider again equation (12.9). If T and U are small quantities of the first order, and if quantities of the third order of smallness can be neglected, then it follows that

$$y \simeq y_1 + [y_t]_1T + [y_u]_1U + \tfrac{1}{2}[y_{tt}]_1T^2 + [y_{ut}]_1TU + \tfrac{1}{2}[y_{uu}]_1U^2$$

This formula was obtained by first keeping t constant, allowing u to vary, and afterwards keeping u constant and varying t. By the same procedure carried out in reverse order, i.e. first keeping u constant, and later t, we obtain

$$y \simeq y_1 + [y_t]_1T + [y_u]_1U + \tfrac{1}{2}[y_{tt}]_1T^2 + [y_{tu}]_1TU + \tfrac{1}{2}[y_{uu}]_1U^2$$

Subtraction of the second equation from the first then gives

$$\{[y_{ut}]_1 - [y_{tu}]_1\} TU \simeq 0$$

provided that third-order small quantities are neglected. Now this can only be true if $[y_{ut}]_1 - [y_{tu}]_1 = 0$, for TU is only of the *second* order, and if it was multiplied by a non-zero coefficient it would not be negligible. So $[y_{ut}]_1 = [y_{tu}]_1$. But the point (t_1, u_1) at which we evaluate these derivatives is an arbitrarily chosen point, and has no special distinction. Therefore the equation $y_{ut} = y_{tu}$ must be true in general, i.e.

$$D_t D_u y = D_u D_t y \qquad . \quad . \quad . \quad (12.11)$$

In a second-order partial derivative the order of differentiation is immaterial. The differentiation with respect to t can be done either before or after that with respect to u, and will give the same answer. This is illustrated by the example $y = t + u^2t^2$ where $y_{tu} = y_{ut} = 4ut$. The equation (12.11) can also be written $f_{tu}(u, t) = f_{ut}(u, t)$ or

$$\left(\frac{\partial}{\partial u}\right)\left(\frac{\partial}{\partial t}\right)y = \left(\frac{\partial}{\partial t}\right)\left(\frac{\partial}{\partial u}\right)y$$

in other notations. The same result applies to functions of three or more variables: if y is a function of t, u, and v, then $y_{tu} = y_{ut}$. For in calculating either $D_u D_t y$ or $D_t D_u y$ we have to suppose that v is held fixed, so that y can effectively be considered as a function of t and u only. For higher orders

$$\begin{aligned} D_t^2 D_u y &= D_t(D_t D_u y) \\ &= D_t(D_u D_t y) \qquad \text{(by 12.11)} \\ &= D_t D_u(D_t y) \\ &= D_u D_t(D_t y) \qquad \text{(by 12.11)} \\ &= D_u D_t^2 y \end{aligned}$$

i.e. $y_{utt} = y_{tut} = y_{ttu}$, so that *partial differentiations can be performed in any order.*

Note, however, that here the symbol $D_t y$ means the derivative in which all the variables $u, v \ldots$ other than t are held fixed, and $D_u y$ means the derivative with t, v, etc., held fixed, so that written in full the relation would be

$$D_{t|u,v\ldots} D_{u|t,v\ldots} y = D_{u|t,v\ldots} D_{t|u,v\ldots} y.$$

(The relation is no longer necessarily true when partial derivatives are taken with *other* quantities held constant, and then the order of differentiation may be important.)

The result $D_t D_u y = D_u D_t y$ enters into the study of the "exact" differential equation (Case *J*, Section 10.9). The equation $My_t + N = 0$ is said to be "exact" if M is the partial derivative of a function

f with respect to y, and N the partial derivative of f with respect to t, that is, if $M = D_y f$, $N = D_t f$. It follows that

$$D_t M = D_t D_y f = D_y D_t f = D_y N.$$

This shows that the condition (10.16) $D_t M = D_y N$ must be satisfied.

12.13 Maxima and minima of functions of several variables

Sometimes it is important to know when a function of two variables is a maximum or a minimum. For example, a craftsman might be ordered to make a rectangular box of given volume V; and if it is to be made of expensive material, he might be further ordered to make it of a shape for which the surface area is as small as possible. Now here the length L and breadth B can be varied independently, the height H being then given by $LBH = V$, i.e. $H = V/LB$, and he has the problem of minimizing the area considered as a function of L and B.

Exactly as in the one-variable case (Section 12.2) such maxima and minima can be classified as "absolute" or "local" and "terminal" or "intermediate". The most important types are the local intermediate maxima and minima, and we shall now consider these.

Suppose therefore that $y = f(t, u)$ is a function which has a maximum when $t = t_1$ and $u = u_1$. This means that $y_1 \geqslant y$ for all values y in the neighbourhood of y_1. It follows that y_1 must still be a maximum even if t alone is varied, u being held fixed: and so from the one-variable case we see that the partial derivative $D_t y = \partial y/\partial t = y_t$ must be zero at (t_1, u_1), i.e. $[y_t]_1 = 0$ (except for rare cases where there is a sharp corner in the graph). Similarly by keeping t fixed and allowing only u to vary we can show that $[y_u]_1 = 0$. The same results follow if y_1 is a minimum. Thus we have the first result:

A necessary condition for $y = f(t, u)$ to be a maximum or minimum when $t = t_1$ and $u = u_1$ is that $[y_t]_1 = 0$ and that $[y_u]_1 = 0$.

Possible values of t_1 and u_1 can accordingly be found by solving these two simultaneous equations: but further investigation is needed to confirm whether the value obtained is in fact a maximum or minimum.

Now let (t, u) be a pair of values of t and u near t_1 and u_1, and let us put $T = t - t_1$, $U = u - u_1$. If (t_1, u_1) is a stationary point, i.e. if $[y_t]_1 = 0$, $[y_u]_1 = 0$, then from equation (12.9)

$$y \simeq y_1 + \tfrac{1}{2}[y_{tt}]_1 T^2 + [y_{tu}]_1 TU + \tfrac{1}{2}[y_{uu}]_1 U^2$$

as far as the second order of small quantities. Thus if the expression

$$(\tfrac{1}{2}[y_{tt}]_1 T^2 + [y_{tu}]_1 TU + \tfrac{1}{2}[y_{uu}]_1 U^2)$$

is negative for all small values of T, U (excluding the case when both T and U are zero) it follows that $y \leqslant y_1$ for all points y near y_1, and so y_1 is a maximum. Conversely if the expression above is always positive then y_1 must be a minimum. If it is sometimes positive and sometimes negative we have neither a maximum nor a minimum. To see how this works out in practice let us take a few examples.

(1) $y = t^2 + u^2$.

Every stationary point must be a solution of the equations $[y_t]_1 = 0$, $[y_u]_1 = 0$, that is, $2t_1 = 0$, $2u_1 = 0$. The only stationary point is therefore $(t_1, u_1) = (0, 0)$.

Second differentiation gives $[y_{tt}]_1 = 2$, $[y_{tu}]_1 = 0$, $[y_{uu}]_1 = 2$. Thus the approximation near (0, 0) is given by

$$\begin{aligned} y &= y_1 + \tfrac{1}{2}[y_{tt}]_1 T^2 + [y_{tu}]_1 TU + \tfrac{1}{2}[y_{uu}]_1 U^2 \\ &= 0 + T^2 + 0 + U^2 = t^2 + u^2 \end{aligned}$$

since $T = t - t_1 = t$, $U = u - u_1 = u$. Here this "approximation" turns out to be exact. Now t^2 is positive except when $t = 0$, and u^2 is positive, except when $u = 0$, and therefore $(t^2 + u^2)$ is positive, except when $t = u = 0$. Thus y is always positive except for its value 0 at the stationary point (0, 0), which will accordingly be both a local and an absolute minimum.

(2) Consider an open (lidless) box of length x, breadth y, and height z. The volume of the box will be $V = xyz$, and its outer area $A = xy + 2xz + 2yz$ (the area of the base is xy, and there are two vertical sides of area xz and two of area yz). Suppose that the box is required to have a given volume V; what dimensions shall we give it to make A a minimum?

If V is fixed then only two of its three dimensions x, y, and z can be varied independently. In fact, since $xyz = V$, we have $z = V/xy$, and therefore $A = xy + 2V/y + 2V/x$, when expressed in terms of x and y only. To make A a maximum or minimum we must first solve the equations $[A_x]_1 = [\partial A/\partial x]_1 = 0$, $[A_y]_1 = [\partial A/\partial y]_1 = 0$, i.e.

$$\begin{aligned} y_1 - 2V/x_1^2 &= 0, \quad \text{or } y_1 = 2V/x_1^2 \\ x_1 - 2V/y_1^2 &= 0, \quad \text{or } x_1 = 2V/y_1^2 \end{aligned}$$

Substitute the value of y_1 from the first equation into the second:

$$x_1 = 2V/y_1^2 = 2V/(2V/x_1^2)^2 = x_1^4/2V.$$

One solution of this would be $x_1 = 0$: but that is impossible, since it would make $z_1 = V/x_1y_1$ infinite. Since therefore $x_1 \neq 0$, we can divide both sides of the equation by x_1 and obtain $1 = x_1^3/2V$, i.e. $x_1 = (2V)^{\frac{1}{3}}$. Since $y_1 = 2V/x_1^2$, we have $y_1 = (2V)^{\frac{1}{3}}$, and since $z_1 = V/x_1y_1$, we have $z_1 = (\frac{1}{4}V)^{\frac{1}{3}}$. Thus the only stationary value of A corresponds to the dimensions $x_1 = (2V)^{\frac{1}{3}}$, $y_1 = (2V)^{\frac{1}{3}}$, $z_1 = (\frac{1}{4}V)^{\frac{1}{3}}$, $A_1 = (108V^2)^{\frac{1}{3}}$, i.e. a box with square base and of height equal to half its breadth. Now for values of x and y near the stationary point we have approximately, writing $X = x - x_1$, $Y = y - y_1$,

$$A \simeq A_1 + \tfrac{1}{2}[A_{xx}]_1 X^2 + [A_{xy}]_1 XY + [A_{yy}]_1 Y^2.$$

But $A_x = y - 2V/x^2$, $A_y = x - 2V/y^2$, whence $A_{xx} = 4V/x^3$, $A_{xy} = 1$, $A_{yy} = 4V/y^3$. Substituting the values $x_1 = (2V)^{\frac{1}{3}}$, $y_1 = (2V)^{\frac{1}{3}}$ we obtain $[A_{xx}]_1 = 2$, $[A_{xy}]_1 = 1$, $[A_{yy}]_1 = 2$, and

$$A \simeq A_1 + X^2 + XY + Y^2$$

The appropriate method for dealing with an expression like $X^2 + XY + Y^2$ is to "complete the square" in the same way as for solving a quadratic equation (Section 3.4). The terms $X^2 + XY$ which contain X are the same as the first two terms of $(X + \frac{1}{2}Y)^2 = X^2 + XY + \frac{1}{4}Y^2$, and so we can write

$$X^2 + XY + Y^2 = (X + \tfrac{1}{2}Y)^2 + \tfrac{3}{4}Y^2.$$

But $(X + \frac{1}{2}Y)^2 + \frac{3}{4}Y^2$ is the sum of two squares, and is therefore positive except when both $X + \frac{1}{2}Y = 0$ and $Y = 0$, i.e. except when $X = 0$ and $Y = 0$, i.e. at the stationary point itself. It follows that $X^2 + XY + Y^2$ is positive at all points near the stationary point,

$$A \simeq A_1 + X^2 + XY + Y^2 > A_1,$$

and therefore A_1 is in fact a minimum, and we have indeed found the box of least area.

(3) $y = t^2 - 4tu + u^2$.

The equations $[y_t]_1 = 0$, $[y_u]_1 = 0$ for a stationary point are

$$2t_1 - 4u_1 = 0, \qquad -4t_1 + 2u_1 = 0$$
$$\text{i.e.} \quad t_1 = 2u, \qquad u_1 = 2t_1$$
$$\text{whence} \quad t_1 = 2(2t_1) = 4t_1,$$
$$\text{i.e.} \quad 0 = 3t_1, \; t_1 = 0, \text{ and } u_1 = 2t_1 = 0.$$

The only stationary point is (0, 0), and at this point $y_1 = 0$.

Now for any point (t, u) near this

$$y \simeq y_1 + \tfrac{1}{2}[y_{tt}]_1T^2 + [y_{tu}]_1TU + \tfrac{1}{2}[y_{uu}]_1U^2$$

where $T = t - t_1 = t$, $U = u - u_1 = u$. But $y_{tt} = 2$, $y_{tu} = -4$, $y_{uu} = 2$, and so the "approximation" becomes

$$y \simeq t^2 - 4tu + u^2.$$

In fact this is our original formula and is exact. To find out how this behaves we again complete the square. The terms containing t are $t^2 - 4tu$, which are the same as the first two terms of $(t - 2u)^2 = t^2 - 4tu + 4u^2$ whence we have

$$y = t^2 - 4tu + u^2 = (t - 2u)^2 - 3u^2.$$

This is now the difference between two squares, not their sum. If $u = 0$ then $y = t^2$ and is positive. But if we choose t and u in such a way as to make $t - 2u = 0$ then $y = -3u^2$ and is negative. Thus there are values of y both greater than the value $y_1 = 0$ and less than it, and we have neither a maximum nor a minimum.

We can look at the problem in this way. Suppose that y is a function of t and u. Then, taking t and u as co-ordinates of points in a plane, we can represent y graphically as the height above this plane (Section 3.6). We shall accordingly get a surface, which will resemble a landscape, with hills and valleys. On this surface there will in general be three kinds of points at which the ground is level (i.e. $y_t = 0$, $y_u = 0$). A maximum point for y corresponds to the highest point of a mountain, where the ground is momentarily level. But as we move away from the peak in any direction the ground begins to slope downwards. The deepest point of a valley is also level, and corresponds to a minimum in y. But on moving away from it we begin to rise. Finally at a pass between two mountain peaks the ground is also level at the head of the pass. But there we have a choice; we can ascend to a peak or descend into a valley, whichever we wish. The function $y = t^2 - 4tu + u^2$ has such a pass or "saddleback point".

Functions of three or more variables are similarly dealt with.

FURTHER EXAMPLES

(4) $y = t^2 + u^2 + v^2$.

The only stationary point is given by $[y_t]_1 = [y_u]_1 = [y_v]_1 = 0$, and is $t_1 = 0$, $u_1 = 0$, $v_1 = 0$, $y_1 = 0$. Now y is the sum of three squares, and is therefore positive except when $t = u = v = 0$: it follows that y_1 is a minimum (both local and absolute).

(5) $y = 6 - 2u + 6v - t^2 - 2u^2 - 3v^2 - 2tu - 2tv$.

To find the stationary point we have to solve the three equations $[y_t]_1 = 0$, $[y_u]_1 = 0$, $[y_v]_1 = 0$, that is

$$\begin{aligned} -2t_1 - 2u_1 - 2v_1 &= 0 \\ -2 - 2t_1 - 4u_1 &= 0 \\ 6 - 2t_1 - 6v_1 &= 0 \end{aligned}$$

The method of solution is as follows. The first equation gives $t_1 = -u_1 - v_1$; we substitute this value in the second and third equations and obtain

$$\begin{aligned} -2 - 2u_1 + 2v_1 &= 0 \\ 6 + 2u_1 - 4v_1 &= 0 \end{aligned}$$

The first equation here gives $u_1 = v_1 - 1$, and on substituting that in the other equation we obtain $4 - 2v_1 = 0$, i.e. $v_1 = 2$. Thus $u_1 = v_1 - 1 = 1$ and $t_1 = -u_1 - v_1 = -3$. The only stationary point is $t_1 = -3$, $u_1 = 1$, $v_1 = 2$, $y_1 = 11$. On putting $T = t - t_1$, $U = u - u_1$, $V = v - v_1$ we see that near this stationary point

$$y \simeq y_1 + \tfrac{1}{2}[y_{tt}]_1 T^2 + \tfrac{1}{2}[y_{uu}]_1 U^2 + \tfrac{1}{2}[y_{vv}]_1 V^2 + [y_{tu}]_1 TU + [y_{tv}]_1 TV + [y_{uv}]_1 UV$$
$$= y_1 - T^2 - 2U^2 - 3V^2 - 2TU - 2TV.$$

We must again try completing the square. The terms containing T are $-T^2 - 2TU - 2TV$. The appropriate square is $-(T+U+V)^2$, so that

$$-T^2-2U^2-3V^2-2TU-2TV = -(T+U+V)^2 - U^2 - 2V^2 + 2UV$$

The remaining terms containing U are therefore $-U^2+2UV$, which are the first two terms of $-(U-V)^2$, so that we can write

$$-T^2-2U^2-3V^2-2TU-2TV = -(T+U+V)^2 - (U-V)^2 - V^2.$$

Now $(T+U+V)^2$, $(U-V)^2$, V^2 are all positive or zero. Thus the expression $-(T+U+V)^2 - (U-V)^2 - V^2$ is always negative, except when $T+U+V=0$ and $U-V=0$ and $V=0$, i.e. except when $T=U=V=0$: and therefore

$$y \simeq y_1 - T^2 - 2U^2 - 3V^2 - 2TU - 2TV < y_1$$

and y_1 is a maximum.

(6) $y = tu + tv + uv$.

The stationary point is $t_1 = u_1 = v_1 = 0$, and the appropriate "approximation" is the exact formula $y = tu + tv + uv$. It is not possible to complete the square in the usual way, since there are no square terms. However, we notice that t, u, and v are independent variables and can be given any values we like. Let us put $v = 0$, then $y = tu$, and clearly this can be made either positive or negative by choosing suitable values of t and u. Thus the stationary point cannot be a maximum or a minimum, and must be a "saddleback".

CHAPTER THIRTEEN

SERIES

13.1 Geometric series

In Section 11.4 we discussed "arithmetic series" such as 1, 2, 3, 4 . . ., or 2, 4, 6, 8 . . ., or $A, A + \delta, A + 2\delta$ . . . in which each term is got from the last by adding on a fixed number or "common difference" δ.

A series in which each term is obtained by multiplying the preceding one by a fixed number r is called a "geometric series" or "geometric progression", and r is called the "common ratio".

Thus suppose a freely multiplying colony contains originally N bacteria. If each bacterium doubles itself in half an hour, then after $\frac{1}{2}$ hour there will be $2N$ bacteria, after 1 hour $2^2N = 4N$, after $1\frac{1}{2}$ hours $2^3N = 8N$, after 2 hours $2^4N = 16N$, and so on. These numbers form a geometric series.

Again consider the metamorphosis of the locust. This passes through five larval stages or instars, all much alike, and acquires wings in a final sixth stage. The head length increases from stage to stage in the following manner (A. J. Duarte, "Growth of the Migratory Locust", *Bull. Ent. Res.*, **29** (1938), 425, quoted by d'Arcy W. Thompson)—

Table 13.1—Head lengths in locusts

Stage	Head length (mm)	Geometric series
I	1·44	1·44
II	1·94	1·96
III	2·70	2·66
IV	3·71	3·62
V	4·89	4·93
Adult	5·59	6·70

The geometric series given for comparison has common ratio 1·36, i.e. each term is obtained by multiplying the preceding one by 1·36. It will be seen that it agrees reasonably well with the observed values, except for the adults. A similar result holds for the lengths of the femurs of the locust, which increase from one stage to the next in approximately the ratio 1 : 1·43.

Another example of a geometric series occurs in the survival of the lapwing (*Vanellus vanellus*) (D. Lack, *British Birds*, **39**, 258–264). A number of these birds were ringed and set free. The numbers of ringed birds found dead in successive years are given in the following table—

Table 13.2—Mortality in Vanellus vanellus

Year	Number recovered	Geometric series
1	141	155·3
2	109	103·5
3	70	69·0
4	42	46·0
5	40	30·7
6	19	20·5
7	19	13·7
8	7	9·1
9	6	6·1
10	5	4·1
11	6	2·7
12	1	1·8
13	0	1·2
14	1	·8
15	0	·6
16	0	·4
17	0	·3
18	0	·2
Total	466	466·0

Each term in the geometric series given for comparison is $\frac{2}{3}$ of the preceding term (rounded off to one place of decimals), and the first term of the series is chosen so as to give approximately the same total number, 466, as the observations. Allowing for the fact that the numbers recovered must always be whole numbers, not fractions, and must be subject to a certain amount of chance variation, there is quite reasonable agreement. (A method of testing the goodness of this agreement is explained in Section 21.17.) A credible explanation of these figures would be that in any year 1 bird in 3 is liable to be killed. For then the population in each successive year would be reduced to $\frac{2}{3}$ of its value in the year immediately preceding, and the number killed would accordingly also be reduced in the ratio $1 : \frac{2}{3}$.

In general if a is the first term of a geometric series, and r is the common ratio, then the second term is ar, the third term ar^2, and the nth term ar^{n-1}.

PROBLEMS

Find the first term a, common ratio r, and nth term of the following geometric series

(1) 1, ·1, ·01, ·001, . . .

(2) 3·6, 2·4, 1·6, . . .

(3) 32, −24, 18, −13·5, . . .

13.2 Sum of a geometric series

Sometimes it is useful to have a formula for the sum $S_n = a + ar + ar^2 + \ldots + ar^{n-1}$ of the first n terms of a geometric series.

We find by direct multiplication (see equation 3.7)

$$(1 - r)(1 + r + r^2 + \ldots + r^{n-1}) = 1 - r^n.$$

If $r \neq 1$ we can divide both sides of this equation by $(1 - r)$, obtaining

$$1 + r + r^2 + \ldots + r^{n-1} = (1 - r^n)/(1 - r) \,. \tag{13.1}$$

Multiplying both sides by a we have

$$\left.\begin{aligned} S_n = a + ar + \ldots + ar^{n-1} &= a(1 - r^n)/(1 - r) \\ &= a(r^n - 1)/(r - 1) \end{aligned}\right\} \tag{13.2}$$

This is the required formula.

EXAMPLES

(1) Sum the series $1 + 3 + 3^2 + \ldots$ to n terms.

Here $a = 1$, $r = 3$, and

$$S_n = a(r^n - 1)/(r - 1) = \tfrac{1}{2}(3^n - 1)$$

(2) Sum the series $\frac{1}{2} + \frac{1}{4} + \frac{1}{8} + \frac{1}{16} + \ldots$ to n terms.

Here $a = \frac{1}{2}$, $r = \frac{1}{2}$, and

$$S_n = a(1 - r^n)/(1 - r) = 1 - (\tfrac{1}{2})^n$$

PROBLEMS

(1) A population consists originally of 10 animals introduced into a new and favourable habitat. It can be expected to double its size each year. How many animals will there be after n years? If each animal consumes 1,000 square metres of grain per year, estimate the total area of crops eaten in n years, and in particular the case when $n = 10$.

(2) Sum the series $1 - \cdot 1 + \cdot 01 - \cdot 001 + \ldots$ to n terms.

(3) Establish the formula for the sum of a geometric series in the following way. Imagine a large population of birds (such as lapwings) consisting initially of N birds. Suppose that in each year a proportion r

of the total survive, and a proportion $k = 1 - r$ are killed. Then in the first year Nk birds will be killed, and Nr will survive; in the second year, of the Nr survivors, Nrk will die and $Nrr = Nr^2$ survive. In the third year, Nr^2k will die and Nr^3 survive. In the first n years, therefore, the number killed will be $Nk + Nkr + Nkr^2 + \ldots + Nkr^{n-1} = k(N + Nr + Nr^2 + \ldots + Nr^{n-1})$, and the number surviving will be Nr^n. But these two numbers added together must equal the original number of birds N. Divide the resulting equation through by $k = 1 - r$, and so obtain a formula for $N + Nr + Nr^2 + \ldots + Nr^{n-1}$.

13.3 Arithmetico-geometric series

The series $1r, 2r^2, 3r^3, 4r^4, \ldots$ is a combination of the arithmetic series $1, 2, 3, 4, \ldots$ and the geometric series $r, r^2, r^3, \ldots$. Its mth term is mr^m. It is called an "arithmetico-geometric series". Occasionally we need to know the sum of such a series to n terms. It can be obtained by the following trick.

Take equation (13.1) and multiply it throughout by r. It becomes

$$r + r^2 + r^3 + \ldots + r^n = (r - r^{n+1})/(1 - r) \quad . \quad . \quad . \quad (13.3)$$

Now this is an identity: it holds for all values of r (except $r = 1$), irrespective of whether r is a fixed or variable quantity. So suppose that r is variable. Since the left- and right-hand sides of this equation are identically equal they must have the same rates of change; that is

$$D_r(r + r^2 + \ldots + r^n) = D_r[(r - r^{n+1})/(1 - r)]$$

or on applying the rule for differentiation of a quotient we find (after some simplification)

$$1 + 2r + 3r^2 + 4r^3 + \ldots + nr^{n-1} = \frac{1 - (n + 1)r^n + nr^{n+1}}{(1 - r)^2}$$

On multiplying both sides of this equation by r we obtain

$$r + 2r^2 + 3r^3 + 4r^4 + \ldots + nr^n = \frac{r - (n + 1)r^{n+1} + nr^{n+2}}{(1 - r)^2} \quad . . \quad (13.4)$$

which is the required formula.

We can go further. We know that the rates of change of both sides of equation (13.4) must be equal, i.e. on differentiating with respect to r

$$1 + 2^2r + 3^2r^2 + 4^2r^3 + \ldots + n^2r^{n-1}$$
$$= \frac{1 + r - (n + 1)^2 r^n + (2n^2 + 2n - 1)r^{n+1} - n^2r^{n+2}}{(1 - r)^3}$$

and on multiplying both sides by r

$$1^2r + 2^2r^2 + 3^2r^3 + \ldots + n^2r^n$$
$$= \frac{r + r^2 - (n + 1)^2 r^{n+1} + (2n^2 + 2n - 1)r^{n+2} - n^2r^{n+3}}{(1 - r)^3} \quad (13.5)$$

This gives the sum of the first n terms of the series whose general (mth) term is m^2r^m, which can be considered as an arithmetico-geometric series of a more complicated type. A further differentiation enables us to sum $r + 2^3r^2 + 3^3r^3 + 4^3r^4 + \ldots$ to n terms. We can also combine these results to sum series whose mth term is of the type $(A + Bm + Cm^2 + ..)r^m$ where A, B, and C are given constants: this is a general "arithmetico-geometric" series. We merely have to add together A times the sum of the series whose mth term is r^m, B times the sum of mr^m, and C times the sum of m^2r^m (and so on if there are further terms). From equations (13.3), (13.4) and (13.5) the sum to n terms is accordingly

$$A\frac{r - r^{n+1}}{1 - r} + B\frac{r - (n + 1)r^{n+1} + nr^{n+2}}{(1 - r)^2}$$

$$+ C\frac{r + r^2 - (n + 1)^2 r^{n+1} + (2n^2 + 2n - 1)r^{n+2} - n^2 r^{n+3}}{(1 - r)^3} + \ldots$$

In the derivation of equations (13.3), (13.4), and (13.5) r has been assumed to be a variable. But inasmuch as these equations are true for all values of r (except $r = 1$, when the right-hand sides of the equations become indeterminate) they are identities; and so the final results will still be true even when r is constant.

PROBLEMS

Find the sum to n terms of the series

(1) $2r + 6r^2 + 12r^3 + \ldots + n(n + 1)r^n$

(2) $6r + 24r^2 + 60r^3 + \ldots + n(n + 1)(n + 2)r^n$.

13.4 The geometric rate of growth

One of the most striking properties of a geometric series is its rapid rate of growth when the common ratio exceeds 1. For example, if we consider the series 1, 2, 4, 8, . . . of powers of 2, then by the time we get to the 11th term it has become 1024, i.e. more than 1000. In each further 10 terms it must again multiply itself a thousandfold: the 21st term must be greater than 1,000,000, and the 31st term greater than 1,000,000,000. We are already approaching astronomical magnitudes although each term is no more than double the preceding one.

In fact any geometric series with common ratio $r > 1$ will ultimately grow very large, and will eventually exceed any arithmetic series, no matter how rapidly the latter increases. Compare for example the arithmetic series 0, 1000, 2000, 3000, . . . with common difference 1000, and the geometric series 1·023, 1·047, 1·072, 1·097 . . . with common ratio 1·023 ($= 10^{\cdot 01}$). At first sight the arithmetic series seems an easy winner. Its tenth term is 10,000, whereas the geometric series has only got to 1·259 by the tenth term, and must, one would think, be feeling pretty small. The hundredth term of the arithmetic series is 100,000, while that of the geometric series is only 10. Each time we go on another 100 terms, we add 100,000 to the arithmetic term and multiply the geometric term by 10: so the 200th terms are respectively 200,000 and 100, the 300th terms are 300,000 and 1000, the 400th are 400,000 and 10,000. The geometric series is now evidently catching up. When we get to the thousandth term the arithmetic series has mounted to 1 million, but the geometric term is ten thousand million, and so an easy winner.

Conversely if r is less than 1 the terms of a geometric series rapidly fade away to vanishing point. The powers of $\frac{1}{4}$ are $(\frac{1}{4})^1 = \cdot 25$, $(\frac{1}{4})^2 = \cdot 0625$, $(\frac{1}{4})^3 = \cdot 0156$, $(\frac{1}{4})^4 = \cdot 0039$, $(\frac{1}{4})^5 = \cdot 0010$, $(\frac{1}{4})^6 = \cdot 0002$, $(\frac{1}{4})^7 = \cdot 0001$, and all subsequent powers are zero to 4 places of decimals. If we take a value of r nearer 1 the diminution may not be quite so dramatic, but even so it will not usually take very long for the terms to become imperceptible. In the data for *Vanellus vanellus* quoted above, although a bird has apparently 2 chances out of 3 of surviving any one year, not one bird has been observed to survive more than 14 years.

13.5 The sum to infinity

Consider the series $\frac{1}{2}, \frac{1}{4}, \frac{1}{8}, \frac{1}{16} \ldots$ with common ratio $\frac{1}{2}$. The successive sums S_n are $S_1 = \frac{1}{2}$, $S_2 = \frac{1}{2} + \frac{1}{4} = \frac{3}{4}$, $S_3 = \frac{1}{2} + \frac{1}{4} + \frac{1}{8} = \frac{7}{8}$, $S_4 = \frac{1}{2} + \frac{1}{4} + \frac{1}{8} + \frac{1}{16} = \frac{15}{16}$, and so on. Inspection of these figures shows that they are approaching more and more nearly to 1. The general sum to n terms is $S_n = 1 - (\frac{1}{2})^n$, and as n increases the second term $(\frac{1}{2})^n$ soon becomes imperceptible, and the sum S_n practically 1. Thus

$$\lim_{n \to \infty} S_n = 1 \qquad . \qquad . \qquad (13.6)$$

It is natural to say that 1 is the "sum of an infinite number of terms" or "sum to infinity" of the series $\frac{1}{2} + \frac{1}{4} + \frac{1}{8} + \frac{1}{16} + \frac{1}{32} + \ldots$ and we can write $\sum_{a=1}^{\infty} (\frac{1}{2})^a = 1$. This form of words is not intended to imply any more than that the more terms we take of the series the nearer the sum approaches 1.

The sum to n terms of the general series $a + ar + ar^2 + ar^3 + \ldots$ is $S_n = a(1 - r^n)/(1 - r)$. If $|r| < 1$ the power r^n rapidly tends to zero as $n \to \infty$, so that S_n approaches the limit $S_\infty = a/(1 - r)$.

$$a + ar + ar^2 + ar^3 + \ldots \textit{ to infinity} = a/(1 - r) \qquad . \qquad (13.7)$$

If however $|r| > 1$ the successive terms of the series increase indefinitely in size, and there is no "sum to infinity".

EXAMPLES

(1) The series $\frac{1}{10} + \frac{1}{100} + \frac{1}{1000} + \ldots$, i.e. $\cdot 1 + \cdot 01 + \cdot 001 + \cdot 0001 + \ldots$ has first term $a = \frac{1}{10}$ and common ratio $r = \frac{1}{10}$. Its sum

to infinity is therefore $\frac{1}{10}/(1 - \frac{1}{10}) = \frac{1}{9}$. This justifies the representation of the fraction $\frac{1}{9}$ by the infinite decimal ·111111. . . .

(2) Achilles runs at 10 metres per second, but the Tortoise at only ·1 metre per second. Nevertheless the Tortoise challenges Achilles to a race, giving himself 10 metres start. He says "Achilles will take 1 second to cover the 10 metres advantage I have: but by that time I shall have gained an additional ·1 metre. Achilles will take ·01 second to cover this, but I shall then have gained another ·001 metre. Achilles will take ·0001 second to run this, but I shall still be in front: we can go on for an infinite number of steps of this kind, and at every point I shall always be a little ahead. So Achilles cannot ever catch me up." But unfortunately for the tortoise the total time involved in this infinite succession is only $1 + \cdot 01 + (\cdot 01)^2 + (\cdot 01)^3 + \ldots$ seconds, $= 1/(1 - \cdot 01)$ $= 100/99$ seconds, and is therefore finite. In fact, Achilles will pass the tortoise after this interval of 100/99 seconds. We can deduce that the infinite decimal 1·010101 . . . = 100/99.

13.6 Convergence and divergence

Let $x_1, x_2, x_3, x_4 \ldots$ be any series of numbers, and let $S_n = x_1 + x_2 + x_3 + \ldots + x_n = \sum_{\alpha=1}^{n} x_\alpha$ be the sum of the first n terms. If S_n tends to a definite limit S_∞ as n tends to ∞, then the series is said to be "convergent" and S_∞ is called the "sum to infinity" or simply the "sum" $\Sigma_1^\infty x_\alpha$ of the series. We also say that the series "converges to S_∞". If there is no such limit the series is "divergent".

Query: Can an arithmetic series be convergent?

Sometimes the solution to a problem can be expressed as a convergent series, but not in any simpler form. This is often quite a satisfactory solution for numerical calculation. For by summing a sufficient number of terms of the series we can get as near as we wish to the true answer. But there is one important qualification: the convergence must be reasonably rapid. A series may have a sum (to infinity) and yet converge to it so slowly that we have to add thousands of terms to get a reasonable approximation. Such a series would be useless for calculation—for example it can be shown that the series $1 - \frac{1}{2} + \frac{1}{3} - \frac{1}{4} + \frac{1}{5} - \ldots$ has the sum ln 2, but it would be practically impossible to use it in that form. The convergence is too slow. On the other hand a geometric series does converge rapidly, provided that the common ratio r is substantially less than 1 in absolute value. Hence any series which tends to its final limit as rapidly as a geometric series will be ideal for the purposes of calculation.

We shall therefore say that a series $x_1 + x_2 + x_3 + \ldots$ is "geometrically convergent" if each term is smaller in absolute value than the corresponding term of some convergent geometric series; e.g. if $|r| < 1$ then the series $r/1 + r^2/2 + r^3/3 + \ldots$ is "geometrically convergent", because each term is smaller in magnitude than the corresponding term of the series $r + r^2 + r^3 + \ldots$. Similarly, if we can find two positive numbers A and r such that $|r| < 1$ and $|x_m| < Ar^m$ for all values of m, then the series $x_1 + x_2 + \ldots$ is, by definition, geometrically convergent.

There are two points to notice in connection with this definition. The first is that the property of convergence or divergence of a series depends on the "tail end" of the series: what the first few terms do is of no importance so long as they are finite. Thus it is convenient to widen the definition:

Formal definition of geometric convergence

If there are positive numbers A, r, and N such that

(i) $|r| < 1$

(ii) for every term x_m for which $m \geqslant N$ (i.e. for every term from x_N onwards) the inequality $|x_m| \leqslant Ar^m$ is satisfied,

then the series is called "geometrically convergent". Such a series can be conveniently used for numerical calculation.

The second point is that we are to some extent begging the question in calling the series "geometrically *convergent*", since it is not immediately obvious from the definition that such a series necessarily converges, i.e. that the sum $S_n = x_1 + x_2 + \ldots + x_n$ of the first n terms does tend to a limit S_∞ as $n \to \infty$. Now commonsense suggests that this must be so. For the essence of "convergence" is that by taking n sufficiently large we can make S_n approximate to S_∞ as closely as we like; i.e. the addition of any further terms to S_n will not alter its value appreciably. Now we know that this is true for the geometric series $Ar + Ar^2 + Ar^3 + \ldots$. Not only do the terms themselves rapidly become negligibly small, but also the sum of any number of successive terms from the nth onwards is negligibly small when n is large. But if this is true for the geometric series $Ar + Ar^2 + Ar^3 + \ldots$ it will be true *a fortiori* for the original series $x_1 + x_2 + x_3 + \ldots$, whose terms are even smaller in magnitude, at least from a certain point onwards. So the series Σx_a would seem to converge, and so indeed it does: but a complete formal proof requires techniques a little beyond the scope of this book.

But, taking this for granted, how can we determine most easily whether a given series is convergent or divergent? There are two simple and useful tests.

(I) Let $\rho_m = x_{m+1}/x_m$ be the ratio of two successive terms of the series. Then if there is a positive number r less than 1 such that all these ratios ρ_m are less in magnitude than r, or at any rate all from a certain term onwards, then the series Σx_a converges at least as rapidly as a certain geometric series of common ratio r. For this condition means that each term from a certain point onwards is reduced in absolute value by at least as much as the ratio r when compared with the preceding term, and so the terms decrease at least as rapidly as if each was r times the one before it. A formal proof runs as follows. Expressed in symbols the condition becomes

$$|\rho_m| = |x_{m+1}/x_m| \leqslant r < 1 \quad \text{when } m \geqslant N \quad . \qquad (13.8)$$

For $|x_{m+1}/x_m| = |x_{m+1}|/|x_m|$. Let us put $A = |x_N|/r^N$; then $|x_N| = Ar^N$. For the next term we have

$$\begin{aligned} |x_{N+1}| &= |x_N| \cdot \frac{|x_{N+1}|}{|x_N|} \\ &= Ar^N\,|\rho_N| \\ &\leqslant Ar^N r \qquad \text{(by 13.8)} \\ &\leqslant Ar^{N+1}. \end{aligned}$$

For the $(N + 2)$nd term,

$$\begin{aligned} |x_{N+2}| &= |x_{N+1}|\,\frac{|x_{N+2}|}{|x_{N+1}|} \\ &\leqslant Ar^{N+1} \cdot |\rho_{N+1}| \\ &\leqslant Ar^{N+1} \cdot r \qquad \text{(by 13.8)} \\ &\leqslant Ar^{N+2}. \end{aligned}$$

In the same way $|x_{N+3}| = |x_{N+2}|\,|\rho_{N+2}| \leqslant Ar^{N+3}$, and so we show for every successive term from x_N onwards that $|x_m| \leqslant Ar^m$. Thus the series Σx_a converges more rapidly than, or at least as rapidly as the geometric series ΣAr^a.

Examples of the use of this test will be found in the next section.

(II) Suppose that the ratio $\rho_m = x_{m+1}/x_m$ tends to a definite limit L as m tends to infinity. Then if $|L| < 1$ the series Σx_a is geometrically convergent.

For since $\rho_m \to L$ we know that $|\rho_m| \to |L|$. Now take r to be any positive number between $|L|$ and 1. Then since the ratios $|\rho_m|$ approach $|L|$ as a limit, and $|L| < r$, there must be some value N of m beyond which all subsequent $|\rho_m|$ are less than r. (Otherwise they would not be approaching nearer and nearer to $|L|$.) But this is simply condition (I) above for the series to converge geometrically. This is very reasonable, since the condition $x_{m+1}/x_m \to L$ simply means that the further we go along the series the more nearly it behaves like a geometric series of common ratio L.

13.7 The Taylor series

In Section 12.7 we have seen that a function y of a variable t can be approximated to by the formula

$$y = y_1 + [y_t]_1 T + [y_{tt}]_1 T^2/\underline{|2} + \ldots + [D_t^n y]_1 T^n/\underline{|n}.$$

Here t_1 is a fixed value of t and $T = t - t_1$ is supposed small. The suffix 1 signifies that t is to be put equal to t_1 after the differentiation has been performed. Here we mean that $y_1 + [y_t]_1 T$ is a first approximation, valid when the terms containing T^2 can be neglected; $y_1 + [y_t]_1 T + \frac{1}{2}[y_{tt}]_1 T^2$ is a second approximation, and so on. In general the more terms we take the closer is the approximation, and it is natural to conjecture that if we extend the series to an infinite number of terms we may get a perfectly exact value.

$$\begin{aligned} y &= y_1 + [y_t]_1 T + [y_{tt}]_1 T^2/\underline{|2} + \ldots \\ &= \sum_0^\infty [D_t^a y]_1 T^a/\underline{|a} \qquad (13.9) \end{aligned}$$

This series is known as the "Taylor series" for the function y. Alternative ways of writing it are

$$y = y_1 + \left(\frac{dy}{dt}\right)_1 (t - t_1) + \frac{1}{\underline{|2}}\left(\frac{d^2y}{dt^2}\right)_1 (t - t_1)^2 + \ldots$$

and

$$\begin{aligned} f(t) = f(t_1 + T) &= f(t_1) + f'(t_1) \,.\, (t - t_1) + \frac{1}{\underline{|2}} f''(t_1) \,.\, (t - t_1)^2 + \ldots \\ &= f(t_1) + f_t(t_1) \,.\, T + \frac{1}{\underline{|2}} f_{tt}(t_1) \,.\, T^2 + \ldots \end{aligned}$$

according to whichever notation is convenient.

Thus supposing this series to be valid, examples (1), (2) and (3) of Section (12.7) would give

$$\begin{aligned} 1/t &= 1 - (t - 1) + (t - 1)^2 - (t - 1)^3 + \ldots \text{ to infinity} \\ e^t &= 1 + t/\underline{|1} + t^2/\underline{|2} + t^3/\underline{|3} + \ldots \text{ to infinity} \\ \text{antilog } t &= 1 + t/M + t^2/M^2\,\underline{|2} + t^3/M^3\,\underline{|3} + \ldots \text{ to infinity.} \end{aligned}$$

The first of these series is certainly true if $|1 - t| < 1$, for it is a simple geometric series of common ratio $(1 - t)$ and first term 1, and therefore by (13.7) its sum to infinity is $1/[1 - (1 - t)] = 1/t$. This being so we may be encouraged to think that (13.9) may be true under fairly general conditions.

There is an important special case of (13.9), when $t_1 = 0$. Then $T = t - t_1 = t$, and the series can be written

$$y = f(t) = y_0 + [y_t]_0 t + [y_{tt}]_0 t^2/\underline{|2} + [y_{ttt}]_0 t^3/\underline{|3} + \dots$$

$$= y_0 + \left(\frac{dy}{dt}\right)_0 t + \left(\frac{d^2y}{dt^2}\right)_0 \frac{t^2}{\underline{|2}} + \left(\frac{d^3y}{dt^3}\right)_0 \frac{t^3}{\underline{|3}} + \dots$$

$$= f(0) + f'(0) \cdot t + f''(0) \cdot t^2/\underline{|2} + f'''(0) \cdot t^3/\underline{|3} + \dots \tag{13.10}$$

where here the suffix $_0$ means that t must be set equal to 0 after differentiation. [Sometimes series (13.9) is called "Taylor's series" and the special case (13.10) "Maclaurin's series". The two series are, however, merely different ways of writing what is essentially the same theorem.]

Two questions arise which require answering. Suppose we take any given function y and expand it in a Taylor series according to equation (13.9) or (13.10). Will the series be convergent? And if it is convergent is its sum equal to the original function y?

For the moment we shall take it on trust that if the series has a sum (to infinity) that sum is equal to y. But it is interesting and useful to find if the series does converge for several simple and important functions y. We shall use in every case the form (13.10) with $t_1 = 0$.

EXAMPLES

(1) $y = e^t$. Here all derivatives y_t, y_{tt}, y_{ttt} etc. are equal to e^t, and therefore $y_0 = [y_t]_0 = [y_{tt}]_0 = \dots = 1$. The series becomes

$$e^t = 1 + t + t^2/\underline{|2} + t^3/\underline{|3} + \dots \text{ to infinity.}$$

The mth term $x_m = t^{m-1}/\underline{|m-1}$, and $x_{m+1} = t^m/\underline{|m}$. Thus $\rho_m = x_{m+1}/x_m = t/m$; and for any given value of t this tends to 0 as m tends to infinity. Therefore by condition (II) of the preceding section this series converges for all values of t without restriction.

(2) $y = \ln(1 + t)$. The successive derivatives are $y_t = 1/(1 + t)$, $y_{tt} = -1/(1 + t)^2$, $y_{ttt} = \underline{|2}/(1 + t)^3$, $y_{tttt} = -\underline{|3}/(1 + t)^4$, and in general $D_t^m y = (-1)^{m-1} \underline{|m-1}/(1 + t)^m$. Thus when $t = 0$ we obtain $y_0 = 0$, $[y_t]_0 = 1$, $[y_{tt}]_0 = -\underline{|1}$, $[y_{ttt}]_0 = \underline{|2}$, etc. and the series becomes

$$\ln(1 + t) = t/\underline{|1} - \underline{|1}\, t^2/\underline{|2} + \underline{|2}\, t^3/\underline{|3} - \dots$$

$$= t - t^2/2 + t^3/3 - t^4/4 + \dots \text{ (by 12.6)}$$

The mth term is $x_m = (-1)^{m-1} t^m/m$, and the $(m + 1)$th is $x_{m+1} = (-1)^m t^{m+1}/(m + 1)$. So $\rho_m = x_{m+1}/x_m = -mt/(m + 1)$, and $|\rho_m| = m|t|/(m + 1) < |t|$. If therefore $|t| < 1$ this series converges geometrically by condition I of the previous section.

(3) $y = \sin t$. Here the successive derivatives are $y_t = \cos t$, $y_{tt} = -\sin t$, $y_{ttt} = -\cos t$, $y_{tttt} = \sin t$, and thereafter cyclically, $\cos t$, $-\sin t$, $-\cos t$, $\sin t$. On putting $t = 0$ we obtain the values $y_0 = 0$, $[y_t]_0 = 1$, $[y_{tt}]_0 = 0$, continuing with $-1, 0, 1, 0, -1, 0, \ldots$ and so on, and the Taylor series is

$$y = \sin t = t - t^3/\underline{|3} + t^5/\underline{|5} - t^7/\underline{|7} + \ldots$$

The mth term of this series is $x_m = (-1)^{m-1}\, t^{2m-1}/\underline{|2m-1}$, and the $(m+1)$th is $(-1)^m\, t^{2m+1}/\underline{|2m+1}$, whence the ratio

$$\rho_m = x_{m+1}/x_m = -\,t^2\,\underline{|2m-1}\,/\,\underline{|2m+1}.$$

But $$\underline{|2m+1} = (2m+1)\,\underline{|2m} = 2m(2m+1)\,\underline{|2m-1},$$

so that $\rho_m = -t^2/[2m(2m+1)]$, and this tends to zero as m tends to infinity. Thus by condition II this series converges geometrically for all values of t without restriction.

This series can be used to calculate the value of $\sin t$. But it must be remembered that here t is measured in radians. If we want $\sin \theta°$, we must use the relation $1° = \pi/180$ radians $= H$ radians (say), so that $\theta° = H\theta$ radians, and on replacing t by $H\theta$ in the series this gives

$$\sin \theta° = H\theta - H^3\theta^3/\underline{|3} + H^5\theta^5/\underline{|5} - H^7\theta^7/\underline{|7} + \ldots$$

PROBLEMS

Show that the following series are given by (13.10), and write down the mth term of each:

(1) $e^{-t} = 1 - t + t^2/\underline{|2} - t^3/\underline{|3} + t^4/\underline{|4} + \ldots$ (for all t).

(2) $\text{antilog}\, t = 1 + t/M + t^2/M^2\,\underline{|2} + t^3/M^3\,\underline{|3} + \ldots$ (for all t).

(3) $\cos t = 1 - t^2/\underline{|2} + t^4/\underline{|4} - t^6/\underline{|6} + \ldots$ (for all t).

(4) $\sinh t = t + t^3/\underline{|3} + t^5/\underline{|5} + t^7/\underline{|7} + \ldots$ (for all t).

(5) $\cosh t = 1 + t^2/\underline{|2} + t^4/\underline{|4} + t^6/\underline{|6} + \ldots$ (for all t).

(6) $\ln(1-t) = -t - t^2/2 - t^3/3 - t^4/4 - \ldots$ (for $|t| < 1$).

(7) $(1-t)^{-2} = 1 + 2t + 3t^2 + 4t^3 + \ldots$ (for $|t| < 1$).

13.8 The binomial series

A specially important series is that for $y = (u+t)^n$.
Differentiation gives $y_t = n(u+t)^{n-1} = nu^{n-1}$ when $t = 0$;

$$y_{tt} = n(n-1)(u+t)^{n-2} = n(n-1)u^{n-2} \quad \text{when} \quad t = 0;$$

$$y_{ttt} = n(n-1)(n-2)(u+t)^{n-3} = n(n-1)(n-2)u^{n-3} \quad \text{if} \quad t = 0;$$

Therefore by (13.10),

$$y = (u+t)^n$$
$$= u^n + nu^{n-1}t + \frac{n(n-1)}{\underline{|2}} u^{n-2}t^2 + \frac{n(n-1)(n-2)}{\underline{|3}} u^{n-3}t^3 + \ldots \tag{13.11}$$

This is Newton's "binomial series". When does it converge? The mth term is

$$x_m = \frac{n(n-1)(n-2)\ldots(n-m+2)}{\underline{|m-1}} u^{n-m+1} t^{m-1}$$

and the $(m+1)$th term is

$$x_{m+1} = \frac{n(n-1)(n-2)\ldots(n-m+2)(n-m+1)}{\underline{|m}} u^{n-m} t^m.$$

The ratio $\rho_m = x_{m+1}/x_m$ is therefore

$$\frac{n(n-1)(n-2)\ldots(n-m+2)(n-m+1)\,\underline{|m-1}\, u^{n-m} t^m}{n(n-1)(n-2)\ldots(n-m+2)\,\underline{|m}\, u^{n-m+1} t^{m-1}} = \frac{n-m+1}{m}\frac{t}{u}$$

But as $m \to \infty$, $(n-m+1)/m \to -1$ and therefore $\rho_m \to -t/u$. It follows by condition II above (Section 13.6) that if $|-t/u| < 1$, i.e. if $|t|/|u| < 1$, i.e. if $|t| < |u|$, the series converges geometrically, irrespective of the value of n. For example, if we put $n = -1$ we get

$$1/(u+t) = u^{-1} - u^{-2}t + \frac{(-1)(-2)}{\underline{|2}} u^{-3}t^2 - \ldots$$
$$= u^{-1} - u^{-2}t + u^{-3}t^2 - u^{-4}t^3 + \ldots$$

provided that $|t| < |u|$. This is in fact a geometric series with common ratio $r = u^{-1}t$ so that $|r| = |t|/|u| < 1$. Again if we put $n = \frac{1}{2}$ we have a series for the square root

$$\sqrt{(u+t)} = u^{\frac{1}{2}} + \tfrac{1}{2}u^{-\frac{1}{2}}t + \frac{\frac{1}{2}(-\frac{1}{2})}{\underline{|2}} u^{-\frac{3}{2}}t^2 + \frac{\frac{1}{2}(-\frac{1}{2})(-\frac{3}{2})}{\underline{|3}} u^{-\frac{5}{2}}t^3 + \ldots$$
$$= \sqrt{u}[1 + \tfrac{1}{2}(t/u) - \tfrac{1}{8}(t^2/u^2) + \tfrac{1}{16}(t^3/u^3) - \ldots]$$

In the special case when n is a positive integer or zero the series is simplified. We know that the mth term is

$$x_m = n(n-1)(n-2)\ldots(n-m+2)u^{n-m+1}t^{m-1}/\underline{|m-1}$$

and if $m \geqslant n+2$ this must contain a factor $(n-n) = 0$, and so must be zero. That is, the series becomes a finite one of $(n+1)$ terms only, all subsequent terms vanishing, and therefore it converges for all values of u and t without restriction.

For example:

$$
\begin{aligned}
(u+t)^0 &= 1\\
(u+t)^1 &= u+t\\
(u+t)^2 &= u^2+2ut+t^2\\
(u+t)^3 &= u^3+3u^2t+3ut^2+t^3\\
(u+t)^4 &= u^4+4u^3t+6u^2t^2+4ut^3+t^4\\
(u+t)^5 &= u^5+5u^4t+10u^3t^2+10u^2t^3+5ut^4+t^5
\end{aligned}
$$

which can also be found by direct multiplication.

The coefficients in these powers have a number of simple properties. Before we discuss them, however, it is helpful to introduce a distinctive notation. Unfortunately there are a number of different symbols in use, and no general agreement on notation. But the simplest one seems to be due to H. E. Soper, and runs as follows.

The expressions n, $n(n-1)$, $n(n-1)(n-2)$, and so on, which occur in the binomial series bear a distinct resemblance to powers. The ordinary power n^2 means $n \times n$, the product of two equal factors: whereas $n \times (n-1)$ is the product of two factors in which the second is diminished by 1 compared with the first. Similarly $n^3 = n \times n \times n$, with 3 equal factors, whereas in $n \times (n-1) \times (n-2)$ the factors decrease by 1 at each step. We call an expression like $n(n-1)(n-2)$ a "descending factorial", and to emphasize its analogy with the ordinary power n^3 we shall write it as $(n-)^3$. Thus $(n-)^1 = n$; $(n-)^2 = n(n-1)$, $(n-)^3 = n(n-1)(n-2)$ and in general

$$(n-)^r = n(n-1)(n-2)\ldots(n-r+1) \quad . \quad (13.12)$$

[Other symbols often used for $(n-)^r$ are ${}^nP_r$, $(n)_r$, $(n)^{(r)}$, and $n_{(r)}$.]

We can therefore write the binomial series as

$$(u+t)^n = u^n + \frac{(n-)^1}{\underline{|1}}\, u^{n-1}\, t^1 + \frac{(n-)^2}{\underline{|2}}\, u^{n-2}\, t^2 + \ldots$$

We can also clearly have an "ascending factorial" $(n+)^r$ where the factors rise by 1 at each step:

$$
\begin{aligned}
(n+)^2 &= n(n+1),\\
(n+)^3 &= n(n+1)(n+2),\\
(n+)^4 &= n(n+1)(n+2)(n+3) \text{ and in general}\\
(n+)^r &= n(n+1)(n+2)\ldots(n+r-1) \quad . \quad (13.13)
\end{aligned}
$$

Such an ascending factorial occurs in the expansion

$$(u-t)^{-n} = u^{-n} + \frac{(n+)^1}{\underline{|1}}\, u^{-n-1}\, t + \frac{(n+)^2}{\underline{|2}}\, u^{-n-2}\, t^2 + \ldots$$

obtained from (13.11) by writing $-t$, $-n$ in place of t and n. The ascending factorial also occurs in the theory of generalized arithmetic series (Section 11.6), where for example we saw that

$$(1+)^m + (2+)^m + (3+)^m + \ldots + (n+)^m = (n+)^{m+1}/(m+1)$$

Another way of writing $(n+)^r$ is $n^{[r]}$.

Other important expressions, frequently used in higher mathematics, are $(n-)^r/\underline{|r}$, $n^r/\underline{|r}$, and $(n+)^r/\underline{|r}$. Convenient symbols for these are $(n-)_r$, $(n)_r$, and $(n+)_r$ respectively, and they are named "reduced descending factorials", "reduced powers", and "reduced ascending factorials". Thus, by definition,

$$\left.\begin{aligned}(n-)_r &= (n-)^r/\underline{|r} = n(n-1)(n-2)\ldots(n-r+1)/\underline{|r}\\ (n)_r &= n^r/\underline{|r} = n \,.\, n \,.\, n \ldots n/\underline{|r}\\ (n+)_r &= (n+)^r/\underline{|r} = n(n+1)(n+2)\ldots(n+r-1)/\underline{|r}\end{aligned}\right\} \qquad (13.14)$$

Other symbols for $(n-)_r$ are ${}^nC_r$, $\binom{n}{r}$, $n^{[r]}$ and $n_{(r)}$, and other symbols for $(n+)_r$ are ${}^nH_r$ and $n_{[r]}$. The alternative symbols ${}^nC_r$ and $\binom{n}{r}$ for the reduced descending factorial $(n-)_r$ are of fairly common occurrence, and should be specially noted.

The binomial series can therefore also be written in the general form, valid for any value of n,

$$(u+t)^n = u^n + (n-)_1 u^{n-1} t + (n-)_2 u^{n-2} t^2 + \ldots$$

We have not yet assigned any meaning to the symbols $(n-)^0$ and $(n+)^0$. However $(n-)^r$ and $(n+)^r$ obey the recurrence relations

$$\begin{aligned}(n-)^{r+1} &= (n-r)(n-)^r\\ (n+)^{r+1} &= (n+r)(n+)^r\end{aligned}$$

for all values of r greater than zero. If we define $(n-)^0$, $(n+)^0$ in such a way as to make these true even when $r = 0$, we find $(n-)^1 = n(n-)^0$, $(n+)^1 = n(n+)^0$: but $(n-)^1 = (n+)^1 = n$, so that $(n-)^0 = (n+)^0 = 1$. Since $\underline{|0} = 1$ we therefore have

$$(n-)^0 = (n)^0 = (n+)^0 = (n-)_0 = (n)_0 = (n+)_0 = 1.$$

Using this convention we can compress the binomial series into the form

$$(u+t)^n = \sum_{a=0}^{\infty} (n-)_a u^{n-a} t^a \qquad (13.15)$$

and the series for e^t in the form

$$e^t = (t)_0 + (t)_1 + (t)_2 + (t)_3 + \ldots = \sum_0^{\infty} (t)_a$$

The values of $(n-)_r$ are given for positive integers n in the following table, known as "Pascal's triangle".

Table 13.3—Pascal's triangle

Values of the binomial coefficients

$$(n-)_r = n(n-1)(n-2)\ldots(n-r+1)/\underline{|r}$$

$r =$	0	1	2	3	4	5	6	7	8	9	10
$n =$ 0	1										
1	1	1									
2	1	2	1								
3	1	3	3	1							
4	1	4	6	4	1						
5	1	5	10	10	5	1					
6	1	6	15	20	15	6	1				
7	1	7	21	35	35	21	7	1			
8	1	8	28	56	70	56	28	8	1		
9	1	9	36	84	126	126	84	36	9	1	
10	1	10	45	120	210	252	210	120	45	10	1

From each row of this table we can read off at once the corresponding binomial series. Thus taking the row $n = 6$ we have

$$(u+t)^6 = u^6 + 6u^5t + 15u^4t^2 + 20u^3t^3 + 15u^2t^4 + 6ut^5 + t^6.$$

The most striking property of this table is that each row is symmetrical: it reads exactly the same starting from the end and working backwards. In symbols this means that

$$(n-)_{n-r} = (n-)_r \quad . \quad . \quad . \qquad (13.16)$$

This property follows from the identity $(u+t)^n = (t+u)^n$; e.g.

$$(u+t)^3 = u^3 + 3u^2t + 3ut^2 + t^3 = (t+u)^3 = t^3 + 3t^2u + 3tu^2 + u^3$$

we reverse the series but have the same coefficients. It can also be proved as follows. If n is an integer then

$$(n-)^r = n(n-1)(n-2)\ldots(n-r+1) \text{ (by definition)}$$

$$= \frac{n(n-1)\ldots(n-r+1)(n-r)(n-r-1)\ldots 2.1}{(n-r)(n-r-1)\ldots 2.1}$$

$$= \underline{|n}/\underline{|n-r}$$

whence

$$(n-)_r = (n-)^r/\underline{|r} = \frac{\underline{|n}}{\underline{|r}\,\underline{|n-r}} \qquad (13.17)$$

From this formula

$$(n-)_{n-r} = \frac{\underline{|n}}{\underline{|n-r}\,\underline{|n-(n-r)}} = \frac{\underline{|n}}{\underline{|n-r}\,\underline{|r}} = (n-)_r$$

By the use of equation (13.17) the binomial series can be expressed in another rather elegant form. We have

$$(u+t)^n = u^n + (n-)_1\, u^{n-1} t^1 + (n-)_2\, u^{n-2} t^2 + \ldots + t^n$$

$$= \frac{\underline{|n}}{\underline{|n}\,\underline{|0}}\, u^n t^0 + \frac{\underline{|n}}{\underline{|n-1}\,\underline{|1}}\, u^{n-1} t^1 + \frac{\underline{|n}}{\underline{|n-2}\,\underline{|2}}\, u^{n-2} t^2 + \ldots + \frac{\underline{|n}}{\underline{|0}\,\underline{|n}}\, u^0 t^n$$

Divide this equation throughout by $\underline{|n}$. It becomes

$$\frac{(u+t)^n}{\underline{|n}} = \frac{u^n\, t^0}{\underline{|n}\,\underline{|0}} + \frac{u^{n-1}\, t^1}{\underline{|n-1}\,\underline{|1}} + \frac{u^{n-2}\, t^2}{\underline{|n-2}\,\underline{|2}} + \ldots + \frac{u^0\, t^n}{\underline{|0}\,\underline{|n}}$$

or, on using "reduced powers" $(u)_r = u^r/\underline{|r}$,

$$(u+t)_n = (u)_n\,(t)_0 + (u)_{n-1}\,(t)_1 + (u)_{n-2}\,(t)_2 + \ldots + (u)_0\,(t)_n \quad (13.18)$$

provided that n is an integer.

The other property of Pascal's triangle which is worth noting is that each number in it is obtained by adding the number immediately above it to the number immediately above it and one place to the left. (This property enables us to construct the triangle very quickly.) That is, in symbolic form,

$$(n+1-)_r = (n-)_r + (n-)_{r-1} \quad . \quad . \quad . \quad (13.19)$$

The proof runs as follows. By (13.17)

$$(n-)_r + (n-)_{r-1} = \frac{\underline{|n}}{\underline{|r}\,\underline{|n-r}} + \frac{\underline{|n}}{\underline{|r-1}\,\underline{|n-r+1}}$$

We now use the relations $\underline{|r} = r\underline{|r-1}$, $\underline{|n-r+1} = (n-r+1)\,\underline{|n-r}$.

Thus

$$(n-)_r + (n-)_{r-1} = \frac{\underline{|n}}{r\underline{|r-1}\,\underline{|n-r}} + \frac{\underline{|n}}{\underline{|r-1}\,(n-r+1)\,\underline{|n-r}}$$

$$= \frac{\underline{|n}}{\underline{|r-1}\,\underline{|n-r}}\left(\frac{1}{r} + \frac{1}{n-r+1}\right)$$

$$= \frac{\underline{|n}}{\underline{|r-1}\,\underline{|n-r}}\;\frac{n+1}{r(n-r+1)}$$

$$= \frac{(n+1)\,\underline{|n}}{r\underline{|r-1}\,.\,(n-r+1)\,\underline{|n-r}}$$

$$= \frac{\underline{|n+1}}{\underline{|r}\,\underline{|n-r+1}} = (n+1-)_r$$

PROBLEMS

(1) Show that $(-n-)^r = (-1)^r (n+)^r$

(2) Show that, if n is a positive integer, $(n+)^r = \underline{|n+r-1}/\underline{|n-1}$

(3) Show that each number in Pascal's triangle is the sum of all the numbers above it in the column immediately to the left.

13.9 Correctness of the Taylor series

We now return to the problem of determining whether the sum of the Taylor series for a function y is in fact equal to y, as it should be.

For certain functions y the series cannot converge. The function $y = 1/t$ cannot have a Taylor series expansion about $t_1 = 0$, for the first term of the series, $y_1 = 1/0$ is infinite. The same applies to ln t. There can be no series for $y = \sqrt{t}$ about $t_1 = 0$, for when $t = 0$ the first derivative $y_t = 1/2\sqrt{t}$ is infinite. Similarly $|t|$ cannot be expressed by such a series, for it has no derivative when $t = 0$.

However when the series converges it is almost always true that its sum is equal to y. This is a consequence of a very general theorem (see Section 14.25). But in many simple cases it is possible to give a direct proof. We make use of the more precise form of the approximation given in Section 12.9 (the "hare, tortoise, and snail" argument) which shows that y must lie in value between

$$(y_1 + [y_t]_1 T + [y_{tt}]_1 T^2/\underline{|2} + \ldots + [D_t^n y]_{min} T^n/\underline{|n})$$
$$\text{and } (y_1 + [y_t]_1 T + [y_{tt}]_1 T^2/\underline{|2} + \ldots + [D_t^n y]_{max} T^n/\underline{|n})$$

where as usual $T = t - t_1$ and $[D_t^n y]_{min}$ and $[D_t^n y]_{max}$ are respectively the least and greatest values of $D_t^n y$ in the range of values we are considering. We can express this fact in the form

$$y = S_n + R_n$$

where $S_n = y_1 + [y_t]_1 T + \ldots + [D_t^{n-1} y] T^{n-1}/\underline{|n-1}$ = the sum of the first n terms of the Taylor series for y, and R_n lies between

$$[D_t^n y]_{min} T^n/\underline{|n} \quad \text{and} \quad [D_t^n y]_{max} T^n/\underline{|n}.$$

Now what we want to show is that y is the sum to infinity of the series, which by definition means that $S_n \to y$ as $n \to \infty$, i.e. since $R_n = y - S_n$, that $R_n \to 0$.

Let $|D_t^n y|_{max}$ denote the greatest absolute value of $D_t^n y$; both $[D_t^n y]_{min}$ and $[D_t^n y]_{max}$ cannot exceed this in absolute value, and so $[D_t^n y]_{min} T^n/\underline{|n}$ and $[D_t^n y]_{max} T^n/\underline{|n}$ cannot exceed $|D_t^n y|_{max} |T|^n/\underline{|n}$ in absolute value. Therefore R_n, which lies between these values, must satisfy

$$|R_n| < |D_t^n y|_{max} |T|^n/\underline{|n}.$$

So whenever we can prove that

$$|D_t^n y|_{max} |T|^n / \lfloor\underline{n} \to 0 \quad \text{as} \quad n \to \infty \quad (13.19)$$

it follows that $R_n \to 0$, i.e. $y - S_n \to 0$, i.e. $S_n \to y$, i.e. that the series has in fact the correct sum y.

This condition looks somewhat formidable, but in practice is usually not too difficult to apply. Only a few simple cases will be considered here: others can be covered by very similar arguments.

(i) The series $1 + t + t^2/\lfloor\underline{2} + t^3/\lfloor\underline{3} + \ldots$ for e^t, considered for values of t lying between -1 and 1.

Here $y = e^t$, and $t_1 = 0$, so that $T = t$. Also we have $D_t^n y = e^t$, and lies between $1/e$ and e for values of t between -1 and 1. Thus $|D_t^n y|_{max} = e$. So in order to show that the series $1 + t + t^2/\lfloor\underline{2} + \ldots$ does in fact converge to the correct value $y = e^t$ we have by (13.19) to show that $e|t|^n/\lfloor\underline{n} \to 0$ as $n \to \infty$. But this is clearly the case, since $|t| \leqslant 1$, and therefore $e|t|^n/\lfloor\underline{n} \leqslant e/\lfloor\underline{n}$ which $\to 0$ as $n \to \infty$.

A slightly more complicated version of this argument can be used to show that the series sums to e^t for values of t lying between -2 and 2, or between -3 and 3, and so on, finally including any desired value of t.

(ii) The series $\ln(1 + t) = t - t^2/2 + t^3/3 - \ldots$ for values of t lying between 0 and 1.

If $y = \ln(1 + t)$ then $D_t^n y = (-1)^{n-1} \lfloor\underline{n-1}\, (1 + t)^{-n}$. The smallest value of t is 0, so the smallest value of $(1 + t)$ is 1, and the greatest value of $(1 + t)^{-n}$ is $1^{-n} = 1$. Therefore $|D_t^n y|_{max} = \lfloor\underline{n-1}$. To prove that the series has the correct sum we must show by (13.19) that

$$\lfloor\underline{n-1}\, |t|^n / \lfloor\underline{n} = |t|^n/n \to 0$$

as $n \to \infty$. But $|t| \leqslant 1$, so that $|t|^n/n \leqslant 1/n$, which certainly tends to zero, and *a fortiori* $|t|^n/n$ must also tend to zero.

(iii) The series for $1/(1 - t)$. This turns out to be $1 + t + t^2 + t^3 + \ldots$, i.e. a geometric series, and we have shown above that if $|t| < 1$ this has the correct sum $1/(1 - t)$.

In such a way we can verify, though rather clumsily, that the Taylor series for a particular function y does in fact have y as its sum to infinity. But as we have said there is a more advanced theorem which shows that this must be true for all cases which arise in practice.

13.10 Manipulations with series

We can find the series for cosh t and sinh t in the following way. For all values of t we have

$$e^t = 1 + t + t^2/\lfloor\underline{2} + t^3/\lfloor\underline{3} + t^4/\lfloor\underline{4} + \ldots.$$

Replace t by $(-t)$ throughout to obtain

$$e^{-t} = 1 - t + t^2/\underline{|2} - t^3/\underline{|3} + t^4/\underline{|4} - \ldots .$$

Now by definition $\cosh t$ is $\frac{1}{2}(e^t + e^{-t})$, and $\sinh t$ is $\frac{1}{2}(e^t - e^{-t})$. Let us therefore first add the series for e^t and e^{-t} by adding together the constant terms, adding the terms containing t, and so on. We find $e^t + e^{-t} = 2 + 0t + 2t^2/\underline{|2} + 0t^3 + 2t^4/\underline{|4} + \ldots$. Finally on division by 2 we get

$$\cosh t = 1 + t^2/\underline{|2} + t^4/\underline{|4} + \ldots .$$

In the same way a subtraction of the series for e^{-t} from that for e^t and a division by 2 gives

$$\sinh t = t + t^3/\underline{|3} + t^5/\underline{|5} + \ldots .$$

But are we justified in treating the series in this way? If the series were finite we should be merely following the ordinary laws of algebra in such a calculation. But here we are dealing with infinite series, and what is referred to as the "sum" of such a series is strictly speaking not an ordinary sum, but the limit of a sum. So it is not at once obvious that our operations here are logically justifiable. However, that is so. We would expect them to be so, since the convergence of the series means that all terms are negligible after a certain point, so that we can almost treat the series as if they had only a finite number of terms, neglecting the rest. This can be expressed symbolically as follows. Let S_m be the sum of the first m terms of the series for e^t, and S'_m the sum of m terms for e^{-t}. Then if we include the zero terms in the series we have obtained, which we hope has the sum $\cosh t$, and write it as

$$(\text{presumed } \cosh t) = 1 + 0t + t^2/\underline{|2} + 0t^3 + t^4/\underline{|4} + \ . \qquad (13.20)$$

the sum of the first m terms of this series is $S''_m = \frac{1}{2}(S_m + S'_m)$, because of the way it has been obtained. But $S_m \to e^t$ as $m \to \infty$, since the first series has sum e^t to infinity, and $S'_m \to e^{-t}$, and so $S''_m = \frac{1}{2}(S_m + S'_m)$, which tends in the limit to $\frac{1}{2}(e^t + e^{-t}) = \cosh t$, i.e. the series (13.20) does in fact sum to $\cosh t$. In the same way the sum S'''_m of the first m terms of the series

$$0 + t + 0t^2 + t^3/\underline{|3} + 0t^4 + \ldots \qquad . \qquad (13.21)$$

is $\frac{1}{2}(S_m - S'_m)$, and so tends to $\frac{1}{2}(e^t - e^{-t}) = \sinh t$ as $m \to \infty$. Thus (13.21) has the sum $\sinh t$.

A slightly more complicated argument shows that we can also multiply series. For example, suppose we wanted to calculate, for $|t| < 1$,

$$\begin{aligned}(1 - t)^{-2} &= (1 - t)^{-1}(1 - t)^{-1}\\ &= (1 + t + t^2 + t^3 + \ldots)(1 + t + t^2 + t^3 + \ldots)\end{aligned}$$

Multiplying the term "1" in the first series by the "1" of the second gives us the constant term "1" in the product. A term containing t can arise by multiplication in two ways, either by multiplying 1 in the first series by t in the second, or by multiplying t in the first series by 1 in the second; this gives us $1 . t + t . 1 = 2t$ in the product series for $(1 - t)^{-2}$. There will be three products containing t^2, viz. $1 . t^2$, $t . t$, $t^2 . 1$, giving $3t^2$ in the $(1 - t)^{-2}$ series. Thus by multiplying the series in this way we find $(1 - t)^{-2} = 1 + 2t + 3t^2 + 4t^3 + \ldots$ and that is correct, for it is precisely the expansion we get using the binomial series.

We can prove this result by a different argument. Write $(1 - t)^{-1} = 1 + t + t^2 + t^3 + \ldots$. and differentiate it, using the rule for differentiation of a sum.

$$D_t(1 - t)^{-1} = D_t 1 + D_t t + D_t t^2 + D_t t^3 + \ldots$$

This gives once more

$$(1 - t)^{-2} = 1 + 2t + 3t^2 + 4t^3 + \ldots$$

Again let us take the geometric series

$$(1 + t^2)^{-1} = 1 - t^2 + t^4 - t^6 + \ldots$$

and integrate it, using the rule for integration of a sum. We find

$$\int (1 + t^2)^{-1}\, dt = \int 1 dt - \int t^2 dt + \int t^4 dt - \int t^6 dt + \ldots$$

that is

$$\tan^{-1} t = C + t - t^3/3 + t^5/5 - t^7/7 + \ldots \qquad (13.22)$$

Putting $t = 0$ shows that $C = 0$.

All these operations are what one would expect to be able to do with series. It can be shown that they are valid for all, or almost all, series one is likely to meet with in practice, and in particular they are valid for all geometrically convergent Taylor series. For an exact discussion the reader is referred to books on Analysis (the theory of real and complex functions).

PROBLEMS

(1) Multiply the series $e^t . e^{-t}$ and show that the product is 1.

(2) Taking the series $t - t^3/\underline{|3} + t^5/\underline{|5} - \ldots$ for sin t, obtain the series for cos t by differentiation.

(3) Multiply the series $(1 - t)^2$ and $(1 - t)^{-2}$ and show that the product is 1.

(4) By integrating the series for $(1 + t)^{-1}$ obtain the series for $\ln(1 + t)$.

13.11 Uniqueness of Taylor series

We can now show that a function y of t can be expressed as a Taylor series in powers of $(t - t_1)$ in only one way. For convenience of demonstration we shall take the case $t_1 = 0$. Suppose then that we find two possible series for y, say

$$y = a_0 + a_1 t + a_2 t^2 + \dots = b_0 + b_1 t + b_2 t^2 + \dots. \quad (13.23)$$

By putting $t = 0$ we see at once that $a_0 = b_0$. Now differentiate (13.23): we obtain

$$y_t = a_1 + 2a_2 t + \dots = b_1 + 2b_2 t + \dots$$

If we put $t = 0$ we see that $a_1 = b_1$. Again differentiate:

$$y_{tt} = 2a_2 + 6a_3 t + \dots. = 2b_2 + 6b_3 t + \dots$$

whence on putting $t = 0$, $2a_2 = 2b_2$, i.e. $a_2 = b_2$. Proceeding in this way we see that $a_m = b_m$ for all m, i.e. the series are identical.

As an example of the use of this result consider the identity $(1 + t)^m \times (1 + t)^n = (1 + t)^{m+n}$. If $|t| < 1$ we can use the binomial series

$$(1 + t)^m = (m-)_0 + (m-)_1 t + (m-)_2 t^2 + (m-)_3 t^3 + \dots$$
$$(1 + t)^n = (n-)_0 + (n-)_1 t + (n-)_2 t^2 + (n-)_3 t^3 + \dots$$

whence on multiplication

$$\begin{aligned}(1 + t)^m (1 + t)^n &= (m-)_0 (n-)_0 \\ &+ [(m-)_1 (n-)_0 + (m-)_0 (n-)_1] t \\ &+ [(m-)_2 (n-)_0 + (m-)_1 (n-)_1 + (m-)_0 (n-)_2] t^2 \\ &+ [(m-)_3 (n-)_0 + (m-)_2 (n-)_1 + (m-)_1 (n-)_2 + (m-)_0 (n-)_3] t^3 \\ &+ \dots\end{aligned}$$

whereas by the binomial series

$$(1 + t)^{m+n} = (m+n-)_0 + (m+n-)_1 t + (m+n-)_2 t^2 + (m+n-)_3 t^3 + \dots$$

These series must be identical, so we see that

$$\begin{aligned}(m+n-)_0 &= (m-)_0 (n-)_0 \\ (m+n-)_1 &= (m-)_1 (n-)_0 + (m-)_0 (n-)_1 \\ (m+n-)_2 &= (m-)_2 (n-)_0 + (m-)_1 (n-)_1 + (m-)_0 (n-)_2 \\ (m+n-)_3 &= (m-)_3 (n-)_0 + (m-)_2 (n-)_1 + (m-)_1 (n-)_2 + (m-)_0 (n-)_3\end{aligned}$$

and in general for any positive integer r

$$(m+n-)_r = (m-)_r (n-)_0 + (m-)_{r-1} (n-)_1 + \dots + (m-)_0 (n-)_r \quad (13.24)$$

This is known as van der Monde's theorem.

Similarly by multiplying $(1 - t)^{-m} (1 - t)^{-n} = (1 - t)^{-m-n}$ we get

$$(m+n+)_r = (m+)_r (n+)_0 + (m+)_{r-1} (n+)_1 + \dots + (m+)_0 (n+) \quad (13.25)$$

and for comparison the binomial series can be written (equation 13.18)

$$(m+n)_r = (m)_r (n)_0 + (m)_{r-1} (n)_1 + \dots. + (m)_0 (n)_r.$$

All these relations are true for any values of m and n.

PROBLEM

(1) What relations does one get by expressing $e^{At} \cdot e^{Bt} = e^{(A+B)t}$ in series form?

13.12 The solution of differential equations by series

Suppose we wish to solve the differential equation $y_t = Ky$. Let us suppose that the solution y has a Taylor series $p_0 + p_1t + p_2t^2 + p_3t^3 + \ldots$ where the p_m's are numbers we wish to determine. Then

$$\begin{aligned} y_t &= p_1 + 2p_2t + 3p_3t^2 + 4p_4t^3 + \ldots \\ Ky &= Kp_0 + Kp_1t + Kp_2t^2 + Kp_3t^3 + \ldots \end{aligned}$$

and the equation $y_t = Ky$ shows that these two series must be identical, i.e.

$$\begin{aligned} p_1 &= Kp_0 \\ 2p_2 &= Kp_1, \text{ that is, } p_2 = Kp_1/2 = K^2p_0/\underline{|2} \\ 3p_3 &= Kp_2, \text{ that is, } p_3 = Kp_2/3 = K^3p_0/\underline{|3} \\ 4p_4 &= Kp_3, \text{ that is, } p_4 = Kp_3/4 = K^4p_0/\underline{|4} \end{aligned}$$

The solution is therefore

$$\begin{aligned} y &= p_0 + p_1t + p_2t^2 + p_3t^3 + \ldots \\ &= p_0 + Kp_0t + K^2p_0t^2/\underline{|2} + K^3p_0t^3/\underline{|3} + \ldots \\ &= p_0e^{Kt} \end{aligned}$$

This solution introduces one new constant p_0, as indeed it should, since we know that the solution of a first-order differential equation contains an arbitrary constant.

This process can be applied to almost any differential equation. It will be successful except when there are solutions which do not have such a Taylor series (e.g. solutions like $y = 1/t$ or $y = \sqrt{t}$).

EXAMPLES

(1) Solve the second-order equation $y_{tt} = -\omega^2y$, where ω is a constant.

Let $y = p_0 + p_1t + p_2t^2 +$ etc. Then $y_{tt} = 1.2.p_2 + 2.3.p_3t + 3.4.p_4t^2 + \ldots$ and $-\omega^2y = -\omega^2p_0 - \omega^2p_1t - \omega^2p_2t^2 - \ldots$ These series must be identical:

$$\begin{aligned} 1.2\, p_2 &= -\omega^2p_0, \text{ that is, } p_2 = -\omega^2p_0/\underline{|2} \\ 2.3\, p_3 &= -\omega^2p_1, \text{ that is, } p_3 = -\omega^2p_1/\underline{|3} \\ 3.4\, p_4 &= -\omega^2p_2, \text{ that is, } p_4 = -\omega^2p_2/3.4 = \omega^4p_0/\underline{|4} \\ 4.5\, p_5 &= -\omega^2p_3, \text{ that is, } p_5 = -\omega^2p_3/4.5 = \omega^4p_1/\underline{|5} \\ 5.6\, p_6 &= -\omega^2p_4, \text{ that is, } p_6 = -\omega^2p_4/5.6 = -\omega^6p_0/\underline{|6} \end{aligned}$$

and $y = p_0 + p_1t + p_2t^2 + \ldots$

$$= p_0 + p_1 t - \frac{\omega^2 p_0 t^2}{\underline{|2}} - \frac{\omega^2 p_1 t^3}{\underline{|3}} + \frac{\omega^4 p_0 t^4}{\underline{|4}} + \frac{\omega^4 p_1 t^5}{\underline{|5}} - \ldots$$

$$= p_0 \left(1 - \frac{\omega^2 t^2}{\underline{|2}} + \frac{\omega^4 t^4}{\underline{|4}} - \frac{\omega^6 t^6}{\underline{|6}} + \ldots\right) + \frac{p_1}{\omega}\left(\omega t - \frac{\omega^3 t^3}{\underline{|3}} + \frac{\omega^5 t^5}{\underline{|5}} - \ldots\right)$$

$$= p_0 \cos \omega t + (p_1/\omega) \sin \omega t.$$

This contains two new constants, p_0 and p_1, as we would expect, and is the general solution.

(2) The equation $t^2 y_{tt} + t y_t + (t^2 - n^2)y = 0$ is known as "Bessel's equation". If n is an integer then there is a Taylor series solution $y = C\mathcal{J}_n(t)$ where C is an arbitrary constant and $\mathcal{J}_n(t)$, known as "Bessel's function of order n" is given by

$$\mathcal{J}_n(t) = \frac{(\frac{1}{2}t)^n}{\underline{|n}} - \frac{(\frac{1}{2}t)^{n+2}}{\underline{|1}\,\underline{|n+1}} + \frac{(\frac{1}{2}t)^{n+4}}{\underline{|2}\,\underline{|n+2}} - \frac{(\frac{1}{2}t)^{n+6}}{\underline{|3}\,\underline{|n+3}} + \ldots \qquad (13.26)$$

This cannot, however, be the complete solution, since it only contains one arbitrary constant, whereas the original equation involves the second derivative y_{tt} and so is of the second order. It follows that there must be another solution $Y_n(t)$ to the equation which has no Taylor series about $t_1 = 0$. Such a solution can be found, but has a very complicated form (see E. T. Whittaker and G. N. Watson, *A course of modern analysis*, Cambridge U.P., 4th edn, 1935). A biological application of these functions to population spread is given by J. G. Skellam, "Random dispersal in theoretical populations", *Biometrika*, **38** (1951), 196.

13.13 Multiple Taylor series

The results of Section 12.11 suggest that a function $y = f(t, u)$ of two variables can be expressed by an infinite series

$$y = [y]_1 + [y_t]_1 T + [y_u]_1 U + \frac{[y_{tt}]_1 T^2 U^0}{\underline{|2}\,\underline{|0}} + \frac{[y_{tu}]_1 T^1 U^1}{\underline{|1}\,\underline{|1}} + \frac{[y_{uu}]_1 T^0 U^2}{\underline{|0}\,\underline{|2}} + \text{etc.}$$

the general term being $\dfrac{[D_t{}^p D_u{}^q y]_1 T^p U^q}{\underline{|p}\,\underline{|q}}$, where as usual T means $(t - t_1)$, U means $(u - u_1)$ and the suffix $_1$ means that we must put $t = t_1$ and $u = u_1$ after the differentiations have been performed. This is generally true, provided that the series converges. We can also extend this to functions of 3 or more variables. The most important application is the "multinomial theorem" which enables us to convert an expression like $(t + u + v + w)^n$ into series form. This again is really only needed in practice when n is a positive integer. We then get a finite series valid for all values of t, u, v, and w,

$$(t + u + v + w)^n = \sum_{\alpha+\beta+\gamma+\delta=n} \frac{\underline{|n}\, t^\alpha u^\beta v^\gamma w^\delta}{\underline{|\alpha}\,\underline{|\beta}\,\underline{|\gamma}\,\underline{|\delta}} \quad . \quad . \qquad (13.27)$$

We take all possible combinations of integers $\alpha \geqslant 0$, $\beta \geqslant 0$, $\gamma \geqslant 0$ and $\delta \geqslant 0$ such that $\alpha + \beta + \gamma + \delta = n$, obtain one term

$$\underline{|n}.t^{\alpha}u^{\beta}v^{\gamma}w^{\delta}/\underline{|\alpha}\,\underline{|\beta}\,\underline{|\gamma}\,\underline{|\delta}$$

for each such combination, and add all these terms together.

13.14 The sigma notation for double sums

Suppose we have 8 numbers arranged in a rectangle as follows—

$$\begin{matrix} x_{11} & x_{12} & x_{13} & x_{14} \\ x_{21} & x_{22} & x_{23} & x_{24} \end{matrix} \qquad (13.28)$$

Such an arrangement could for example occur if we had samples from two different nationalities, and had classified them into their blood-groups O, A, B, or AB—

Table 13.4—Blood-group classification

Nationality	Numbers in blood-groups				Total
	O	A	B	AB	
British ..	$x_{11}=$ 108	$x_{12}=$ 111	$x_{13}=$ 22	$x_{14}=$ 9	250
Latvians ..	$x_{21}=$ 55	$x_{22}=$ 69	$x_{23}=$ 33	$x_{24}=$ 15	172
Total ..	163	180	55	24	422

These numbers are taken for illustrative purposes from Ruth Sanger and R. R. Race, "Blood Groups in a Sample of 250 People", *Ann. Eugen. Lond.*, **15** (1949), 77, and R. R. Race, Ruth Sanger, Sylvia D. Lawler and D. V. Keetch, "Blood Groups of Latvians", *Ann. Eugen. Lond.*, **14** (1948), 134.

Such a systematic arrangement of numbers in a rectangular array is known as a "matrix". Here we have (ignoring the totals) an array of two rows and four columns: it is called a "2×4 matrix". The use of two suffixes x_{rs} provides a systematic way of writing the numbers in the array. The first suffix r refers to the row, i.e. the nationality, 1 denoting British and 2 Latvian, while the second suffix s refers to the column, i.e. the blood-group, 1 denoting group O, 2 A, 3 B and 4 AB. Thus x_{13} means "the number of British persons of blood-group B".

Now we can add all the numbers x_{rs}, obtaining the "grand total" $x_{11} + x_{12} + x_{13} + x_{14} + x_{21} + x_{22} + x_{23} + x_{24} = 422$, as shown in the bottom right-hand corner of Table 13.4. This can be written as $\sum\limits_{\alpha,\beta} x_{\alpha\beta}$ or as $\Sigma x_{\alpha\beta}$ meaning the sum of all the numbers x_{rs}.

Table 13.4 also gives "row totals" (right-hand column) such as $x_{11} + x_{12} + x_{13} + x_{14} = \Sigma x_{1\beta}$, meaning the total number of persons of British nationality, and $x_{21} + x_{22} + x_{23} + x_{24} = \Sigma x_{2\beta} =$ the total number of Latvians. Addition of these again gives us the grand total $\Sigma x_{1\beta} + \Sigma x_{2\beta} = \Sigma x_{\alpha\beta}$, (i.e. 250 + 172 = 422). In Σ notation this can be written even more concisely:

$$\sum_{\alpha}\left(\sum_{\beta} x_{\alpha\beta}\right) = \sum_{\alpha,\beta} x_{\alpha\beta} \quad . \quad . \quad . \quad (13.29)$$

i.e. if we first add all the $x_{\alpha\beta}$ for a given value of α but different values of β we get the total $\sum_{\beta} x_{\alpha\beta}$ of row α; a further summation of all these row totals gives the grand total $\sum_{\alpha,\beta} x_{\alpha\beta}$.

In the same way we can find column totals, $x_{11} + x_{21} = \Sigma x_{\alpha 1}$, the total number of group O, $x_{12} + x_{22} = \Sigma x_{\alpha 2}$, the total group A, $\Sigma x_{\alpha 3}$, group B, and $\Sigma x_{\alpha 4}$, group AB. These column totals are shown at the foot of Table 13.4. The sum of these column totals must be the grand total: 163 + 180 + 55 + 24 = 422, or in general

$$\sum_{\beta}\left(\sum_{\alpha} x_{\alpha\beta}\right) = \sum_{\alpha,\beta} x_{\alpha\beta}$$

Thus it does not matter whether we first sum the columns, and then add the column totals together, or first sum the rows, and then add the row totals together.

This discussion can readily be generalized to three-way, four-way, and other more complicated classifications. Suppose we add a third suffix, such as a 1 for male sex, and 2 for female. Then x_{241} would mean the number of persons of Latvian nationality, group AB, and male sex. $\sum_{\alpha} x_{\alpha 41}$ would be the total number of any nationality of group AB and male sex, and $\sum_{\alpha,\beta} x_{\alpha\beta 1} = \sum_{\alpha}\left(\sum_{\beta} x_{\alpha\beta 1}\right) = \sum_{\beta}\left(\sum_{\alpha} x_{\alpha\beta 1}\right)$ would be the total number of males. The grand total would be $\Sigma x_{\alpha\beta\gamma}$, and the summation can be done in any order, e.g. $\sum_{\alpha}\left[\sum_{\beta}\left(\sum_{\gamma} x_{\alpha\beta\gamma}\right)\right]$, or $\sum_{\beta}\left(\sum_{\alpha,\gamma} x_{\alpha\beta\gamma}\right)$.

Another point worth noting is the product of two sums, such as $(x_1 + x_2 + x_3 + x_4)(y_1 + y_2)$. By direct multiplication this becomes

$$(x_1y_1 + x_2y_1 + x_3y_1 + x_4y_1 + x_1y_2 + x_2y_2 + x_3y_2 + x_4y_2),$$

i.e. $\sum_{\alpha,\beta} x_\alpha y_\beta$ summed over all possible combinations of suffixes α and β. Thus we can write

$$(\Sigma x_\alpha)(\Sigma y_\beta) = \Sigma x_\alpha y_\beta \quad . \quad . \quad (13.30)$$

It would be incorrect to write $(\Sigma x_\alpha)(\Sigma y_\alpha) = \Sigma x_\alpha y_\alpha$, for this would mean for instance $(x_1 + x_2 + x_3 + x_4)(y_1 + y_2) = (x_1y_1 + x_2y_2)$ which in general is not true.

Similar reasoning shows that $(\Sigma x_\alpha)(\Sigma y_{\beta\gamma}) = \Sigma x_\alpha y_{\beta\gamma}$, and that $(\Sigma x_\alpha)(\Sigma y_\beta)(\Sigma z_\gamma) = \Sigma x_\alpha y_\beta z_\gamma$, and other formulas of this kind. Such formulas are often useful, especially in the study of matrices (Chapter 18).

These results can equally well be applied to infinite series, i.e. series such as $\Sigma x_{\alpha\beta}$ where α can take all values 1, 2, 3, 4, . . . and so on indefinitely, and similarly β can take all values 1, 2, 3. . . . They will be valid if the series is "geometrically convergent", that is, if we can find two positive numbers A and R with the properties (i) $R < 1$, (ii) $|x_{rs}| \leqslant AR^{r+s}$ for all values of r and s. However, such double infinite series are not very frequently needed in practice.

CHAPTER FOURTEEN

DIRECTED MAGNITUDES

14.1 Vectors and scalars

Forces, velocities, electric fields, and electric currents all have this in common, that they have a direction as well as a magnitude. If a body is being acted upon by a force it is important to know not only what the magnitude of the force is, but also whether it is urging the body to move north, south, east, west, upwards, or downwards. A physical quantity which has direction as well as magnitude is known as a "vector" (or more precisely a "spatial vector"). In modern textbooks it is usually distinguished by heavy type, e.g. $\boldsymbol{v}$ for a velocity or $\boldsymbol{F}$ for a force. A quantity which has only magnitude, such as temperature, volume, mass, or electric charge, is known as a "scalar".

Now we usually consider two measurements of length or mass as being "equal" if they have the same magnitude regardless of position, time, or other external conditions. A mass of 1 kilogram in London is regarded as equal to a mass of 1 kilogram in Paris, whether measured in the year 1900 or 1950: that is the normal way of using the word "equal". In the same way we shall regard two forces or other vectors as "equal" if they have the same magnitudes and the same direction, irrespective of where and when we measure them. It is convenient to speak in that way. But if they are equal in magnitude but different in direction we shall regard them as unequal forces. There is one exception to this. If the magnitude of a vector is zero, its direction is of no importance. A body acted upon by zero force is not urged one way or another, and it is futile to ask what direction the zero force is acting in. A body which is standing still can be imagined is moving with zero velocity in any direction. A vector whose magnitude is zero we shall call a "zero vector" $\boldsymbol{O}$, and regard its direction as arbitrary or undetermined.

The word "vector" is Latin for "carrier", the point being that anything which is carried must be carried a certain distance in a certain direction. The word "scalar" is related to "scale", since two scalar quantities can be compared on a single scale of magnitudes without reference to direction.

14.2 Geometric representation of a vector

From the definition of a vector as a quantity with magnitude and direction we see that it can be represented by a directed line AB, the direction of the line (from A to B) corresponding to the direction of the vector, and the length of the line being equal to the magnitude of the vector (in suitable units) (Fig. 14.1). Any point A can be chosen as starting-

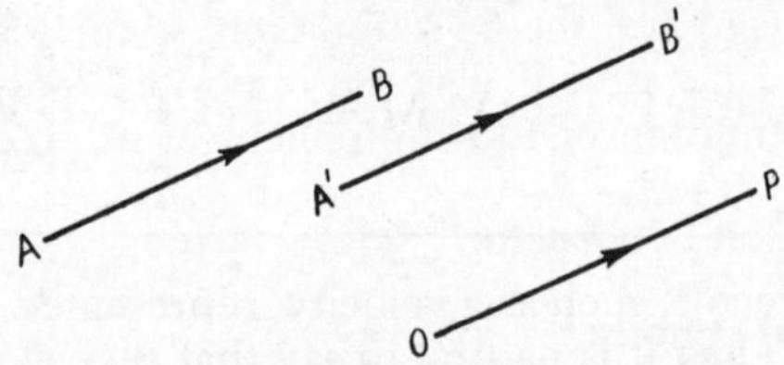

Fig. 14.1—Vectors represented by lines

point for the line. If A' is any other point, and the line $A'B'$ equal in length and parallel to AB, then $A'B'$ will equally well represent the same vector. However, suppose we choose arbitrarily any one fixed point or "origin" O in space (Fig. 14.1). Then we can draw a line OP equal and parallel to AB, and representing the same vector $\boldsymbol{v}$. So for each vector $\boldsymbol{v}$ there will be just one such line, and just one such end-point P. Conversely for each point P there will be just one vector OP. In this sense we can speak of the *point* P as "representing the vector $\boldsymbol{v} = OP$". The advantage is that in a complicated diagram it is easier to draw a single point P to represent a vector, rather than a line OP or AB. The representation of a vector by a line is the more natural one, and we shall use it wherever possible. The representation by a point P has the disadvantage that it is necessary to choose a fixed point O as origin, and such a choice is arbitrary. For when we say that the vector $\boldsymbol{v}$ is represented by a point P, we really mean that it is represented by the line OP. But the use of a point P in this way is sometimes helpful.

Note that the zero vector $\boldsymbol{O}$ is represented by any line AA of zero length, or by the point O.

14.3 Components of vectors

Let AB (Fig. 14.2) be a directed line representing a vector $\boldsymbol{v}$, and let HK be another line in a direction D making an angle θ with AB. Draw

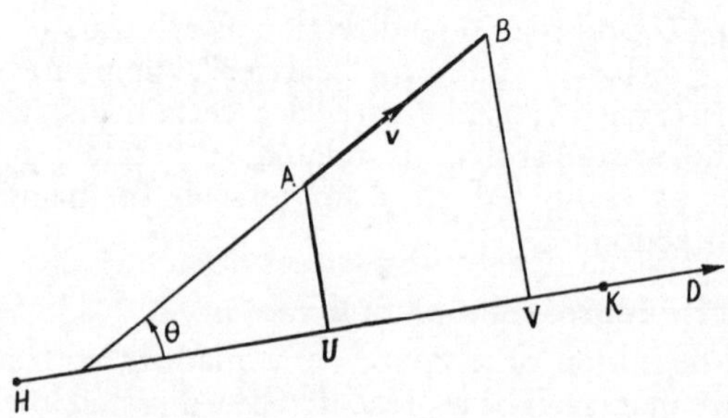

Fig. 14.2—The component UV of a vector AB along a line HK

AU, *BV* perpendicularly onto *HK*. Then the distance *UV* is called the "component of $\boldsymbol{v}$ along *HK*" and is equal to the magnitude of $\boldsymbol{v}$ times the cosine of θ (Section 5.6). Here *UV* is supposed to be measured with its proper sign, positive if the direction from *U* to *V* is the same as that from *H* to *K*, and negative if it is the opposite direction. This agrees with the usual definition of the component of a force in mechanics.

14.4 Magnification of a vector

Let $\boldsymbol{v}$ be any vector, such as a velocity, represented by a directed line *AB* (Fig. 14.3). Then it is natural to say that the velocity is doubled if its magnitude is doubled, its direction remaining the same. Thus if

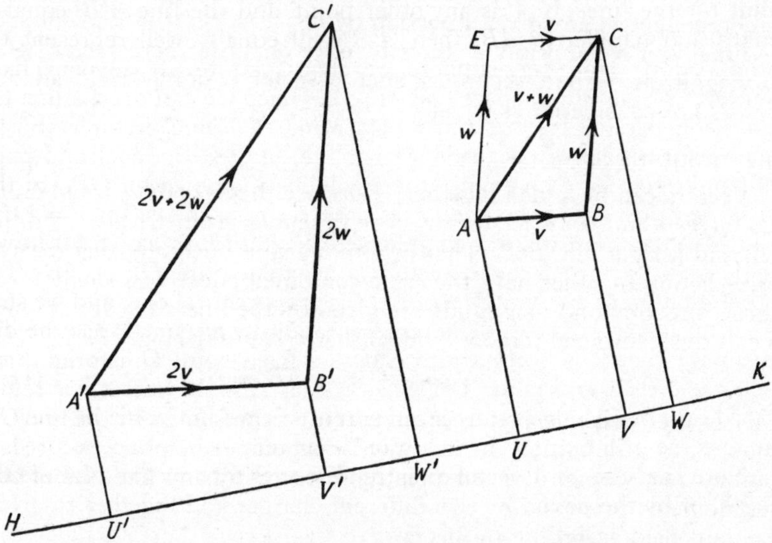

Fig. 14.3—Multiplication of a vector by a number, and addition of vectors

A'B' is drawn parallel to *AB* but of twice its length, then *A'B'* represents $\boldsymbol{v}$ multiplied by 2, or $2\boldsymbol{v}$. Similarly the product $k\boldsymbol{v}$ where k is any positive number k, will be taken to mean the vector with the same direction as $\boldsymbol{v}$, but k times greater in magnitude. The vector $-k\boldsymbol{v}$ will naturally signify the vector $k\boldsymbol{v}$ with its direction reversed, e.g. if $A'B' = 2\boldsymbol{v}$, then $B'A' = -2\boldsymbol{v}$, by definition. The vector *BA* is the reverse of *AB*, and will be denoted by $-\boldsymbol{v}$, or $(-1)\boldsymbol{v}$.

In this way we can define the product of an ordinary number, positive or negative, and a vector. We can show that this product obeys the law

$$h(k\boldsymbol{v}) = (hk)\boldsymbol{v} \quad . \quad . \quad . \quad (14.1)$$

For $(hk)\boldsymbol{v}$ means the original vector multiplied hk times in magnitude, but with its direction unaltered (or reversed, if hk is negative). $k\boldsymbol{v}$ means that the magnitude has been multiplied by k, and the direction unchanged (except for a reversal if k is negative). If then we multiply $(k\boldsymbol{v})$ by h we have in all multiplied the magnitude by hk, and we have left the direction unaltered, or reversed it, according to the usual rule of signs in algebra (+ve × +ve = +ve, +ve × −ve = −ve, etc.) So $h(k\boldsymbol{v}) = (hk)\boldsymbol{v}$.

Furthermore, if HK is any line, and we draw AU, BV, $A'U'$, $B'V'$ perpendicularly onto HK, then since AB and $A'B'$ make the same angle with HK, the ratio of UV to AB must be the same as the ratio of $U'V'$ to $A'B'$, each being the cosine of this angle (Fig. 14.3). So if $A'B' = kAB$, then $U'V' = kUV$. If we multiply a vector by k, we also multiply its component in any given direction by k.

Finally, if we multiply a vector by 0 we obtain a "zero vector" $\boldsymbol{O}$ of no magnitude. The direction of such a vector is immaterial: all zero vectors are considered as equal.

14.5 Addition of vectors

Vectors can be added together. Suppose that a body is acted on by two forces $\boldsymbol{v}$ and $\boldsymbol{w}$. $\boldsymbol{v}$, let us say, is represented geometrically by a line AB, and $\boldsymbol{w}$ by a line BC. Then we know experimentally that the two forces acting together have the same combined effect as a single force, whose direction and magnitude are given by the line AC. It is natural to call this combined force $\boldsymbol{v} + \boldsymbol{w}$, so that *vectorially* (Fig. 14.3)

$$AC = \boldsymbol{v} + \boldsymbol{w} = AB + BC \qquad . \qquad (14.2)$$

This is the "triangle of forces" theorem: from our point of view, it amounts to a definition of "addition" for vectors. It can be used to combine many other directed quantities besides forces: for example the electric fields produced by two different charges add together to give a resultant field according to this law.

This method of addition of vectors is very similar to ordinary addition in its properties. The laws of operation for ordinary addition and multiplication have been set out in Section 3.2 (which the reader is recommended to re-read). Let us write out those of vectors for comparison.

Firstly let us draw AU, BV, CW perpendicularly onto the line HK (Fig. 14.3). Then UV is the component of $\boldsymbol{v} = AB$ along HK, VW is the component of $\boldsymbol{w} = BC$, and UW the component of $\boldsymbol{v} + \boldsymbol{w}$. Since $UV + VW = UW$ (taking signs into account) we see that the component of $\boldsymbol{v} + \boldsymbol{w}$ along HK is the sum of the components of $\boldsymbol{v}$ and $\boldsymbol{w}$.

Secondly, let us complete the parallelogram $ABCE$ by drawing AE parallel to BC and EC parallel to AB. Since opposite sides of a parallelogram are equal in length, we see that vectorially $AE = BC = \boldsymbol{w}$, $EC = AB = \boldsymbol{v}$, and the relation $AE + EC = AC$ becomes

$$\boldsymbol{w} + \boldsymbol{v} = AC = \boldsymbol{v} + \boldsymbol{w} \qquad . \qquad . \qquad (14.3)$$

(the "commutative law of addition"). In the same way if we are given three vectors, $AB = \boldsymbol{v}$, $BC = \boldsymbol{w}$, and $CD = \boldsymbol{x}$, and we complete

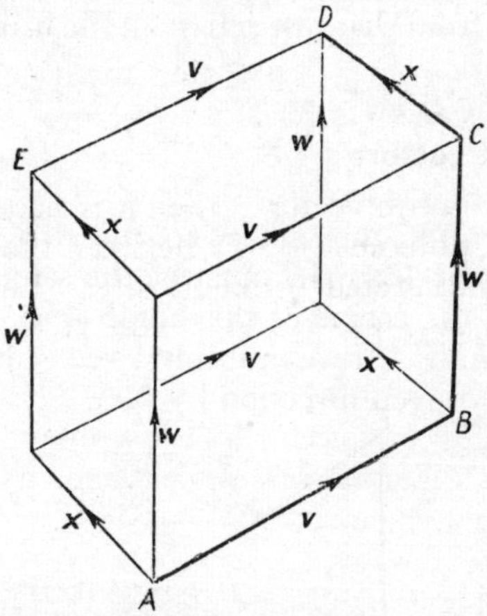

Fig. 14.4—The sum of three vectors

the parallelopipedon, as shown in Fig. 14.4, we see that vectorially

$$\begin{aligned} AD &= AB + BD = \boldsymbol{v} + (\boldsymbol{w} + \boldsymbol{x}) \\ &= AC + CD = (\boldsymbol{v} + \boldsymbol{w}) + \boldsymbol{x} \\ &= AE + ED = (\boldsymbol{w} + \boldsymbol{x}) + \boldsymbol{v}, \text{ etc.} \end{aligned} \qquad (14.4)$$

i.e. *vector additions can be performed in any order.*

The addition of $\boldsymbol{O}$ to any vector leaves it unchanged, for

$$\boldsymbol{O} + \boldsymbol{v} = AA + AB = AB = \boldsymbol{v} \qquad (14.5)$$

This is only natural, for if we interpret the vectors as forces then $\boldsymbol{O}$ is a zero force, and does not change any other force on combination with it.

Now let ABC (Fig. 14.3) be the triangle of vectors representing the addition $\boldsymbol{v} + \boldsymbol{w} = (\boldsymbol{v} + \boldsymbol{w})$. Draw a second triangle $A'B'C'$ similar and parallel to ABC, but of twice the size. Then $A'B'$ represents $2\boldsymbol{v}$, $B'C'$ represents $2\boldsymbol{w}$, and $A'C'$ $2(\boldsymbol{v} + \boldsymbol{w})$. But vectorially $A'B' + B'C' = A'C'$, i.e. $2\boldsymbol{v} + 2\boldsymbol{w} = 2(\boldsymbol{v} + \boldsymbol{w})$. In the same way

$$k\boldsymbol{v} + k\boldsymbol{w} = k(\boldsymbol{v} + \boldsymbol{w}) \qquad (14.6)$$

for any number k (the so-called "distributive law").

Finally we know that two vectors acting along the same line (i.e. in the same direction, or in directly opposite directions) combine like ordinary positive or negative numbers. A force of 2 kilograms weight acting downwards combines with a force of 3 kilograms weight, also acting downwards, to give one of 5 kilograms. We can write this

$$h\boldsymbol{v} + k\boldsymbol{v} = (h + k)\boldsymbol{v} \qquad (14.7)$$

In short, vectors obey exactly the same laws of addition and multiplication as ordinary algebraic symbols, except that so faı we have given no method of multiplying 2 vectors together, but only an ordinary number and a vector. Apart from that restriction all the usual algebraic formulas hold good.

14.6 Subtraction of vectors

Let $\boldsymbol{v}$ and $\boldsymbol{w}$ be any two vectors; then it is natural to call the vector sum $\boldsymbol{w} + (-\boldsymbol{v})$ the "difference $\boldsymbol{w} - \boldsymbol{v}$ between the vectors $\boldsymbol{w}$ and $\boldsymbol{v}$".

This is easily interpreted geometrically. Let $\boldsymbol{v} = AB$, $\boldsymbol{w} = AE$ (Fig. 14.5). Then $\boldsymbol{w} - \boldsymbol{v} = (-\boldsymbol{v}) + \boldsymbol{w} = BA + AE = BE$. Since

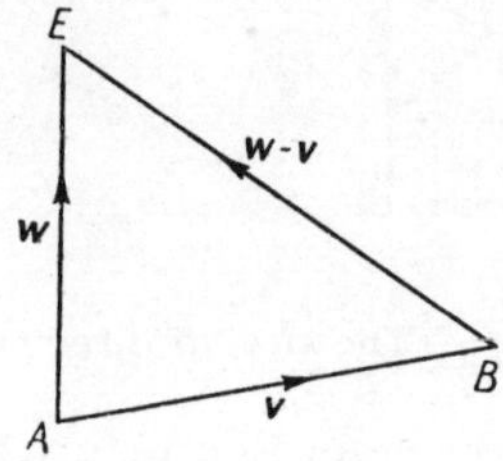

Fig. 14.5—Vector subtraction

vectorially $AB + BE = AE$ we have

$$\boldsymbol{v} + (\boldsymbol{w} - \boldsymbol{v}) = \boldsymbol{w} \quad . \quad . \quad . \quad (14.8)$$

Thus $(\boldsymbol{w} - \boldsymbol{v})$ is the vector which has to be added to $\boldsymbol{v}$ to make it equal to $\boldsymbol{w}$, which corresponds to the usual idea of subtraction as the reverse of addition.

Also we have from this definition

$$\boldsymbol{v} - \boldsymbol{v} = BB = \boldsymbol{O} \quad . \quad . \quad . \quad (14.9)$$

Thus subtraction also obeys all the ordinary algebraic laws.

EXAMPLE

(1) To show that the three medians of a triangle meet in a point.

A "median" of a triangle ABC is defined as the line joining a vertex to the mid-point of an opposite side; e.g. AP, where P is the mid-point of BC (Fig. 14.6).

Now choose an origin O, and call the vectors OA, OB, OC, $\boldsymbol{a}$, $\boldsymbol{b}$ and $\boldsymbol{c}$ respectively. Then by the subtraction law, $AB = \boldsymbol{b} - \boldsymbol{a}$, $BC = \boldsymbol{c} - \boldsymbol{b}$. But $BP = \frac{1}{2}BC = \frac{1}{2}(\boldsymbol{c} - \boldsymbol{b})$, so that $AP = AB + BP = (\boldsymbol{b} - \boldsymbol{a}) + \frac{1}{2}(\boldsymbol{c} - \boldsymbol{b}) = \frac{1}{2}\boldsymbol{b} + \frac{1}{2}\boldsymbol{c} - \boldsymbol{a}$. Now choose a point G on AP

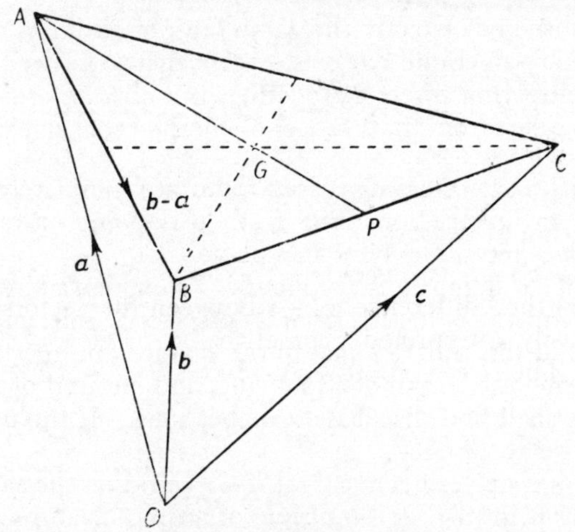

Fig. 14.6—The centroid G of a triangle ABC is the intersection of the three medians

and $\frac{2}{3}$ of the distance along AP, so that $AG = \frac{2}{3}\,AP = \frac{1}{3}\boldsymbol{b} + \frac{1}{3}\boldsymbol{c} - \frac{2}{3}\boldsymbol{a}$. It follows that

$$OG = OA + AG = \boldsymbol{a} + (\tfrac{1}{3}\boldsymbol{b} + \tfrac{1}{3}\boldsymbol{c} - \tfrac{2}{3}\boldsymbol{a})$$
$$= \tfrac{1}{3}\boldsymbol{a} + \tfrac{1}{3}\boldsymbol{b} + \tfrac{1}{3}\boldsymbol{c}.$$

Thus we have shown that G lies on the median AP through A. But the vector OG is expressed symmetrically in terms of $\boldsymbol{a}$, $\boldsymbol{b}$, and $\boldsymbol{c}$, so it must equally well be true that G lies on the medians through B and through C. The point G is called the "centroid" of the triangle ABC.

PROBLEM

(1) Show that in any triangular pyramid (or "tetrahedron") the four lines joining the vertices to the centroids of opposite faces all pass through a certain point G. Also show that the three lines joining the mid-points of opposite edges also pass through this point.

14.7 The magnitude of a vector

It is customary to denote the magnitude of a vector $\boldsymbol{v}$, irrespective of its direction, by the symbol $|\boldsymbol{v}|$ (read as "mod $\boldsymbol{v}$" and called the modulus or absolute magnitude of $\boldsymbol{v}$). Thus $|\boldsymbol{v}|$ is an ordinary positive number. It obeys the law

$$|k\boldsymbol{v}| = |k|\,|\boldsymbol{v}| \quad . \quad . \quad . \quad (14.10)$$

Also, since no side of a triangle can exceed in length the sum of the other 2 sides,

$$|AC| \leqslant |AB| + |BC|,$$
$$\text{i.e. } |\boldsymbol{v} + \boldsymbol{w}| \leqslant |\boldsymbol{v}| + |\boldsymbol{w}| \quad . \quad . \quad . \quad (14.11)$$

PROBLEM

(1) Prove that $|\boldsymbol{v} - \boldsymbol{w}| \leqslant |\boldsymbol{v}| + |\boldsymbol{w}|$.

14.8 Vectors in a plane

All properties of vectors discussed so far are completely general, and true without any restriction. But from this point onwards we shall restrict ourselves to vectors lying in a plane.

The reason for doing so is as follows. To express a physical quantity in numerical form it is necessary first to select a unit, and then to say what multiple of this unit is equal to the quantity in question. Thus if we are measuring the length of a foot rule, and our unit of measurement is an inch, we shall find that it is 12 inches long. If the unit is a centimetre, it will be 30·48 cm.

Now it would be very convenient if we could do the same for a vector, expressing it in turn as a multiple of a unit. Naturally this cannot be done in terms of ordinary numbers alone, for a velocity of (say) 2 metres per second eastwards is not a multiple of a velocity of 1 metre per second northwards. But it can be done by allowing a generalization of the idea of "number". And this is not an unreasonable device. For the idea of number has been generalized several times in the course of history. It began with positive integers, used in counting. But men soon discovered that it was useful to supplement these with fractions, which were used in measuring. Later it was found convenient to invent negative numbers, and irrational numbers such as $\sqrt{2}$ which cannot be expressed as an exact fraction. If the reader is prepared to accept yet a further generalization it will be possible to include the measurement of vectors in a plane. Unfortunately it is more difficult to do the same for general vectors in three-dimensional space; and that is why we have to confine our attention henceforward to a single plane.

Choose arbitrarily any vector $OI = \boldsymbol{I}$ (say) in the plane, and call it the "unit vector". Now consider any other vector $\boldsymbol{v} = OP$, of length r. This in general differs from $\boldsymbol{I}$, the unit, both in magnitude and direction: it will be r times greater in magnitude, and it will differ in direction by some angle, say θ (Fig. 14.7; θ will be measured as positive if the

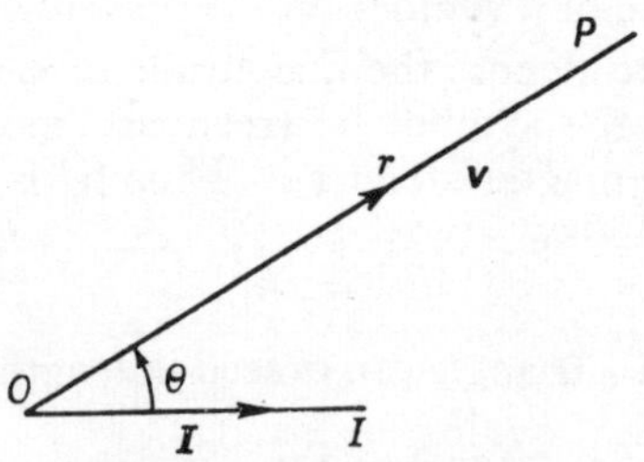

Fig. 14.7—Complex numbers

The measure $\{r, \theta\}$ of a vector $\boldsymbol{v}$ in terms of a unit $\boldsymbol{I}$

direction of rotation from OI to OP is anticlockwise and negative if clockwise, according to the usual convention). We shall then say that the measure of $\boldsymbol{v}$ in terms of the unit $\boldsymbol{I}$ is $\{r, \theta\}$. We shall look upon $\{r, \theta\}$ as a *single* number of a new type, called a "complex number" z ("complex" because it is composed of two parts or ordinary real numbers r and θ). Probably the simplest way of thinking of the complex number $\{r, \theta\}$ is to consider it as the sign for an operation, namely the operation of magnifying r times and then rotating through an angle θ. The performance of this operation on the vector $\boldsymbol{I}$ gives us the vector $\boldsymbol{v}$. We look upon $\boldsymbol{v}$ as the unit vector $\boldsymbol{I}$ "multiplied by" the complex number $z = \{r, \theta\}$, or in symbols

$$\boldsymbol{v} = z\boldsymbol{I} = \{r, \theta\}\boldsymbol{I} \quad . \quad . \quad . \quad (14.12)$$

In contrast to this an ordinary positive number h can be taken simply as a sign of magnification alone. For when we say that 1 foot has the measure 12 when expressed in units of 1 inch, we mean that a foot is an inch magnified 12 times. If $\boldsymbol{v}$ is a vector, then the product $h\boldsymbol{v}$ signifies the vector $\boldsymbol{v}$ magnified h times, but unaltered in direction. In the same way a multiplication of a vector $\boldsymbol{v}$ by a complex number $\{r, \theta\}$ means, by definition, a magnification r times and a rotation through the angle θ. In particular a multiplication by the complex number $\{h, 0\}$ is simply an h-times magnification, and has exactly the same effect as a multiplication by the positive number h. Thus the complex number $\{h, 0\}$ is equivalent to the real number h in its effect on vectors—which is how a complex number is defined. For all practical purposes we can consider h and $\{h, 0\}$ as identical.

Similarly to multiply $\boldsymbol{v}$ by $(-h)$ means to magnify it h times and reverse its direction. Multiplication of $\boldsymbol{v}$ by the complex number $\{h, \pi\}$ magnifies it h times, and rotates it through π radians, $= 180°$, i.e. this also reverses its direction. Thus the complex number $\{h, \pi\}$ is exactly equivalent to the real negative number $(-h)$. In short we can consider the ordinary real numbers we have so far dealt with as particular cases of complex numbers, those in which $\theta = 0$ or π.

14.9 Multiplication of complex numbers

Complex numbers will hardly be worthy of the name of "numbers" unless we can add, subtract, multiply and divide them. How we do so is of course a matter of definition, not of fact. We are free in principle to define "addition", "subtraction", "multiplication" and "division" for this new type of number in any way that takes our fancy. But since complex numbers have been shown to include real numbers as a particular case, it is clearly desirable that whatever definitions we choose for the addition, etc., of complex numbers agree with the familiar ones for the special case when the numbers are real. It is also desirable that the ordinary properties of the signs $+$, $-$, $\times$ and $:$ should be preserved as far as possible, and that the definitions should sound plausible and reasonable.

The operation which has the simplest definition is that of multiplication. Let us first go back to the idea of multiplication for real numbers. Suppose that a magnitude M_1 is x_1 times a magnitude M_2, and M_2 is x_2 times M_3. Then M_1 is $x_1 x_2$ times M_3. For example, 1 yard = 3 feet; 1 foot = 12 inches; so 1 yard = 3 × 12 inches.

This suggests a rule for multiplying complex numbers. Suppose that a vector $\boldsymbol{v}_1$ is $z_1 = \{r_1, \theta_1\}$ times a second vector $\boldsymbol{v}_2$, i.e. $\boldsymbol{v}_1$ is obtained from $\boldsymbol{v}_2$ by magnifying it r_1 times and rotating it through the angle θ_1. Suppose further that $\boldsymbol{v}_2$ is z_2 times $\boldsymbol{v}_3$ i.e. is obtained by a magnification r_2 and rotation θ_2. Then it is natural to say that $\boldsymbol{v}_1$ is $(z_1 z_2)$ times $\boldsymbol{v}_3$ That is, if $\boldsymbol{v}_1 = z_1 \boldsymbol{v}_2$ and $\boldsymbol{v}_2 = z_2 \boldsymbol{v}_3$ then

$$\boldsymbol{v}_1 = z_1 (z_2 \boldsymbol{v}_3) = (z_1 z_2)\boldsymbol{v}_3 \quad . \quad . \quad (14.13)$$

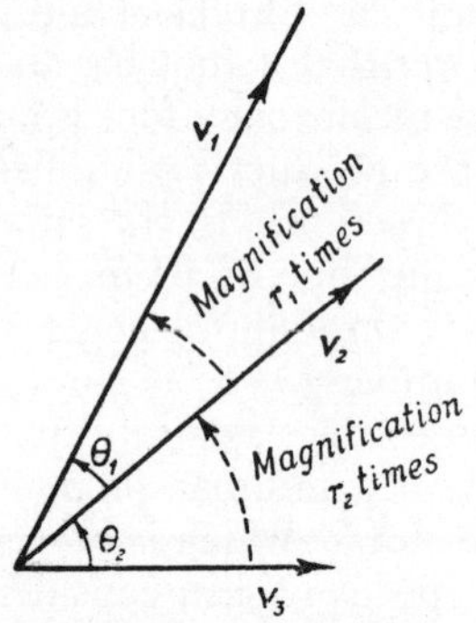

Fig. 14.8—Multiplication of complex numbers

But (Fig. 14.8) $\boldsymbol{v}_1$ is obtained from $\boldsymbol{v}_3$ by magnifying it $r_1 r_2$ times and rotating it through $\theta_1 + \theta_2$, so that

$$z_1 z_2 = \{r_1, \theta_1\}\{r_2, \theta_2\} = \{r_1 r_2, \theta_1 + \theta_2\} \quad (14.14)$$

This is the definition of multiplication. If we interpret z_1 and z_2 as operations of magnification and rotation, then the product $z_1 z_2$ will be interpreted as the effect of successive application of the two operations. First (to change $\boldsymbol{v}_3$ into $\boldsymbol{v}_2$) we magnify r_2 times and rotate through an angle θ_2; secondly (to change $\boldsymbol{v}_2$ into $\boldsymbol{v}_1$) we magnify r_1 times, and rotate through θ_1. The total effect is a magnification $r_1 r_2$ and a rotation $\theta_1 + \theta_2$.

From (14.14) we see also that

$$\begin{aligned} z_2 z_1 = \{r_2, \theta_2\}\{r_1, \theta_1\} &= \{r_2 r_1, \theta_2 + \theta_1\} \\ &= \{r_1 r_2, \theta_1 + \theta_2\} \\ &= z_1 z_2 \quad . \quad . \quad (14.15) \end{aligned}$$

so that multiplication of complex numbers can be done in either order just as for real numbers. Also if $z_3 = \{r_3, \theta_3\}$ is a third complex number

$$\begin{aligned} z_1(z_2 z_3) &= \{r_1, \theta_1\}\{r_2 r_3, \theta_2 + \theta_3\} = \{r_1 r_2 r_3, \theta_1 + \theta_2 + \theta_3\} \\ (z_1 z_2) z_3 &= \{r_1 r_2, \theta_1 + \theta_2\}\{r_3, \theta_3\} = \{r_1 r_2 r_3, \theta_1 + \theta_2 + \theta_3\} \\ (z_2 z_3) z_1 &= \{r_2 r_3, \theta_2 + \theta_3\}\{r_1, \theta_1\} = \{r_1 r_2 r_3, \theta_1 + \theta_2 + \theta_3\} \end{aligned}$$

i.e. $z_1(z_2 z_3) = (z_1 z_2) z_3 = (z_2 z_3) z_1$, so that multiplications of any number of complex numbers can be done in any order, just as for real numbers ("commutative" and "associative" laws, in technical language).

There is one special case of interest. Multiplication by $\{0, \theta\}$ means by definition "magnification 0 times, and rotation through angle θ". But magnification by 0 results in everything being reduced to zero, independently of the angle θ. Thus it is convenient to consider $\{0, \theta\}$ as equal to the real number 0 whatever the value of θ. We then have

$$0z = 0 \quad . \quad . \quad . \quad (14.16)$$

whatever z may be, exactly as for real numbers.

14.10 Powers of complex numbers

Since multiplication of complex numbers obeys laws similar to ordinary multiplication it is natural to denote the product zz of z into itself by the symbol z^2, and call it the "square of z". Similarly zzz is denoted by z^3 and so on. Such powers obey the usual laws, e.g. $z^2z^3 = (zz)(zzz) = zzzzz = z^5$, and in general $z^m z^n = z^{m+n}$.

In terms of the $\{r, \theta\}$ notation we have by the rule (14.14) for multiplication

$$z^2 = \{r, \theta\}\{r, \theta\} = \{rr, \theta + \theta\} = \{r^2, 2\theta\}$$
$$z^3 = \{r, \theta\}\{r, \theta\}\{r, \theta\} = \{rrr, \theta + \theta + \theta\} = \{r^3, 3\theta\}$$

and in general $z^n = \{r, \theta\}^n = \{r^n, n\theta\}$.

14.11 Multiplication of real numbers

We have now to show that this definition of multiplication for complex numbers agrees with the usual definition when the numbers are real.

Consider first the case of two positive numbers h and k. As we have already seen, h is effectively equal to the complex number $\{h, 0\}$, and similarly $k = \{k, 0\}$. The product of the two real numbers h and k is hk, while the product of the two complex numbers $\{h, 0\}$ and $\{k, 0\}$ is $\{hk, 0 + 0\} = \{hk, 0\}$. If the two definitions are to agree, these two products must be equal, i.e. $hk = \{hk, 0\}$. But that is true since hk is positive.

Now consider two negative numbers $-h$ and $-k$. According to the ordinary rule for multiplication their product is positive and equal to hk. Now when we interpret $-h$ as a complex number it becomes $\{h, \pi\}$, i.e. a magnification by h, and a rotation through π radians $= 180°$. Similarly $-k = \{k, \pi\}$. The complex product of $\{h, \pi\}$ and $\{k, \pi\}$ is by definition $\{hk, \pi + \pi\} = \{hk, 2\pi\}$, that is, a magnification by hk and a rotation through $180° + 180° = 360°$. But a rotation through 360° is equivalent in its effect to no rotation at all, i.e.

$$\{hk, 2\pi\} = \{hk, 0\}.$$

This is equivalent to the real number hk, so the two definitions of multiplication again give the same answer. Incidentally this provides an interpretation of the usual rule that the product of two negative numbers is positive: it means that two rotations through 180° combine to bring the vector back to its original direction.

We leave the reader to verify in a similar way that the agreement still holds in the remaining case of the multiplication of a positive and a negative number.

14.12 The square root of minus one

By our definition, multiplication by a number $\{1, \theta\}$ means rotation through the angle θ without change in magnitude. Thus $\{1, 0\}$ or $\{1, 2\pi\}$ signifies no change, and is equivalent to multiplication by 1, while $\{1, \pi\}$ merely reverses directions, and is equivalent to -1.

Now consider $\{1, \frac{1}{2}\pi\}$. This means rotation through a right angle in an anticlockwise direction: we call it "i". Then

$$i^2 = ii = \{1, \tfrac{1}{2}\pi\}\,\{1, \tfrac{1}{2}\pi\} = \{1, \pi\} = -1.$$

i.e. a rotation through 90° repeated is equivalent to one through 180°, or a simple reversal. i is therefore the famous "square root of minus one", or rather i is one such square root: another is $-i = \{1, \frac{3}{2}\pi\}$, for $(-i)(-i) = \{1, 3\pi\}$, and a rotation through 3π radians is equivalent to one through π, i.e. $\{1, 3\pi\} = \{1, \pi\} = -1$.

When mathematicians first began considering the square root of minus one, they could not see any simple interpretation of it, for they knew that the square of any ordinary number was positive. So they called the square root of a negative number, such as i, an "imaginary number", and ordinary numbers such as 2, π, or $-\frac{3}{2}$ were called "real" numbers. Nowadays we know that i has a simple interpretation: it is a rotation through 90°. The equation $i^2 = -1$ means simply that two successive rotations through 90° are equivalent to a reversal. So i is no more imaginary than any other number. But the old names still remain, and we talk of 2, π, and $-\frac{3}{2}$ as "real" numbers, and i, $2\pi i$ and $-\frac{3}{2}i$ as "imaginary" numbers. These terms are now purely technical, and no longer imply that a "real" number really exists, or that an "imaginary" one doesn't.

Consider further the number $\omega = \{1, \frac{2}{3}\pi\}$. This means a rotation through $\frac{2}{3}\pi$ radians $= 120°$. Thus if we perform the rotation three times we come back to the original position, $\omega^3 = \{1, 2\pi\} = 1$. Thus ω is a complex cube root of 1. Another cube root is $\omega^2 = \{1, \frac{4}{3}\pi\}$, for $(\omega^2)^3 = \{1, 4\pi\} = 1$; and of course 1 is also a cube root of 1. So 1 has 3 complex cube roots, 1, ω, and ω^2. It has 4 fourth roots, 1, i, -1, and $-i$, and we can show that in general it has n nth roots.

PROBLEM

(1) What are the n nth roots of 1?

14.13 Division of complex numbers

In ordinary arithmetic the quotient x/y means that number which when multiplied by y gives x. We shall therefore take the quotient of two complex numbers z_1/z_2 to mean that number which gives z_1 on multiplication by z_2. Now if $z_1 = \{r_1, \theta_1\}$ and $z_2 = \{r_2, \theta_2\}$, then to get from z_2 to z_1 we have to multiply the magnitude of z_2 by r_1/r_2, and rotate it through an angle $\theta_1 - \theta_2$; i.e.

$$z_1/z_2 = \{r_1, \theta_1\}/\{r_2, \theta_2\} = \{r_1/r_2, \theta_1 - \theta_2\} \tag{14.17}$$

This is the definition of division of complex numbers. It has the property we require, for

$$(z_1/z_2)z_2 = \{r_1/r_2, \theta_1 - \theta_2\}\{r_2, \theta_2\} = \{r_1, \theta_1\} = z_1$$

by the rule (14.14) for multiplication. We also see that the definition (14.17) fails to give a definite quotient only if $r_2 = 0$, i.e. $z_2 = 0$, so we can divide any complex number by any other complex number except 0.

In particular the number $1/z = \{1, 0\}/\{r, \theta\} = \{1/r, -\theta\}$ is called the "reciprocal" of z.

PROBLEMS

(1) Show that $1/i = -i$.
(2) Show that $i^3 = -i$.
(3) Show that $1/\omega = \omega^2$.
(4) Show that $z_2/z_1 = z_2 \cdot (1/z_1)$.
(5) Show that $(z_1z_2)^2 = z_1^2 \cdot z_2^2$.
(6) Show that $(1/z_1)(1/z_2) = 1/(z_1z_2)$.
(7) What are the cube roots of -1?
(8) What are the square roots of ω?
(9) Show that every complex number $z = \{r, \theta\}$ has exactly two square roots, except the number 0 which has only one.

14.14 Geometrical representation of a complex number

A complex number is the measure of a vector $\boldsymbol{v}$ in terms of a unit vector $\boldsymbol{I}$. If therefore we choose arbitrarily any vector OI in the plane as unit vector $\boldsymbol{I}$, we can represent the complex number $z = \{r, \theta\}$ by the vector $\boldsymbol{v} = z\boldsymbol{I}$. $\boldsymbol{v}$ has length r and makes an angle θ with $\boldsymbol{I}$. The number 1 corresponds to the vector $\boldsymbol{I}$ itself $(= 1\boldsymbol{I})$, and the number 0 to the vector $\boldsymbol{O}$. Various complex numbers are drawn in this way in Fig. 14.9.

If the number $z_2 = \{r_2, \theta_2\}$ is represented by a vector $\boldsymbol{v}_2$, then the product $z_1z_2 = \{r_1, \theta_1\}\{r_2, \theta_2\}$ is represented by the vector $z_1\boldsymbol{v}_2$, i.e. by $\boldsymbol{v}_2$ magnified r_2 times and rotated through θ_2.

Furthermore the absolute value or length or magnitude of the vector $\boldsymbol{v} = z\boldsymbol{I}$ is simply $|\boldsymbol{v}| = r$. So we call r the "absolute value",

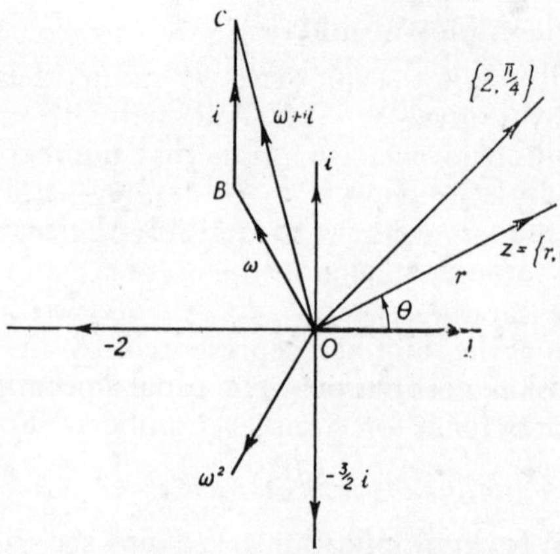

Fig. 14.9—Complex numbers represented by vectors

"magnitude", or "modulus" $|z|$ of the complex number $z = \{r, \theta\}$ Since $\{r_1, \theta_1\}\{r_2, \theta_2\} = \{r_1 r_2, \theta_1 + \theta_2\}$ we have

$$|z_1 z_2| = r_1 r_2 = |z_1|\,|z_2| \quad . \quad . \quad . \quad (14.18)$$

For a positive number $|h| = |\{h, 0\}| = h$, for a negative number $|-h| = |\{h, \pi\}| = h$, so that for real numbers this agrees with our former definition of the modulus as magnitude irrespective of sign.

The angle θ is the "phase" or "amplitude" of the complex number $z = \{r, \theta\}$. We write $\theta = \text{ph}\, z$. Unfortunately this is not unique: the number $\{r, \theta\}$ is the same as $\{r, \theta + 2\pi\}$, or $\{r, \theta + 4\pi\}$, since a complete revolution has no effect. So if θ is one value of the phase, $\theta + 2n\pi$ is another value, where n stands for any integer.

14.15 Complex addition and subtraction

So far we have not given any definition of "addition" for complex numbers. But we have a definition of addition of vectors, and we have a method of representing complex numbers by vectors. So it is natural to combine these facts to supply the required definition. Let z_1, z_2 be any two complex numbers, represented by the vectors $\mathbf{v}_1$ and $\mathbf{v}_2$. Let $\mathbf{v}_3 = \mathbf{v}_1 + \mathbf{v}_2$ be the sum of $\mathbf{v}_1$ and $\mathbf{v}_2$, and let z_3 be the corresponding complex number. Then it is natural to call z_3 the "sum of z_1 and z_2", and to write $z_3 = z_1 + z_2$. Such an addition can be performed graphically by using the triangle law of vector addition. Thus in Fig. 14.9 the vector OB represents $\omega = \{1, \frac{2}{3}\pi\}$, and the vector BC represents $i = \{1, \frac{1}{2}\pi\}$. The sum $OB + BC = OC$ therefore (by definition) represents the number $\omega + i$. By measurement or trigonometric calculation we find the length OC to be $2 \cos 15°$ and the angle $\angle IOC$ to be $\frac{7}{12}\pi$, so that

$$\omega + i = \{2 \cos 15°, \tfrac{7}{12}\pi\}.$$

This rule for addition must obey all the ordinary laws of addition, since vectors do so. For instance, since $\mathbf{v}_1 + \mathbf{v}_2 = \mathbf{v}_2 + \mathbf{v}_1$ it follows that $z_1 + z_2 = z_2 + z_1$, and since $|\mathbf{v}_1 + \mathbf{v}_2| \leqslant |\mathbf{v}_1| + |\mathbf{v}_2|$ (formula 14.11) it follows that

$$|z_1 + z_2| \leqslant |z_1| + |z_2| \qquad (14.19)$$

It also agrees with the general rule for the addition of real numbers, which correspond to vectors lying in the same straight line as $\mathbf{I}$.

We can define the difference $z_1 - z_2$ of the two complex numbers z_1 and z_2 as being the number represented by the vector $\mathbf{v}_1 - \mathbf{v}_2$. Since subtraction of vectors obeys all the usual algebraic rules, it follows that the same will be true for complex numbers. For instance, since $(\mathbf{v}_1 - \mathbf{v}_2) + \mathbf{v}_2 = \mathbf{v}_1$ we see that $(z_1 - z_2) + z_2 = z_1$.

PROBLEMS

Prove the following results:

(1) $\omega + \omega^2 = -1$.
(2) $\omega - \omega^2 = \sqrt{3} \cdot i$.
(3) For any number z, $z - z = 0$.

14.16 The distributive law

We have now defined all four arithmetical operations for complex numbers. We have shown that as far as multiplication and division are concerned the rules of manipulation are exactly the same as for ordinary real numbers. The same applies to addition and subtraction. One further property remains to be proved: the so-called "distributive law" $z(z_1 + z_2) = zz_1 + zz_2$ which involves both multiplication and addition.

But this is easy to establish. Let z_1 be represented by a vector AB, and z_2 by a vector BC: then $z_1 + z_2 = AC$ (Fig. 14.10). Now if

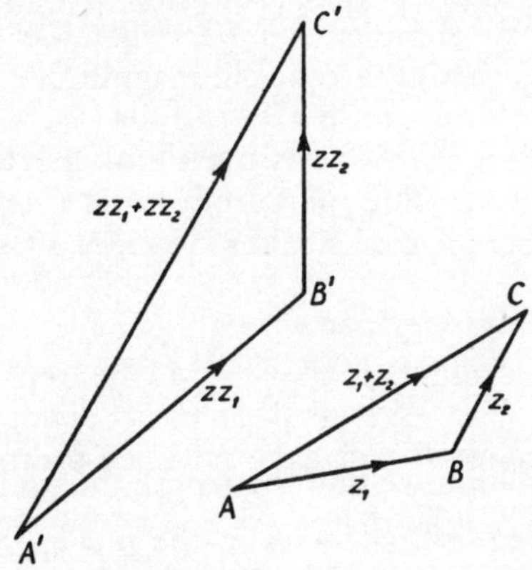

Fig. 14.10—The distributive law for complex numbers

$z = \{r, \theta\}$ then zz_1 is represented by a vector $A'B'$ obtained from AB by magnification by r and rotation through θ. The vector $B'C'$ representing zz_2 is similarly obtained from BC. Thus $A'C' = A'B' + B'C' = zz_1 + zz_2$. But all we have done is to magnify the whole triangle ABC r times, and rotate it through θ; so $A'C' = \{r, \theta\}\, AC$, i.e.

$$zz_1 + zz_2 = z(z_1 + z_2) \quad . \quad . \quad . \quad (14.20)$$

We can now prove that any ordinary algebraic formula is true for complex numbers. For example,

$$(z_1 + z_2)^2 = (z_1 + z_2)(z_1 + z_2) = (z_1 + z_2)z_1 + (z_1 + z_2)z_2 \quad \text{(by 14.20)}$$
$$= z_1(z_1 + z_2) + z_2(z_1 + z_2)$$
$$= z_1z_1 + z_1z_2 + z_2z_1 + z_2z_2$$
$$= z_1^2 + z_1z_2 + z_1z_2 + z_2^2$$
$$= z_1^2 + 2z_1z_2 + z_2^2.$$

In general we can do with complex numbers anything that can be done with real numbers, and also many other things that were previously impossible. Thus we are no longer prevented from taking the square root of a number because that number is negative. Other ways in which we have greater freedom of action will appear presently.

As a rule therefore an argument will gain in generality if we use complex numbers instead of real ones, for real numbers can be considered as a special case. Before we illustrate this further it may be as well to point out the few exceptions. If z is a complex number we can no longer assert that z^2 is positive or zero, as is the case for real numbers. Neither can we compare complex numbers. If $\boldsymbol{v}_1$ represents a force of 1 unit pointing eastwards, and $\boldsymbol{v}_2$ a unit force northwards, then $\boldsymbol{v}_1$ and $\boldsymbol{v}_2$ are certainly unequal, but it would make no sense to say that one was greater than or less than the other: relations such as "greater" or "less" apply only to real numbers. It is true that we can say of two vectors or complex numbers that the first is greater than (or equal to, or less than) the other in magnitude. But such a relation can be written $|z_1| > |z_2|$ (or $|z_1| = |z_2|$, or $|z_1| < |z_2|$) and is merely a comparison between the two positive numbers $|z_1|$ and $|z_2|$ with no new principle involved. But apart from these rules, everything which is true of real numbers is true also for complex ones.

14.17 Real and imaginary parts

Let the complex number $z = \{r, \theta\}$ be represented in a plane by the vector $OP = \boldsymbol{v}$. We shall as usual denote the unit vector $\boldsymbol{I}$ by OI. Then the point P will have polar co-ordinates $\{r, \theta\}$ relative to OI.

Now draw two co-ordinate axes in the plane, the x-axis OX along OI, and the y-axis OY at right angles to OX (in an anti-clockwise

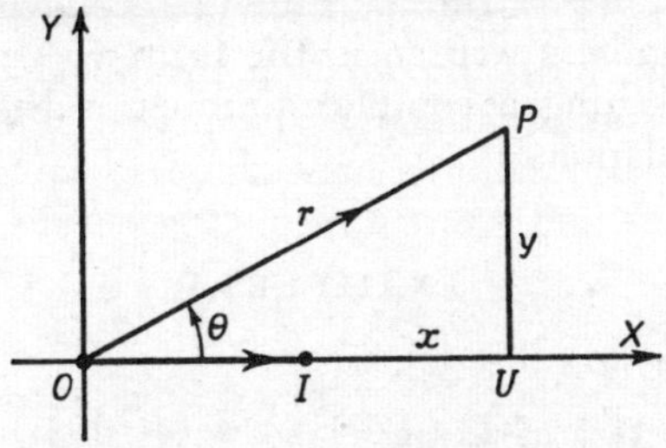

Fig. 14.11—Real and imaginary parts of a complex number
$z = \{r, \theta\} = x + yi$

direction). Then if P has cartesian co-ordinates (x, y) (Fig. 14.11), and we draw PU perpendicularly on to the x-axis, the length OU will be x, and the length UP will be y. Now vectorially $OP = OU + UP$. Here OP represents the number z. $OU = x \,.\, OI$, and therefore represents the real number x, and UP is y times a unit vector pointing in the direction of the y-axis, and so represents yi, where $i = \{1, \frac{1}{2}\pi\}$. Thus $OP = OU + UP$ is equivalent to

$$z = x + yi \quad . \quad . \quad . \quad . \quad (14.21)$$

where x and y are ordinary real numbers.

This notation "$x + yi$" gives us an alternative way of writing a complex number. It is related to the $\{r, \theta\}$ notation in the way that cartesian co-ordinates are related to polars, that is

$$x = r\cos\theta \qquad y = r\sin\theta$$
$$|z| = r = \sqrt{(x^2 + y^2)} \quad . \quad . \quad . \quad (14.22)$$

Thus

$$z = \{r, \theta\} = x + iy = r\cos\theta + ir\sin\theta$$
$$= r(\cos\theta + i\sin\theta) \quad . \quad (14.23)$$

For example, $\omega = \{1, \frac{2}{3}\pi\} = \cos\frac{2}{3}\pi + i\sin\frac{2}{3}\pi = -\frac{1}{2} + \frac{1}{2}\sqrt{3}i$.

x is called the "real part" of the complex number z ($x = \text{Rl}\, z$), and y

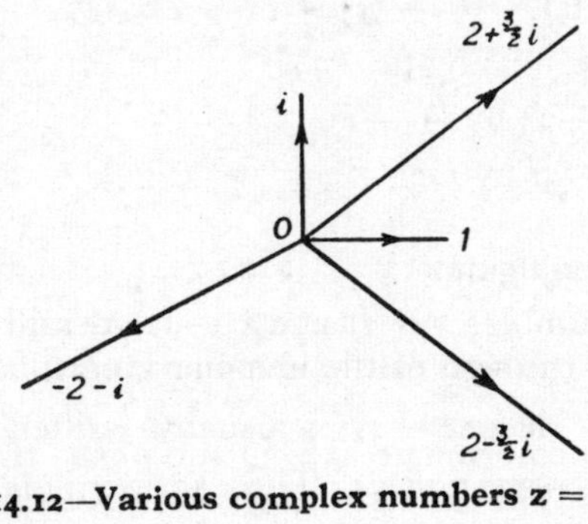

Fig. 14.12—Various complex numbers $z = x + yi$

the "imaginary part" ($y = \text{Im } z$). Fig. 14.12 shows diagrammatically several complex numbers written in the form $x + yi$.

This form of writing a complex number is the most convenient one for most calculations.

EXAMPLES

(1) Add $(3 + 4i)$ to $(2 + 5i)$.

$$(3 + 4i) + (2 + 5i) = (3 + 2) + (4i + 5i) = 5 + 9i$$

(2) Subtract $(2 + 3i)$ from $(6 + 7i)$.

$$(6 + 7i) - (2 + 3i) = (6 - 2) + (7i - 3i) = 4 + 4i$$

(3) Multiply $(1 + i)$ by i.

$$i(1 + i) = i + i^2 = i - 1$$

(4) Multiply $(2 + i)$ by $(1 + 2i)$.

$$(2 + i)(1 + 2i) = 2(1 + 2i) + i(1 + 2i)$$
$$= 2 + 4i + i + 2i^2 = 5i$$

(5) Divide $3 + 4i$ by $2 + 3i$.

Here the trick is to multiply numerator and denominator of the fraction $(3 + 4i)/(2 + 3i)$ by $(2 - 3i)$, to make the denominator real:

$$\frac{3 + 4i}{2 + 3i} = \frac{(3 + 4i)(2 - 3i)}{(2 + 3i)(2 - 3i)} = \frac{3(2 - 3i) + 4i(2 - 3i)}{(2)^2 - (3i)^2}$$
$$= \frac{6 - 9i + 8i - 12i^2}{4 + 9}$$
$$= \frac{18 - i}{13} = \frac{18}{13} - \frac{1}{13}i$$

PROBLEMS

(1) Add $(3 - 2i) + (6 + 7i) + (1 - i)$.
(2) Multiply $(1 + i)$ by $(i - 1)$.
(3) Divide $(1 + i)$ by $(1 - i)$.
(4) Divide $2 + 3i$ by i.

14.18 Complex conjugates

The complex numbers $x + iy$ and $x - iy$ are said to be "conjugate": they differ only in the sign of the imaginary part.

If $z = x + iy$, then $x - iy$ is usually written $\bar{z}$ or z^* (read as "z bar" or "z star" respectively). Thus, for example, $\overline{2 + 3i} = 2 - 3i$.

These conjugates have several important properties. Let $z_1 = x_1 + iy_1$ and $z_2 = x_2 + iy_2$ be any two complex numbers. Then $z_1 + z_2 = (x_1 + iy_1) + (x_2 + iy_2) = (x_1 + x_2) + i(y_1 + y_2)$. Now by definition $\bar{z}_1 = x_1 - iy_1$, $\bar{z}_2 = x_2 - iy_2$, so that

$$\bar{z}_1 + \bar{z}_2 = (x_1 + x_2) - i(y_1 + y_2) = \overline{z_1 + z_2} \quad \ldots \quad (14.24)$$

i.e. the conjugate of a sum is the sum of the conjugates.

The following similar results are left as exercises for the reader.

PROBLEMS

Prove that:

(1) The conjugate of $\{r, \theta\}$ is $\{r, -\theta\}$.

(2) If z is real, then $\bar{z} = z$, and conversely.

(3) $\bar{i} = -i$.

(4) $\bar{\omega} = \omega^2$.

(5) $\overline{z_1 - z_2} = \bar{z}_1 - \bar{z}_2$.

(6) $\overline{z_1 z_2} = \bar{z}_1 \bar{z}_2$.

(7) $\bar{z} + z = 2x$ (where $z = x + iy$).

(8) $z - \bar{z} = 2yi$.

(9) $z\bar{z} = |z|^2$.

(10) $\overline{z^2} = \bar{z}^2$.

(11) $\overline{1 + 2z + 3z^2} = 1 + 2\bar{z} + 3\bar{z}^2$.

(12) $\overline{z^n} = \bar{z}^n$.

(13) $\overline{1/z} = 1/\bar{z}$.

(14) Interpret $\bar{z}$ geometrically.

There is one other important property. Suppose we have an equation such as $2 + 2z + z^2 + z^3 = 0$ with real coefficients. This equation can have solutions or "roots" which are real: it can also have roots which are complex and not real: e.g. $2 + 2z + z^2 + z^3 = 0$ has the roots -2, i, and $-i$. We can prove that if z is a root then $\bar{z}$ is also a root: and if z is not real $\bar{z}$ will be different from z. We do that as follows. Let the given equation be $A + Bz + Cz^2 + Dz^3 = 0$, z being any root. (A cubic equation is chosen for the sake of illustration, but the argument remains valid for any polynomial.) It follows, on taking the complex conjugate, that $\overline{A + Bz + Cz^2 + Dz^3} = \bar{0} = 0$ also. But by the rules we have given above

$$\overline{A + Bz + Cz^2 + Dz^3} = \bar{A} + \overline{Bz} + \overline{Cz^2} + \overline{Dz^3}$$
$$= \bar{A} + \bar{B}\bar{z} + \bar{C}\overline{z^2} + \bar{D}\overline{z^3}$$
$$= \bar{A} + \bar{B}\bar{z} + \bar{C}\bar{z}^2 + \bar{D}\bar{z}^3.$$

Since A, B, C, D are real, $\bar{A} = A$, $\bar{B} = B$, $\bar{C} = C$ and $\bar{D} = D$, and therefore $A + B\bar{z} + C\bar{z}^2 + D\bar{z}^3 = 0$, i.e. $\bar{z}$ is also a root.

14.19 Complex logarithms

We now consider the problem of extending the ideas of powers, exponentials, trigonometric functions, and so on, to complex numbers.

Can we find a logarithm of a complex number? We speak of "finding" such a logarithm: strictly speaking of course it is a question of definition: we are free to define a logarithm in any way we choose. But such a definition will seem arbitrary unless the new kind of logarithm so defined has the same kind of properties as the old one, and unless it agrees in value with the old when z is real. Now the most important property of the logarithm function is that it reduces multiplication to addition, i.e. that $\log (ab) = \log a + \log b$. Is there a similar function $\log z$ of a complex number z such that

$$\text{if } z_3 = z_1 z_2 \quad \text{then} \quad \log z_3 = \log z_1 + \log z_2? \quad . \; . \; (14.25)$$

Let $z_1 = \{r_1, \theta_1\}$, $z_2 = \{r_2, \theta_2\}$, $z_3 = \{r_3, \theta_3\}$: then since $z_1 z_2 = z_3$ we know that $r_3 = r_1 r_2$, i.e.

$$\log r_3 = \log r_1 + \log r_2 \text{ (with ordinary logs)}$$

and $$\theta_3 = \theta_1 + \theta_2$$

Thus if A and B are any numbers whatever (real or complex),

$$(A \log r_3 + B\theta_3) = A(\log r_1 + \log r_2) + B(\theta_1 + \theta_2)$$
$$= (A \log r_1 + B\theta_1) + (A \log r_2 + B\theta_2)$$

Thus any function of the form $A \log r + B\theta$ (with $\log r$ according to the old definition), will satisfy the fundamental property of logarithms. However, we shall also require our new definition to be equivalent to the old definition when z is an ordinary positive number, so that $\log \{r, 0\}$ according to the new definition must equal $\log r$ according to the old. That is, $A \log r + B.0$ must equal $\log r$; clearly this will be true if we take $A = 1$. We have thus extended the idea of logarithms to complex numbers, $\log \{r, \theta\} = \log r + B\theta$.

However, there is still one further condition which it is desirable to satisfy. Consider natural logarithms. If x is small and real we know that $\ln (1 + x) = x - \frac{1}{2}x^2 + \frac{1}{3}x^3 - \ldots \simeq x$ to the first order of small quantities. Can we arrange for this to be true also for the complex logarithm, i.e. if $|z|$ is small can we arrange our definition so that $\ln (1 + z) \simeq z$, neglecting small quantities of the second order?

Let OI be the unit vector (Fig. 14.13) and let IP represent a small complex number $z = x + iy$. Draw PU perpendicularly onto OI produced: then $IU = x$ and the length $UP = y$.

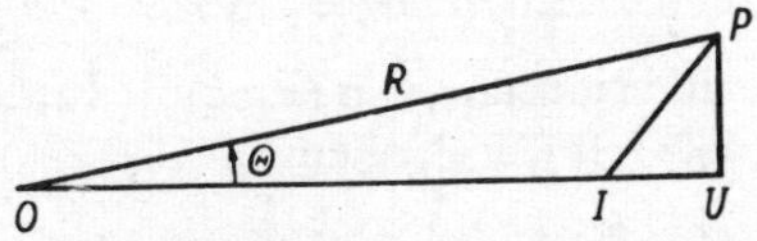

Fig. 14.13—The value of ln (1 + z) when z is small

Now the vector OI represents the number 1, and IP represents z, and vectorially $OP = OI + IP = 1 + z$. Suppose that $1 + z = \{R, \Theta\}$; then R means the length OP, and Θ the angle $\angle IOP$.

Since $z = IP$ is small we see that Θ is a small angle. So $OU = OP \cos \Theta \simeq OP = R$ to the first order of small quantities (see Section 6.12). Therefore

$$R \simeq OU = OI + IU = 1 + x.$$

Also $y = PU = R \sin \Theta$

$\simeq R\Theta$ to the first order, since Θ is small

$\simeq (1 + x)\Theta$

$\simeq \Theta$, neglecting the product $x\Theta$ which is of the second order of smallness.

Thus $1 + z = \{R, \Theta\} = \{1 + x, y\}$.

According to our new definition of logarithms (suitable for complex numbers) we must take $\ln \{R, \Theta\}$ to mean $\ln R + B\Theta$; that is

$$\begin{aligned} \ln (1 + z) &\simeq \ln (1 + x) + By \\ &\simeq x + By \quad \text{since } x \text{ is small.} \end{aligned}$$

But we wish to arrange our new definition so that $\ln (1 + z)$ is approximately equal to $z = x + iy$ when z is small, and this will be so if we take B to be i.

Thus it seems appropriate and reasonable to define the natural logarithm of the complex number $z = \{r, \theta\}$ to be $\ln r + i\theta$, where $\ln r$ stands for the usual real natural logarithm of the positive number r. However, it will be convenient to make a distinction in the next few sections between the definition for real numbers, as given in Chapter 6, and the new definition for complex numbers. For the time being, therefore, the complex natural logarithm will be denoted by the symbol $\text{Ln}\, z$, with a capital letter, and is accordingly defined as

$$\text{Ln}\, z = \text{Ln}\, \{r, \theta\} = \ln r + i\theta \quad . \quad . \quad (14.26)$$

(This convention concerning the use of the capital is not generally adopted, but seems helpful here.)

The following properties accordingly follow from this definition:

(i) It agrees with the usual definition for real logarithms. For any positive number r can be considered also as a complex number $\{r, 0\}$; and from (14.26)

$$\text{Ln}\ \{r, 0\} = \ln r$$

(ii) It obeys the addition law, $\text{Ln}\,(z_1 z_2) = \text{Ln}\ z_1 + \text{Ln}\ z_2$. For if $z_1 = \{r_1, \theta_1\}$, and $z_2 = \{r_2, \theta_2\}$, then

$$\begin{aligned} z_1 z_2 &= \{r_1 r_2, \theta_1 + \theta_2\}. \text{ So} \\ \text{Ln}\ z_1 &= \ln r_1 + i\theta_1, \\ \text{Ln}\ z_2 &= \ln r_2 + i\theta_2, \text{ and} \\ \text{Ln}\ (z_1 z_2) &= \ln (r_1 r_2) + i\,(\theta_1 + \theta_2) \\ &= \ln r_1 + \ln r_2 + i\theta_1 + i\theta_2 = \text{Ln}\ z_1 + \text{Ln}\ z_2. \end{aligned}$$

(iii) If z is small, $\text{Ln}\ (1 + z) \simeq z$, as shown above.

There is, however, one new feature about the complex logarithm which is not shared by the real logarithm: it is many-valued. For the angle θ associated with the complex number z is not uniquely determined; we could equally well take it to be $\theta + 2\pi$, or $\theta + 4\pi$, or $\theta - 2\pi$. For 2π radians $=$ 1 complete turn, and we can add any number of complete turns to an angle without altering the final result. It follows from (14.26) that $\ln r + i(\theta + 2\pi)$, $\ln r + i(\theta + 4\pi)$ and in general $\ln r + i(\theta + 2n\pi)$ are all possible values of $\text{Ln}\ z$. Thus $\text{Ln}\ z$ is a function with an infinite number of different values, any two of which differ by a multiple of $2\pi i$. It may be compared in that respect with the real function $\tan^{-1}x$, which also has an infinite number of values. For if $\theta = \tan^{-1}x$, i.e. $x = \tan\,\theta$, then x is also equal to $\tan\,(\theta + \pi)$, $\tan\,(\theta + 2\pi)$, etc., so that $\theta + \pi$, $\theta + 2\pi, \ldots$ and in general $\theta + n\pi$ are all possible values of $\tan^{-1}x$.

It follows that if we wish to be strictly accurate we should modify the statement of the properties (i), (ii) and (iii) above as follows:

(i)′ $\ln r$ is one possible value of $\text{Ln}\ r = \text{Ln}\ \{r, 0\}$.

(ii)′ Whatever z_1 and z_2 may be (other than 0), $\text{Ln}\,z_1 + \text{Ln}\,z_2$ is one value of $\text{Ln}\,(z_1 z_2)$.

(iii)′ If z is small, there is a value of $\text{Ln}\ (1 + z)$ approximately equal to z.

In Chapter 6 we defined and studied the logarithms of *positive* numbers, but did not find it possible to give any definition of a logarithm of a negative number. The new complex definition provides every number other than zero with a logarithm. For example, $-1 = \{1, \pi\}$ and therefore one logarithm of -1 is $\ln 1 + \pi i = \pi i$. Other possible values are $3\pi i$, $5\pi i$, and in general $(2n + 1)\pi i$.

Note—When logarithms are being used for numerical calculations involving negative numbers it is best to use the special device explained in Section 6.8.

PROBLEMS

(1) Show that one value of Ln i is $\frac{1}{2}\pi i$.

(2) Show that one value of Ln ω is $\frac{2}{3}\pi i$.

(3) Find Ln $(1 + i)$.

(4) Find Ln $(-e)$.

(5) Show that $\overline{\text{Ln } z} = \text{Ln } \bar{z}$.

14.20 Complex exponentials

For real numbers the function $X = \exp x$ was defined as the natural antilogarithm; i.e. the relations $X = \exp x$ and $x = \ln X$ mean the same. The same definition can be applied to complex numbers: if $z = \text{Ln } Z$, then we shall write $Z = \text{Exp } z$, using for the moment a capital letter to distinguish this from the real exponential function. To find what this definition means, let us write z in the form $x + iy$, and Z in the form $\{R, \Theta\}$. Then the relation $z = \text{Ln } Z$ becomes

$$x + iy = \ln R + i\Theta$$

i.e. $x = \ln R$, $y = \Theta$, or $R = \exp x$, $\Theta = y$, and by (14.23)

$$\begin{aligned} \text{Exp}\, z &= \{R, \Theta\} \\ &= \{\exp x, y\} \\ &= \exp x (\cos y + i \sin y) \qquad . \quad . \quad (14.27) \end{aligned}$$

Given any complex number $z = x + iy$ this formula gives us a convenient way of calculating $Z = \text{Exp } z$.

PROBLEMS

Find the values of (1) Exp i, (2) Exp πi, (3) Exp $(1 + \pi i)$.

The alternative form of this relation, $\text{Exp } (x + iy) = \{\exp x, y\}$ provides a simple geometric construction for the exponential or natural logarithm of a complex number.

Let OI be the unit vector. Draw through I an equiangular spiral with centre O and angle $\phi = \frac{1}{4}\pi = 45°$ (Fig. 14.14). As we showed in Chapter 6, this spiral is the one defining the ordinary exponential function, in the sense that if we draw a line OP' making an angle x (radians) with OI and meeting the spiral at P', then the length OP' is $\exp x$. Now rotate OP' about O, keeping its length constant, until it becomes OP where $\angle IOP = y$. Then the vector OP represents the complex number $\{\exp x, y\}$, i.e. the required exponential.

Conversely, suppose that a complex number Z is given, and is represented by the vector OP in terms of the unit vector OI. To find

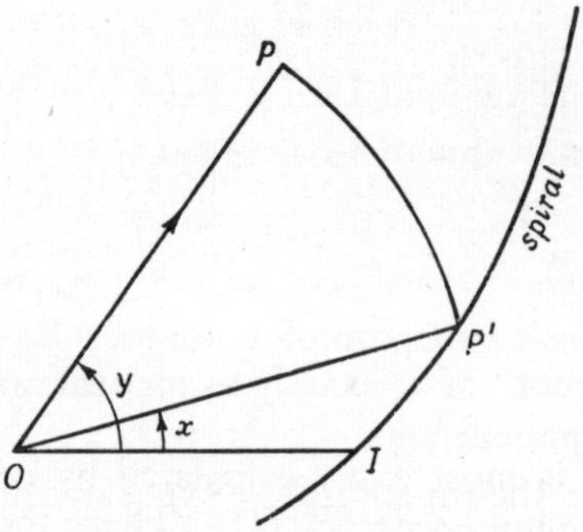

Fig. 14.14—The exponential of a complex number

the natural logarithm of Z, draw a circle with centre O through P to meet the spiral at P'. Then

$$\text{Ln}\, Z = \angle IOP' + i\, \angle IOP \qquad (14.28)$$

provided that the angles are measured in radians.

The relation $\text{Exp}\ z = \exp x\ (\cos y + i \sin y) = \{\exp x, y\}$ shows that the complex exponential is a single-valued function: given the value of z, that of $\text{Exp}\ z$ is completely determined. Furthermore if z is real (so that $z = x$, $y = 0$) we have $\text{Exp}\ z = \{\exp x, 0\} = \exp x = \exp z$. Thus the complex and real exponential functions agree exactly in value whenever z is real, which is of course the only case in which the real function is defined. So there is really no point in distinguishing between the two functions, which are always equal when both are applicable, and we shall henceforth use the same symbol $\exp z$ to mean either function, as the context allows. Furthermore the definition of $Z = \exp z$ as meaning the same as the relation $z = \text{Ln}\ Z$ shows that

$$Z = \exp z = \exp \text{Ln}\ Z \qquad (14.29)$$

for all Z except $Z = 0$, and irrespective of what particular value we choose for $\text{Ln}\ Z$.

Query—Is it equally true that $z = \text{Ln} \exp z$ for all values of z?

We also find from the definition (14.27) that

$$\begin{aligned} \exp [z_1 + z_2] &= \exp [(x_1 + iy_1) + (x_2 + iy_2)] \\ &= \exp [(x_1 + x_2) + i(y_1 + y_2)] \\ &= \{\exp (x_1 + x_2), (y_1 + y_2)\} \\ &= \{(\exp x_1 \exp x_2), (y_1 + y_2)\} \quad \text{(since } x_1, x_2 \text{ are real)} \\ &= \{\exp x_1, y_1\} \{\exp x_2, y_2\} \quad \text{(by 14.14)} \\ &= \exp z_1 . \exp z_2 \qquad (14.30) \end{aligned}$$

whatever the values of z_1 and z_2 may be.

In the same way it can be shown that

$$\exp [z_1 - z_2] = \exp z_1/\exp z_2 \qquad (14.31)$$

Thus the complex exponential obeys the same laws as the real one.

14.21 Complex roots

If z is the square of the complex number Z, then it is natural to call Z the "square root" of z, exactly as for real numbers, and to write $Z = \sqrt{z}$.

The value of Z is most easily calculated by using the polar form, writing $z = \{r, \theta\}$ and $Z = \{R, \Theta\}$. Then the relation $Z = \sqrt{z}$ means by definition the same as $z = Z^2 = Z \times Z$, that is

$$\begin{aligned}\{r, \theta\} &= \{R, \Theta\}\,\{R, \Theta\} \\ &= \{RR, \Theta + \Theta\} \\ &= \{R^2, 2\Theta\}.\end{aligned}$$

This implies that $r = R^2$, i.e. $R = \sqrt{r}$, the ordinary positive square root of the positive number r. It does not, however, necessarily mean (as one might at first be tempted to suppose) that $\theta = 2\Theta$, but merely that θ and 2Θ must differ by an integral number of complete turns, $2\pi n$, so that

$$2\Theta = \theta + 2\pi n, \quad \text{or} \quad \Theta = \tfrac{1}{2}\theta + \pi n$$

Thus the possible values of Θ are $\frac{1}{2}\theta$, $\frac{1}{2}\theta \pm \pi$, $\frac{1}{2}\theta \pm 2\pi$, . . . etc., and the possible values of $Z = \sqrt{z} = \{R, \Theta\}$ are

$$\{\sqrt{r}, \tfrac{1}{2}\theta\}, \{\sqrt{r}, \tfrac{1}{2}\theta \pm \pi\}, \{\sqrt{r}, \tfrac{1}{2}\theta \pm 2\pi\}, \ldots \text{etc.}$$

But $\{\sqrt{r}, \frac{1}{2}\theta\}$, $\{\sqrt{r}, \frac{1}{2}\theta \pm 2\pi\}$, $\{\sqrt{r}, \frac{1}{2}\theta \pm 4\pi\}$, . . . etc., are different ways of writing the same complex number; and the same holds for $\{\sqrt{r}, \frac{1}{2}\theta \pm \pi\}$, $\{\sqrt{r}, \frac{1}{2}\theta \pm 3\pi\}$, . . . etc.

So in fact a complex number has just two square roots:

$$\sqrt{z} = \{\sqrt{r}, \tfrac{1}{2}\theta\} \text{ or } \{\sqrt{r}, \tfrac{1}{2}\theta + \pi\}.$$

The second root $\{\sqrt{r}, \frac{1}{2}\theta + \pi\}$ is simply the first rotated through π, i.e. reversed in direction, or multiplied by -1. Thus the two roots can alternatively be written as $\sqrt{z} = \pm \{\sqrt{r}, \frac{1}{2}\theta\}$, and are calculated by the simple rule "take the square root of the modulus r, and halve the polar angle θ". But unfortunately it is no longer possible to distinguish one of them as positive and the other as negative, as in the real case, since in general they will be complex numbers.

These square roots have similar properties to those of ordinary real square roots. For example, we can show that the usual process of halving the logarithm can be used for calculating a square root; that is to say, the number Z whose natural logarithm is $\frac{1}{2}$ Ln z will be a square root of Z^2. For this number Z is by definition $\exp(\frac{1}{2}\,\mathrm{Ln}\,z)$, and therefore

$$\begin{aligned} Z^2 &= \exp(\tfrac{1}{2}\,\mathrm{Ln}\,z) \times \exp(\tfrac{1}{2}\,\mathrm{Ln}\,z) \\ &= \exp(\tfrac{1}{2}\,\mathrm{Ln}\,z + \tfrac{1}{2}\,\mathrm{Ln}\,z) \quad \text{(by 14.30)} \\ &= \exp \mathrm{Ln}\,z \\ &= z \quad \text{(by 14.29)}. \end{aligned}$$

PROBLEMS

(1) Show that $\sqrt{(z^3)} = \pm(\sqrt{z})^3$.

(2) Show that $\sqrt{(z_1 z_2)} = \pm\sqrt{z_1} \cdot \sqrt{z_2}$.

(3) Show that $\sqrt{(z_1/z_2)} = \pm\sqrt{z_1}/\sqrt{z_2}$.

(4) What number has only one square root?

(5) What are the values of $\sqrt{(2i)}$?

(6) How many cube roots does a complex number $z = \{r, \theta\}$ have, and what are they?

(7) Show that $\sqrt{(\cos\theta + i\sin\theta)} = \pm(\cos\frac{1}{2}\theta + i\sin\frac{1}{2}\theta)$.

14.22 Complex powers

For any real number x we know that $x^2 = \exp(2\ln x)$, $x^3 = \exp(3\ln x)$, $x^4 = \exp(4\ln x)$, and in general $x^n = \exp(n\ln x)$, whenever n is a positive integer. For these relations are by definition equivalent to $\ln(x^2) = 2\ln x$, $\ln(x^3) = 3\ln x, \ldots \ln(x^n) = n\ln x$, that is, to the rule that the nth power of x is that number whose logarithm is n times the logarithm of x (see Section 6.5).

It is not difficult to show that a similar relation holds for complex numbers, using the complex logarithm and exponential functions:

$$z^n = \exp(n\,\mathrm{Ln}\,z) \quad . \quad . \quad . \quad (14.32)$$

whenever n is a positive integer. For starting with equation (14.29), we apply equation (14.30) repeatedly.

$$\begin{aligned} z^2 = z \times z &= \exp(\mathrm{Ln}\,z)\exp(\mathrm{Ln}\,z) \\ &= \exp(\mathrm{Ln}\,z + \mathrm{Ln}\,z) \\ &= \exp(2\,\mathrm{Ln}\,z) \\ z^3 = z \times z^2 &= \exp(\mathrm{Ln}\,z)\exp(2\,\mathrm{Ln}\,z) \\ &= \exp(\mathrm{Ln}\,z + 2\,\mathrm{Ln}\,z) \\ &= \exp(3\,\mathrm{Ln}\,z) \\ z^4 = z \times z^3 &= \exp(\mathrm{Ln}\,z)\exp(3\,\mathrm{Ln}\,z) \\ &= \exp(\mathrm{Ln}\,z + 3\,\mathrm{Ln}\,z) \\ &= \exp(4\,\mathrm{Ln}\,z) \end{aligned}$$

and so on up to any desired power.

From the definition of the exponential function as the natural antilogarithm it follows that equation (14.32) can equally well be stated in the form "If Z and z are such that (some value of) Ln Z is n times (some value of) Ln z, then $Z = z^n$". Thus powers of complex numbers can be found by taking logarithms in the usual way.

Furthermore, since exp 0 = 1, we have by (14.31)

$$\begin{aligned}\exp(-n \operatorname{Ln} z) &= \exp(0 - n \operatorname{Ln} z)\\ &= (\exp 0)/(\exp n \operatorname{Ln} z)\\ &= 1/z^n\end{aligned}$$

So, just as with real numbers, it is natural and convenient to write $1/z^n$ as z^{-n}. All the usual rules of indices hold, such as that $z^m z^n = z^{m+n}$, $z^m/z^n = z^{m-n}$, and so on. Since we have also shown in the previous section that $\exp(\frac{1}{2} \operatorname{Ln} z)$ is a square root of z, it is natural to write $\sqrt{z}$ as $z^{\frac{1}{2}}$, and also $\sqrt{z^3}$ as $z^{\frac{3}{2}}$, $1/\sqrt{z}$ as $z^{-\frac{1}{2}}$, and so on, exactly as for real numbers. Again all the usual laws will hold, with proofs exactly similar to those of Section 6.10; except that in the complex case there will be some caution necessary since $\sqrt{z}$ is now a two-valued function. Thus we may have $z^{\frac{3}{2}} = z \,.\, z^{\frac{1}{2}}$ or $z^{\frac{3}{2}} = -z \,.\, z^{\frac{1}{2}}$, according to the values chosen for the square root.

This gives a definition of z^n whenever n is an integer or a simple fraction such as $\frac{1}{2}$ or $-\frac{3}{2}$. It is tempting to generalize this definition to complex powers, and to write $\exp(u \operatorname{Ln} z)$ as z^u for any complex values of z and u. This is indeed the standard definition of z^u. But this notation is rarely used in practice because in general it has an infinite number of different values corresponding to the different values of Ln z, and they are difficult to keep under control.

It is, however, customary and convenient to use the symbol k^z, where k is a positive number, for the quantity

$$k^z = \exp(z \ln k) \qquad . \quad . \quad . \quad (14.33)$$

with the convention that we choose the ordinary real natural logarithm for ln k. (This is only possible if k is positive.) With this convention k^z becomes a *single-valued* function of z and k, and all the usual laws of indices hold without any restriction. In particular if we take k to be e we have

$$e^z = \exp(z \ln e) = \exp z$$

So that the function exp z can also be written as e^z, exactly as for the real function. The equations (14.29), (14.30), (14.31), and (14.27) can then be written in the form

$$\left.\begin{aligned} e^{\operatorname{Ln} Z} &= Z\\ e^{z_1+z_2} &= e^{z_1} \,.\, e^{z_2}\\ e^{z_1-z_2} &= e^{z_1}/e^{z_2}\\ e^{x+iy} &= \{e^x, y\} = e^x(\cos y + i \sin y)\end{aligned}\right\} \quad . \quad (14.34)$$

From these equations we can deduce that

$$\left.\begin{aligned} e^{2\pi i} &= e^{4\pi i} = e^{6\pi i} = \ldots = 1 \\ e^{\pi i} &= e^{3\pi i} = e^{5\pi i} = \ldots = -1 \\ e^{\frac{1}{2}\pi i} &= \{e^0, \tfrac{1}{2}\pi\} = \{1, \tfrac{1}{2}\pi\} = i \\ e^{z+\pi i} &= e^z \cdot e^{\pi i} = -e^z \\ e^{z+2\pi i} &= e^z \cdot e^{2\pi i} = e^z \end{aligned}\right\} \quad . \quad . \quad (14.35)$$

The watchful reader will notice that the definition (14.33) of k^z, where k is positive, agrees exactly with the old definition (6.12) when z is real. But it does not necessarily agree with the definition (14.32) when n is fractional. Thus $2^{\frac{1}{2}}$ would mean the positive square root of 2 if interpreted according to definition (14.33), but according to (14.32) it could mean either $\sqrt{2}$ or $-\sqrt{2}$. This is an unfortunate ambiguity, but on any occasion in which such a symbol arises the context will usually make the meaning clear.

14.23 Complex trigonometric functions

In Section 6.14 the so-called "hyperbolic functions" cosh x and sinh x were defined as $\frac{1}{2}(e^x + e^{-x})$ and $\frac{1}{2}(e^x - e^{-x})$ respectively. This definition can readily be extended to any number z, writing

$$\left.\begin{aligned} \cosh z &= \tfrac{1}{2}(e^z + e^{-z}) \\ \sinh z &= \tfrac{1}{2}(e^z - e^{-z}) \end{aligned}\right\} \quad . \quad . \quad (14.36)$$

PROBLEM

(1) Express cosh $(x + iy)$ and sinh $(x + iy)$ in terms of x and y.

Now if we put $x = 0$ in (14.34) we obtain the remarkable formula

$$e^{iy} = \cos y + i \sin y.$$

Replacement of y by $-y$ in this equation gives us

$$\begin{aligned} e^{-iy} &= \cos(-y) + i\sin(-y) \\ &= \cos y - i \sin y \end{aligned}$$

whence

$$\begin{aligned} e^{iy} + e^{-iy} &= 2\cos y \\ e^{iy} - e^{-iy} &= 2i \sin y \end{aligned}$$

i.e. $\cos y = (e^{iy} + e^{-iy})/2, \quad \sin y = (e^{iy} - e^{-iy})/2i$

FURTHER PROBLEM

(2) Using the construction for exp $z = e^z$ given in Section 14.20 show that these results are true.

These equations hold for all real values of y. It is tempting to generalize them to complex numbers, defining

$$\cos z = (e^{iz} + e^{-iz})/2, \quad \sin z = (e^{iz} - e^{-iz})/2i \quad . \quad (14.37)$$

for all values of z. Since e^{iz} and e^{-iz} are one-valued functions of z, it follows that cos z and sin z are also one-valued functions; and these definitions will agree with the usual ones when z is real. They can no longer be directly interpreted as trigonometric ratios: but we can use the geometric construction for an exponential given in Section (14.20) to find the values of sin z and cos z for any value of z. Another method is to use equation (14.34):

$$e^{iz} = e^{i(x+iy)} = e^{(-y+ix)} = e^{-y}(\cos x + i \sin x)$$
$$e^{-iz} = e^{-i(x+iy)} = e^{(y-ix)} = e^{y}(\cos x - i \sin x)$$

whence

$$\begin{aligned}\cos z = \cos(x+iy) &= \tfrac{1}{2}(e^{iz} + e^{-iz}) \\ &= \tfrac{1}{2}(e^{-y} + e^{y}) \cos x + \tfrac{1}{2}i(e^{-y} - e^{y}) \sin x \\ &= (\cosh y \,.\, \cos x) - i(\sinh y \,.\, \sin x)\end{aligned}$$

and similarly

$$\sin z = (e^{iz} - e^{-iz})/2i = (\cosh y \,.\, \sin x) + i(\sinh y \,.\, \cos x)$$

and if we know the numerical values of x and y we can use these equations to calculate $\cos(x + iy)$ and $\sin(x + iy)$.

Notice that equation (14.37) can be written as $\cos z = \cosh iz$; $\sin z = (\sinh iz)/i$ which shows how the trigonometric and hyperbolic functions are related. This explains the remarkable similarity of their properties, which seems very mysterious when we consider real numbers only.

The remaining trigonometric functions can be defined as follows:

$$\tan z = (\sin z)/(\cos z) \qquad \sec z = 1/(\cos z)$$
$$\cot z = (\cos z)/(\sin z) \qquad \operatorname{cosec} z = 1/(\sin z)$$

exactly as in the real case, and similarly for the hyperbolic functions:

$$\tanh z = (\sinh z)/(\cosh z), \text{ etc.}$$

PROBLEMS

(3) Find $\cos(1 + i)$, $\sin(\frac{1}{2} + \frac{1}{2}\pi i)$.

(4) Show that $\tan z = \dfrac{e^{iz} - e^{-iz}}{i(e^{iz} + e^{-iz})}$, and give similar formulas for cot z, sec z and cosec z.

(5) Show that $\tan z = (\tanh iz)/i$, and give similar formulas for the remaining functions.

Now the definitions $\cos z = (e^{iz} + e^{-iz})/2$ and $\sin z = (e^{iz} - e^{-iz})/2i$ can be used to deduce the properties of the general cosine and sine functions, and any properties so deduced will automatically hold for the ordinary real cosine and sine functions. That is the chief practical importance of these definitions. We scarcely ever need them to calculate cos z or sin z for complex values of z; but we can use them to deduce the properties of cos z and sin z, and they are often easier to manipulate than the geometrical constructions of Chapter 5.

For instance, we have immediately from the definitions (remembering that $e^{2\pi i} = e^{-2\pi i} = 1$ and $e^{\pi i} = e^{-\pi i} = -1$).

$$\begin{aligned}
\cos 0 &= (e^0 + e^0)/2 = (1 + 1)/2 = 1 \\
\sin 0 &= (e^0 - e^0)/2i = 0 \\
\cos \pi &= (e^{\pi i} + e^{-\pi i})/2 = (-1 - 1)/2 = -1 \\
\sin \pi &= (e^{\pi i} - e^{-\pi i})/2i = (-1 - [-1])/2 = 0 \\
\cos 2\pi &= (e^{2\pi i} + e^{-2\pi i})/2 = (1 + 1)/2 = 1 \\
\cos (z + 2\pi) &= (e^{zi+2\pi i} + e^{-zi-2\pi i})/2 \\
&= (e^{zi}e^{2\pi i} + e^{-zi}e^{-2\pi i})/2 \\
&= (e^{zi} + e^{-zi})/2 = \cos z \\
\cos (z + \pi) &= (e^{zi+\pi i} + e^{-zi-\pi i})/2 \\
&= (e^{zi}e^{\pi i} + e^{-zi}e^{-\pi i})/2 \\
&= (-e^{zi} - e^{-zi})/2 = -\cos z
\end{aligned}$$

$$\begin{aligned}
(\cos z)^2 + (\sin z)^2 &= \left(\frac{e^{iz} + e^{-iz}}{2}\right)^2 + \left(\frac{e^{iz} - e^{-iz}}{2i}\right)^2 \\
&= \frac{(e^{iz})^2 + 2e^{iz}e^{-iz} + (e^{-iz})^2}{4} - \frac{(e^{iz})^2 - 2e^{iz}e^{-iz} + (e^{-iz})^2}{4} \\
&= 4e^{iz}e^{-iz}/4 = 4e^{iz-iz}/4 = 1
\end{aligned}$$

$$\begin{aligned}
(\cos z)^2 &= \frac{(e^{iz})^2 + 2e^{iz}e^{-iz} + (e^{-iz})^2}{4} \\
&= \tfrac{1}{4}(e^{2iz} + 2 + e^{-2iz}) \\
&= \tfrac{1}{4}(2 \cos 2z + 2) = \tfrac{1}{2}(\cos 2z + 1)
\end{aligned}$$

We also have directly from the definitions

$$\begin{aligned}
\cos z + i \sin z &= e^{iz} \\
\cos z - i \sin z &= e^{-iz} \qquad . \quad . \quad . \quad (14.38)
\end{aligned}$$

whence

$$e^{i(z_1+z_2)} = e^{iz_1} e^{iz_2} = (\cos z_1 + i \sin z_1)(\cos z_2 + i \sin z_2)$$

$$= (\cos z_1 . \cos z_2 - \sin z_1 . \sin z_2) + i (\cos z_1 . \sin z_2 + \sin z_1 . \cos z_2)$$

and similarly

$$e^{-i(z_1+z_2)} = e^{-iz_1} e^{-iz_2}$$

$$= (\cos z_1 . \cos z_2 - \sin z_1 . \sin z_2) - i (\cos z_1 . \sin z_2 + \sin z_1 . \cos z_2)$$

whence

$$\begin{aligned}
\cos (z_1 + z_2) &= [e^{i(z_1+z_2)} + e^{-i(z_1+z_2)}]/2 \\
&= \cos z_1 . \cos z_2 - \sin z_1 . \sin z_2 \\
\sin (z_1 + z_2) &= [e^{i(z_1+z_2)} - e^{-i(z_1+z_2)}]/2i \\
&= \cos z_1 . \sin z_2 + \sin z_1 . \cos z_2.
\end{aligned}$$

FURTHER PROBLEMS

Using the definitions (14.37) prove the following relations:

(6) $\cos(-z) = \cos z$.

(7) $\tan(-z) = -\tan z$.

(8) $\cos(\frac{1}{2}\pi - z) = \sin z$.

(9) $\sin(z + \frac{1}{2}\pi) = \cos z$.

(10) $\cos(z_1 - z_2) = \cos z_1 \,.\, \cos z_2 + \sin z_1 \,.\, \sin z_2$.

(11) $\sin \frac{1}{2}\pi = 1$.

14.24 Complex limits

Suppose that $\boldsymbol{v} = OP$ is a variable vector, and $\boldsymbol{V} = OQ$ a fixed one. Then it is natural to say that "$\boldsymbol{v}$ tends in the limit to $\boldsymbol{V}$" if the point P tends to Q, i.e. if the distance PQ becomes smaller and smaller and tends to zero. Now the vector $PQ = OQ - OP = \boldsymbol{V} - \boldsymbol{v}$, so that the distance PQ is by definition $|\boldsymbol{V} - \boldsymbol{v}|$; to say that "$\boldsymbol{v}$ tends to $\boldsymbol{V}$" or that "$\boldsymbol{v} \to \boldsymbol{V}$" will therefore mean by definition that $|\boldsymbol{V} - \boldsymbol{v}| \to 0$. Now $|\boldsymbol{V} - \boldsymbol{v}|$ is not a vector but an ordinary positive number: so this definition reduces the idea of the limit of a vector to that of the limit of an ordinary number. There is no need to introduce any really new idea.

Similarly a complex number z will be said to tend to a limit Z if $|Z - z| \to 0$.

It is now a fairly simple matter to show that these limits obey the laws we should expect. As an example we shall prove that if $z_1 \to$ a limit Z_1, and $z_2 \to$ a limit Z_2, then $z_1 + z_2 \to Z_1 + Z_2$. For

$$|(Z_1 + Z_2) - (z_1 + z_2)| = |(Z_1 - z_1) + (Z_2 - z_2)|$$
$$\leqslant |Z_1 - z_1| + |Z_2 - z_2| \quad \text{[by (14.19)]}$$

But $|Z_1 - z_1| \to 0$ and $|Z_2 - z_2| \to 0$ by hypothesis, and therefore $|(Z_1 + Z_2) - (z_1 + z_2)|$, which is less than their sum, must also tend to 0, i.e. $z_1 + z_2$ tends to $Z_1 + Z_2$. Similarly (with a little more trouble) it can be shown, for instance, that $z_1 z_2 \to Z_1 Z_2$.

14.25 Differentiation of complex functions

Let w be a function of a complex number z; e.g. $w = z^2$, or $w = \sin z$.

Then if w_1 is the value of w when $z = z_1$, and w_2 the value when $z = z_2$, we shall call $\delta w = w_2 - w_1$ the "change in w", $\delta z = z_2 - z_1$ the "change in z", and the ratio $\delta w/\delta z$ the "average rate of change of w". If this ratio tends to a definite limit L as $\delta z \to 0$ we shall say that "w is a differentiable function of z" and call L the "derivative". "derivate" or "differential coefficient" of w with respect to z. We write L as dw/dz, or $D_z w$, or Dw, or w_z, or w', exactly as for real variables,

EXAMPLES

(1) $w = z$. In this case $w_1 = z_1$, $w_2 = z_2$, $\delta w = w_2 - w_1 = z_2 - z_1 = \delta z$, and therefore $\delta w/\delta z = 1$. But as $\delta z \to 0$ the number 1 tends to the limit 1, since it is a constant anyway. Therefore $D_z z = 1$.

(2) $w = e^z = \exp z$. In this case $w_1 = e^{z_1}$, $w_2 = e^{z_2} = e^{z_1+\delta z}$, so that

$$\begin{aligned}\delta w = w_2 - w_1 &= e^{z_1+\delta z} - e^{z_1} \\ &= e^{z_1}e^{\delta z} - e^{z_1} \\ &= e^{z_1}(e^{\delta z} - 1)\end{aligned}$$

But when δz is small, $e^{\delta z} \simeq 1 + \delta z$ (equation 14.29), so that $\delta w \simeq e^{z_1}\delta z$. Thus $\delta w/\delta z \simeq e^{z_1}$, and in the limit we have $D_z w = D_z e^z = e^{z_1} = e^z$ (where the suffix $_1$ is omitted as no longer necessary when z_1 and z_2 become indistinguishable).

(3) $w = \text{Ln } z$. Here $w_1 = \text{Ln } z_1$, $w_2 = \text{Ln } z_2 = \text{Ln } (z_1 + \delta z) = \text{Ln } [z_1(1 + \delta z/z_1)] = \text{Ln } z_1 + \text{Ln } (1 + \delta z/z_1)$, so that

$$\delta w = w_2 - w_1 = \text{Ln } (1 + \delta z/z_1).$$

Now when δz is small, $\delta z/z_1$ is small (provided that $z_1 \neq 0$), and by the definition of the complex logarithm we know that one value of $\text{Ln } (1 + \delta z/z_1) \simeq \delta z/z_1$. (Here we should be a bit careful, since the logarithm has many values differing by multiples of $2\pi i$. But it goes almost without saying that in differentiating $w = \text{Ln } z$ we shall take for $w_2 = \text{Ln } z_2$ that value of the logarithm which differs only slightly from $w_1 = \text{Ln } z_1$, so that the difference $\delta w = \text{Ln } (1 + \delta z/z_1)$ will be a small quantity without any multiples of $2\pi i$ added on. It must then be true that $\text{Ln } (1 + \delta z/z_1) \simeq \delta z/z_1$). We have therefore $\delta w \simeq \delta z/z_1$, or $\delta w/\delta z \simeq 1/z_1$. The smaller δz is, the more accurate this will be, and in the limit as $\delta z \to 0$ we shall have $D_z \text{ Ln } z = 1/z$.

Now all the usual rules for differentiation will hold, since the proofs already given for real variables can be translated almost word for word to apply to complex variables. So, for example, by the product rule,

$$D_z z^2 = D_z(zz) = (D_z z)\, z + z\,(D_z z) = z + z = 2z$$

By using the rules for products, sums, and functions of functions, we see that

$$\begin{aligned}D_z \cos z &= D_z[\tfrac{1}{2}(e^{zi} + e^{-zi})] \\ &= \tfrac{1}{2}D_z e^{zi} + \tfrac{1}{2}D_z e^{-zi} \\ &= \tfrac{1}{2}ie^{zi} - \tfrac{1}{2}ie^{-zi} \\ &= -(e^{zi} - e^{-zi})/2i = -\sin z\end{aligned}$$

and similarly $D_z \sin z = \cos z$, just as in the real case. Any other function which can be built up by addition, subtraction, multiplication, division, exponentials, logarithms, sines and cosines can be differentiated in the usual way: and that means almost all functions which occur in practice. We can go on to second, third, and further derivatives. We cannot however speak of "maxima" and "minima" of complex functions, since as a rule we cannot say of two complex numbers that one is greater than or less than the other.

It is also possible to consider series of complex numbers, and the definitions of "convergence", "geometric convergence" and the tests for convergence will be the same as for real series. In particular we can have a Taylor series

$$w = w_1 + [w_z]_1 (z - z_1) + [w_{zz}]_1 (z - z_1)^2/|\underline{2} + \ldots$$

where the suffix ($_1$) means that we have to put $z = z_1$ after differentiation: and as a special case if $z_1 = 0$ we have

$$w = w_0 + [w_z]_0 z + [w_{zz}]_0 z^2/|\underline{2} + \ldots$$

where the suffix ($_0$) means that we put $z = 0$ after differentiation. Now one could try to prove that this series converges and gives the correct answer by the same methods as for real variables. But fortunately there is a general theorem which states in effect that if the function w can be differentiated once, to give $w_z = D_z w$, then it can be repeatedly differentiated (to give $w_{zz} = D_z^2 w$, w_{zzz}, etc.), and that the Taylor series must converge for sufficiently small values of w and must give the correct sum. In fact to be more precise there must either be a positive number R (depending on the particular function w and particular value z_1 that we choose) such that the series is geometrically convergent when $|z - z_1| < R$, or else (as for e^z, sin z, and cos z) the series is geometrically convergent for all values of z. R is called the "radius of convergence". This powerful theorem is unfortunately not very easy to prove. (See any standard text-book on complex variables.) But since all the functions met with in practice can be differentiated, this theorem implies that their Taylor series will be valid, and in particular this covers the geometric, exponential, logarithmic series [for ln $(1 + z)$] and binomial theorem exactly in the form we have obtained for real numbers. Thus

$$e^z = 1 + z + z^2/|\underline{2} + z^3/|\underline{3} + \ldots$$

In this series put firstly iz and secondly $-iz$ in place of z: we get

$$e^{iz} = 1 + iz - z^2/|\underline{2} - iz^3/|\underline{3} + z^4/|\underline{4} + \ldots$$

$$e^{-iz} = 1 - iz - z^2/|\underline{2} + iz^3/|\underline{3} + z^4/|\underline{4} - \ldots$$

respectively, whence by addition and subtraction

$$\cos z = (e^{iz} + e^{-iz})/2 = 1 - z^2/|\underline{2} + z^4/|\underline{4} - \ldots$$

$$\sin z = (e^{iz} - e^{-iz})/2i = z - z^3/|\underline{3} + z^5/|\underline{5} - \ldots$$

This shows how the cosine and sine series arise.

If the function is not differentiable when $z = z_1$ the Taylor series will fail. Thus $1/z$ and Ln z cannot be differentiated when $z = 0$, since they are then infinite, and so they have no Taylor series. It is perhaps also worth noting that $w = \bar{z}$, the complex conjugate of z, cannot be differentiated for any value of z according to our definition, and so will not be expressible as a Taylor series. For if $z = x + iy$, then $\bar{z} = x - iy$. Thus

$$\delta z = z_2 - z_1 = (x_2 + iy_2) - (x_1 + iy_1) = (x_2 - x_1) + i(y_2 - y_1) = \delta x + i\delta y$$
$$\delta \bar{z} = \bar{z}_2 - \bar{z}_1 = (x_2 - iy_2) - (x_1 - iy_1) = (x_2 - x_1) - i(y_2 - y_1) = \delta x - i\delta y$$
$$\delta \bar{z}/\delta z = (\delta x - i\delta y)/(\delta x + i\delta y).$$

Now the essence of differentiability is that when δz is small the quotient $\delta w/\delta z$ will approximate to a certain limiting value, the derivative. But one possible way of making δz small is to take δx small and $\delta y = 0$ exactly, when the formula above shows that $\delta \bar{z}/\delta z = 1$. Another way is to take $\delta x = 0$ and δy small, when $\delta \bar{z}/\delta z = -1$. So, however small δz may be, the quotient $\delta \bar{z}/\delta z$ can take the values 1 and -1, as well as other possible values, and therefore does not approach any definite limit. It follows that any function which involves $\bar{z}$, such as sin $\bar{z}$ or $\bar{z}^2$ or $|z| = \sqrt{(z\bar{z})}$ will not as a rule be differentiable, and cannot be expressed as a Taylor series.

14.26 Integration of complex functions

If a function $W = F(z)$ has its derivative $D_z W = F_z(z)$ equal to a given function, $w = f(z)$, then W is said to be an "indefinite integral" of w, and we write

$$W = \int w \, dz.$$

Since the rules for differentiation are the same as for a real variable it follows conversely that the rules for integration will be the same: e.g. $\int z \, dz = \frac{1}{2}z^2 + C$, $\int \frac{dz}{z} = \text{Ln } z + C$, and so on. It can be shown that if w is a differentiable function of z then it has an indefinite integral $W = F(z) = \int w \, dz$. In fact, on taking the Taylor series for w, say

$$w = a_0 + a_1 z + a_2 z^2 + \dots.$$

and integrating it we obtain

$$W = C + a_0 z + a_1 z^2/2 + a_2 z^3/3 + \dots.$$

It is clear that if the series for W converges, then we obtain that for w by differentiation, and therefore W is the indefinite integral of w. It is not difficult to show that the W series does converge for sufficiently small z*.

* Actually the formal proof of this theorem to be found in more advanced books follows quite different lines. But the reasons for that need not concern us here.

This method allows us to integrate any function which is likely to arise in practice. One instance is the "normal integral" $\frac{1}{\sqrt{2\pi}}\int e^{-\frac{1}{2}z^2}\,dz$ which is of interest in statistical theory. By the exponential series

$$e^{-\frac{1}{2}z^2} = 1 - \tfrac{1}{2}z^2 + \frac{1}{4\lfloor 2}z^4 - \frac{1}{8\lfloor 3}z^6 + \ldots + \frac{(-1)^r}{2^r\lfloor r}z^{2r} + \ldots$$

and therefore on integration,

$$\int e^{-\frac{1}{2}z^2}\,dz = C + z - \frac{1}{6}z^3 + \frac{1}{40}z^5 - \ldots + \frac{(-1)^r}{2^{r+1}r\lfloor r}z^{2r+1} + \ldots$$

The normal integral is obtained by multiplying this series by $(2\pi)^{-\frac{1}{2}}$. The product is geometrically convergent for all values of z, and can therefore be used for calculation.

It is also possible to define a complex definite integral $\int_{z_1}^{z_2} w\,dz = F(z_2) - F(z_1)$, where $F(z) = W$ is the indefinite integral. This again will obey all the usual rules for definite integrals.

In short we see that almost every process used on real numbers is a particular case of a more general process concerning complex numbers. It is true that complex numbers as such rarely occur in practical problems where all quantities normally occurring are real. (But complex numbers as such can be used to study vibrations, or in plane geometry.) Yet, even when all the quantities concerned are real, it can be helpful to use complex quantities as an intermediate stage in the calculation. We proceed to demonstrate this in the next chapter.

CHAPTER FIFTEEN

SOME USEFUL INTEGRALS

15.1 Trigonometric integrals

Certain integrals can be readily evaluated by the use of complex numbers. Many arise from the formula $\int e^{Ax}\,dx = e^{Ax}/A + C$, applicable whether x is real or complex, provided that A is not zero. (If $A = 0$ the integral becomes $\int dx = x + C$.)

Consider for example the integral of $e^x \cos x$. By formula (14.37),

$$\begin{aligned}\int e^x \cos x \,.\, dx &= \int e^x \,.\, \tfrac{1}{2}(e^{ix} + e^{-ix})\,dx\\ &= \tfrac{1}{2}\int (e^x e^{ix} + e^x e^{-ix})\,dx\\ &= \tfrac{1}{2}\int [e^{(1+i)x} + e^{(1-i)x}]\,dx\\ &= \frac{e^{(1+i)x}}{2(1+i)} + \frac{e^{(1-i)x}}{2(1-i)} + C\end{aligned}$$

This is a valid answer, but rather awkward to use in this form owing to the presence of complex functions. But it can be reduced to a real form. To do this we first bring the fractions to a common denominator $2(1+i)(1-i) = 2(1^2 - i^2) = 4$. Thus

$$\begin{aligned}\int e^x \cos x\,dx &= \frac{(1-i)\,e^{(1+i)x} + (1+i)\,e^{(1-i)x}}{2(1+i)(1-i)} + C\\ &= \tfrac{1}{4}[(1-i)\,e^{x+ix} + (1+i)\,e^{x-ix}] + C\\ &= \tfrac{1}{4}[(1-i)\,e^x e^{ix} + (1+i)\,e^x e^{-ix}] + C\\ &= \tfrac{1}{4}e^x[(1-i)\,e^{ix} + (1+i)\,e^{-ix}] + C\\ &= \tfrac{1}{4}e^x[(e^{ix} + e^{-ix}) - i\,(e^{ix} - e^{-ix})] + C\\ &= \tfrac{1}{4}e^x[(2\cos x) - i\,(2i\sin x)] + C\\ &= \tfrac{1}{2}e^x[\cos x + \sin x] + C\end{aligned}$$

This can be readily checked: on differentiating the right-hand side we do in fact get $e^x \cos x$.

PROBLEMS

(1) Show similarly that $\int e^x \sin x \,.\, dx = \tfrac{1}{2}e^x[\sin x - \cos x] + C$.

(2) Find $\int e^x \cos 2x \,.\, dx$.

Now consider $\int \cos Ax \,.\, \cos Bx \,.\, dx$.

Firstly we have from (14.37)

$$\begin{aligned}\cos Ax \,.\, \cos Bx &= \tfrac{1}{2}(e^{Aix}+e^{-Aix}) \,.\, \tfrac{1}{2}(e^{Bix}+e^{-Bix}) \\ &= \tfrac{1}{4}[e^{Aix}e^{Bix}+e^{Aix}e^{-Bix}+e^{-Aix}e^{Bix}+e^{-Aix}e^{-Bix}] \\ &= \tfrac{1}{4}[e^{(A+B)ix}+e^{(A-B)ix}+e^{-(A-B)ix}+e^{-(A+B)ix}] \\ &= \tfrac{1}{4}[e^{(A+B)ix}+e^{-(A+B)ix}]+\tfrac{1}{4}[e^{(A-B)ix}+e^{-(A-B)ix}] \\ &= \tfrac{1}{2}\cos(A+B)x+\tfrac{1}{2}\cos(A-B)x \quad . \quad . \quad (15.1)\end{aligned}$$

Now if $K \neq 0$, $\int \cos Kx \,.\, dx = (\sin Kx)/K + C$, as follows either from our standard formulas, or from

$$\begin{aligned}\int \cos Kx \,.\, dx &= \int \tfrac{1}{2}(e^{Kix} + e^{-Kix})\, dx \\ &= \tfrac{1}{2}e^{Kix}/Ki - \tfrac{1}{2}e^{-Kix}/Ki + C \\ &= [e^{Kix} - e^{-Kix}]/2iK + C \\ &= (\sin Kx)/K + C.\end{aligned}$$

Therefore if $A \neq B$ and $A \neq -B$ we have

$$\int \cos Ax \,.\, \cos Bx \,.\, dx = \frac{\sin(A+B)x}{2(A+B)} + \frac{\sin(A-B)x}{2(A-B)} + C \quad . \quad (15.2)$$

The exceptional cases in which $A = \pm B$ are not difficult. If $A = B \neq 0$ then (15.1) gives us $(\cos Ax)^2 = \frac{1}{2}\cos 2Ax + \frac{1}{2}$, so that in this case

$$\int (\cos Ax)^2 dx = (\sin 2Ax)/4A + x/2 + C \quad . \quad (15.3)$$

If $A = -B \neq 0$ we get the same result. Finally if $A = B = 0$ the integral becomes $\int 1\, dx = x + C$.

Similarly we find that if $A \neq B$ and $A \neq -B$,

$$\int \sin Ax \,.\, \sin Bx \,.\, dx = \frac{\sin(A-B)x}{2(A-B)} - \frac{\sin(A+B)x}{2(A+B)} + C$$

$$\int \sin Ax \,.\, \cos Bx \,.\, dx = -\frac{\cos(A+B)x}{2(A+B)} - \frac{\cos(A-B)x}{2(A-B)} + C$$

and the exceptional cases can be derived from the formulas ($A \neq 0$):

$$\left.\begin{aligned}\int (\sin Ax)^2\, dx &= x/2 - (\sin 2Ax)/4A + C \\ \int \sin Ax \,.\, \cos Ax \,.\, dx &= -(\cos 2Ax)/4A + C\end{aligned}\right\} \quad (15.4)$$

In general this method enables us to evaluate any integral of the form $\int e^{Ax} \cos B_1x \,.\, \cos B_2x \,.\, \sin B_3x \,.\, dx$ with any number of cosines and sines.

We can also integrate xe^{Kx}. For since $\int e^{Kx}\,dx = K^{-1}e^{Kx}$, we have by the "integration by parts" method

$$\int xe^{Kx}\,dx = x\,(K^{-1}e^{Kx}) - \int (D_x x)(K^{-1}e^{Kx})\,dx$$
$$= K^{-1}xe^{Kx} - K^{-2}e^{Kx} + C$$

A further integration by parts gives us the integral

$$\int x^2e^{Kx}\,dx = K^{-1}x^2e^{Kx} - 2K^{-2}\,xe^{Kx} + 2K^{-3}e^{Kx} + C$$

and in general if n is any positive integer,

$$\int x^n\,e^{Kx}\,dx$$
$$= K^{-n-1}\,\underline{|n}\,.\,e^{Kx}\left[\frac{(Kx)^n}{\underline{|n}} - \frac{(Kx)^{n-1}}{\underline{|n-1}} + \frac{(Kx)^{n-2}}{\underline{|n-2}} - \ldots + (-1)^n\right] + C \qquad (15.5)$$

as can be verified by direct differentiation. This allows us to integrate expressions like $x^2 \sin x$, $x \cos x$, $x\,e^x \cos x$, and more generally of a form like $x^ne^{-Ax} \cos B_1x\,.\,\cos B_2x\,.\,\sin B_3x$.

EXAMPLE

(1) Find $\int x \cos Ax\,.\,dx$.

$$\int x \cos Ax\,.\,dx = \int \tfrac{1}{2}x\,(e^{Aix} + e^{-Aix})\,dx$$
$$= \tfrac{1}{2}\,[(Ai)^{-1}x - (Ai)^{-2}]\,e^{Aix} + \tfrac{1}{2}\,[(-Ai)^{-1}x - (-Ai)^{-2}]\,e^{-Aix} + C$$
$$= x\,\frac{e^{Aix} - e^{-Aix}}{2Ai} + \frac{e^{Aix} + e^{-Aix}}{2A^2} + C$$
$$= (x \sin Ax)/A + (\cos Ax)/A^2 + C \quad . \quad . \quad . \quad . \quad (15.6)$$

Similarly $\int x \sin Ax\,.\,dx = -\,(x \cos Ax)/A + (\sin Ax)/A^2 + C$.

FURTHER PROBLEMS

(3) Find $\int x^2 \sin x\,.\,dx$.

(4) Find $\int x \sin x\,.\,\cos x\,.\,dx$.

(5) Find $\int x\,e^x \cos x\,.\,dx$.

15.2 Vibrations

Consider a body near a state of equilibrium. Denote its position measured from the equilibrium point by y (we use the term "position" for convenience; but a similar analysis can be applied to any state of a system in equilibrium, e.g. y could mean temperature, or volume, or any other convenient property, provided that the equilibrium point is taken as the origin, $y = 0$).

If the equilibrium is stable there will be a force tending to restore the system to equilibrium whenever it is slightly disturbed. The simplest assumption that we can make, and one that will often be nearly true in practice, is that the restoring force will be proportional to the displacement y. So the acceleration of the body will be proportional to y, but in the opposite direction, and we can write

$$D_t^2 y = y_{tt} = -Ky$$

where t stands for the time and K is a positive constant. It will be convenient to denote $\sqrt{K}$ by ω, so that this equation becomes

$$y_{tt} = -\omega^2 y \quad . \quad . \quad . \quad . \quad (15.7)$$

Now it has been shown in Section 13.12 that the solution of this equation is $y = p_0 \cos \omega t + (p_1/\omega) \sin \omega t$, where p_0 and p_1 are arbitrary constants. (More precisely we have shown that this is the only solution with a Taylor series about $t_1 = 0$; in Section 16.26 it will be shown that there are no other solutions.) Since p_1 is arbitrary, (p_1/ω) is also an arbitrary constant, and we can equally well write this general solution in the form

$$y = a \cos \omega t + b \sin \omega t \quad . \quad . \quad . \quad (15.8)$$

where a and b are arbitrary constants. ($a = p_0$ and $b = p_1/\omega$). A motion of the form specified by (15.8) is known as "simple harmonic". There is another way of writing it. Draw a right-angled triangle with sides a and b (Fig. 15.1). Let A be the hypotenuse, and ϕ the angle opposite b, so that $A = \sqrt{(a^2 + b^2)}$, $a = A \cos \phi$, $b = A \sin \phi$. Then

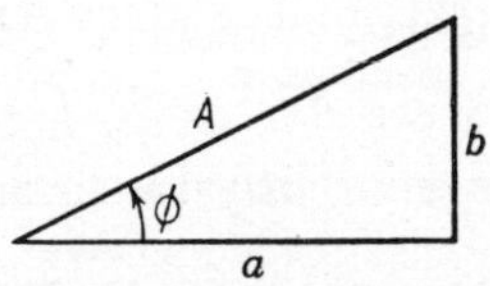

Fig. 15.1—Amplitude A and phase ϕ of harmonic motion

$$\begin{aligned} y &= a \cos \omega t + b \sin \omega t \\ &= A \cos \phi \,.\, \cos \omega t + A \sin \phi \,.\, \sin \omega t \\ &= A \cos (\omega t - \phi) \quad . \quad . \quad . \quad . \quad . \quad (15.9) \end{aligned}$$

If $\phi = 0$ this gives an equation $y = A \cos \omega t$. This means that when we plot y against t we obtain a wave-form for the graph, like that of $\cos t$ (Fig. 5.9). The constant A is the greatest value which y can take, and is called the "amplitude" of the vibration. If ϕ is not zero the effect is to delay the vibration by a constant interval of time ϕ/ω, since $\omega t - \phi = \omega(t - \phi/\omega)$: that is to say, the whole graph is shifted along the t-axis by the amount ϕ/ω but otherwise unaffected. The constant ϕ is called the "phase".

When t is increased by $2\pi/\omega$ the displacement y returns to its former value, for then $A \cos [\omega (t + 2\pi/\omega) - \phi] = A \cos [\omega t - \phi + 2\pi] = A \cos [\omega t - \phi]$. The vibration has an exact period $2\pi/\omega$, or frequency $\omega/2\pi$. The constant ω obtained by multiplying the frequency by 2π is sometimes called the "pulsatance".

Many mechanical vibrations are of this form: the waves on a pond, the vibrations of a tuning fork, the note of an organ. Now our calculation depends on two assumptions. The first is that a displacement from equilibrium produces a restoring *acceleration* or second derivative. This is true for mechanical oscillations. But it will not be true for a change in temperature, for which Newton's law of cooling states that the displacement in temperature produces a proportional rate of change, or *first* derivative, tending to restore equilibrium. In that case the temperature difference falls off according to an exponential law, without oscillations. The second assumption is that the restoring acceleration is exactly proportional to the displacement, according to the law $y_{tt} = -\omega^2 y$. Consider here a pendulum of length L subject to a small displacement y (Fig. 15.2). Then the angle θ from the vertical is given by $y = L\theta$, or $\theta = y/L$. If the mass of the bob is m, it is acted on by a downward force mg, and the component of this perpendicular to the string is $-mg \sin \theta$, or $-mg\theta$ approximately, since θ is by hypothesis small. The acceleration of the bob $=$ force/mass $= -mg\theta/m = -g\theta = -gy/L$, the minus sign indicating that it is towards the equilibrium position. That is,

Fig. 15.2—Simple pendulum

$$y_{tt} = -gy/L = -\omega^2 y \quad \text{where} \quad \omega = \sqrt{(g/L)}.$$

Thus, by our previous argument, the pendulum undergoes simple harmonic motion with period $2\pi/\omega = 2\pi\sqrt{(L/g)}$. (But if the displacement is large this approximation fails, and the motion is no longer of this type.)

In many cases of periodic motion the proportionality between displacement and acceleration breaks down. When a note is played

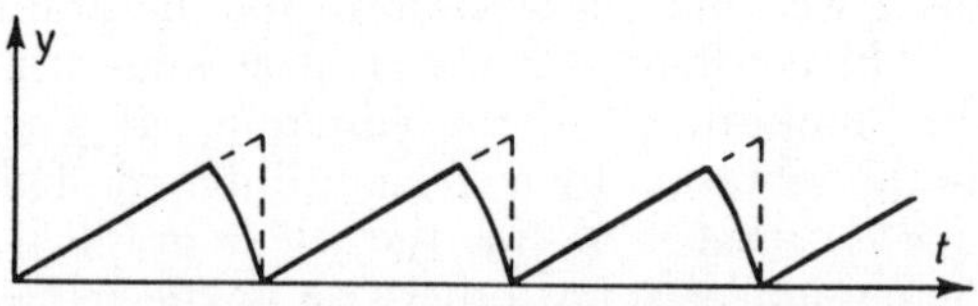

Fig. 15.3—The motion of a violin string

on a violin the string tends to stick to the bow for a certain time, and then return very rapidly to its former position. The motion will be of the form shown graphically in Fig. 15.3, with a straight ascent, and a sharp peak, but nevertheless exactly periodic. Now when we listen to the playing of a single note on a violin (or an organ, or a piano) we hear not only a note of period equal to the period of vibration, called the "fundamental", but also "overtones" or "harmonics" with periods $\frac{1}{2}$, $\frac{1}{3}$, $\frac{1}{4}$ . . . of the fundamental period, i.e. with frequencies exactly multiples of the fundamental frequency. Thus it seems that such a note can be thought of as made up of a mixture of the fundamental and overtones in varying proportions, and the human ear has a mechanism for separating these out.

Now it is fairly clear why the overtones or harmonics must have frequencies which are exact multiples of the fundamental. Suppose for instance that a note has an exact frequency of 500 cycles per second. Then a component of frequency 1000 or 1500 per second will necessarily repeat itself exactly every 1/500th of a second, and so will fit into this period; but a component of 700 cycles per second could not do so. In general let us suppose that the fundamental note has frequency $\omega/2\pi$. It is then represented by $a_1 \cos \omega t + b_1 \sin \omega t$, where a_1 and b_1 are constants, and t stands for the time. A note of twice this frequency is represented by $a_2 \cos 2\omega t + b_2 \sin 2\omega t$, one of three times the frequency by $a_3 \cos 3\omega t + b_3 \sin 3\omega t$, and so on. So if the observed displacement y can be broken up into a combination of harmonics it must be of the form

$$y = a_0 + a_1 \cos \omega t + b_1 \sin \omega t + a_2 \cos 2\omega t + b_2 \sin 2\omega t + \ldots \text{ etc.}$$

We can readily separate these terms by using integral calculus. To simplify the calculations we shall take the unit of time to be such that $\omega = 1$, i.e. that the period is 2π. (If not, this can be arranged by a suitable change of units.) Now suppose first that y can be analysed into a finite number N of vibrations: then we can write

$$y = a_0 + a_1 \cos t + b_1 \sin t + \ldots + a_N \cos Nt + b_N \sin Nt \quad . \quad . \quad (15.10)$$

The device which solves the problem of finding the a's and b's is as follows. From formulas (15.2) to (15.4) above we see that, if m and n are zero or positive integers, but $m \neq n$,

$$\begin{aligned} \int_0^{2\pi} \cos mx \,.\, \cos nx \,.\, dx &= \int_0^{2\pi} \cos mx \,.\, \sin nx \,.\, dx \\ &= \int_0^{2\pi} \sin mx \,.\, \sin nx \,.\, dx \\ &= 0 \end{aligned}$$

since $\sin 2\pi(m+n) = \sin 2\pi(m-n) = 0$ and also $\cos 2\pi(m+n) = \cos 2\pi(m-n) = 1$. Also if $m = n \neq 0$ we have

$$\int_0^{2\pi} (\cos nx)^2 \, dx = \int_0^{2\pi} (\sin nx)^2 \, dx = \pi,$$

$$\int_0^{2\pi} \cos nx \,.\, \sin nx \,.\, dx = 0.$$

The series (15.10) for y becomes, on multiplication throughout by $\cos nt$,

$$y \cos nt = a_1 \cos nt + a_1 \cos t . \cos nt + b_1 \sin t . \cos nt + \\ \ldots + a_n \cos nt . \cos nt + b_n \sin nt . \cos nt + \ldots \\ + a_N \cos Nt . \cos nt + b_N \sin Nt . \cos nt.$$

On integrating this from 0 to 2π we shall therefore obtain a zero integral from every term on the right-hand side except $a_n \cos nt . \cos nt$, which has the integral πa_n. Thus (for $n \neq 0$)

$$\int_0^{2\pi} y \cos nt . dt = \int_0^{2\pi} a_n \cos nt . \cos nt . dt = \pi a_n$$

i.e.

$$a_n = \pi^{-1} \int_0^{2\pi} y . \cos nt . dt \quad . \quad . \quad . \quad (15.11)$$

Similarly if we multiply equation (15.10) through by $\sin nt$ and integrate we shall find that every term will give a zero integral except $b_n \sin nt . \sin nt$, which has integral $b_n \pi$;

$$\int_0^{2\pi} y \sin nt . dt = \int_0^{2\pi} b_n \sin nt . \sin nt . dx = \pi b_n$$

i.e.

$$b_n = \pi^{-1} \int_0^{2\pi} y \sin nt . dt \quad . \quad . \quad . \quad (15.12)$$

Finally to find a_0 we simply integrate y as it stands from 0 to 2π; every term but the first has a zero integral, and therefore

$$\int_0^{2\pi} y \, dt = \int_0^{2\pi} a_0 \, dt = 2\pi a_0$$

i.e.

$$a_0 = (2\pi)^{-1} \int_0^{2\pi} y . dt \quad . \quad . \quad . \quad (15.13)$$

Thus each coefficient a_r or b_r can be expressed in the form of an integral.

However, in general, a periodic wave-form y will not be expressible exactly as the sum of a finite number of harmonics. But it is natural to suppose that it can be expressed as the sum of a convergent infinite series: the first few terms would represent the fundamental and simplest harmonics, while the tail end of the series would consist of higher harmonics of small amplitude and negligible effect.

$$y = a_0 + a_1 \cos t + b_1 \sin t + a_2 \cos 2t + b_2 \sin 2t + \ldots \text{ to infinity} \quad (15.14)$$

It is also natural to suppose that the coefficients a_n and b_n can be calculated from the integrals (15.11), (15.12) and (15.13). In fact these assumptions are true for a very large class of functions y: this is known as "Fourier's Theorem". But unfortunately it requires considerable ingenuity to prove the result properly, as the series (15.14), known as Fourier's series, can often be very slowly convergent. We shall simply have to accept the result for our purposes.

To take a concrete instance, let us consider the motion of a violin string as shown in Fig. 15.3. To simplify the calculations the graph will be slightly idealized by assuming that the return motion is instantaneous, as suggested by the dotted lines. We then have the relation $y = Kt$ for values of t from 0 to 2π, and therefore

$$a_0 = (2\pi)^{-1} \int_0^{2\pi} y\, dt = (2\pi)^{-1} \int_0^{2\pi} Kt\, dt$$
$$= (2\pi)^{-1} [\tfrac{1}{2}Kt^2]_0^{2\pi} = \pi K$$
$$a_n = \pi^{-1} \int_0^{2\pi} y \cos nt \,.\, dt = \pi^{-1} \int_0^{2\pi} Kt \,.\, \cos nt \,.\, dt$$
$$= \pi^{-1} K [n^{-1} t \sin nt + n^{-2} \cos nt]_0^{2\pi} = 0$$
$$b_n = \pi^{-1} \int_0^{2\pi} y \sin nt \,.\, dt = \pi^{-1} \int_0^{2\pi} Kt \,.\, \sin nt \,.\, dt$$
$$= \pi^{-1} K [-n^{-1} t \cos nt + n^{-2} \sin nt]_0^{2\pi} = -2Kn^{-1}$$

So y is represented by the series

$$y = K[\pi - 2 \sin t - \sin 2t - \frac{2}{3} \sin 3t - \,.\,.\, - \frac{2}{r} \sin rt - \,.\,.\,.\,]$$

and the successive harmonics do in fact decrease in amplitude, but only very slowly.

If the value of y is not given by an algebraic expression, but by a graph or set of numerical values instead, it is still possible to find a_r and b_r by the use of numerical integration. A suitable way of doing this is suggested by the following interpretation of the integrals (15.11), (15.12) and (15.13). As t varies in value between 0 and 2π, a_0 is the average value of y, a_n is twice the average value of $y \cos nt$, and b_n is twice the average value of $y \sin nt$. Imagine then the interval from 0 to 2π divided into twelve equal parts by points $T_1 = 0$, $T_2 = \frac{\pi}{6}$ $(= 30°)$, $T_3 = \frac{2\pi}{6}$ $(= 60°)$, $T_4 = \frac{3\pi}{6}$ $(= 90°)$, and so on up to $T_{12} = \frac{11\pi}{6}$ $(= 330°)$. The final point $T_{13} = 2\pi$ will simply repeat the first, since y is periodic, and can therefore be ignored. Let $Y_1, Y_2, \ldots Y_{12}$ be the corresponding observed values of y. Then a_0 will be approximately the average value $\frac{1}{12}(Y_1 + Y_2 + \ldots + Y_{12})$; a_1 will be approximately the doubled average $\frac{2}{12}(Y_1 \cos 0° + Y_2 \cos 30° + Y_3 \cos 60° + \ldots + Y_{12} \cos 330°)$, and b_1 will be the doubled average $\frac{2}{12}(Y_1 \sin 0° + Y_2 \sin 30° + \,.\,.\, + Y_{12} \sin 330°)$. a_2 will be twice the average of $(Y_r \cos 2T_r)$ for values of r from 1 to 12, and so on. If greater accuracy is required the number of points must be increased, and various schemes have been worked out to reduce the labour of computation (see E. T. Whittaker & G. Robinson, *The Calculus of Observations*, Blackie, 1937, Chapter 10).

The ability of the human ear to separate a note into fundamental and overtones suggests that it acts in a similar way to a Fourier analyser. But comparisons between sounds as heard and as mechanically recorded show that it does not, or at least not very accurately. The fundamental pitch can sometimes be heard very clearly in a note which exact analysis shows to consist almost entirely of overtones: this is probably due to distortion of the sound in the auditory channels.

It has been many times suggested that a Fourier analysis may be helpful in finding the causes of a vibration. This may be so in the study of musical notes, and in X-ray crystallography, or in tracking sea-waves, or even in analysing the components of an electrocardiogram. But it is doubtful whether it is helpful in studying biological periodic phenomena in general, since the mechanism which causes fluctuations in the number of a species from year to year, or causes the nervous shaking of the hand when held out, may not be at all analogous to the process which causes the swing of a pendulum. The decomposition of such a process into harmonic vibrations may form a simple means of description, and is always mathematically possible: but it can be irrelevant to the explanation.

PROBLEMS

(1) A periodic function y has the value 0 when t lies between 0 and π, and the value 1 when t lies between π and 2π, and thereafter repeats. What is its Fourier series? (This example is not so artificial as it might seem at first sight. If a constant pressure is applied to a body at regular intervals, the behaviour of the body may depend on its resonance with the various harmonics. The same will apply to an electrical system transmitting a signal composed of a series of Morse dashes.)

(2) A periodic function $y = f(t)$ has the following values:

t	0	$\frac{\pi}{6}$	$\frac{2\pi}{6}$	$\frac{3\pi}{6}$	$\frac{4\pi}{6}$	$\frac{5\pi}{6}$	$\frac{6\pi}{6}$	$\frac{7\pi}{6}$	$\frac{8\pi}{6}$	$\frac{9\pi}{6}$	$\frac{10\pi}{6}$	$\frac{11\pi}{6}$
y	1·3	1·666	·8	−·5	−1·1	−·966	−·7	−·466	−·2	−·1	−·1	·366

Find its components up to the terms in $\cos 3t$ and $\sin 3t$.

15.3 Partial fractions

A "rational function" is one such as

$$(2x^3 - 7x^2 - 4x + 37)/(x + 2) = P_1(x)/P_2(x),$$

which is the quotient of one polynomial $P_1(x)$ divided by another $P_2(x)$. We sometimes wish to integrate such a function. The method is as follows.

The obvious first step is to divide $P_1(x)$ by $P_2(x)$ algebraically (see Section 3.8). Thus, on dividing $(2x^3 - 7x^2 - 4x + 37)$ by $(x + 2)$ we find the quotient $(2x^2 - 11x + 18)$ and remainder 1, so that

$$2x^3 - 7x^2 - 4x + 37 = (2x^2 - 11x + 18)(x + 2) + 1$$

and

$$\frac{2x^3 - 7x^2 - 4x + 37}{x + 2} = 2x^2 - 11x + 18 + \frac{1}{x + 2}$$

whence

$$\int \frac{2x^3 - 7x^2 - 4x + 37}{x + 2}\,dx = \tfrac{2}{3}x^3 - \frac{11}{2}x^2 + 18x + \int \frac{dx}{x + 2}. \qquad (15.15)$$

But by changing the variable from x to $x + 2 = z$, so that $dx/dz = 1$, we have

$$\int \frac{dx}{x + 2} = \int \frac{1}{z}\frac{dx}{dz}\,dz = \ln z + C = \ln(x + 2) + C \qquad (15.16)$$

and on substituting this in (15.15) we get the complete integral. This method will give an integral at once, whenever the denominator $P_2(x)$ is of the first degree. More generally, a division of $P_1(x)$ by $P_2(x)$ will give a polynomial quotient $Q(x)$ and a polynomial remainder $R(x)$, so that

$$P_1(x) = P_2(x)Q(x) + R(x)$$

and

$$P_1(x)/P_2(x) = Q(x) + R(x)/P_2(x).$$

For example

$$\frac{2x^3 - 7x^2 - 4x + 37}{x^2 - x - 6} = 2x - 5 + \frac{3x + 7}{x^2 - x - 6}.$$

The quotient $Q(x)$ is easily integrated; in this case the integral of $2x - 5$ is $x^2 - 5x + C$. There remains the last term $R(x)/P_2(x)$, or in this case $(3x + 7)/(x^2 - x - 6)$, still to be dealt with. The numerator $R(x)$ will always be of lower degree than the denominator.

Now in this case the denominator $P_2(x) = x^2 - x - 6$ has factors $(x - 3)(x + 2)$. Divide the numerator $R(x) = 3x + 7$ by one of these, say by $(x + 2)$, to get

$$3x + 7 = 3(x + 2) + 1$$

so that

$$\frac{3x + 7}{x^2 - x - 6} = \frac{3(x + 2) + 1}{(x - 3)(x + 2)} = \frac{3}{x - 3} + \frac{1}{(x - 3)(x + 2)} \qquad (15.17)$$

Now the last term on the right-hand side can be decomposed as follows:

$$\frac{1}{(x-3)(x+2)} = \frac{1}{5(x-3)} - \frac{1}{5(x+2)}. \qquad (15.18)$$

This is easily verified by bringing the right-hand side to common denominator $5(x-3)(x+2)$. On substituting this in (15.17) we get

$$\frac{3x+7}{x^2-x-6} = \frac{16}{5(x-3)} - \frac{1}{5(x+2)} \qquad (15.19)$$

This is "the decomposition of $(3x+7)/(x^2-x-6)$ into partial fractions". (A useful check is to substitute particular values of x in (15.19). Thus on putting $x=0$ we get $-\frac{7}{6} = -\frac{16}{15} - \frac{1}{10}$, or, with common denominator 30, this is $-\frac{35}{30} = -\frac{32}{30} - \frac{3}{30}$, which is correct. Similarly on putting $x=1$ we get $-\frac{10}{6} = -\frac{16}{10} - \frac{1}{15}$, or $-\frac{50}{30} = -\frac{48}{30} - \frac{2}{30}$). We can now integrate (15.19), getting

$$\int \frac{3x+7}{x^2-x-6}\,dx = \frac{16}{5}\ln(x-3) - \frac{1}{5}\ln(x+2) + C.$$

The same process applies wherever the denominator $P_2(x)$ has two unequal factors $(x-a)(x-b)$. The key identity is

$$\frac{1}{(x-a)(x-b)} = \frac{1}{(a-b)(x-a)} + \frac{1}{(b-a)(x-b)} \qquad (15.20)$$

Thus, for example,

$$\begin{aligned}\frac{2x+1}{(x-1)(x-4)} &= \frac{2(x-1)+3}{(x-1)(x-4)} \\ &= \frac{2}{x-4} + 3 \times \frac{1}{(x-1)(x-4)} \\ &= \frac{2}{x-4} + 3\left[-\frac{1}{3(x-1)} + \frac{1}{3(x-4)}\right] \\ &= -\frac{1}{x-1} + \frac{3}{x-4}\end{aligned}$$

Again, if we put $x=0$ we get $\frac{1}{4} = 1 - \frac{3}{4}$, which checks. The general formula

$$\int \frac{dx}{x-a} = \ln(x-a) + C \qquad (15.21)$$

enables us to integrate the expression

$$\int \frac{(2x+1)\,dx}{(x-1)(x-4)} = -\ln(x-1) + 3\ln(x-4) + C.$$

An expression like $1/[(x-a)(x-b)(x-c)]$ can be split up into partial fractions by repeated use of (15.20). For example

$$\frac{1}{(x-3)(x+2)(x+1)}$$

$$= \left[\frac{1}{(x-3)(x+2)}\right]\frac{1}{x+1}$$

$$= \left[\frac{1}{5(x-3)} - \frac{1}{5(x+2)}\right]\frac{1}{x+1}$$

$$= \frac{1}{5(x-3)(x+1)} - \frac{1}{5(x+2)(x+1)}$$

$$= \frac{1}{5}\left[\frac{1}{4(x-3)} - \frac{1}{4(x+1)}\right] - \frac{1}{5}\left[\frac{-1}{x+2} + \frac{1}{x+1}\right]$$

$$= \frac{1}{20(x-3)} + \frac{1}{5(x+2)} - \frac{5}{20(x+1)} \quad . \quad (15.22)$$

Expressions such as $(3x^2+x+2)/[(x-3)(x+2)(x+1)]$ can now be decomposed into partial fractions. We first apply Theorem 3.1 to put the numerator into the form

$$3x^2 + x + 2 = \alpha + \beta(x-3) + \gamma(x-3)(x+2) \quad (15.23)$$

This can be done by first dividing $(3x^2+x+2)$ by $(x-3)$ to give quotient $(3x+10)$ and remainder 32, so that

$$3x^2 + x + 2 = 32 + (x-3)(3x+10).$$

The quotient $(3x+10)$ is now in turn divided by $(x+2)$ to give quotient 3 and remainder 4, so that $(3x+10) = 4 + 3(x+2)$, or

$$\begin{aligned} 3x^2 + x + 2 &= 32 + (x-3)[4 + 3(x+2)] \\ &= 32 + 4(x-3) + 3(x-3)(x+2). \end{aligned}$$

Alternatively, one can find α, β, γ in order by substituting in turn the particular values $x = 3$, $x = -2$, and $x = 0$ in (15.23). We therefore have

$$\frac{3x^2+x+2}{(x-3)(x+2)(x+1)}$$

$$= \frac{32}{(x-3)(x+2)(x+1)} + \frac{4}{(x+2)(x+1)} + \frac{3}{x+1} \quad . \quad (15.24)$$

The first term on the right-hand side can be decomposed by (15.22):

$$\frac{32}{(x-3)(x+2)(x+1)} = \frac{8}{5(x+3)} + \frac{32}{5(x+2)} - \frac{8}{x+1}$$

the second term can be decomposed by (15.20):

$$\frac{4}{(x+2)(x+1)} = -\frac{4}{x+2} + \frac{4}{x+1}$$

When these values are substituted in (15.24) and terms with the same denominator are added together we get finally

$$\frac{3x^2+x+2}{(x-3)(x+2)(x+1)} = \frac{8}{5(x-3)} + \frac{12}{5(x+2)} - \frac{1}{x+1} \qquad (15.25)$$

(which can be checked by substituting, e.g., $x = 0$). A similar argument holds in general. Any fraction of the form $R(x)/[(x-a)\ldots(x-g)]$ can be decomposed by repeated applications of (15.20) into the form

$$\frac{R(x)}{(x-a)(x-b)\ldots(x-g)} = \frac{A}{x-a} + \frac{B}{x-b} + \ldots + \frac{G}{x-g}$$

provided that the numerator $R(x)$ is of lower degree than the denominator, and provided that all the factors in the denominator are unequal. To find the values of $A, B, \ldots G$, it is not necessary to go through the whole algebraic process of decomposition, as there is a short-cut method which we illustrate on the case (15.25) already discussed. Let us assume that

$$\frac{3x^2+x+2}{(x-3)(x+2)(x+1)} = \frac{A}{x-3} + \frac{B}{x+2} + \frac{C}{x+1} \qquad (15.26)$$

and try to find the values of the constants A, B, C. Multiply this equation through by $(x-3)$; it becomes

$$\frac{3x^2+x+2}{(x+2)(x+1)} = A + \frac{B(x-3)}{x+2} + \frac{C(x-3)}{x+1} \quad . \qquad (15.27)$$

Now put $x = 3$; we get at once $\frac{32}{20} = A$, or $A = \frac{8}{5}$. Similarly if we multiply (15.26) through by $(x+2)$ and then put $x = -2$ we get $B = \frac{12}{5}$; and on multiplying (15.26) by $(x+1)$ and putting $x = -1$ we get $C = -1$. There is an important logical point here. The equation (15.26) is an identity, which will hold for all values of x which do not make any denominator zero, i.e. for x different from 3, -2, and -1. The equation (15.27) which we derive from it will also be true when x is different from 3, -2 and -1; it is not immediately obvious that it will also hold when $x = 3$. However, we can get over this difficulty by taking the limit of each side of the equation (15.27) as x tends to 3. These limits must be equal, since the equation certainly holds when x is near 3 but not equal to 3; and since there is no denominator $(x-3)$ anywhere in this equation, and the functions involved are of a simple form, these limits will be the actual values of the expressions when $x = 3$. Hence the equation still holds when $x = 3$, justifying the procedure. A similar remark holds in other cases discussed later.

We therefore get from (15.25) the integral

$$\int \frac{3x^2+x+2}{(x-3)(x+2)(x+1)}\,dx = \frac{8}{5}\ln(x-3) + \frac{12}{5}\ln(x+2) - \ln(x+1) + C.$$

As another example we derive (15.19). Try

$$\frac{3x+7}{(x-3)(x+2)} = \frac{E}{x-3} + \frac{F}{x+2}$$

Multiply through by $(x-3)$ to get

$$\frac{3x+7}{x+2} = E + F\frac{x-3}{x+2}$$

Put $x = 3$ (or, more strictly, take the limit as x tends to 3); this gives $E = \frac{16}{5}$. Similarly, a multiplication by $(x+2)$ followed by a substitution $x = -2$ give $F = -\frac{1}{5}$.

PROBLEMS

Decompose the following expressions into partial fractions:

(1) $1/[(x-2)(x+2)]$
(2) $1/(x^2-9)$
(3) $(2x+3)/(x^2-1)$
(4) $(x+4)/(x^2+x-2)$
(5) $x/(x^2-2)$
(6) $(x+3)/(x^2+2x)$
(7) $(3x+2)/[(x-1)(x-2)(x-3)]$
(8) $(4x^2+12x-14)/[(x+1)(x-1)(x-3)]$
(9) $(7x+8)/[(x-1)(x+2)(x+5)]$
(10) $(3x^2-2x)/[(x-1)(x^2-3)]$

15.4 Partial fractions with repeated factors

The procedure described above for splitting $R(x)/P_2(x)$ fails if two or more of the factors of $P_2(x)$ are equal. For example,

$$\frac{3y^2+2y+1}{y^3} = \frac{3}{y} + \frac{2}{y^2} + \frac{1}{y^3} \qquad (15.28)$$

can be considered as a decomposition into a new type of "partial fractions", with the denominators of degrees 1, 2, and 3; and this particular decomposition cannot be further simplified to one in which all denominators are of the first degree. From it we get the integral

$$\int \frac{3y^2+2y+1}{y^3}\,dy = -3\ln y - 2y^{-1} - \tfrac{1}{2}y^{-2} + C. \qquad (15.29)$$

If we make (for example) the substitution $y = x - 2$ in (15.28) and (15.29), we get (since $dy = dx$)

$$\frac{3(x-2)^2 + 2(x-2) + 1}{(x-2)^3}$$

$$= \frac{3x^2 - 10x + 9}{(x-2)^3} = -\frac{3}{x-2} + \frac{2}{(x-2)^2} + \frac{1}{(x-2)^3} \quad (15.30)$$

$$\int \frac{3x^2-10x+9}{(x-2)^3}\,dx = -3\ln(x-2) - \frac{2}{x-2} - \frac{1}{2(x-2)^2} + C. \quad (15.31)$$

However, so long as the denominator of a fraction contains two different factors $(x-a)(x-b)$, it is possible to separate these by (15.20). One can therefore anticipate that if a fraction contains a factor $(x-a)^n$, the effect of this will be to produce in general a collection of terms of the form

$$+ \frac{A_1}{x-a} + \frac{A_2}{(x-a)^2} + \ldots + \frac{A_n}{(x-a)^n} + \ldots$$

in the decomposition into partial fractions. To take a simple example, we would expect to have a decomposition of the form

$$\frac{2x^2 - 4x - 5}{(x-3)(x+2)^2} = \frac{A}{x-3} + \frac{B_1}{x+2} + \frac{B_2}{(x+2)^2} \cdot \quad (15.32)$$

where we get partial fractions with denominators $(x+2)$ and $(x+2)^2$ in consequence of the denominator $(x+2)^2$ in the original expression $R(x)/P_2(x)$. This is true, provided, as usual, that $R(x)$ is of lower degree than $P_2(x)$. We will not give a general proof here, but show how the decomposition arises in the particular case (15.32), which will indicate the general argument. By using Theorem 3.1 we have

$$2x^2 - 4x - 5 = 1 - 2(x-3) + 2(x-3)(x+2)$$

whence

$$\frac{2x^2 - 4x - 5}{(x-3)(x+2)^2} = \frac{1}{(x-3)(x+2)^2} - \frac{2}{(x+2)^2} + \frac{2}{x+2} \cdot \quad (15.33)$$

By repeated use of (15.18)

$$\frac{1}{(x-3)(x+2)^2} = \left[\frac{1}{5(x-3)} - \frac{1}{5(x+2)}\right]\frac{1}{x+2}$$

$$= \frac{1}{5(x-3)(x+2)} - \frac{1}{5(x+2)^2}$$

$$= \frac{1}{25(x-3)} - \frac{1}{25(x+2)} - \frac{1}{5(x+2)^2}$$

and on substituting this in (15.33) we get

$$\frac{2x^2 - 4x - 5}{(x-3)(x+2)^2} = \frac{1}{25(x-3)} + \frac{49}{25(x+2)} - \frac{11}{5(x+2)^2} \quad (15.34)$$

$$\int \frac{2x^2-4x-5}{(x-3)(x+2)^2}\,dx = \frac{1}{25}\ln(x-3) + \frac{49}{25}\ln(x+2) + \frac{11}{5(x+2)} + C\,. \quad (15.35)$$

The problem now arises: can we find the values of A_1, B_1, and B_2 directly from (15.32) without going through all the intermediate steps? By multiplying (15.32) by $(x-3)$ and then putting $x = 3$ we get immediately $A = \frac{1}{25}$; and by multiplying by $(x+2)^2$ and setting $x = -2$ we get $B_2 = -\frac{11}{5}$. It is not quite so easy to find B_1, but there are several possible ways. We can take the fractions already found on to the left-hand side of (15.32), getting

$$\frac{2x^2 - 4x - 5}{(x-3)(x+2)^2} - \frac{1}{25(x-3)} + \frac{11}{5(x+2)^2} = \frac{B_1}{x+2} \quad (15.36)$$

If the calculations have been done correctly, the left-hand side should now simplify to an expression of the form $B_1/(x+2)$, and this will give us the value of B_1. In fact, the left-hand side, expressed with common denominator $25(x-3)(x+2)^2$, becomes after a little calculation

$$\frac{49x^2 - 49x - 294}{25(x-3)(x+2)^2} = \frac{49(x-3)(x+2)}{25(x-3)(x+2)^2} = \frac{49}{25(x+2)}$$

whence $B_1 = \frac{49}{25}$. An alternative method is to try substituting one or more values of x in (15.36); if we put $x = 0$ we get $+\frac{5}{12} - \frac{1}{75} + \frac{11}{20} = \frac{1}{2}B_1$, whence again $B_1 = \frac{49}{25}$. Other methods for simplifying the calculation are given in H. Kestelman and H. S. W. Massey, *Ancillary Mathematics*, Pitman, 1959.

EXAMPLE

(1) Find the partial fractions for

$$\frac{x^2+1}{x^3(x-1)^2} = \frac{A_1}{x} + \frac{A_2}{x^2} + \frac{A_3}{x^3} + \frac{B_1}{x-1} + \frac{B_2}{(x-1)^2}$$

Multiply by x^3 and set $x = 0$; this gives $A_3 = 1$. Multiply by $(x-1)^2$ and set $x = 1$; this gives $B_2 = 2$. Take the terms thus found on to the left-hand side and simplify; this gives

$$\frac{x^2+1}{x^3(x-1)^2} - \frac{1}{x^3} - \frac{2}{(x-1)^2} = \frac{A_1}{x} + \frac{A_2}{x^2} + \frac{B_1}{x-1}$$

and this reduces to

$$\frac{-2x^3 + 2x}{x^3(x-1)^2} = \frac{-2(x+1)x(x-1)}{x^3(x-1)^2} = \frac{-2x-2}{x^2(x-1)} = \frac{A_1}{x} + \frac{A_2}{x^2} + \frac{B_1}{x-1}$$

Multiply by x^2 and put $x = 0$; this gives $A_2 = 2$. Multiply by $(x - 1)$ and put $x = 1$; this gives $B_1 = -4$. Take the terms thus found on to the left-hand side and simplify; this gives finally $4/x = A_1/x$, whence $A_1 = 4$. Thus finally we get

$$\frac{x^2 + 1}{x^3(x - 1)^2} = \frac{4}{x} + \frac{2}{x^2} + \frac{1}{x^3} - \frac{4}{x - 1} + \frac{2}{(x - 1)^2}$$

PROBLEMS

Find the following integrals:

(1) $\int \frac{dx}{x^2(x+2)}$ (2) $\int \frac{(1+x)dx}{(1-x)^2}$ (3) $\int \frac{(2x+1)dx}{(x-2)^2(x+1)}$

(4) $\int \frac{dx}{(x+1)^2(x+4)}$ (5) $\int \frac{(x^3+3x+5)dx}{(x+1)^4}$.

15.5 Complex roots

The method we have explained for decomposition into partial fractions can only be used if we can factorize the original denominator. In Section 3.10 we discovered a theorem which helps us to do this: it states that the polynomial

$$P(x) = a + bx + cx^2 + \ldots + hx^n$$

has a factor $(x - \lambda)$ if and only if λ is a root of the equation $P(x) = 0$, i.e. if and only if

$$a + b\lambda + c\lambda^2 + \ldots + h\lambda^n = 0.$$

Now if the numbers used are restricted to real values there may be no such roots: e.g. the equation $1 + x^2 = 0$ has no real roots, and therefore $1 + x^2$ has no real factors. But there are two complex roots, $x = i$ and $x = -i$, and therefore two complex factors of $1 + x^2$, namely, $x - i$ and $x + i$. Similarly we have found that the equation $x^3 = 1$, or $x^3 - 1 = 0$, has two complex roots $x = \omega$ and $x = \omega^2$ in addition to its real root $x = 1$, so that $x^3 - 1$ has the factors $(x - 1)$, $(x - \omega)$ and $(x - \omega^2)$. This is a particular example of a general theorem, that an equation of the nth degree

$$a + bx + cx^2 + \ldots + hx^n = 0 \qquad (h \neq 0)$$

has always at least one root, which may be complex; and as a rule it has n roots. This is true whether $a, b, c \ldots h$ are real or complex.

The explanation of this theorem is perhaps best illustrated by considering a special case, such as a cubic $a + bx + cx^2 + hx^3 = 0$ where x can represent either a real or complex number. $h \neq 0$, for otherwise the equation would not be cubic. Now it will be convenient to consider the special case $h = 1$, i.e. to take the equation in the standard form

$$A + Bx + Cx^2 + x^3 = 0. \qquad (15.37)$$

Since $h \neq 0$ the equation is reduced to this form by division by h: from

$$a + bx + cx^2 + hx^3 = 0$$

we get
$$(a/h) + (b/h)x + (c/h)x^2 + x^3 = 0$$

which is of the form (15.22) with $A = a/h$, $B = b/h$, $C = c/h$.

Now let us put $y = A + Bx + Cx^2 + x^3$. Suppose that the complex number x is represented by a vector OP, and y by a vector OQ (Fig. 15.4). (Note that x and y represent distinct numbers here, not

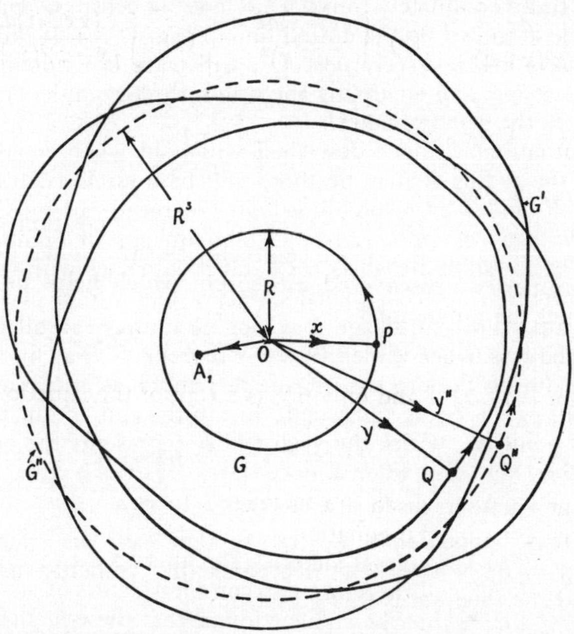

Fig. 15.4—Proof of the existence of a root of an algebraic equation

the real and imaginary parts of a single complex number z, as in the last chapter.) Thus to each point P there corresponds a number x; to each number x there corresponds $y = A + Bx + Cx^2 + x^3$; and to each y there corresponds a vector OQ, and therefore a point Q. In short we can say that P represents x, and Q represents y.

Now imagine that P is moved to trace out a continuous curve G: then Q, being determined in position by the position of P, will move so as to trace out another continuous curve G'. In particular we shall imagine P to move round a circle G of centre O and radius R. What can we then say about G'?

First let us suppose that $R = |x|$ is large. Then we know that in the expression $y = A + Bx + Cx^2 + x^3$ the term x^3 will far outweigh the others. That means that y will be nearly equal to x^3, in the sense that the difference between $y = A + Bx + Cx^2 + x^3$ and y'' (say) $= x^3$ will be negligible in comparison with their absolute magnitudes. Let y'' be represented by a vector OQ'': then Q'' will not be far from Q, and as P traces out the circle G, Q'' will trace out a curve G'' not very different from the curve G' traced out by Q.

Now since P lies on a circle of radius R, the vector $x = OP$ can be written $\{R, \theta\}$: and as P traces out the circle, θ will change continuously in value from 0 to 2π. But $y'' = x^3 = \{R, \theta\}^3 = \{R^3, 3\theta\}$; consequently 3θ changes in value from 0 to 6π, and the point Q'' representing y'' will run three times completely round a circle G'' of centre O and radius R^3. (This circle is shown by the dotted line in Fig. 15.4.) It follows that Q which always lies relatively near Q'', will trace out a loop G', nearly circular in shape, and encircling the origin three complete times before returning to the starting point.

The situation we have described will hold when R is large. But whatever the radius R may be there will be a circle G, traced out by the point P, and a corresponding loop G' traced out by Q. Suppose we start with a very large circle G, and imagine it gradually shrunk until finally its radius becomes zero and it coincides with the origin O. Then the loop G' will also shrink, rather like a loop of string being tightened up. This shrinkage may not be uniform in all parts of the loop: indeed it is conceivable that some parts of G' may move outwards for a time during the process—again very much as a loop of string can move about as it is being tightened. But in the end G' must contract to the single point A_1, where the vector OA_1 represents the coefficient A which is the value of y when $x = 0$.

We now have two cases to consider:

(i) It may happen that $A = 0$. In that case the equation $y = 0$ becomes $y = Bx + Cx^2 + x^3 = 0$, and this certainly has the root $x = 0$, and the theorem is therefore true.

(ii) The other case is that $A \neq 0$. But $A =$ the vector OA_1, so that the point A_1 is different from O. Now on inspection of Fig. 15.4 we see that it is inconceivable that the loop G' could shrink to the single point A_1 without passing at least once over O. In fact, since G' is a triple loop, we shall expect it to pass three times over O during the shrinking process: but in exceptional cases all three loops might pass over together at the same point and at the same moment. In saying that the contrary is inconceivable we are appealing to our common experience of strings and loops: but it is possible to give a strict though quite complicated mathematical proof of this point. Now G' is by definition the loop traced out by Q, where OQ is the vector representing $y = A + Bx + Cx^2 + x^3$. So if G' passes over O, there must be some point $Q = O$, that is to say, there must be some value of $y = OQ = 0$. The corresponding value of x is therefore a root of the equation $y = 0$. If the loop G' passes at least 3 times over O, as will usually be the case, then there will be at least 3 corresponding roots of the equation $y = 0$. But a cubic equation cannot have more than 3 roots (Section 3.10), so usually it will have 3, and always at least 1 root. By replacing the number "3" by "n" throughout we obtain a similar proof of the theorem for the nth degree equation

$$y = A + Bx + Cx^2 + \ldots + x^n = 0.$$

Returning to the cubic equation, let α be a root. Then

$$y = A + Bx + Cx^2 + x^3$$

has a factor $x - \alpha$; let us say $y = (x - \alpha)(A' + B'x + x^2)$. But the equation $A' + B'x + x^2 = 0$ must in turn have at least one root β, which may or may not be different from α, and therefore

$$A' + B'x + x^2 = (x - \beta)(A'' + x),$$

and $$y = (x - \alpha)(x - \beta)(A'' + x).$$

Finally the equation $A'' + x = 0$ has the root $\gamma = -A''$, and so

$$y = A + Bx + Cx^2 + x^3 = (x - \alpha)(x - \beta)(x - \gamma),$$

i.e. any cubic expression can be completely factorized into linear factors. The same sort of argument holds equally for an nth-degree polynomial

$$A+Bx+Cx^2+ \ldots +x^n = (x-\alpha)(x-\beta)(x-\gamma) \ldots (x-\lambda) \quad (15.38)$$

there being n factors in all, though not all the factors need be unequal. If the coefficient of $x^n = H$ is not unity, then we shall have

$$A+Bx+Cx^2+ \ldots +Hx^n = H(x-\alpha)(x-\beta) \ldots (x-\lambda). \quad (15.39)$$

Thus any polynomial can in theory be completely factorized. It may not be very easy in practice to find these factors: that is a point we shall discuss later. But it follows that any "rational function" or ratio of two polynomials

$$f(x) = (a + bx + cx^2 + \ldots + kx^m)/(A + Bx + Cx^2 + \ldots + Hx^n)$$

can be expressed in the form $\dfrac{a + bx + cx^2 + \ldots + kx^m}{H(x - \alpha)(x - \beta) \ldots (x - \lambda)}$.

It can therefore be decomposed into partial fractions by the methods we have explained above, and so we can find $\int f(x)\,dx$.

PROBLEMS

Integrate (1) $\displaystyle\int \frac{dx}{(x + i)(x - i)}$ (2) $\displaystyle\int \frac{x\,dx}{(x + \omega)(x - \omega)}$

(3) $\displaystyle\int \frac{(x^3 + x)\,dx}{(x + i)^2(x - i)^2}$

15.6 Real factors

We have now obtained in theory a general solution of the problem of integrating any rational function, i.e. the ratio of any two polynomials. In fact, this solution is not only a theoretical one, it is also quite practicable. But when we wish to integrate a real expression it may be rather inconvenient to have to manipulate a number of complex roots of an equation, and it may be useful to rearrange the results in a form involving real numbers only.

Suppose that $y = P(x) = A + Bx + Cx^2 + \ldots + x^n$ is the polynomial we wish to factorize, where the coefficients $A, B, C \ldots$ are all real. If all the roots of the equation $P(x) = 0$ are real, the previous section gives us a real factorization. If not, let α_1 be a non-real root: then $P(x)$ will have the non-real factor $x - \alpha_1$. It was, however, shown in Section 14.18 that the complex conjugate $\bar{\alpha}_1$ will also be a root, and $P(x)$ will have the factor $x - \bar{\alpha}_1$. This is different from $x - \alpha_1$, since α_1 is not real, and therefore $P(x)$ has both factors $(x - \alpha_1)$ and $(x - \bar{\alpha}_1)$, so that

$$P(x) = (x - \alpha_1)(x - \bar{\alpha}_1)P_1(x)$$

where $P_1(x)$ is the quotient, a polynomial of degree 2 less than that of $P(x)$. Let us write α_1 as the sum $a_1 + ib_1$ of real and imaginary parts; by definition $\bar{\alpha}_1 = a_1 - ib_1$ and

$$\begin{aligned}(x - \alpha_1)(x - \bar{\alpha}_1) &= (x - a_1 - ib_1)(x - a_1 + ib_1)\\ &= x^2 - 2a_1x + (a_1{}^2 + b_1{}^2)\\ &= x^2 - 2a_1x + c_1 \quad \text{(say)}\end{aligned}$$

where $c_1 = a_1{}^2 + b_1{}^2 = |\alpha_1|^2$. So

$$\begin{aligned}P(x) &= (x^2 - 2a_1x + c_1)P_1(x), \quad \text{or}\\ P_1(x) &= P(x)/(x^2 - 2a_1x + c_1).\end{aligned}$$

But the numbers a_1 and c_1 are real, and $P(x)$ has real coefficients: the polynomial $P_1(x)$ obtained by division must therefore also have real coefficients. We can now repeat the argument with P_1: if it has a non-real factor $(x - \alpha_2) = (x - a_2 - ib_2)$ it will also be divisible by $(x - \bar{\alpha}_2) = (x - a_2 + ib_2)$ and therefore by $(x - \alpha_2)(x - \bar{\alpha}_2) = x^2 - 2a_2x + c_2$. Continuing in this way we can reduce any real polynomial to real factors:

$$P(x) = (x^2 - 2a_1x - c_1)(x^2 - 2a_2x - c_2) \ldots (x - \beta_1)(x - \beta_2) \quad . \quad (15.40)$$

where the quadratic expressions $(x^2 - 2a_1x - c_1)$, $(x^2 - 2a_2x - c_2)$, etc., are obtained by combining pairs of complex conjugate factors, while the linear expressions $(x - \beta_1)$, $(x - \beta_2)$, etc., are obtained from the real roots of the equation $P(x) = 0$. These factors need not be all different. A quadratic factor can be repeated: we might have

$$x^2 - 2a_1x - c_1 = x^2 - 2a_2x - c_2;$$

and the same holds for linear factors.

Thus since the roots of the equation $x^3 = 1$ are ω, ω^2 and 1 [for $\bar{\omega} = \omega^2$] the polynomial $x^3 - 1$ can be factorized in complex form as $(x - \omega)(x - \omega^2)(x - 1)$, or in real form as $(x^2 + x + 1)(x - 1)$.

15·7 Partial fractions with quadratic factors

Since two complex factors can be combined into a single real (quadratic) factor, it will be natural to do the same with the corresponding partial fractions. Thus

$$\frac{1}{x^3 - 1} = \frac{\omega}{3(x - \omega)} + \frac{\omega^2}{3(x - \omega^2)} + \frac{1}{3(x - 1)}$$

by the ordinary decomposition. Addition of the first two partial fractions gives

$$\frac{\omega}{3(x - \omega)} + \frac{\omega^2}{3(x - \omega^2)} = \frac{\omega(x - \omega^2) + \omega^2(x - \omega)}{3(x - \omega)(x - \omega^2)}$$

$$= \frac{-x - 2}{3(x^2 + x + 1)}$$

Similarly

$$\frac{2x}{x^3+x^2+x+1} = \frac{2x}{(x+1)(x+i)(x-i)} = \frac{1+i}{2(x-i)} + \frac{1-i}{2(x-i)} - \frac{1}{x+1}$$

and by adding together the first two fractions we obtain

$$\frac{2x}{x^3 + x^2 + x + 1} = \frac{x + 1}{x^2 + 1} - \frac{1}{x + 1}$$

In general, if the denominator of a real rational function has a quadratic factor $(x^2 - 2a_1x + c_1)$ which is the product of two conjugate complex factors $(x - \alpha_1)(x - \bar{\alpha}_1)$, then the corresponding conjugate partial fractions can be combined together to make a real partial fraction with quadratic denominator. Thus

$$\frac{A_1}{x - \alpha_1} + \frac{\bar{A}_1}{x - \bar{\alpha}_1} = \frac{B_1x + C_1}{x^2 - 2a_1x + c_1}$$

where B_1 and C_1 are real. Suppose we wished to find the real partial fraction decomposition of $2x/(x^3 + x^2 + x + 1)$ directly, without first splitting it into complex fractions and then recombining them. We try the decomposition

$$\frac{2x}{(x^2 + 1)(x + 1)} = \frac{Bx + C}{x^2 + 1} + \frac{D}{x + 1}.$$

The value of D is found as usual by multiplying through by $(x + 1)$ and then putting $x = -1$; this gives $D = -1$.

If we similarly multiply by $(x^2 + 1)$, and then put $x = i$ (which makes $x^2 + 1 = 0$), we get

$$2i/(i + 1) = Bi + C.$$

The left-hand side is simplified by multiplying numerator and denominator by $-i + 1$, the conjugate of the denominator. This gives

$$\frac{2i(-i + 1)}{(i + 1)(-i + 1)} = \frac{2 + 2i}{-i^2 + 1} = 1 + i = Bi + C.$$

Since B and C are real, it follows that $B = 1$, $C = 1$, and this completes the decomposition. As another example, consider the decomposition

$$\frac{x^2 + 9x + 2}{(x^2 - 2x + 4)(x + 2)} = \frac{Bx + C}{x^2 - 2x + 4} + \frac{D}{x + 2}.$$

To find D, multiply through by $(x + 2)$ and set $x = -2$: this gives $D = -1$. Let ρ, $\bar{\rho}$ be the roots of $x^2 - 2x + 4 = 0$. These can be found to be $1 \pm \sqrt{(-3)}$. Multiply through by $(x^2 - 2x + 4)$ and then set $x = \rho$. This shows that $B\rho + C$ is equal to the value of

$$(x^2 + 9x + 2)/(x + 2)$$

when $x = \rho$. Now there is a quick method of finding the value of a real polynomial $P(x)$ when x takes a particular complex value, such as ρ. We know that ρ is a solution of the real equation $x^2 - 2x + 4 = 0$. Now divide $P(x)$ by $x^2 - 2x + 4$ (which we can legitimately do so long as $x^2 - 2x + 4 \neq 0$) to give quotient $Q(x)$ and remainder $R(x)$, so that

$$P(x) = (x^2 - 2x + 4)Q(x) + R(x).$$

Take the limit of this equation as $x \to \rho$; since all the parts are polynomials in x, there is no difficulty about zero denominators, so that the effect is the same as setting $x = \rho$. This gives $P(\rho) = R(\rho)$.

For example, on dividing $(x^2 + 9x + 2)$ by $(x^2 - 2x + 4)$ we get

$$x^2 + 9x + 2 = (x^2 - 2x + 4) \times 1 + (11x - 2).$$

Hence, on putting $x = \rho = 1 + \sqrt{(-3)}$ we have

$$\rho^2 + 9\rho + 2 = 11\rho - 2 = 9 + 11\sqrt{(-3)}.$$

But we have shown that

$$B\rho + C = (\rho^2 + 9\rho + 2)/(\rho + 2)$$

$$(B + C) + B\sqrt{(-3)} = \frac{9 + 11\sqrt{(-3)}}{3 + \sqrt{(-3)}}$$

$$= \frac{[9 + 11\sqrt{(-3)}][3 - \sqrt{(-3)}]}{[3 + \sqrt{(-3)}][3 - \sqrt{(-3)}]}$$

$$= [60 + 24\sqrt{(-3)}]/12$$

$$= 5 + 2\sqrt{(-3)}.$$

Since B and C are real, it follows that $B + C = 5$, $B = 2$, whence $C = 3$, and the final decomposition is

$$\frac{x^2 + 9x + 2}{(x^2 - 2x + 4)(x + 2)} = \frac{2x + 3}{x^2 - 2x + 4} - \frac{1}{x + 2}.$$

It can be shown that a repeated quadratic factor must be dealt with as follows:

$$\frac{x^3 - 32}{x(x^2 + 4)^2} = \frac{A}{x} + \frac{Bx + C}{x^2 + 4} + \frac{Dx + E}{(x^2 + 4)^2}.$$

To find A, multiply through by x and put $x = 0$; this gives $A = -2$. To find D and E, multiply through by $(x^2 + 4)^2$ and put $x = 2i$; this gives

$$[(2i)^3 - 32]/2i = -4 + 16i = 2iD + E,$$

whence $D = 8$, $E = -4$. There are various ways of finding B and C analogous to those discussed in Section 15.4. For example, we can take the terms already found over to the left-hand side, to give

$$\frac{x^3 - 32}{x(x^2 + 4)^2} + \frac{2}{x} + \frac{-8x + 4}{(x^2 + 4)^2} = \frac{Bx + C}{x^2 + 4}.$$

On simplification of the left-hand side we find $B = 2$, $C = 1$.

We now have to face the problem of integrating an expression of the form $(Bx + C)/(x^2 - 2ax + c)^n$. To illustrate the process, consider the particular case $\int \frac{(x + 1)\,dx}{(x^2 - x + \frac{1}{2})^2}$. The first step is to "complete the square" in the denominator. Here $(x^2 - x + \frac{1}{2}) = (x - \frac{1}{2})^2 + (\frac{1}{2})^2$: in general $x^2 - 2ax + c = (x - a)^2 + (c - a^2) = (x - a)^2 + b^2$, where $b = \sqrt{(c - a^2)}$. ($b^2 = c - a^2$ will always be positive, since $x^2 - 2ax + c$ has non-real factors.)

Now change the variable from x to $X = x - a$; the integral becomes

$$\int \frac{Bx + C}{(x^2 - 2ax + c)^n}\,dx = \int \frac{B(X + a) + C}{(X^2 + b^2)^n}\frac{dx}{dX}\,dX$$
$$= \int \frac{BX + C'}{(X^2 + b^2)^n}\,dX$$

where $C' = Ba + C$; and in our particular example, $X = x - \frac{1}{2}$,

$$\int \frac{(x + 1)\,dx}{(x^2 - x + \frac{1}{2})^2} = \int \frac{X + \frac{3}{2}}{(X^2 + \frac{1}{4})^2}\,dX$$
$$= \int \frac{X\,dX}{(X^2 + \frac{1}{4})^2} + \int \frac{\frac{3}{2}\,dX}{(X^2 + \frac{1}{4})^2}$$

The first integral is easily dealt with by the substitution $t = X^2 + \frac{1}{4}$, for then $dt/dX = 2X$, i.e. $X = \frac{1}{2}dt/dX$, and the integral becomes

$$\int \frac{\frac{1}{2}dt/dX \,.\, dX}{t^2} = \int \frac{dt}{2t^2} = -\tfrac{1}{2}t^{-1} + K$$

$$= -\tfrac{1}{2}(X^2 + \tfrac{1}{4})^{-1} + K = -\tfrac{1}{2}(x^2 - x + \tfrac{1}{2})^{-1} + K,$$

where K is the constant of integration. (We cannot use the customary letter C, as it has already been used in another connection.) In general the integral $\int \frac{BX\,dX}{(X^2 + b^2)^n}$ can be evaluated by substituting $X^2 + b^2 = t$; it then becomes $\int \frac{1}{2}Bt^{-n}\,dt$.

The second integral will be $\int \frac{\frac{3}{2}\,dX}{(X^2 + \frac{1}{4})^2}$, or in general $\int \frac{C'\,dX}{(X^2 + b^2)^n}$. If $n = 1$ this can be directly integrated, for

$$\int \frac{C'\,dX}{X^2 + b^2} = C'\,b^{-1}\tan^{-1}(X/b)$$

by a standard formula. If $n > 1$ then we invoke the following ***reduction formula***:

$$\int \frac{dX}{(X^2+b^2)^n} = \frac{X}{2(n-1)b^2(X^2+b^2)^{n-1}} + \frac{(2n-3)}{2(n-1)b^2}\int \frac{dX}{(X^2+b^2)^{n-1}} \quad (15.41)$$

which is easily checked by differentiation. This expresses the integral $\int \frac{dx}{(X^2 + b^2)^n}$ in terms of the integral $\int \frac{dX}{(X^2 + b^2)^{n-1}}$; and by repeated use of the formula we can finally simplify it to $\int \frac{dX}{(X^2 + b^2)}$ which is known. Thus in our case $n = 2$.

$$\int \frac{\frac{3}{2}dX}{(X^2 + \frac{1}{4})^2} = \frac{\frac{3}{2}X}{2.\frac{1}{4}.(X^2 + \frac{1}{4})} + \frac{\frac{3}{2}.1}{2.1.\frac{1}{4}}\int \frac{dX}{X^2 + \frac{1}{4}}$$

$$= \frac{3X}{X^2 + \frac{1}{4}} + 6\tan^{-1}(2X)$$

$$= \frac{3(x - \frac{1}{2})}{x^2 - x + \frac{1}{2}} + 6\tan^{-1}(2x - 1)$$

An addition of the integrals $\int \frac{XdX}{(X^2 + \frac{1}{4})^2}$ and $\int \frac{\frac{3}{2}dX}{(X^2 + \frac{1}{4})^2}$ finally gives

$$\int \frac{(x+1)\,dx}{(x^2-x+\frac{1}{2})^2} = -\frac{1}{2x^2-2x+1} + \frac{3(x - \frac{1}{2})}{x^2-x+\frac{1}{2}} + 6\,\tan^{-1}(2x-1) + K$$

$$= \frac{3x-2}{x^2-x+\frac{1}{2}} + 6\tan^{-1}(2x-1) + K$$

Any other real rational function can be similarly dealt with.

PROBLEMS

Integrate the following functions:

(1) $\dfrac{1}{(x+3)(x-4)}$ (2) $\dfrac{1}{(x^2+4)(x+3)}$ (3) $\dfrac{x+3}{(x^2+4)(x^2+1)}$

(4) $\dfrac{1}{x(x^2+1)^2}$ (5) $\dfrac{2x-4}{(x^2+1)(x^2+x+1)}$

15.8 Rational functions of sines and cosines

Any rational function of $\sin x$ and $\cos x$, as, for example, $[1 + \sin x] \div [\cos x + \sin x]$ or $[(\cos x)^2 + 3(\sin x)^3]/[\sin x + 1]$, can always be integrated. The general method is to make the substitution $t = \tan \frac{1}{2}x$. Then $\sin x = 2t/(1+t^2)$, $\cos x = (1-t^2)/(1+t^2)$, and $dx/dt = 2/(1+t^2)$. The integral is therefore reduced to that of a rational function of t, and can be evaluated by the methods given in the preceding sections.

Similarly any rational function of $\sinh x$ and $\cosh x$ can be integrated by the substitution $t = \tanh \frac{1}{2}x$, for then $\sinh x = 2t/(1-t^2)$, $\cosh x = (1+t^2)/(1-t^2)$ and $dx/dt = 2/(1-t^2)$.

In particular cases there may be a short cut. For example, an expression of the form $\int f(\sin x).\cos x.dx$ can be integrated by the substitution $X = \sin x$, for $\cos x = dX/dx$ and the integral becomes $\int f(X)\,dX$.

EXAMPLES

(1) $$\int (\sin x)^3 \cos x.dx = \int X^3\,dX = \tfrac{1}{4}X^4 + C = \tfrac{1}{4}(\sin x)^4 + C.$$

(2) $$\begin{aligned}\int (\sin x)^3(\cos x)^3\,dx &= \int (\sin x)^3[1-(\sin x)^2]\cos x.dx \\ &= \int X^3(1-X^2)\,dX \\ &= \tfrac{1}{4}X^4 - \tfrac{1}{6}X^6 + C \\ &= \tfrac{1}{4}(\sin x)^4 - \tfrac{1}{6}(\sin x)^6 + C.\end{aligned}$$

Similarly any integral of the form $\int f(\cos x).\sin x.dx$ becomes $-\int f(X)\,dX$ on substituting $X = \cos X$. An integral of the form $\int (\sin x)^m (\cos x)^n\,dx$, where m and n are positive integers or zero, can always be found by using the identities $\cos x = \frac{1}{2}(e^{ix} + e^{-ix})$, $\sin x = (e^{ix} - e^{-ix})/2i$.

EXAMPLES

(1) $\int (\cos x)^2 dx = \int \frac{1}{4}(e^{2ix} + 2 + e^{-2ix})\, dx$

$$= \frac{e^{2ix}}{8i} + \frac{x}{2} - \frac{e^{-2ix}}{8i} + C$$

$$= \tfrac{1}{4}\sin 2x + \tfrac{1}{2}x + C.$$

(2) $\int (\cos x)^2(\sin x)^2 dx = \int -\frac{1}{16}(e^{ix} + e^{-ix})^2 (e^{ix} - e^{-ix})^2\, dx$

$$= -\frac{1}{16}\int (e^{2ix} - e^{-2ix})^2\, dx$$

$$= -\frac{1}{16}\int (e^{4ix} - 2 + e^{-4ix})\, dx$$

$$= -\frac{e^{4ix}}{64i} + \frac{x}{8} + \frac{e^{-4ix}}{64i} + C.$$

$$= -\tfrac{1}{32}\sin 4x + \tfrac{1}{8}x + C$$

If either m or n or both are negative integers a (possibly repeated) application of the reduction formulas (verifiable by differentiation)

$\int (\sin x)^m (\cos x)^n\, dx$

$$= \frac{m+n+2}{n+1}\int (\sin x)^m (\cos x)^{n+2}\, dx - \frac{(\sin x)^{m+1}(\cos x)^{n+1}}{n+1}$$

$$= \frac{m+n+2}{m+1}\int (\sin x)^{m+2} (\cos x)^{n}\, dx - \frac{(\sin x)^{m+1}(\cos x)^{n+1}}{m+1}$$

will reduce the integral to a simpler form.

If a function contains only even powers of $\sin x$ and $\cos x$ it can be integrated by substituting $X = \tan x$, $(\sin x)^2 = X^2/(1 + X^2)$, $(\cos x)^2 = 1/(1 + X^2)$, $dx/dX = 1/(1 + X^2)$.

15.9 Integrals containing irrational functions

A rational function of x and $\sqrt{(Ax + B)}$, where A and B are constants, can always be integrated. Examples of such functions are $x + \sqrt{(x + 1)}$, $(x^3 + 3x)/[1 + \sqrt{(2x - 3)}]$ and indeed any expression which can be obtained from x and $\sqrt{(Ax + B)}$ by using only combinations of additions, subtractions, multiplications and divisions, including additions and multiplications by ordinary real or complex numbers. The substitution $X = \sqrt{(Ax + B)}$ transforms the integral of any such function into the integral of a rational function of X.

EXAMPLES

(1) Find $\int x\sqrt{(x+1)}.dx$.

Put $X = \sqrt{(x+1)}$, then $x = X^2 - 1$, and $dx/dX = 2X$. The integral therefore becomes

$$\begin{aligned}\int (X^2-1)X.2X\,dX &= \int (2X^4 - 2X^2)\,dX \\ &= \tfrac{2}{5}X^5 - \tfrac{2}{3}X^3 + C \\ &= \tfrac{2}{5}(x+1)^{5/2} - \tfrac{2}{3}(x+1)^{3/2} + C.\end{aligned}$$

(2) Find $\int (2x+3)/\sqrt{(x-2)}.dx$.

Put $X = \sqrt{(x-2)}$: then $x = X^2 + 2$, and $dx/dX = 2X$. Thus

$$\begin{aligned}\int (2x+3)/\sqrt{(x-2)}.dx &= \int (2X^2+7).2X/X.dX \\ &= \tfrac{4}{3}X^3 + 14X \\ &= \tfrac{4}{3}(x-2)^{3/2} + 14(x-2)^{1/2}.\end{aligned}$$

More generally any rational function of x and $\left(\dfrac{Ax+B}{Cx+D}\right)^{\frac{1}{n}}$ can be integrated, where n is any positive integer. It is enough to substitute $X = \left(\dfrac{Ax+B}{Cx+D}\right)^{\frac{1}{n}}$ when the integral is transformed to the integral of a rational function of X.

In particular this implies that we can integrate any rational function of x and $\sqrt{(x^2+Bx+C)}$. For x^2+Bx+C can be factorized as, say, $(x-\alpha)(x-\beta)$; and then we can write $\sqrt{(x^2+Bx+C)} = (x-\alpha)\sqrt{\dfrac{x-\beta}{x-\alpha}}$. So by the substitution $X = \sqrt{\dfrac{x-\beta}{x-\alpha}}$ we can reduce the integral to one of a rational function of X. However, if α and β are complex this substitution is rather awkward, and it is then better to proceed by "completing the square", writing $x^2+Bx+C = (x+\tfrac{1}{2}B)^2 + (C-\tfrac{1}{4}B) = X^2+b^2$, where $X = x+\tfrac{1}{2}B$ and $b = \sqrt{(C-\tfrac{1}{4}B)}$. We then substitute $X = b\cos\theta$ and reduce the integral to a trigonometric one of the forms considered in Sections 15.1 and 15.8.

15.10 Integrals containing the square root of a cubic or quartic polynomial

In general an integral containing the square root of a third or fourth degree polynomial cannot be solved by the methods of the preceding section. Such integrals are called "elliptic integrals", and require the introduction of new functions, called "elliptic functions", which are very similar to the ordinary trigonometric functions in many ways. Unfortunately to present the theory in its simplest and most lucid form involves delving rather more deeply into the theory of complex functions than is possible here, and the reader is referred to text-books on the subject. A particularly simple and elegant account will be found in *Jacobian Elliptic Functions*, by E. H. Neville (2nd edn, 1951, O.U.P.) and in *The Elliptic Functions as they should be*, by A. Eagle (Galloway & Porter, 1958); there is a useful collection of formulas and tables in *Jacobian Elliptic Function Tables*, by L. M. Milne-Thomson (Dover Publications, 1950).

Integrals involving roots of polynomials of degree 5 or more are best dealt with by numerical methods (Section 11.16).

CHAPTER SIXTEEN

PHYSICAL AND CHEMICAL MAGNITUDES

16.1 Time

There are two natural units of time, the day and the year.

The day, or more accurately the "mean solar day", is measured as the average interval of time between successive times at which the sun is due south of any given point. It is perhaps worth pointing out that this is not the same as the time of rotation of the earth, or "sidereal day": for in consequence of the earth's revolution round the sun the earth has to do more than a complete revolution to catch up with the sun. But it is the ordinary solar day which concerns us here since it corresponds to the regular alternation of darkness and light; the sidereal day is of interest only to astronomers. For convenience the day is divided into 24 hours, the hours into 60 minutes, and the minutes into 60 seconds each, in a way which is both familiar and universal. That is, 1 day = 86,400 seconds.

The year, or more exactly the "tropical year", is defined as the average interval between successive spring equinoxes (times when the plane of the equator passes through the sun). Again this is not quite identical with the time of revolution of the earth round the sun, or "sidereal year", because the plane of the equator is slowly changing its direction. That phenomenon is known as "precession". Nevertheless it is the tropical year which is of the greatest interest, since it corresponds to the regular recurrence of the four seasons. It has a length of 365·2422 days.

In practice we replace the exact tropical year by a "civil year" of 365 or 366 days. The alternation of leap and non-leap years is so arranged by the Gregorian calendar that the average length of the civil year over a long period is 365·2425 days: this differs quite inappreciably from the length 365·2422 days of the exact tropical year.

Besides these units we have the conventional ones of the week = 7 days, and the month, originally no doubt derived from the moon, but now a conventional and irregular unit. This leads to such problems as to find the number of days between two given dates, such as Sept. 30, 1949, and June 21, 1950. Such problems are easily dealt with as follows. First consider a non-leap year. Instead of numbering the days in the usual way, we shall simply number them consecutively, beginning with January 1 as day 1, and ending with December 31 as day 365. Thus Jan. 26, 1949, will be day 26 of year 1949, and can be written simply as 49D26; March 3, 1950, will be day 62, or 50D62 in brief. Any date can be converted by the use of the following table.

Table 16.1—Month numbers

January (non-leap)	0	June	151
,, (leap)	−1	July	181
February (non-leap)	31	August	212
,, (leap)	30	September	243
March	59	October	273
April	90	November	304
May	120	December	334

To find the day number, add the day of the month to the "month number" given in this table: e.g. Sept. 30 = day (243 + 30) = day 273. June 21 = day (151 + 21) = day 172. To find the time between any two dates in the same year it is then only necessary to subtract the smaller day number from the larger: e.g. from noon on June 21 to noon on Sept. 30 is 273 − 172 = 101 days.

To find the number of days between two dates in different years requires two new definitions. We shall restrict ourselves for simplicity to the 20th century: the "year number" of a given date will then be the last 2 figures of the year, omitting the prefix 19: so 1949 has year number $y = 49$, and 1950 has year number $y = 50$. The date June 21, 1949 = 49D172 has year number $y = 49$ and day number $d = 172$; Sept. 30, 1950 = 50D273, with $y = 50$, $d = 273$. We also introduce the "quarter year number" q, which is the quotient obtained by dividing the year number by 4, ignoring the remainder. Thus for 1917, $y = 17$, $q = 4$; for 1949, $y = 49$, $q = 12$. The number of days between day d_1 of year y_1 (quarter year q_1) and day d_2 of year y_2 (quarter year q_2) is then

$$n = 365\,(y_2 - y_1) + (q_2 - q_1) + (d_2 - d_1) \quad . \quad . \quad (16.1)$$

For example, from noon on June 21, 1949 = 49D172 to noon on Sept. 30, 1950 = 50D273 is

$$365\,(50 - 49) + (12 - 12) + (273 - 172) = 466 \text{ days}:$$

from Sept. 30, 1946 = 46D273 to June 21, 1949 = 49D172 is

$$365\,(49 - 46) + (12 - 11) + (172 - 273) = 995 \text{ days}.$$

These numbers are also useful in finding the day of the week for any given date. The rule is to divide the sum $(y + q + d)$ of the year, quarter year, and day numbers by 7; the remainder then determines the day as in Table 16.2.

Table 16.2—Determination of the day of the week

Remainder	*Day*	*Remainder*	*Day*
0	Sunday	4	Thursday
1	Monday	5	Friday
2	Tuesday	6	Saturday
3	Wednesday		

This is equivalent to the continental numbering in which the week begins on Monday. Thus July 3, 1951 = 51D184 by Table 16.1. $y + q + d = 51 + 12 + 184 = 247$. On division by 7 this gives remainder 2, showing that it is a Tuesday.

For leap years the scheme is slightly modified by counting Jan. 1 as day 0 instead of day 1, Jan. 2 as day 1, and so on consecutively to Dec. 31 = day 365. This has two advantages. Firstly, the month numbers have to be adjusted only for the months of January and February, as shown in Table 16.1. So any date after February will have the same day number whether the year is leap or not: Christmas day will be day 359 in any year. Secondly, the rules for finding the number of days between two given dates and for finding the day of the week, apply without any change. If we had counted Jan. 1 as "day 1" in a leap year these rules would have had to be modified.

EXAMPLES

(1) Find the number of days from June 13, 1948, to Feb. 2, 1952. From Table 16.1 these dates are 48D164 and 52D32. The number of days is $n = 365\,(52 - 48) + (13 - 12) + (32 - 164) = 1329$.

(2) What day of the week will Feb. 4, 1960, be? $60 + 15 + 34 = 109$: this gives remainder 4 on division by 7, i.e. this will be Thursday.

16.2 Length

For measures other than those of time there are, of course, two competing systems of units, the metric system, used internationally and almost universal in scientific work, and the British system.

The metric system has the advantage of having systematic subdivisions of its units by powers of 10, according to the prefixes deci- (d) = 10^{-1}, centi- (c) = 10^{-2}, milli- (m) = 10^{-3}, micro- (μ) = 10^{-6}, nano- (n) = 10^{-9}, pico- (p) = 10^{-12}, femto- (f) = 10^{-15}, atto- (a) = 10^{-18}. Also we have the multiples deka- (dk- or da-) = 10, hecto- (h) = 10^2, kilo-(k) = 10^3, mega (M) = 10^6, giga- (G) = 10^9 and tera- (T) = 10^{12}.

In contrast to this the British system has irregular subdivisions, such as 12 inches to 1 foot, 3 feet to 1 yard, and in addition the units employed in Britain are slightly different from the American ones. For these reasons it seems doomed to eventual extinction, in spite of certain advantages over the metric system. But for various reasons medical and biometrical data are still often collected in British units, and while that is so it is useful to have conversion factors.

The standard British inch is 2·54 cm, formerly 2·5399956 centimetres. The American inch is defined by the relation 1 metre = 39·37 inches; that is, 1 inch = 2·5400051 cm. For practical purposes all these definitions are equivalent. Further measures in common use are

1 mil = 10^{-3} inches.
1 foot = 12 inches.
1 yard = 3 feet = 36 inches.
1 mile = 8 furlongs = 1760 yards = 5280 feet = 63,360 inches.
1 fathom = 2 yards = 72 inches.

The principal conversion factors are therefore as follows (the common logarithms are also given):

			Logarithms
1 inch	=	2·54 cm	·40483
1 foot	=	·30480 m	$\bar{1}$·48402
1 yard	=	·91440 m	$\bar{1}$·96114
1 mile	=	1·6093 km	·20665
1 metre	=	39·370 inch	1·59517
	=	3·2808 feet	·51598
	=	1·0936 yards	·03886
1 km	=	·62137 miles	$\bar{1}$·79335

16.3 Area

This is simply the product of two lengths, and is accordingly measured in square inches, square centimetres, square metres, etc. The metric *are* = 100 square metres. The British system has the additional units 1 acre = 4 roods = 4840 square yards = $\frac{1}{640}$ square mile. Thus the principal conversion factors are:

			Logarithms
1 yd^2	=	·83613 m^2	$\bar{1}$·92227
1 ft^2	=	·092903 m^2	$\bar{2}$·96803
1 in^2	=	6·4516 cm^2	·80967
1 m^2	=	1·1960 yd^2	·07773
	=	10·764 ft^2	1·03197
1 cm^2	=	·15500 in^2	$\bar{1}$·19033
1 $mile^2$	=	2·5900 km^2	·41330
1 km^2	=	·38610 $mile^2$	$\bar{1}$·58670

16.4 Volume

A volume is measured as the product of length, breadth, and height, and will therefore be most naturally measured in cubic inches, cubic centimetres (cc), or cubic metres. The cubic metre is sometimes called the "stere".

The metric system has a second unit of volume, the litre, defined as the volume of 1 kilogram of distilled water at 4°C and 760 mm of mercury pressure. The exact relation is 1 litre = 1000·028 cc, or 1 cc = ·000999972 litres. For most purposes, therefore, a litre is equivalent to 1000 cc.

The British system has a whole range of extra units, based on the imperial gallon, which is defined as the volume of 10 pounds avoirdupois of distilled water, weighed in air against brass weights, at 62° F and 30 inches of mercury pressure. The derived units, still used in medicine, are related thus:

60 minims	=	1 fluid drachm
8 fluid drachms	=	1 fluid ounce
20 fluid ounces	=	4 gills = 1 pint
2 pints	=	1 quart
4 quarts	=	8 pints = 160 fluid ounces = 1 gallon.

The following are the conversion factors:

			Logarithms
1 pint	=	·56826 dm^3	$\bar{1}$·75455
1 gallon	=	4·5461 dm^3	·65764
1 in^3	=	16·387 cc	1·21450
1 ft^3	=	28·317 dm^3	1·45205
1 dm^3	=	·035315 ft^3	$\bar{2}$·54795
1 cc	=	·061024 in^3	$\bar{2}$·78550
1 gallon	=	·16054 ft^3	$\bar{1}$·20558
1 ft^3	=	6·2288 gallons	·79442

In the U.S.A. the gallon is defined as 231 cubic inches, i.e. as 3·7854 dm^3.

The above relations will hold for litres in place of dm^3, except that the last significant figure may be slightly changed.

16.5 Mass

The metric unit of mass is the kilogram. 1 gram = ·001 kg, 1 tonne = 1000 kg.

The usual British unit is the pound avoirdupois = ·45359243 kg. This is divided thus: 1 dram = $\frac{1}{256}$ lb.; 1 ounce (av.) = $\frac{1}{16}$ lb; 1 stone = 14 lb.; 1 quarter = 28 lb.; 1 hundredweight = 112 lb.; 1 ton = 2240 lb. In America the "short hundredweight" = 100 lb. and "short ton" = 2000 lb are also in use.

There is also a second system of weights, the "Troy" and "Apothecary's" system, used in medicine. The basic unit is the grain $= \frac{1}{7000}$ lb avoirdupois. The other units are the scruple = 20 grains, the drachm = 60 grains, the pennyweight = 24 grains, the Troy ounce = 480 grains, and the Troy pound (rarely used) = 5760 grains = 12 Troy ounces. The principal conversion factors are:

			Logarithms
1 grain	=	·064799 gm	$\bar{2}$·81157
1 drachm	=	3·8879 gm	·58972
1 troy ounce	=	31·104 gm	1·49282
1 av. ounce	=	28·350 gm	1·45255
1 av. pound	=	·45359 kg	$\bar{1}$·65667
1 stone	=	6·3503 kg	·80279
1 ton	=	1·0161 tonne	·00691
1 gm	=	·035274 oz av.	$\bar{2}$·54745
1 kg	=	2·2046 lb av.	·34333
1 tonne	=	·98421 ton	$\bar{1}$·99309

16.6 Velocity

Velocity is measured by distance divided by time. The natural units are accordingly feet per second, miles per hour, metres per second and kilometres per hour.

		Logarithms
1 ft/sec	$= \frac{15}{22}$ m.p.h. = ·68182 m.p.h.	$\bar{1}$·83367
1 m.p.h.	$= \frac{22}{15}$ ft/sec = 1·4667 ft/sec	·16633
1 m/sec	= 3·6 km/hour	·55630
1 km/hour	$= \frac{5}{18}$ m/sec = ·27778 m/sec	$\bar{1}$·44370
1 ft/sec	= ·3048 m/sec	$\bar{1}$·48402
1 m/sec	= 3·2808 ft/sec	·51598

16.7 Derived units: the M.K.S. and C.G.S. systems

From now on we deal with more complicated quantities, such as accelerations, forces, pressures, etc.: and fortunately the units are considerably simplified. For it is usual to express such quantities in a consistent system of units, derived from a single unit of time, a single unit of mass, and a single unit of length.

In all systems the unit of time is the second. The standard units of mass and length are the kilogram and metre respectively, and the derived units form the metre-kilogram-second or M.K.S. or Giorgi system. Sometimes the gram is chosen as the unit of mass and the centimetre as unit of length, and we obtain the centimetre-gram-second or C.G.S. system. The M.K.S. system has the advantage that it includes the common electrical units, such as the volt, the ampere, and the watt. The C.G.S. system does not agree with these electrical units, which in itself is inconvenient, and moreover it gives units which are of rather awkward sizes for most purposes, e.g. the "erg" or energy unit is too small to be convenient. Fortunately the two systems differ only by powers of 10, and so are readily interconverted by a shift in the decimal point. Thus a velocity of 1 cm/sec = ·01 m/sec, and an acceleration of 1 cm/sec^2 = ·01 m/sec^2.

16.8 Change of units

Almost always we measure a velocity by dividing the distance covered by the time: and to express this division we write the unit of velocity symbolically as the quotient, "unit of distance/unit of time", as "metres/sec" or "miles/hour". Such a choice is very natural, and reduces the arithmetic to a simple division. But it is not inevitable. It would be possible to measure distances in feet, time in seconds, and velocity in miles per hour. In that case, a distance y feet covered in a time t seconds corresponds to a velocity of $v = \frac{15}{22}\, y/t$ miles per hour. Such a choice of units means that the relation $v = y/t$ no longer holds: but for some very special purposes, such as the timing of cars or trains, it might be more convenient than the usual choice of feet per second. A similar situation could arise in metric units if we measured the distance y in metres, the time t in seconds, and the velocity v in kilometres per hour. The relation would then be $v = 3{\cdot}6\, y/t$. We can say that when we take 1 metre as the unit of distance, and 1 second as the unit of time, then 1 metre/sec is a "derived" unit of velocity, or to use an alternative phrase, a "germane" unit. It can be defined as that unit which simplifies a particular formula. Thus if a length L is measured in metres, the natural derived unit of volume V is the cubic metre or stere, since the formula for the volume of a cube is then $V = L^3$. We could if we liked choose another unit as the unit of volume; for example we could take the volume of a sphere of radius 1, which might be called a "spherical metre". The volume V of a sphere of radius R would then be $V = R^3$ spherical metres, instead of $V = \frac{4}{3}\pi R^3$ cubic metres. But here experience shows that the cubic metre is a much more useful unit than the spherical metre—not by divine command, but simply because we find it so. If natural bodies were commonly of spherical shape it might be otherwise. Both the cubic and spherical metres could be considered as derived units, but here experience shows us which one to take.

Sometimes it may be convenient to change the principal units: it is then important to know what happens to the derived units. Suppose for instance that a velocity has been estimated to be 8 yards/minute, and we wish to change the units to inches and seconds. What then will the velocity be expressed in inches/sec?

Now a velocity of 8 yards/minute means one in which 8 yards = 8×36 inches are covered in 1 minute = 60 seconds. Since the velocity is calculated by dividing the distance gone by the time taken —for that is the meaning of our derived units—this velocity will be $\frac{8\times 36}{60}$ inch/sec = 4·8 inch/sec. But we can also obtain this answer by treating the symbolical division "yard/min." as if it was an actual division:

$$8 \text{ yards/min.} = \frac{8 \text{ yards}}{1 \text{ min.}} = \frac{8 \times 36 \text{ inches}}{60 \text{ sec}} = 4{\cdot}8 \text{ inch/sec.}$$

Similarly a volume of 1 metre³ means the volume of a cube of side 1 metre, i.e. 100 cm, and so by the formula $V = L^3$, where V is in cc and L in cm, this is 100^3 cm³. But we could also get this result by writing $(1 \text{ metre})^3 = (100 \text{ cm})^3 = 100^3 \text{ cm}^3$, treating the symbol "cm" exactly like any ordinary algebraic symbol. This pleasing property holds throughout the whole range of derived units, and makes their interconversion an easy matter.

16.9 Force, energy and power

In the metre-kilogram-second system the unit of force is the *newton*, defined as the force needed to give a kilogram mass an acceleration of 1 metre/sec². In general a force f newtons acting on a mass m kilograms produces an acceleration $D_t^2 y$ metres/sec² according to the law "force = mass × acceleration"

$$f = m\, D_t^2 y \qquad . \qquad . \qquad . \qquad . \qquad (16.2)$$

$D_t^2 y$, the second derivative with respect to the time, is often denoted by $\ddot{y}$ (read as "y double dot"). Similarly the first derivative $D_t y$ of the position y, or in ordinary language the velocity, is often written as $\dot{y}$ (read, "y dot").

In the C.G.S. system the unit of force is the *dyne*, which gives a mass of 1 gram an acceleration of 1 cm/sec². In the British (foot-pound-second) system the corresponding unit is the *poundal*, which gives one pound an acceleration of 1 ft/sec²: but this is very rarely used in practice. The conversion factors are

$$1 \text{ dyne} = 10^{-5} \text{ newtons}$$
$$1 \text{ poundal} = {\cdot}13825 \text{ newtons}$$

The work done by a force in moving an object is measured by the product of the distance gone times the component of the force in the direction of motion. The M.K.S. unit of work is the *joule* = newton × metre = kg . m²/sec². The C.G.S. unit is the *erg* = dyne × cm = gm . cm²/sec². Thus 1 erg = 10^{-7} joule.

Energy has the same physical nature as work, and is accordingly also measured in joules or ergs. The moment of a force, though not similar to "work" in its conception, is also measured quantitatively as a product of a length and a force, and so can also be expressed in joules.

Power is defined as the rate of doing work. If δW is the amount of work done in a time δt, the average power is measured as $\delta W/\delta t$, and the instantaneous power as the limit of this, or dW/dt, or $\dot{W}$. The M.K.S. unit of power is accordingly the *watt*, or joule per second ($=$ kg . m^2/sec^3). The C.G.S. unit of 1 erg/sec $= 10^{-7}$ watts seems to have no distinctive name.

16.10 Gravitational units

Sometimes a force is measured not in absolute units, but as a weight: thus 1 kg weight = the pull of the earth on 1 kg mass. If g is the acceleration due to gravity, then the weight of a mass m is mg absolute units, or m "gravitational units". Unfortunately g varies from point to point on the earth's surface, so that these gravitational units are not constants, but depend on the place selected for measurement. At Greenwich $g = 9{\cdot}8117$ m/sec $= 981{\cdot}17$ cm/sec $= 32{\cdot}1906$ ft/sec^2, so that 1 kg weight = 9·8117 newtons (logarithm ·99174), 1 gm weight = 981·17 dynes, 1 lb weight = 32·191 poundals = 4·4505 newtons. A newton is therefore equivalent to the weight of $\frac{1}{10}$ kg or $3\frac{1}{2}$ ounces approximately. If we take the usual standard value of g at latitude 45°, namely 9·80665 m/sec^2, or 32·1740 ft/sec^2, we find

		Logarithms
1 kg wt	= 9·8067 newtons	·99152
1 lb wt	= 4·4482 newtons	·64819

The corresponding units of work and power are

1 kg metre	= 9·8067 joules	·99152
1 ft lb	= 1·3558 joules	·13220
1 metric horse-power	= (75 kg metres/sec) = 735·50 watts	2·86658
1 British horse-power	= (550 ft lb/sec) = 745·70 watts	2·87256

16.11 Density

If a volume V cubic metres of any substance contains a mass m kilograms, we say that the average density is $\rho = m/V$ kilograms/cubic metre. This is the M.K.S. unit; in the C.G.S. system the unit will be 1 gm/cc = 1000 kg/m^3.

Thus the density of water is about 1000 (at 4° C; at 15° C it is 999·1). Most other solids and liquids have densities in the range 500 to 10,000 (or in gm/cc, ·5 to 10). Lead has a density $1{\cdot}137 \times 10^4$ kg/m^3. The density of air at 0° C and 760 mm of mercury pressure is 1·293 kg/m^3, and most other common gases are of a similar density.

Now let us consider a *small* volume: to indicate that it is small we shall call it δV instead of V, meaning "a little bit of volume". Then the mass δm contained within it will also become small, and common experience suggests that the density will be practically uniform throughout δV. In calling δV "small" we mean that it must be small in length, in breadth, and in height: a very long thin body could have a very small volume but vary considerably in its density. One can further imagine this element δV of volume shrinking ever smaller and smaller, so that the ratio $\delta m/\delta V$ or average density tends to a definite limit ρ which can be called the "density at the point" and denoted by dm/dV. Strictly speaking this is nonsense, for below a certain size we should notice the presence of atoms, and the density will not be uniform. But it is a convenient idealization to speak as if the limit did exist, and will not do any harm provided that its limitations are clearly recognized.

Actually the density so defined is only a special kind of density. If the mass is concentrated along a line whose thickness can be neglected, such as a thin wire, we can speak of the "line density" λ measured in kilograms per metre, or grams per centimetre. λ will be the limit of the ratio $\delta m/\delta L =$ mass/length for a small length δL of the wire, as $\delta L \to 0$. If the mass is concentrated on a surface it will have a "surface density" σ kg/m^2, measured as mass per unit area. We can also speak of densities of other quantities than mass. An electric charge can have a line, surface, or volume density; a surface density of a force is called a "pressure", and so on.

16.12 Multiple integrals

Imagine a curve C (Fig. 16.1) on which the line-density of a substance is λ (λ need not be constant, but can vary from one point of the curve

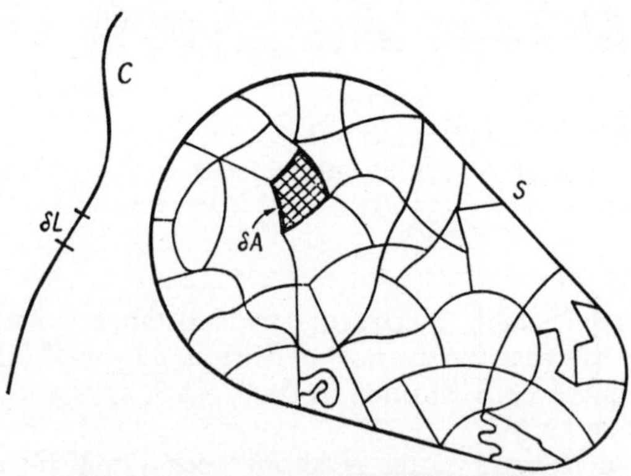

Fig. 16.1—Single and double integrals

to another.) If then we take a small length δL of the curve, the mass δm contained in this will be approximately $\lambda\,\delta L$, and therefore the total mass on the curve will be approximately $M = \Sigma\, \delta m \simeq \Sigma\, \lambda\, \delta L$. If we suppose these small lengths δL to be cut up into still smaller lengths, the formula will approximate even more closely to the total mass: and in the limit, as all the δL tend to zero, we shall get the exact formula $M = \int_C \lambda\, dL$. This is an ordinary definite integral: the suffix C means that the integral is taken along the curve, with the two ends of the curve as limits of integration.

Now suppose we have a surface S on which the surface density is σ (Fig. 16.1) (σ may vary from point to point of the surface). Let us cut this surface up into small areas δA; that is, these areas must be small both in length and breadth. Then the mass contained in the area δA is approximately $\delta m \simeq \sigma \,.\, \delta A$, and the total mass is $M = \Sigma\, \delta m \simeq \Sigma\, \sigma \,.\, \delta A$. A division of these areas into still smaller ones will improve the approximation: and as all the small bits of area δA tend to zero the sum $\Sigma\, \sigma \,.\, \delta A$ will tend to the total mass M as a limit. Now this is analogous to an ordinary definite integral, but is of a new kind. It is called the "double" or "surface" integral $\int_S \sigma \,.\, dA$.

In the same way the mass of a three-dimensional body B of volume-density ρ (possibly varying from point to point) can be found by cutting it up into small elements of volume δV. Each of these will contain a mass $\delta m \simeq \rho\, \delta V$, and the total mass M will be approximated to by the sum $\Sigma\, \rho\, \delta V$ which will tend to the limit M as all the volume elements δV tend to zero. This limit is called the "triple" or "volume" integral $M = \int_B \rho\, dV$.

If the surface S is plane there is a simple interpretation of the double integral $\int_S \sigma\, dA$. We can represent σ graphically by taking an axis

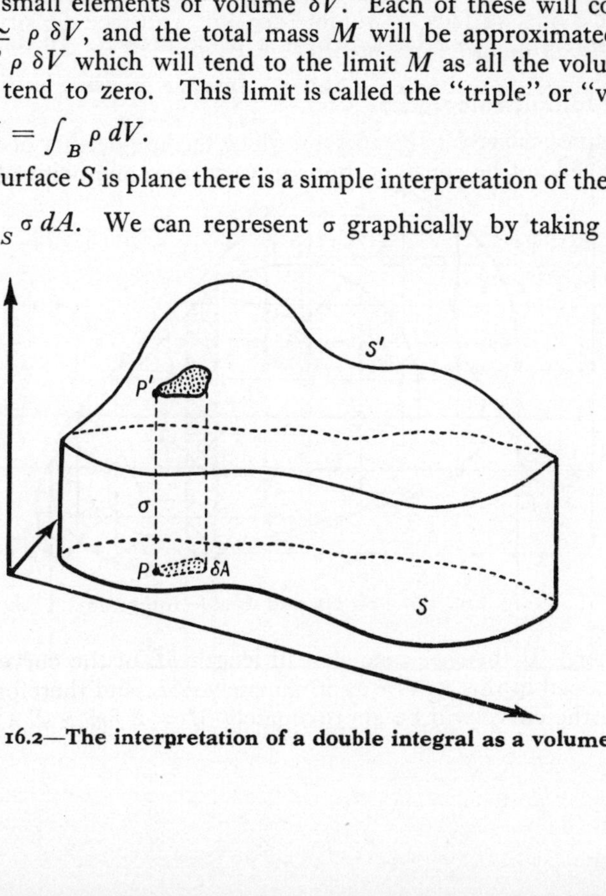

Fig. 16.2—The interpretation of a double integral as a volume

perpendicular to S as an axis of σ: the value of σ at any point P on the surface will then be plotted as a point P' at height $\sigma = PP'$ above P. These points P' will lie on a curved surface S' (Fig. 16.2). The integral $\int_S \sigma\, dA$ will then be equal to the volume between the surfaces S and S'. (This is analogous to the representation of the ordinary integral $\int y\, dx$ as the area between the x-axis and the curve for y.) For if we take any small element of area δA in S and erect above it a column of height σ reaching up to the surface S', the volume contained in this column will be approximately area of base times height, i.e. $\sigma\, \delta A$. The total volume contained in all such columns will be $\Sigma\, \sigma\, \delta A$, approximately: and in the limit this sum will become $\int \sigma\, dA$ by definition. This in effect replaces a surface distribution of variable density σ by a volume distribution of uniform density 1 but variable height σ; and naturally this gives the same integral.

16.13 Repeated integrals

How in practice can we calculate such a double or triple integral? One method is the following. Consider for the sake of simplicity a double integral $\int_S \sigma\, dA$ taken over a plane area S. In this plane we can introduce co-ordinates x and y. Instead of using a division into

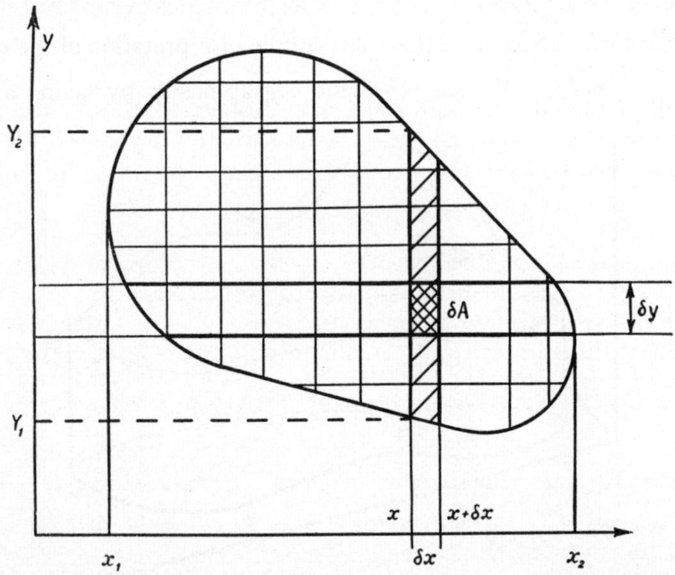

Fig. 16.3—A double integral considered as a repeated integral

small areas δA of arbitrary shapes, we divide the surface into small rectangles by two parallel systems of thin strips, one parallel to the y-axis, the other system parallel to the x-axis. A strip of width δx parallel to the y-axis will intersect one of width δy parallel to the x-axis in a small rectangle of area $\delta A = \delta x \,.\, \delta y$ (Fig. 16.3). The total mass $M = \int_S \sigma \, dA$ will therefore be approximated by the sum $\Sigma \, \delta m = \Sigma \, (\sigma \, \delta A) = \Sigma \, (\sigma \,.\, \delta x \,.\, \delta y)$.

Now concentrate attention on a single strip of width δx parallel to the y-axis. We shall suppose that it lies between the x-co-ordinate values x and $x + \delta x$, and that the lowest point on this strip has y-co-ordinate Y_1, and the highest point has y-co-ordinate Y_2, with all intermediate points belonging to the strip. (That is, we take S to be of a fairly simple shape, with the property that each vertical line cuts it in one range only. A shape like the crescent moon, which can be cut twice by a single vertical line, will not be covered by our argument: but as a rule it is not difficult to make suitable allowance for this complication when it occurs in practice.)

We notice that Y_1 and Y_2 are functions of x which depend on the shape of the boundary curve.

Now the total mass contained within the vertical strip is approximately $\Sigma \, \sigma \, \delta A = \Sigma \, \sigma \, \delta x \,.\, \delta y$, summed for all elements of area contained in this strip. But here δx is constant, and we can write this sum as $\delta x \,.\, \Sigma \, \sigma \, \delta y$, summed over all elements of the strip, i.e. between $y = Y_1$ and $y = Y_2$. Now this sum is the ordinary approximation to a definite integral: so the total mass in the strip is approximately

$$\delta x \,.\, \int_{Y_1}^{Y_2} \sigma \, dy$$

where the integration is performed with respect to y only, keeping x fixed at the value x. To find the total mass we now have to add all these strips together: that is

$$M \simeq \Sigma \left(\int_{Y_1}^{Y_2} \sigma \, dy \right) \delta x.$$

This summation is performed over all possible values of x, say from x_1 to x_2, where x_1 is the least value of x and x_2 the greatest (Fig. 16.3). In the limit when we take both the δx and the δy strips to tend to zero in thickness this sum will again become a definite integral, and we shall expect exact equality:

$$M = \int_S \sigma \, dA = \int_{x_1}^{x_2} \left(\int_{Y_1}^{Y_2} \sigma \, dy \right) dx \quad . \qquad (16.3)$$

The right-hand side here is called a "repeated integral" and is usually written $\int_{x_1}^{x_2} \int_{Y_1}^{Y_2} \sigma \, dy \, dx$ or $\int\int_S \sigma \, dy \, dx$, without the brackets. (Some text-books write it as $\int_{x_1}^{x_2} \int_{Y_1}^{Y_2} \sigma \, dx \, dy$, which does not seem so logical a notation.) The interpretation is that the inside integral is to be performed with respect to y, keeping x constant; and the limits Y_1 and Y_2 are the boundary values of y for the particular value of x.

EXAMPLE

(1) Find the mass contained within the right-angled triangle S with vertices at O, (1, 0), and (1, 1), given that the surface density σ at the point (x, y) is $\sigma = x + y$ (Fig. 16.4).

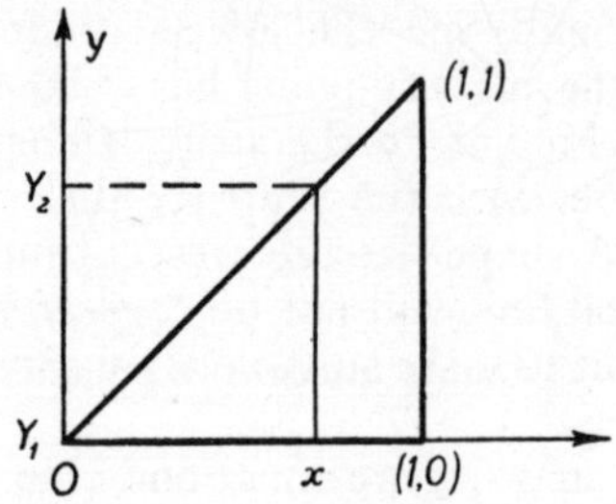

Fig. 16.4—The mass of a triangle of varying density

Here $Y_1 = 0$, $Y_2 = x$, and $x_1 = 0$, $x_2 = 1$. The mass is therefore

$$M = \int_S \sigma\, dA = \int_0^1 \left(\int_0^x (x + y)\, dy\right) dx$$

$$= \int_0^1 [xy + \tfrac{1}{2}y^2]_0^x\, dx$$

$$= \int_0^1 (x^2 + \tfrac{1}{2}x^2)\, dx$$

$$= [\tfrac{1}{3} \cdot \tfrac{3}{2}\, x^3]_0^1 = \tfrac{1}{2}.$$

Similarly a volume integral $M = \int_B \rho\, dV$ can be expressed as a repeated integral

$$M = \int_{x_1}^{x_2} \int_{Y_1}^{Y_2} \int_{Z_1}^{Z_2} \rho\, dz\, dy\, dx \quad . \quad . \quad (16.4)$$

with the following interpretation: x, y, and z are rectangular cartesian co-ordinates. We begin by integrating ρ with respect to z, keeping x and y constant. The limits of integration for ρ are Z_1 and Z_2, the least and greatest values of z for the particular values of x and y; i.e. Z_1 and Z_2 are functions of x and y. This integral is now in turn integrated with respect to y only between the limits Y_1 and Y_2, which are the least and greatest values of y for a given value of x. The final integration is with respect to x.

Sometimes (Fig. 16.5) it is useful to take polar rather than cartesian co-ordinates in evaluating a surface integral. Imagine the area divided

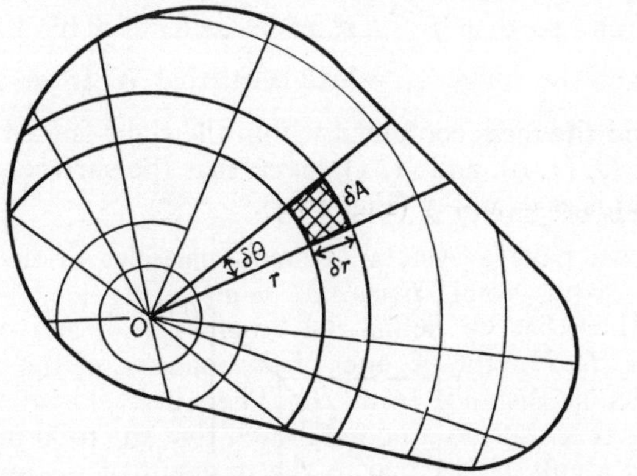

Fig. 16.5—A double integral in polar co-ordinates

up into thin circular strips of width δr and thin radial wedges of angle $\delta\theta$. These intersect in small areas δA which are nearly rectangular. The radial width of such a rectangle is δr; the perpendicular width is $r\,\delta\theta$, by the formula for the length of a circular arc. The area δA is therefore approximately $r\,\delta\theta\,\delta r$, the mass contained within it is approximately $\sigma\,r\,\delta\theta\,\delta r$, and the total mass within S is approximately $\Sigma\,\sigma\,r\,\delta\theta\,\delta r$. In the limit, as the thickness δr of the circular strips and the angle $\delta\theta$ of the radial wedges both tend to zero, this will become a repeated integral; either $\iint_S \sigma r\,d\theta\,dr$, if we integrate first with respect to θ, and afterwards with respect to r, or $\iint_S \sigma r\,dr\,d\theta$ if the first integration is with respect to r. Thus

$$M = \iint_S \sigma\,dx\,dy = \iint_S \sigma r\,d\theta\,dr = \iint_S \sigma r\,dr\,d\theta \qquad (16.5)$$

In each case the integrals are to be taken between the appropriate limits. Thus if we begin by integration of σr with respect to θ, keeping r constant, the limits of integration will be the extreme values of θ possible for that value of r. The second integration will then be over all possible values of r. Conversely, if we begin by integrating with respect to r, keeping θ fixed, the range of integration will be over all possible values of r for that value of θ.

FURTHER EXAMPLE

(2) Find the mass contained within a circle of radius 1, given that the surface density is proportional to the distance from the centre, i.e. $\sigma = Kr$.

Here $M = \iint_S Kr^2\, d\theta\, dr$. The first integration will be with respect to θ: here the polar angle θ will range from 0 to 2π to cover the whole circle, so that $\int_0^{2\pi} Kr^2\, d\theta = 2\pi Kr^2$. This has now to be integrated over the range of values of r, that is, from 0 to 1; $M = \int_0^1 2\pi Kr^2\, dr = [\frac{2}{3}\pi Kr^3]_0^1 = \frac{2}{3}\pi K$.

16.14 Centres of gravity

Suppose we have a system of small particles of masses m_1, m_2, $m_3 \ldots . m_n$, with total mass $M = m_1 + m_2 + \ldots + m_n = \Sigma m_a$. Suppose further that we define the "moment" of any particle about a given plane Π to be the product of the mass m_r of the particle times the perpendicular distance from Π. Then it is shown in text-books on mechanics that there is a point G such that the total moment of the whole system about Π is equal to the moment of a single particle of mass M placed at G: and this is true whatever plane Π we take. G is then known as the "centre of mass" or "centre of gravity" of the system. In particular the moment about any plane through G is zero; it follows that if the system is rigid it will exactly balance if supported at the centre of gravity.

Consider first a distribution of particles in a plane. Let (x_r, y_r) be the co-ordinates of the particle m_r (Fig. 16.6). Then the position of

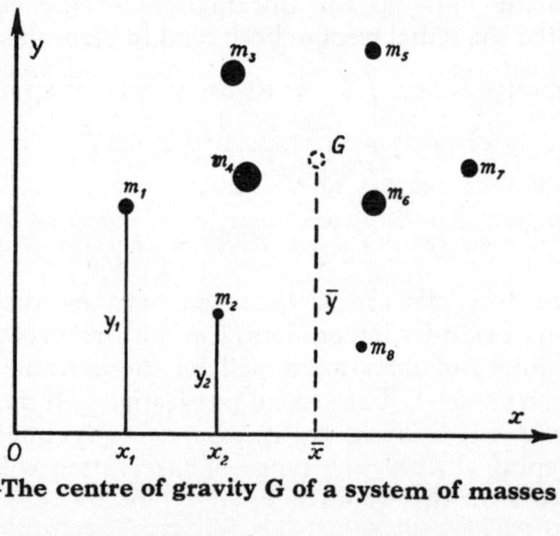

Fig. 16.6—The centre of gravity G of a system of masses in a plane

G can be found in the following way. We shall call the co-ordinates of G $(\bar{x}, \bar{y})$; this is the usual notation—here the bar over the letter means "centre of gravity" and has no connection with the former use of a bar to denote the complex conjugate. Now if we take moments about a plane through the origin O perpendicular to the x-axis, then the moment of the mass m_r is $m_r x_r$, since x_r is its distance from the plane. It follows that the total moment of all the particles is $\Sigma\, m_a x_a$. This must be equal to the moment $M\bar{x}$ of the hypothetical particle of mass M placed at the centre of gravity, whence

$$M\bar{x} = \Sigma\, m_a x_a$$

Similarly taking moments about a plane through O perpendicular to the y-axis gives $M\bar{y} = \Sigma\, m_a y_a$. G is therefore the point with co-ordinates $(\bar{x}, \bar{y})$ given by

$$\bar{x} = \Sigma\, m_a x_a/M, \qquad \bar{y} = \Sigma\, m_a y_a/M.$$

For a set of points in three dimensions we have similarly a centre of gravity G with co-ordinates $(\bar{x}, \bar{y}, \bar{z})$ where

$$\bar{x} = \Sigma\, m_a x_a/M; \qquad \bar{y} = \Sigma\, m_a y_a/M; \qquad \bar{z} = \Sigma\, m_a z_a/M \quad . \quad (16.6)$$

Alternatively we can regard equations (16.6) as the definition of the centre of gravity.

One point perhaps calls for comment. The equation for $\bar{x}$ can be written in the form

$$\bar{x} = \frac{m_1 x_1 + m_2 x_2 + m_3 x_3 + \ldots + m_n x_n}{m_1 + m_2 + m_3 + \ldots + m_n} \quad . \quad (16.7)$$

Now if all the masses $m_1, m_2 \ldots m_n$ are equal this means that $\bar{x}$ is the ordinary average $(x_1 + x_2 + \ldots + x_n)/n$ of the co-ordinates $x_1, x_2, \ldots x_n$. If the m_r's are unequal we therefore call $\bar{x}$ the *weighted average* or *weighted mean* of $x_1, x_2 \ldots x_n$ with weights $m_1, m_2, \ldots m_n$ respectively.

PROBLEMS

(1) A system consists of three particles. One is at (0, 0) and has mass 1, one is at (1, 1) and has mass 2, and one is at (3, 0) and has mass 3. Where is the centre of gravity?

(2) A system consists of a mass 1 at (0, 0, 0), a mass 2 at (1, 1, 1), a mass 5 at (2, 1, 0) and a mass 2 at (1, 2, 1). Where is the centre of gravity?

Now consider a plane area S over which the mass is continuously distributed with surface density σ, instead of being concentrated in a few points. Then each small element of area δA will contain a mass $\sigma\,\delta A$; and in the definition of the centre of gravity according to equation (16.6), instead of writing $\bar{x} = \Sigma\, m_a x_a/M$, we shall have $\bar{x} \simeq \Sigma\, \sigma x\, \delta A/M$, where "$x$" is the x-co-ordinate of the element δA, and the summation is over all such elements. If we divide the area into smaller and smaller pieces δA, whose maximum size tends to zero, this sum will tend to the limit

$$\bar{x} = M^{-1} \int_S \sigma x \, dA \quad . \quad . \quad . \quad (16.8)$$

and similarly $\bar{y} = M^{-1} \int_S \sigma y \, dA$.

These formulas will give the position of the centre of gravity for any plane sheet. For a three-dimensional body B of density ρ the discrete formulas (16.6) will be similarly replaced by

$$\bar{x} = M^{-1} \int \rho x \, dV; \qquad \bar{y} = M^{-1} \int \rho y \, dV;$$

$$\bar{z} = M^{-1} \int \rho z \, dV \quad . \quad . \quad . \quad (16.9)$$

Usually these integrals can be found most easily by turning them into repeated integrals.

EXAMPLES

(1) What is the centre of gravity of the triangle of Fig. 16.4 which has surface density $\sigma = x + y$?

As we have already shown, the mass $M = \int \sigma \, dA = \frac{1}{2}$. So

$$\begin{aligned} \bar{x} = M^{-1} \int \sigma x \, dA &= 2 \int_0^1 \int_0^x (x + y) \, x \, dy \, dx \\ &= 2 \int_0^1 [x^2 y + \tfrac{1}{2} x y^2]_0^x \, dx \\ &= 2 \int_0^1 \tfrac{3}{2} x^3 \, dx \\ &= \tfrac{3}{4} [x^4]_0^1 = \tfrac{3}{4} \end{aligned}$$

$$\begin{aligned} \bar{y} = M^{-1} \int \sigma y \, dA &= 2 \int_0^1 \int_0^x (x + y) \, y \, dy \, dx \\ &= 2 \int_0^1 [\tfrac{1}{2} x y^2 + \tfrac{1}{3} y^3]_0^x \, dx \\ &= 2 \int_0^1 \tfrac{5}{6} x^3 \, dx \\ &= \tfrac{5}{12} [x^4]_0^1 = \tfrac{5}{12} \end{aligned}$$

The C.G. is therefore at $(\frac{3}{4}, \frac{5}{12})$.

(2) Taking the same triangle, where is the centre of gravity if the surface density is uniform, say $\sigma = 1$?

The mass is now

$$\begin{aligned} M = \int \sigma \, dA &= \int_0^1 \int_0^x 1 \, . \, dy \, dx \\ &= \int_0^1 x dx = \tfrac{1}{2}. \end{aligned}$$

(This also follows from the formula: mass = surface density × area $= 1 \times \frac{1}{2} = \frac{1}{2}$.)

$$\bar{x} = M^{-1} \int \sigma x \, dA = 2 \int_0^1 \int_0^x x \, dy \, dx$$
$$= 2 \int_0^1 [xy]_0^x \, dx$$
$$= 2 \int_0^1 x^2 \, dx = \tfrac{2}{3}$$
$$\bar{y} = M^{-1} \int \sigma y \, dA = 2 \int_0^1 \int_0^x y \, dy \, dx$$
$$= 2 \int_0^1 [\tfrac{1}{2}y^2]_0^x \, dx$$
$$= \int_0^1 x^2 \, dx = \tfrac{1}{3}$$

The centre of gravity is therefore at the point $(\frac{2}{3}, \frac{1}{3})$. This is also the centroid, or intersection of the three medians of the triangle—a property which can be shown to be true in general for any triangle of uniform density.

16.15 Moments of inertia

The "moment of inertia" (or "second moment") of a mass m about a line L is defined as the product of m times the square of the distance of m from L. For a system of masses the moment of inertia about L is the sum of the moments of inertia of the separate masses, i.e. if h_r is the distance of the mass m_r from the line L,

$$I = \text{moment of inertia} = \Sigma \, m_a h_a{}^2 \quad . \quad . \quad (16.10)$$

If instead of discrete particles we have a continuous distribution of mass over an area S, then in the above formula the point mass m_r must be replaced by $\sigma \, \delta A$, the mass of the element of area δA; and h_r will be replaced by h, the distance of this element from the line L. The sum $\Sigma \, m_a h_a{}^2$ becomes $\Sigma \, \sigma h^2 \, \delta A$, or in the limit $I = \int_S \sigma h^2 \, dA$. Similarly if the mass is continuously distributed over a volume B with density ρ, the moment of inertia will be $\int_S \rho h^2 \, dV$. If a body is rotated about an axis L with angular velocity ω radians per second, then it can be shown that its kinetic energy is $\frac{1}{2}I\omega^2$ and its angular momentum $I\omega$, where I is the moment of inertia about the axis. (Compare with the formulas $\frac{1}{2}mv^2$ = kinetic energy and mv = momentum for a mass m moving with velocity v.) Also if the body is acted upon by a set of forces with total moment μ about the axis L, the angular acceleration $d\omega/dt$ is given by $\mu = I d\omega/dt$.

As a particular case of our formula the moment of inertia of a plane sheet S about the x-axis is $I_x = \int_S \sigma y^2 \, dA$, and that about the y-axis is $I_y = \int_S \sigma x^2 \, dA$.

EXAMPLE

(1) To find the moment of inertia of a square of side L and surface density σ about an axis through the centre parallel to one of the sides.

Take the centre of the square as origin of co-ordinates, with axes parallel to the sides. Then x varies from $-\frac{1}{2}L$ to $\frac{1}{2}L$, and y from $-\frac{1}{2}L$ to $\frac{1}{2}L$. The moment about the x-axis is therefore

$$\begin{aligned} I = \int_S \sigma y^2 \, dA &= \int_{-\frac{1}{2}L}^{\frac{1}{2}L} \int_{-\frac{1}{2}L}^{\frac{1}{2}L} \sigma y^2 \, dy \, dx \\ &= \int_{-\frac{1}{2}L}^{\frac{1}{2}L} \left[\tfrac{1}{3} \sigma y^3\right]_{-\frac{1}{2}L}^{\frac{1}{2}L} dx \\ &= \int_{-\frac{1}{2}L}^{\frac{1}{2}L} \tfrac{1}{12} \sigma L^3 \, dx = \tfrac{1}{12} \sigma L^4. \end{aligned}$$

For a three-dimensional body the moment of inertia about the x-axis is similarly given by

$$I_x = \int_B \rho (y^2 + z^2) \, dV.$$

16.16 Pressure

By the "pressure" on a surface is meant the force acting perpendicular to the surface per unit area, i.e. the surface density of the perpendicular component. The surface density of the component of force tangential to the surface is called the "shear".

The M.K.S. unit of pressure is therefore the newton/m^2, and the C.G.S. unit the dyne/cm^2 = $\frac{1}{10}$ newton/m^2. A unit which is widely used is the *bar* which = 10^5 newton/m^2 = 10^6 dyne/cm^2. 1 millibar = $\frac{1}{1000}$ bar.

In practice there are a number of other units of pressure sometimes used: the principal one is the *atmosphere*, defined as the pressure of 76 cm of mercury at 0° C. Unfortunately this depends on the value chosen for g: by international agreement g is taken to have the standard value 9·80665 m/sec^2, and the density of mercury at 0° C is taken to be 13595·09 kg/m^3; we therefore have the conversions:

		Logarithms
1 cm mercury	= ·013332 bar	$\bar{2}$·12490
1 atmosphere	= 1·01325 bar	·00572
1 kg wt/cm^2	= ·98066 bar	$\bar{1}$·99152
1 lb wt/in^2	= ·068948 bar	$\bar{2}$·83852

A body subjected to an atmospheric pressure of 1 standard atmosphere, i.e. 1·01325 bar, and at 0° C, is said to be at "normal temperature and pressure" (N.T.P.).

16.17 Viscosity

When different parts of a fluid are moving with different velocities a stress is set up owing to the viscosity of the fluid. If the velocity becomes too great the motion breaks up into eddies, and becomes mathematically very intractable. But if the motion is slow and steady it can be calculated in the following way.

We shall take the simple case of steady parallel motion, as down a tube or channel. Suppose that the motion is in the direction D (Fig. 16.7): and suppose that at a point A the velocity is v, while at a neighbouring point B (where the line AB is perpendicular to the direction

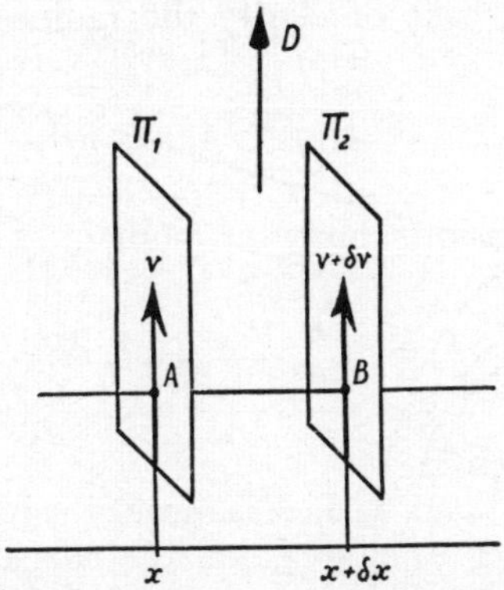

Fig. 16.7—Motion of a viscous fluid

of motion) the velocity is $v + \delta v$. Let us take the x-axis of co-ordinates parallel to AB, so that A has co-ordinate x, B has co-ordinate $x + \delta x$, and therefore the distance $AB = \delta x$. Now draw planes Π_1 and Π_2 through A and B respectively perpendicular to the line AB. The viscosity will cause a shearing force in these planes. If δv is positive, so that the fluid is moving more rapidly upwards at B than at A, there will be an upward shearing force at B acting on the fluid between A and B, tending to accelerate its motion, and a downward shear at A, tending to retardation of the fluid. In other words each portion of fluid tends to pull the neighbouring portions along with it. This shear S will of course be measured as force per unit area, i.e. as newtons/m^2 in M.K.S. units and as dynes/cm^2 in C.G.S. units. Now it is natural to suppose that S will be directly proportional to the velocity difference δv; a little thought will also suggest that it will be inversely proportional to the distance δx, since the further apart the points A and B are the less effect the difference of velocity will have. Putting these suggestions together, we expect S to be a constant multiple of $\delta v/\delta x$, at least when δv and δx are small: or more precisely, we expect the shear at a point A to be proportional to the value of dv/dx at that point. Experiment confirms that this is so: we can write

$$S = \eta\, dv/dx \quad . \quad . \quad . \quad . \quad (16.11)$$

where η is the measure of the viscosity. This can also be written $\eta = S dx/dv$. Now in M.K.S. units S is measured in newtons/m^2, x in metres, and v in metres/sec; η has therefore the unit (newton/m^2) $\times$ m $\times$ (sec/m) $=$ newton . sec/m^2 $=$ kg sec^{-1} m^{-1}. This unit is called the *dekapoise*. The corresponding C.G.S. unit is the *poise* $= \frac{1}{10}$ dekapoise.

We can now find at what rate a liquid will flow down a long thin circular tube. Let the length of the tube be L, and its radius R (see Fig. 16.8). For the sake of clarity the radius of the tube is greatly

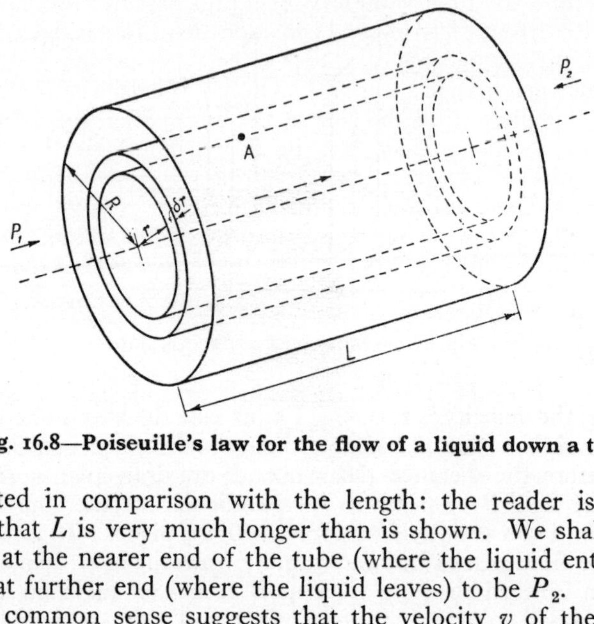

Fig. 16.8—Poiseuille's law for the flow of a liquid down a tube

exaggerated in comparison with the length: the reader is asked to imagine that L is very much longer than is shown. We shall take the pressure at the nearer end of the tube (where the liquid enters) to be P_1, and at further end (where the liquid leaves) to be P_2.

Now common sense suggests that the velocity v of the liquid at any point A within the tube will depend only on its distance r from the axis (supposing that the size of the tube and the pressures P_1 and P_2 are given). Let us draw two cylinders, one of radius r around the axis, and one of slightly greater radius $r + \delta r$; at the surface of the inner cylinder the velocity will be v, and at the surface of the outer one it will be $v + \delta v$. There will therefore be a shear approximately equal to $\eta\, \delta v/\delta r$, or accurately $\eta\, dv/dr$, acting on the surface of the inner cylinder. Since the total area of this curved part of the surface is $2\pi r L$, this means that there will be a total backward force $-2\pi r L \eta\, dv/dr$ due to viscosity on the inner cylinder. For steady motion this must be balanced by the difference of pressure at the two ends. Since the area of each end is πr^2, this forward pressure amounts to $(P_1 - P_2)\pi r^2$, so that $-2\pi r L\eta\, dv/dr = (P_1 - P_2)\pi r^2$, or $dv/dr = -\frac{1}{2}(P_1 - P_2)r/L\eta$. This is a differential equation which can be solved by direct integration; the solution is

$$v = \int (dv/dr)\, dr = -\int \tfrac{1}{2}(P_1 - P_2)\, r dr/L\eta$$
$$= -\tfrac{1}{4}(P_1 - P_2) r_2/L\eta + C$$

The constant C is fixed by the fact that at the boundary of the tube, when $r = R$, the liquid sticks to the boundary wall, and so $v = 0$. Thus $C = \frac{1}{4}(P_1 - P_2)R^2/L\eta$, and on substituting this value we obtain

$$v = \tfrac{1}{4}(P_1 - P_2)(R^2 - r^2)/L\eta.$$

Now let us find the total volume V of liquid passing through the tube in time t. Return to Fig. 16.8 and consider the cylindrical shell between the two cylindrical surfaces of radii r and $r + \delta r$ respectively. Within this shell the velocity will be approximately v (neglecting the correcting term δv). But the area of the ring at the end, which lies between two circles of radii r and $r + \delta r$, will be the circumference of the circle times the width of the ring, i.e. $2\pi r\ \delta r$. So in time t the volume of liquid passing down the shell will be (velocity $\times$ time $\times$ area of end), i.e. $2\pi rvt\ \delta r$. The total volume passing down the whole tube will be the sum of the volumes passing down all such rings, i.e. $\Sigma\ 2\pi rvt\ \delta r$ or more accurately $\int 2\pi rvt\ dr$. Since $v = \frac{1}{4}(P_1 - P_2)(R^2 - r^2)/L\eta$ this integral becomes

$$V = \int_0^R \tfrac{1}{4}(P_1 - P_2)(R^2 - r^2)2\pi rt\ dr/L\eta$$

This is, by the way, less formidable than it seems, as most of the letters denote constants and can be taken outside the integral. Thus

$$V = \tfrac{1}{2}\pi L^{-1}\eta^{-1}t(P_1 - P_2) \int_0^R (R^2 r - r^3)dr$$
$$= \tfrac{1}{2}\pi L^{-1}\eta^{-1}t(P_1 - P_2)\left[\tfrac{1}{2}R^2r^2 - \tfrac{1}{4}r^4\right]_0^R$$
$$= \tfrac{1}{8}\pi L^{-1}\eta^{-1}t(P_1 - P_2)R^4 \qquad . \quad . \quad . \quad (16.12)$$

This is Poiseuille's law. Expressed in the form $\eta = \frac{1}{8}\pi t(P_1 - P_2)R^4/VL$ it can be used to determine the viscosity experimentally. Typical values are: for water at 20° C, ·01006 poise; for ethyl alcohol at 20° C, ·0119; for air at 23° C, 1·830 $\times$ 10^{-4} poise.

EXAMPLE

(1) Brodie has used Poiseuille's law to find the pressure necessary to drive the urine out of the kidneys. He found that urine was discharged from one kidney containing 142,000 glomeruli and tubules at

the rate of 1 cc per minute, i.e. at the rate of 1/(142,000 × 60) cc per second per tubule. Now Poiseuille's law can be written

$$P_1 - P_2 = 8L\eta V/\pi t R^4 \text{ dynes/cm}^2$$
$$= 8L\eta V/1333\pi t R^4 \text{ millimetres of mercury.}$$

Putting $t = 1$, $V = 1/(142{,}000 \times 60)$ cc for each tubule we can make the following calculations:

	Length L (cm)	Radius R (10^{-4} cm)	Loss in pressure (mm Hg)
Proximal convoluted tubule	1·2	6	15
Loop of Henle—			
descending limb	·9	5	23
ascending limb	·9	4·5	35
Distal convoluted tubule	·2	9	½
Collecting tubule	2·2	8	9
Total loss in pressure			82½

Now as the mean aortic blood pressure was 120 mm Hg, and the loss in pressure between the aorta and glomerular capillaries is probably 35 mm Hg, this gives a pressure of 85 mm Hg in the glomerular capillaries, i.e. about enough to account for the secretion of urine. Brodie concluded that the pulsation of the glomerulus does not account for the flow of urine in the tubules, and attributed a secretory action to the glomerular surface. (But the bulk of opinion is against this inference.)

16.18 Heat

Heat is usually measured in units of 1 calorie. This has two varieties, the small or *gram-calorie* (cal.) defined as the heat required to raise 1 gram of water through 1° C; and the large or *kilogram-calorie* (Cal.) which equals 1000 gram-calories. Since the heat required to raise a given mass of water from 0° to 1° C differs from that required from 15° to 16°, or from 50° to 51°, a further specification is necessary to make these units perfectly definite. The most usual one is to take the gram-calorie as the heat required to raise 1 gram of water from 15° to 16° C; this is the "15° gram-calorie". A third kind of calorie is the "international steam table calorie" (I.T. cal) which is approximately 1·0005 15°-gram-calories.

Latent heats of fusion and vaporization are usually expressed in units of 1 gram-calorie per gram, e.g. the latent heat of fusion of ice is about 80 gm-cal/gm, and the latent heat of vaporization of water at 100° C is about 538 gm-cal/gm.

Since heat is a form of energy, it can also be expressed in energy units, i.e. joules in the M.K.S. system or ergs in the C.G.S. system. The conversion factor, known as "Joule's equivalent of heat" is (for a 15°-gm-calorie)

$$J = 4{\cdot}185 \text{ joules/cal} = 4{\cdot}185 \times 10^7 \text{ ergs/cal.}$$

	Logarithms
1 calorie = 4·185 joules	·62170
1 joule = ·23895 cals	$\bar{1}$·37830

Since the definition of the calorie is, as we have seen, both arbitrary and somewhat indefinite, it seems rather surprising that it has not so far been generally replaced by the absolute unit, the joule.

16.19 The method of dimensions

Almost always a physical or chemical equation is expressed in a form which is independent of the system of units. Thus the equation for the volume of a cube of side L is $V = L^3$, and for that of a sphere of radius R is $V = \frac{4}{3}\pi R^3$. We do not need to specify the units employed in these formulas, provided that we understand that if L and R are measured in inches then V is expressed in inch³, if L and R are centimetres then V is in cc, and if L and R are in metres then V is in m³. Similarly the distance a body falls in time t is $\frac{1}{2}gt^2$, it being understood that the appropriate value of g is to be substituted: and in the same way Poiseuille's viscosity equation will hold good in any system of units.

One reason why we write equations in this way is that it is convenient to be able to choose the system of units at will. But the chief reason probably is that our units—such as the metre, the kilogram, and the second—are arbitrarily chosen and have little or no physical significance; and it would be absurd for such arbitrary units to obtrude themselves in an equation expressing a natural law. Thus we quite naturally express our equations in an "absolute" form, without taking any special thought in the matter.

This property of physical equations, that they are true in any system of units, has a very important consequence. It implies that whatever units we use, the unit in which the left-hand side of the equation is expressed must be equal to the unit on the right-hand side. Thus an equation such as $V = L^2$ connecting length and volume can be ruled out at once as impossible: for V will be expressed, say, in metre³, while L^2 will correspondingly be in metre², which is a different unit. Even if we suppose for a moment that we could find a system of units in which the relation $V = L^2$ was true, it would cease to be true when we changed the units. Actually of course we know that there is no system for which $V = L^2$ is true for arbitrary values of V and L: but it does happen to be numerically true for the particular values $L = 1$ (metre), $V = 1$ (metre^3). However, when we change the units to centimetres, L becomes 100 cm, and V becomes 10^6 cm^3, and L^2 is no longer equal to V. Thus the falsity of the equation is demonstrated.

This equality of units provides a very useful check on equations: this check is known as the "method of dimensions". Thus suppose we were uncertain whether the correct equation for the period t of oscillation of a pendulum of length L was $t = 2\pi\sqrt{(L/g)}$ or $t = 2\pi\sqrt{(g/L)}$. We try putting in suitable units, say M.K.S. ones. The time t will be expressed in seconds, g in m/sec^2, and L in m. Thus $2\pi\sqrt{(L/g)}$ will be in units $\sqrt{[\text{m}/(\text{m}/\text{sec}^2)]} = \sqrt{(\text{sec}^2)} = \text{sec}$, whereas $2\pi\sqrt{(g/L)}$ will be in units of $\sqrt{[(\text{m}/\text{sec}^2)/\text{m}]} = \sqrt{(1/\text{sec}^2)} = 1/\text{sec}$. Only the first formula $t = 2\pi\sqrt{(L/g)}$ gives the same units on both sides of the equation, and so it is the correct one. In fact it is possible to say rather more. If it is known that the period of oscillation of a pendulum is a function only of its length L (metres) and of the acceleration due to gravity g (metres/sec^2), it can be deduced that the connecting formula must be $t = K\sqrt{(L/g)}$, where K is a numerical constant. For no other combination of L and g has a second as its unit. Even if one did not know for certain that t depends only on L and g, the study of units would still strongly suggest the relation $t = K\sqrt{(L/g)}$.

We have said above that physical equations are true independently of the system of units. This is not quite true without qualification; it is necessary that similar definitions should be used for derived units in different systems—a condition which is almost always satisfied in practice. To illustrate the point, imagine that a tribe is discovered in Central Africa, called the Ngboglus, who use as their measure of length the radius of a Sacred Jewel, which is perfectly spherical in shape, and that as their unit of volume they use the volume of the Sacred Jewel. In their units the volume of a sphere of radius R is no longer $\frac{4}{3}\pi R^3$, but is now R^3; and the volume of a cube of side L is for them $\frac{3}{4}\pi^{-1}L^3$. The reason is simply that they have a different definition for the derived unit of volume: we use a unit cube, they use a unit sphere, and because of that their equations are different. But since all commonly accepted systems of units use the same methods of defining derived quantities, this particular complication can be ignored (unless of course one wished to do business with the Ngboglus).

16.20 Dimensional constants

We can always select arbitrarily a unit to measure any physical magnitude. Thus we can take the metre, or the inch, or the height of Mount Everest, or the mean distance from the earth to the sun as a unit of length; and we can take the litre, or the gallon, or the fluid ounce, or the Sacred Jewel of the Ngboglus as a unit of volume. Of course a litre is defined officially in terms of a kilogram, but only through the physical properties of water at a special temperature; in effect it has an arbitrary definition. Units chosen in this way we shall call "independent": thus the metre, kilogram, second, litre, calorie and degree centigrade are all independent units. In any system of measurement there will be a number of independent (or "fundamental") units, and other units will be derived from these.

Now if we take independent units for energy E and heat H, we find that heat and energy are interconvertible according to the rule, 1 unit of heat $= J$ units of energy. Thus on taking 1 gram-calorie as the heat unit, and 1 joule as the energy unit, we find 1 gram-calorie $= 4{\cdot}185$ joules. Here the constant J expresses the number of joules for each gram-calorie, and therefore expressed in its correct units it is 4·185 joules/gm-cal. If instead we take 1 kg-cal as unit of heat, and 1 erg as unit of work, we find that $J = 4{\cdot}185 \times 10^{10}$ ergs/kg-cal. This follows from the usual rule for conversion of units

$$\begin{aligned} 4{\cdot}185 \text{ joules/gm-cal} &= 4{\cdot}185 \times (10^7 \text{ ergs})/(10^{-3} \text{ kg-cal}) \\ &= 4{\cdot}185 \times 10^{10} \text{ ergs/kg-cal.} \end{aligned}$$

In general H units of heat are equal to E units of energy, where $E = JH$. Note that this equation is dimensionally correct: E is measured in units of, say, 1 joule, while JH has the units joule/gm-cal $\times$ gm-cal $=$ joule. In this equation J is a universal physical constant. That is to say, it has a fixed value independent of the place, time, and conditions of measurement, or so experimental determinations seem to show. It is also a dimensional constant, i.e. its numerical value depends on the units of measurement.

Now instead of measuring heat in terms of the arbitrary unit of 1 calorie we can use instead the joule. This is a natural unit to take, since we believe that heat is a form of energy, and the joule is the M.K.S. unit of energy. In these units the equation for converting heat into energy becomes simply $E = H$; the constant J has now disappeared, or rather it has become unity. As we have said, this is physically reasonable; but the possibility of making this change follows purely on mathematical grounds, without any appeal to physics. All that is necessary is to replace the arbitrary unit, the calorie, by a new unit equal to $1/J$ calories. In this new unit a quantity of heat $= H$ calories will become $H' = JH$ units, so that the equation $E = HJ$ will become $E = H'$. In general any equation containing a dimensional constant can, by means of a suitable change of units, be replaced by a new equation without the constant: the effect is to replace one independent unit by a new derived unit. Thus, if g, the acceleration of gravity, was an absolute constant on the earth's surface, we could change our units so as to make it equal to 1. The simplest way would be to replace the metre by a new unit equal to g metres. The equation $y = \frac{1}{2}gt^2$ for the distance a body falls in time t would be replaced by $y = \frac{1}{2}t^2$, and many other dynamical calculations would be greatly simplified. If we measure the time t in seconds, the equation $y = \frac{1}{2}t^2$ shows that we should then measure distance in square seconds—a somewhat unfamiliar idea: 1 "square second" would be by definition g metres, or twice the distance a body falls in 1 second. In this system mass and force would be expressed in the same units: 1 kilogram weight would be equal numerically to 1 kilogram. However, since g is not constant this substitution is not possible in practice.

There are two ways of looking on the formula $E = JH$ relating heat and energy. Either we can say that there is a "universal constant of nature" J which relates heat to energy, or we can look upon the constant J as arising because we are so perverse as to measure the same quantity, energy, in two different units. It is at least possible to argue that most dimensional constants arise from such perversity. Thus the velocity of light c is one such constant. If we accept the Special Theory of Relativity we see that time and space are merely two aspects of the same thing, and so distances and times are really quantities of the same kind. Since we use different units, the metre and the second, for distances and times respectively, we can expect the appearance of a universal constant c which is really no more than the ratio of the two units. However, the question of whether two physical quantities arc "really" of the same nature, or "really" different, is a fine philosophical point on which one can argue interminably. The question cannot be settled on the basis of the units customarily used to measure the quantities, which are largely conventional. Heat is usually expressed in different units from other forms of energy, although we believe them to be essentially the same. On the other hand the moment of a force and the work done by a force are both measured by a product force $\times$ distance. But few people would suggest that a moment was a kind of work. The only certainty is that a dimensionless constant, i.e. one independent of units, cannot be eliminated, and must therefore express a real natural law. An example would be the ratio of the electrical force between two electrons to the gravitational force calculated from their masses by means of the universal law of gravitation.

16.21 Atomic weights

Atomic weights measure the relative masses of the atoms of different elements. There were until recently two different scales: on the "chemical scale" the atomic weight of atmospheric oxygen was taken as 16 exactly. This is really the mean atomic weight of the different isotopes contained in it. On the "physical scale" the isotope O^{16} was taken as standard, having an atomic weight of 16 exactly. But by international agreement C^{12} is now the standard, with exact atomic weight 12. This is the old physical scale multiplied by ·999678, and the old chemical scale multiplied by ·99996. For most purposes the differences are negligible.

The atomic weights and atomic numbers of the commoner elements are given in Table 16.3.

Table 16.3—Atomic weights

Element		No.	At. wt.	Element		No.	At. wt.
Aluminium ..	Al	13	26·97	Lithium ..	Li	3	6·94
Antimony ..	Sb	51	121·75	Magnesium ..	Mg	12	24·32
Argon	A	18	39·942	Manganese ..	Mn	25	54·93
Arsenic ..	As	33	74·91	Mercury ..	Hg	80	200·60
Barium ..	Ba	56	137·35	Molybdenum ..	Mo	42	95·95
Beryllium ..	Be	4	9·013	Nickel	Ni	28	58·69
Bismuth ..	Bi	83	208·99	NITROGEN ..	N	7	14·007
Boron	B	5	10·82	OXYGEN ..	O	8	15·999
Bromine ..	Br	35	79·913	PHOSPHORUS ..	P	15	30·974
Calcium ..	Ca	20	40·08	Platinum ..	Pt	78	195·22
CARBON ..	C	6	12·010	POTASSIUM ..	K	19	39·098
CHLORINE ..	Cl	17	35·455	Selenium ..	Se	34	78·96
Chromium ..	Cr	24	52·01	Silicon ..	Si	14	28·06
Cobalt	Co	27	58·94	Silver	Ag	47	1078·876
Copper ..	Cu	29	63·54	SODIUM ..	Na	11	22·990
Fluorine ..	F	9	19·0	Strontium ..	Sr	38	87·63
Gold	Au	79	197·2	SULPHUR ..	S	16	32·065
Helium ..	He	2	4·003	Tin	Sn	50	118·69
HYDROGEN ..	H	1	1·008	Titanium ..	Ti	22	47·90
Iodine	I	53	126·90	Uranium ..	U	92	238·06
Iridium ..	Ir	77	193·1	Vanadium ..	V	23	50·95
Iron	Fe	26	55·85	Wolfram* ..	W	74	183·91
Lead	Pb	82	207·20	Zinc	Zn	30	65·38

* Or Tungsten.

From these atomic weights we obtain the chemical unit of mass, i.e. the *mole* (mol) or gram-molecule = the molecular weight in grams. (Presumably the corresponding M.K.S. unit would be the "kilomole".) The concentration of a solution is usually expressed in units of 1 molar concentration (1 mol/litre = 1 kmol/m³, very nearly) or normal (N) (= 1 gm-equivalent/litre, i.e. molar concentration divided by valency).

16.22 Temperature: specific heats and conductivities

The universal unit of temperature is the degree Celsius, $\frac{1}{100}$ of the difference in temperature between melting ice and boiling water at 76 cm of mercury pressure. Other scales of temperature are the Fahrenheit, Réaumur, and Absolute (or Kelvin) scales: these are connected by the relations that a temperature will be measured as $t_C°$ on the Celsius scale, $t_F°$ Fahrenheit, $t_R°$ Réaumur, and $t_K°$ Absolute, where

$$t_C = \tfrac{5}{9}(t_F - 32) = \tfrac{5}{4}\, t_R = t_K - 273{\cdot}16 \quad . \quad . \quad (16.13)$$

The absolute temperature t_K is usually denoted by the symbol T. It can be exactly defined by means of the second law of thermodynamics (see text-books on the subject) and is very nearly proportional to the pressure of a given mass of hydrogen at constant volume.

The specific heat of a substance is the quantity of heat required to raise unit mass through unit temperature. In the C.G.S. system the unit will be 1 gm-calorie per gram per degree Celsius. Accordingly pure water at 15° C has unit specific heat, by the definition of the gram-calorie. For other substances typical values are: iron, ·10; lead, ·03; glass, ·12 to ·19.

If the temperature within any body is not uniform a conduction of heat will take place from the hotter parts towards the cooler. In particular, consider a slab of uniform thickness b and area A, and let one face be maintained at a temperature T_1 and the opposite face at temperature T_2. The rate of conduction of heat is then directly proportional to the difference in temperature $T_1 - T_2$, and to the area A of the slab, and is inversely proportional to the thickness b. The amount of heat H passing through the slab in time t will therefore be

$$H = c\,(T_1 - T_2)\,t\,A/b \qquad (16.14)$$

where c is a constant, called the " conductivity " of the substance, and measured by the heat passing through a unit cube in unit time when unit difference of temperature is maintained between the opposite faces. The ordinary unit of conductivity is therefore the gm-cal cm^{-1} sec^{-1}. Values of the conductivity for typical substances are (in gm-cal. cm^{-1} sec^{-1} deg^{-1}) copper, ·176, glass ·002, wood ·005, water at 10° C, ·0015.

EXAMPLE

(1) E. G. Glaser (*Nature*, **166** (1950), 1068) performed some experiments on immersion in cold water. The subject of the experiments was a man aged 36, of height 1·80 metres, weight 85 kg, and therefore presumably with surface area 2·05 m^2 by the Dubois formula. It was assumed that his surface tissues would conduct heat at the rate of about 10 kgm-cal. per square metre per hour per °C temperature-difference between his deeper tissues (about 37° C) and the surface. At a water temperature just above freezing point the rate of loss of heat should be about 2·05 $m^2 \times 37° \times 10$ kg-cal. m^{-2} $hour^{-1}$ $°C^{-1}$, i.e. about 12·5 kilogram-calories per minute. Assuming further that his specific heat was about ·83 kilogram-calories per kilogram per degree centigrade, this should result in a fall of temperature of the order of $12{\cdot}5/(85 \times {\cdot}83)$ degrees per minute, i.e. about ·18 degrees per minute. Further experiments showed that while the natural rate of heat production was only 1·2 kilogram calories per minute, this increased to about 12·5 with moderately hard swimming, i.e. sufficient to counterbalance the loss of heat due to the coldness of the water. (So if one falls into cold water the best thing to do is to keep swimming, and so keep warm.)

The law connecting temperature, pressure and volume of a " perfect gas " is $PV = nRT$, where n is the mass of gas expressed in molecular weight units, and R is the " gas constant ". Although real gases do not exactly obey this law, most of them approximate very closely to it at a sufficient dilution. The value of R in various units is

	Logarithms
$8{\cdot}3140 \times 10^7$ erg deg^{-1} mole^{-1}	7·91981
$8{\cdot}3140 \times 10^3$ joule deg^{-1} kmole^{-1}	3·91981
82·053 cm^3 atmos deg^{-1} mole^{-1}	1·91409
1·9864 cal. deg^{-1} mole^{-1}	·29807
623·6 cm^3 cm-Hg deg^{-1} mole^{-1}	3·79491

The degree is of course an arbitrary unit of temperature. If we wish we can obtain a derived unit in which the gas constant $R = 1$; it is only necessary to take $(1/R)^\circ$ as the unit, i.e. the derived temperature unit in the C.G.S. system is $1{\cdot}2028 \times 10^{-8}$ °C, and in the M.K.S. system it is $1{\cdot}2028 \times 10^{-4}$ °C. This unit does not seem to have ever been used in practice. It has the curious property that it makes the specific heat per molecular weight unit a dimensionless constant. In fact twice this specific heat is usually very nearly a small integer, representing, according to the kinetic theory of heat, the number of different ways in which a molecule can move or rotate (the " degrees of freedom " of the molecule). Also in these units the two specific heats of a gas, at constant volume and at constant pressure respectively, differ by 1 for a molecular weight unit (instead of R/J in the usual formula).

16.23 Chemical reactions

If a reaction is of the form molecule of $A +$ molecule of $B \to$ reaction products $R + S$, then Guldberg and Waage's " law of mass action " states that the velocity of the reaction is proportional to the product of the concentrations of A and of B, i.e.

$$\text{reaction velocity} = K[A]\,[B]$$

where $[A]$ and $[B]$ stand for the concentrations of A and B respectively. At any one particular temperature K is constant (and is known as the " reaction constant "). It usually increases exponentially with temperature.

Consider for example the inversion of cane sugar

$$C_{12}H_{22}O_{11} + H_2O \to \underset{\text{Dextrose}}{C_6H_{12}O_6} + \underset{\text{Laevulose}}{C_6H_{12}O_6}$$

Then the reaction velocity $= K[C_{12}H_{22}O_{11}]\,[H_2O]$. If the reaction begins with a very dilute solution the concentration of water will be nearly unity (in suitable units) and will remain so throughout the reaction.

Let $x = [C_{12}H_{22}O_{11}]$ denote the concentration of cane sugar at time t from the start of the reaction. Then the reaction velocity will be equal to the rate of decrease of x, $-dx/dt = -x_t$; and according to the law of mass action this is approximately Kx, i.e.

$$x_t = -Kx.$$

This differential equation has " separable variables " (case A, Section 10.9). It is therefore solved as follows:

Since $dx/dt = -Kx$, $dt/dx = -1/Kx$, and on integration with respect to x, $t = -K^{-1} \ln x + C$. Now if the initial concentration when $t = 0$ is x_0, we have on substitution

$$0 = -K^{-1} \ln x_0 + C, \quad \text{i.e. } C = K^{-1} \ln x_0.$$

So

$$t = -K^{-1} \ln x + K^{-1} \ln x_0$$
$$= K^{-1} \ln (x_0/x).$$

and therefore

$$K = t^{-1} \ln (x_0/x).$$

This can be checked by measuring the concentration x of cane sugar at different times t, and calculating $t^{-1} \ln (x_0/x)$. If our theory is correct, all such values should be equal, and their common value is the reaction constant K.

The following table shows some experimental results:

t (in minutes from start)	x/x_0	$t^{-1} \ln (x_0/x) = K$
0	1	—
1,435	·58240	$3{\cdot}762 \times 10^{-4}$
4,315	·19530	$3{\cdot}785 \times 10^{-4}$
7,070	·07008	$3{\cdot}759 \times 10^{-4}$
11,360	·01484	$3{\cdot}707 \times 10^{-4}$
14,170	·00534	$3{\cdot}692 \times 10^{-4}$
16,935	·00185	$3{\cdot}716 \times 10^{-4}$
19,815	·00069	$3{\cdot}68 \times 10^{-4}$
29,925	·00001	$3{\cdot}8 \times 10^{-4}$

The constancy of K verifies the theory.

Note that for a unimolecular reaction the equation $t = K^{-1} \ln (x_0/x)$ shows that when $x = \frac{1}{2}x_0$, i.e. when the concentration is halved, $t = K^{-1} \ln 2 =$ a constant independent of the initial concentration x_0. This can be used as a test for the unimolecularity of a reaction. For example when diphtheria antitoxin was injected in various strengths into the bodies of animals Bornstein found that the quantity remaining after 4 days was a constant fraction of the original amount, thus suggesting a unimolecular reaction. He confirmed this by plotting log x (logarithm of antitoxin concentration) against t (time elapsed), obtaining a straight line as indicated by the formula $Kt = \ln (x_0/x) = \ln x_0 - \ln x$.

If a reaction is of the form $2A + B \rightarrow R + S$, i.e. $A + A + B \rightarrow R + S$ then the law of mass action indicates that the reaction velocity must be proportional to $[A][A][B] = [A]^2[B]$. Suppose that A is present in a very small concentration $[A] = x$, while that of B can be taken as practically constant. Then the reaction velocity will be $-dx/dt$, and will be of the form kx^2, where k is a constant. We have now to solve the differential equation

$$dx/dt = -kx^2$$
$$\text{i.e. } dt/dx = -k^{-1} x^{-2}$$

whence on integrating with respect to x,

$$t = k^{-1} x^{-1} + C.$$

It follows that for such a reaction we obtain a straight-line graph on plotting x^{-1} against t (see Section 7.3). Similarly, a trimolecular reaction should give a straight line on plotting x^{-2} against t.

Now consider a reversible reaction, say one of the form

$$2A + B \rightleftharpoons 3R + S$$

Then the forward reaction will proceed with velocity $K_1 [A]^2 [B]$, where K_1 is its reaction constant, and the backward reaction will have a velocity $K_2 [R]^3 [S]$. When the equilibrium point is reached these velocities must be equal, i.e.

$$K_1 [A]^2 [B] = K_2 [R]^3 [S]$$

whence $[A]^2 [B]/[R^3] [S] = K_2/K_1 =$ a constant, known as the " dissociation constant " for the reaction. In general if the reaction has the form

$$a \,.\, A + b \,.\, B + \ldots \rightleftharpoons r \,.\, R + s \,.\, S + \ldots$$

(where $a \,.\, A$ means " a molecules of A ") we shall have

$$[A]^a [B]^b \ldots / [R]^r [S]^s \ldots = \text{a constant } k.$$

Thus suppose that the reaction between oxygen and haemoglobin has the form

$$\text{Hb} + n\text{O}_2 \rightleftharpoons \text{HbO}_2$$

Then when equilibrium is established

$$\frac{[\text{HbO}_2]}{[\text{Hb}]\,[\text{O}_2]^n} = \text{dissociation constant } k.$$

Denote the concentration $[O_2]$ of oxygen by x: then this equation gives

$$[\text{HbO}_2]/[\text{Hb}] = kx^n \quad . \quad . \quad . \quad (16.15)$$

Adding 1 to each side we have

$$([\text{HbO}_2] + [\text{Hb}])/[\text{Hb}] = 1 + kx^n \quad . \quad . \quad (16.16)$$

Divide equation (16.15) by (16.16): this gives

$$\frac{[HbO_2]}{[HbO_2] + [Hb]} = \frac{kx^n}{1 + kx^n}$$

But $y = \dfrac{100[HbO_2]}{[HbO_2] + [Hb]}$ represents the percentage saturation of haemoglobin with oxygen. This establishes A. V. Hill's equation

$$y = 100kx^n/(1 + kx^n) \qquad . \qquad . \qquad (16.17)$$

16.24 Electric units

Electrical units can be related to the units of length, time, and mass in three distinct ways—

(*a*) using the force between charges at rest;
(*b*) using the force between currents;
(*c*) using electrolysis, i.e. the decomposition of a chemical substance by the passage of an electric current.

These lead to three possible systems of electric and magnetic units. Those derived from (*a*) are known as " electrostatic units " (e.s.u.), those from (*b*) are " electromagnetic units " (e.m.u.), while (*c*) does not seem to have been ever seriously considered. None of these units have ever been used in practice, as they are of a quite inconvenient order of magnitude—although a hybrid mixture of electrostatic and electromagnetic units (known as Gaussian units) is found in many text-books on theory. To fill the gap a further system of " practical " units has grown up.

The student will find this complicated situation in many books. However as long ago as 1901 G. Giorgi showed that it was possible to obtain a system of units which actually simplifies the theoretical relations and also includes the practical units as a special case. In 1935 this Giorgi system was officially adopted by the International Electrotechnical Commission. A detailed discussion would be outside the scope of this book: a good elementary account will be found in E. G. Cullwick's book *The fundamentals of electro-magnetism* (C.U.P. 1939), (although some of Cullwick's arguments are a little controversial). It is true that electrical theory is not often used in biology, but it may be worth while giving a brief résumé here to show how the different aspects are quite simply linked together. It also illustrates an important type of differential equation (equation 16.30) which may occur in other connections.

The Giorgi system is based on four units: the metre, kilogram, second and coulomb as units of length, mass, time and electrical charge respectively. (There are other ways of choosing the four fundamental units, but they are equivalent to these.) The coulomb can be defined formally by the property:

"Two equal charges of one coulomb placed in empty space at a distance of 1 metre apart will repel one another with a force of $8{\cdot}988 \times 10^9$ newtons. The constant $8{\cdot}988 \times 10^9$ is chosen to be $10^{-7} c^2$, that is, numerically equal to one ten-millionth of the square of the velocity of light measured in metres per second."

In general two charges of q_1 and q_2 coulombs respectively when placed at a distance r metres apart will repel one another with a force

$$f = q_1 q_2 / \kappa_0 r^2 \text{ newtons} \quad . \quad . \quad (16.18)$$

where $\kappa_0 = 1/(8{\cdot}988 \times 10^9) = 10^7/c^2 = 1{\cdot}1126 \times 10^{-10}$ coulomb^2 joule^{-1} m^{-1}. This is the inverse square law for the force between two electric charges. The constant $8{\cdot}988 \times 10^9$ which is used to define the unit of charge, the coulomb, may seem at first sight to be rather oddly chosen, but in practice it turns out to be rather convenient. The reciprocal constant $\kappa_0 = 1{\cdot}1126 \times 10^{-10}$ is known as the "permittivity of free space". From this inverse square law it is possible to deduce the forces acting between any systems of electric charges at rest.

PROBLEM

(1) Show that if any three unknown charges q_1, q_2 and q_3 are given, it is theoretically possible to deduce the value of q_1, q_2, q_3 by measuring the forces between each pair of charges when placed in empty space—except that it is not possible to distinguish between the system of charges q_1, q_2, q_3, and the system $-q_1$, $-q_2$, $-q_3$.

Note—The equation $f = q_1 q_2/\kappa_0 r^2$ will also hold in any other system of units, provided that the appropriate value of κ_0 is taken, according to the usual rule for change of units (Section 16.8). The relation $\kappa_0 = 10^7/c^2$ will no longer necessarily hold.

Electricity can be caused to move by various means, including simple mechanical transportation, electromagnetic forces (as in lightning flashes, dynamos), chemical action (as in cells, accumulators, nerve cells), heat (thermoelectric couple), and light (photoelectric cell). A steady flow of electricity is called a current, and is measured in coulombs per second, or *amperes*. More generally, if q coulombs have been transported down a wire after t seconds, the current will be measured by the rate of change of q, that is, dq/dt amps.

Like everything else electricity will only move if it is urged on by a force, and so as it moves it must do work. This work will generally be dissipated as heat, but can also be used chemically or mechanically. Thus suppose a charge is acted upon by a force with component f along its line of motion; then in moving from a point A to a point B it will gain energy equal to $\int_A^B f\,dL$, where L represents the distance travelled along its path (Section 11.7). In particular, if Q is a charge with fixed position in empty (or nearly empty) space, and q is a movable charge, at a distance r from Q, it follows that the force on q is $f = Qq/\kappa_0 r^2$. It follows that if q moves outward from a distance $r = a$ to a distance $r = b$ it gains energy equal to

$$\int_a^b \frac{Qq}{\kappa_0 r^2}\,dr = \left[-\frac{Qq}{\kappa_0 r}\right]_a^b = \frac{Qq}{\kappa_0}\left(\frac{1}{a} - \frac{1}{b}\right).$$

This energy can be used to do work.

Notice that the energy gained by the charge q (which can be spent in doing external work) is proportional to q. That is only to be expected: 2 coulombs will gain twice as much energy as one coulomb in moving from A to B, provided that the rest of the system is the same in both cases. The amount of energy supplied to a unit charge moving from A to B is called " potential drop " V between A and B: the energy supplied to a charge q will therefore be qV. For example, the potential drop between a point A at distance a from a charge Q and a point B at distance b is (potential of A — potential of B) $= \frac{Q}{\kappa_0}\left(\frac{1}{a} - \frac{1}{b}\right)$. The unit of potential drop is accordingly 1 joule per coulomb, named 1 *volt*.

It follows that if a steady current of I amps $= I$ coulombs per second is passing down a wire along which there is a potential drop of V volts, it will be given energy at the rate of IV joules per second, that is, IV watts. This energy will be dissipated as heat at the rate of $W = IV$ joules per second, or IV/J calories per second.

It is an experimental fact that in most conductors the current I is exactly proportional to the potential drop V (Ohm's law). Thus we can write

$$V = IR \quad . \quad . \quad . \quad . \quad (16.19)$$

The constant of proportionality R is called the "resistance " of the circuit. It is measured in volts per ampere, called *ohms*. Combining this relation with the relation $W = IV$ for the rate of production of heat energy we have

$$W = IV = V^2/R = I^2R \quad . \quad . \quad (16.20)$$

It can be readily shown from Ohm's law that conductors with individual resistances R_1, R_2, R_3, etc. will give a total resistance $R_1 + R_2 + R_3 + \ldots$ when joined in series, and a resistance equal to $(R_1^{-1} + R_2^{-1} + R_3^{-1} + \ldots)^{-1}$ when joined in parallel.

The total drop in potential summed over all the wires in an electric circuit is called the " electromotive force " or " e.m.f." E in the circuit. It follows that if I is the current flowing round the circuit, and R is the total resistance, $R_1 + R_2 + R_3 + \ldots$, then $E = IR$; and the total rate of heat production is $W = EI$.

When a current passes through an electrolytic cell there is a simple proportionality between the amount of substance decomposed and the total quantity of electricity passing through the cell, according to the rule that 96,490 coulombs are required to decompose 1 gram-equivalent, or $9{\cdot}6490 \times 10^7$ coulombs to decompose 1 kg-equiv. This constant of proportionality is known as the *faraday*, F. The values of the fundamental constants can be summarized thus:

	Logarithms
velocity of light $c = 2{\cdot}998 \times 10^8$ m/sec	8·47683
permittivity of vacuum $\kappa_0 = 1{\cdot}1126 \times 10^{-10}$ coulomb^2 joule^{-1} m^{-1}	$\overline{10}{\cdot}04634$
faraday $F = 9{\cdot}649 \times 10^7$ coulombs/kg-equiv.	7·98448

16.25 Dielectrics: electric capacity

The inverse square law $f = q_1 q_2/\kappa_0 r^2$ is valid only in empty space. If the charges q_1 and q_2 are imbedded in a material body then the charges inside the atoms of this body will be disturbed and redistribute themselves. The net effect will be to replace the law by

$$f = q_1 q_2/(K\kappa_0 r^2) \qquad (16.21)$$

where K is a constant characteristic of the material concerned, and known as its " dielectric constant " or " relative permittivity ". It is a pure number: typical values are 1·00058 for air at normal temperature and pressure, 4 to 10 for glass, 1·7 to 2·9 for india-rubber, 4 to 8 for mica. The product $K\kappa_0$ is the " absolute permittivity " κ.

This fact is made use of in constructing condensers or " capacitors ". When a body is raised to a potential V volts, it will acquire a small charge q coulombs, proportional to V. We write

$$q = CV \qquad (16.22)$$

where C is the " capacitance " of the body measured in *farads* or coulombs per volt. A farad is too large a unit for ordinary use, so that most capacities are measured in microfarads (μF) $= 10^{-6}$ farads.

A " capacitor " is a device to hold small quantities of electricity in a compact space. The usual form consists of two parallel plates of area A (square metres) separated by a thin layer of mica, or similar insulator, of thickness δ (metres) and dielectric constant K. Such a capacitor has a capacitance $C = K\kappa_0 A/4\pi\delta$ farads, i.e. when a potential difference of V volts is applied between the two plates, it holds a charge of CV coulombs. (This is a consequence of equation 16.21, but we shall not go through the details here.) In a similar way, if a capacitor is formed by two concentric cylinders of radii R_1 (inner cylinder) and R_2 (outer cylinder) respectively, and of length L (metres), the space between the cylinders being filled with an insulator of dielectric constant K, the capacitance will be approximately $\frac{1}{2}K\kappa_0 L \ln (R_2/R_1)$ farads. This may be relevant to the transmission of a signal along a nerve fibre, which has a very similar construction.

16.26 Magnetic units

The inverse square equation (16.18) represents the force between two stationary charges. When either or both of the charges are in motion there are additional forces between them.

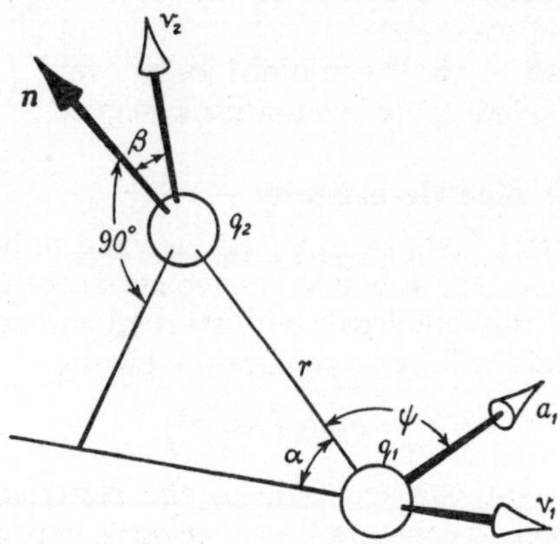

Fig. 16.9—The forces between electric charges

In (Fig. 16.9), let the distance between the two charges be r (metres). Let the charge q_1 move with a velocity v_1 (metres per second) and an acceleration a_1 (metres per second^2), and let q_2 move with a velocity v_2. Let α be the angle between v_1 and the line joining the charges, and further let $\boldsymbol{n}$ be the vector perpendicular to the plane containing the direction v_1 and the two charges, and β the angle between $\boldsymbol{n}$ and the direction of motion v_2.

Then in consequence of the motions of the charges q_2 is acted upon by a force

$$f = q_1 q_2 v_1 v_2 \sin \alpha \sin \beta / r^2 \kappa_0 c^2 \text{ newtons} \qquad (16.23)$$

perpendicular to $\boldsymbol{n}$ and to v_2. Put into simpler language this expression means that this " magnetic force " tends to curve the track of the charge q_2 without altering its speed. Furthermore if q_1 and q_2 are of the same sign, and if when we look at the whole system along the direction $\boldsymbol{n}$ the charge q_1 appears to be moving around q_2 in a clockwise direction, the effect will be to curve the path of q_2 in an anticlockwise direction (and vice versa. If the charges are of opposite signs it will be found that either both will be clockwise, or both anticlockwise.) To take a simple case, consider two charges moving side by side in parallel paths, so that $\alpha = \beta = 90°$; the force equation then reduces to

$$f = \mu_0 q_1 q_2 v_1 v_2 / r^2 \qquad (16.24)$$

where $\mu_0 = 1/\kappa_0 c^2 = 10^{-7}$ newtons sec²/coulomb². Equation (16.24) means that there will be an attraction between charges moving parallel to one another. It follows that two parallel wires carrying currents in the same direction will attract one another, and so will two parallel coils or *electromagnets.* The same effect occurs in a permanent magnet, where the electrons in the iron or other magnetic material all spin parallel to one another, and by such spin behave very similarly to coils carrying currents.

If the charges are not in free space but embedded in a material medium, the forces in equations (16.23) and (16.24) must be multiplied by a factor μ called the " permeability " of the medium.

Besides this " magnetic " force produced by the velocities v_1 and v_2 of the charges, there are still further forces produced by their accelerations. The force on q_2 produced by the acceleration a_1 of q_1 is

$$f = \mu_0 q_1 q_2 a_1 \sin \psi / r \qquad (16.25)$$

where ψ is the angle between the direction of the acceleration a_1 and the line joining the two charges; and f is directed at right angles to this line, in the same plane as a_1 and its component along the direction of a_1 is opposed to the acceleration. In other words, whenever an electron accelerates in a forward direction, all other electrons in the vicinity experience a backward force inversely proportional to their distance, and directly proportional to the acceleration in question. This effect is known as " electromagnetic induction ". (It should perhaps be added that all our equations, such as (16.18), (16.23) and (16.25) are approximations which are very closely obeyed for ordinary velocities, but are subject to correction for velocities approaching that of light. There is also a small additional force due to the combined effect of the velocities and accelerations, but this can normally be neglected.)

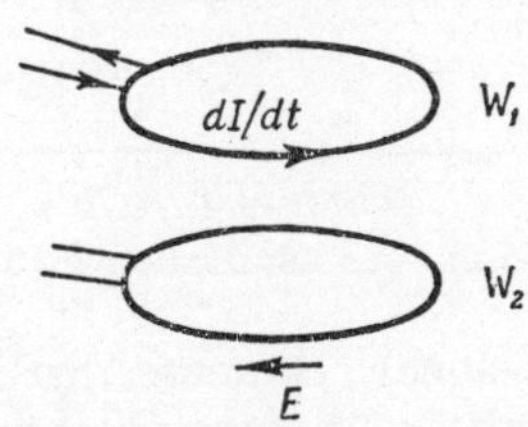

Fig. 16.10—Electromagnetic induction between coils

Induction is the basis of the electrical transformer. Consider two coils of wire, W_1 and W_2 (Fig. 16.10) placed close to one another. Through W_1 we imagine a current I amps to flow. Suppose now that this current is increased: then the electrons in W_1 will be accelerated, and consequently those in W_2 will experience a backward force. Expressed in another way, there will be a back electromotive force E in coil W_2 proportional to the acceleration in W_1, i.e. to the rate of change of current $dI/dt = \dot{I} = I_t$.

Thus we can write

$$E = -MdI/dt \quad . \quad . \quad . \quad (16.26)$$

where M is a constant, called the " coefficient of mutual inductance ". It is measured by the e.m.f. in volts induced in W_2 by a change in current in W_1 of 1 amp per second; i.e. the unit of inductance is 1 volt . sec/amp = 1 *henry*.

We have spoken above of different coils W_1 and W_2: but it is by no means essential to our argument that the coils should be different. The same results apply to a single coil, which has the property of resisting any change in current by producing in itself a back electromotive force E volts given by

$$E = -L\, dI/dt \quad . \quad . \quad . \quad (16.27)$$

Here L is the " coefficient of self-inductance " and forms a sort of electrical inertia of the system, which resists any sudden change of current much as ordinary mass resists any sudden change of velocity.

EXAMPLE

(1) The oscillating circuit (Fig. 16.11). Suppose we have a circuit containing a resistance R (ohms), a capacitor (capacitance C farads) and a

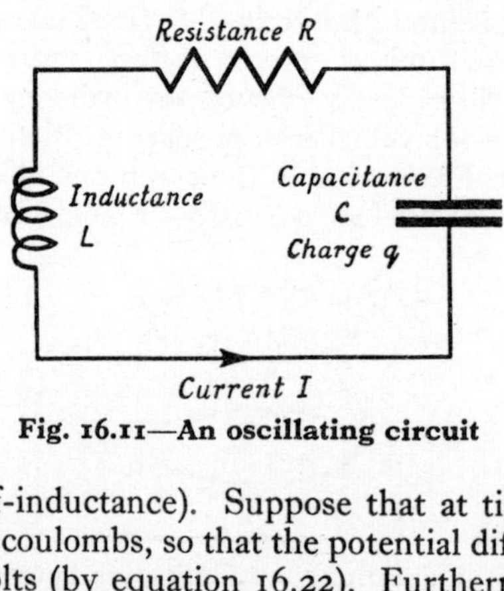

Fig. 16.11—An oscillating circuit

coil (L henries self-inductance). Suppose that at time t the capacitor contains a charge q coulombs, so that the potential difference V between its plates is q/C volts (by equation 16.22). Furthermore the current I amps in the circuit will be produced by the emptying of the capacitor, so that I is minus the rate of change of q, i.e.

$$I = -dq/dt = -D_t q \quad . \quad . \quad . \quad (16.28)$$

The whole system will now act very like a simple pendulum. The current will try to flow from one plate of the capacitor to the other, but in doing so it will be damped by the resistance and hindered by the coil which will oppose any change in the current. By the time the current has really got going the capacitor will be empty: but the current cannot stop itself owing to the presence of the coil, and so the capacitor will be charged up again in the opposite direction before the current stops. The cycle of events then repeats, and we get a series of oscillations.

It is easy to investigate this mathematically. The back e.m.f. due to the coil is $-LD_tI$, i.e. LD_t^2q (sometimes written $L\ddot{q}$, where the two dots denote the second derivative with respect to time). The forward e.m.f. due to the potential difference in the capacitor is q/C, so the net e.m.f. in the circuit is $q/C + L\ddot{q}$ volts. By Ohm's law this must be equal to the current $(-D_tq = -\dot{q})$ times the resistance R, i.e.

$$q/C + L\ddot{q} = -\dot{q}R$$
$$\text{or } L\ddot{q} + \dot{q}R + q/C = 0$$
$$\text{i.e. } LD_t^2q + RD_tq + q/C = 0$$
$$\text{or } Lq_{tt} + Rq_t + q/C = 0 \quad . \quad . \quad . \quad (16.29)$$

according to the notation employed. This is a "linear differential equation of the second order with constant coefficients". We have already considered special cases. If $L = 0$, and we have no coil, the equation reduces to $R\dot{q} + q/C = 0$, i.e. $\dot{q} = q_t = -q/RC$, and this has the solution $q = Ke^{-t/RC}$, an exponential decay like Newton's law of cooling or the decay of a population under unfavourable conditions. If $R = 0$, i.e. there is no resistance, the equation becomes $L\ddot{q} + q/C = 0$, i.e. $\ddot{q} = q_{tt} = -\omega^2 q$ where $\omega = 1/LC$. This corresponds (by Section 15.2) to steady periodic oscillations of period $2\pi/\omega = 2\pi\sqrt{(LC)}$: the solution is $q = K \cos(\omega t + \phi)$ where K and ϕ are arbitrary constants.

In the general case in which $L \neq 0$ we proceed as follows. We first find the values of λ (a constant) for which $q = e^{\lambda t}$ is a possible solution of (16.29). Since in this case $q_t = \lambda e^{\lambda t}$, $q_{tt} = \lambda^2 e^{\lambda t}$, and substitution in (16.29) gives

$$(L\lambda^2 + R\lambda + 1/C)\, e^{\lambda t} = 0$$
$$\text{or } L\lambda^2 + R\lambda + 1/C = 0 \quad . \quad . \quad . \quad . \quad (16.30)$$

since $e^{\lambda t} \neq 0$. Now by hypothesis $L \neq 0$, so this is a quadratic equation for λ with at least one root and probably two distinct roots

$$\lambda_1, \lambda_2 = [-R \pm \sqrt{(R^2 - 4L/C)}]/2L$$

which may however be real or complex. We see that $\lambda_1 + \lambda_2 = -R/L$, i.e.

$$\lambda_2 = \lambda_1 + (\lambda_2 + \lambda_1) - 2\lambda_1$$
$$= \lambda_1 - (2\lambda_1 + R/L) \quad . \quad . \quad . \quad (16.31)$$

so that the roots λ_1, λ_2 are distinct if and only if $2\lambda_1 + R/L \neq 0$. Now let q be *any* solution of the equation (16.29). Write $Q = e^{-\lambda_1 t}q$, so that

$$q = Qe^{\lambda_1 t} \quad . \quad . \quad . \quad . \quad (16.32)$$

Then

$$q_t = Q_t e^{\lambda_1 t} + Q\lambda_1 e^{\lambda_1 t}$$

$$q_{tt} = Q_{tt}e^{\lambda_1 t} + 2Q_t\lambda_1 e^{\lambda_1 t} + Q\lambda_1^2 e^{\lambda_1 t}$$

and substitution of these values in equation (16.29) gives us simply

$$e^{\lambda_1 t} L[Q_{tt} + (2\lambda_1 + R/L)Q_t] = 0$$

as all the other terms cancel, and since $L \neq 0$ and $e^{\lambda_1 t} \neq 0$ this gives us the equation

$$Q_{tt} + (2\lambda_1 + R/L)Q_t = 0 \quad . \quad . \quad (16.33)$$

If now we can solve this equation for Q we can substitute back in (16.32) and find the general value of q. But this is quite simple. There are two cases to consider.

Case I. $2\lambda_1 + R/L = 0$, i.e. the quadratic (16.30) has only one root $\lambda_1 = -R/2L$ (and therefore necessarily real).

Equation (16.33) then reduces to $Q_{tt} = 0$, or on integrating with respect to t,

$$Q_t = \int Q_{tt}\,dt = C_t, \qquad Q = \int Q_t\,dt = C_t t + C_2,$$

where C_1 and C_2 are constants of integration. By (16.32) the solution is

$$q = (C_1 t + C_2)e^{\lambda_1 t}$$

$$= (C_1 t + C_2)e^{-tR/2L} \quad . \quad . \quad . \quad (16.34)$$

Thus the current decays steadily and rapidly to zero with no oscillations.

Case II. $2\lambda_1 + R/L \neq 0$.

Write $\psi = Q_t$, then (16.33) becomes $\psi_t = -(2\lambda_1 + R/L)\psi$, which has the solution $\psi = Q_t = Ke^{-(2\lambda_1 + R/L)t}$. On integrating with respect to t this gives $Q = C_2 e^{-(2\lambda_1 + R/L)t} + C_1$, where $C_2 = -K/(2\lambda_1 + R/L)$ and C_1 is a constant of integration. But by (16.31) $(2\lambda_1 + R/L) = \lambda_1 - \lambda_2$, whence $Q = C_2 e^{(\lambda_2 - \lambda_1)t} + C_1$, and the general solution for q is by (16.32)

$$q = Qe^{\lambda_1 t} = C_2 e^{\lambda_2 t} + C_1 e^{\lambda_1 t} \quad . \quad . \quad . \quad (16.35)$$

provided that the roots λ_1 and λ_2 are distinct. This accordingly gives us the general solution of the differential equation (16.30).

If λ_1 and λ_2 are real, they must be negative in our case, from the formula for the solution of the quadratic equation (16.30). It follows that q is the sum of two components each of which decays exponentially, without oscillation.

If however λ_1 and λ_2 are complex they must be conjugate complex quantities (being solutions of a real equation, Section 14.18) and can be written $-\alpha + i\beta$ and $-\alpha - i\beta$. Since $\lambda_1 + \lambda_2 = -R/L$, we have $\alpha = R/2L$ = a positive quantity. Thus from (16.35)

$$\begin{aligned} q &= C_2 e^{(-\alpha-i\beta)t} + C_1 e^{(-\alpha+i\beta)t} \\ &= e^{-\alpha t}[C_2 e^{-i\beta t} + C_1 e^{i\beta t}] \\ &= e^{-\alpha t}[C_2 (\cos \beta t - i \sin \beta t) + C_1 (\cos \beta t + i \sin \beta t)] \\ &= e^{-\alpha t}[(C_2 + C_1) \cos \beta t + (i\, C_1 - i\, C_2) \sin \beta t] \\ &= e^{-\alpha t}[A \cos \beta t + B \sin \beta t] \quad \text{(say).} \end{aligned}$$

The expression in the square brackets represents an oscillation with period

$$2\pi/\beta = 4\pi i/(\lambda_1 - \lambda_2) = 4\pi\, L/\sqrt{(4L/C - R^2)}.$$

The expression outside the bracket, $e^{-\alpha t} = e^{-tR/2L}$, shows that this oscillation decays in an exponential or geometric manner, and can be shortly described as a " damped oscillation ".

Similar conclusions can be expected whenever we have a " linear second-order equation with constant coefficients " such as (16.29), even if the interpretation of the quantities concerned is in no way electrical. It might also be expected that a similar approach would be helpful in the study of the transmission of impulses in nerves; but the situation will then be more complicated, since besides the capacity, resistance and inductance of the fibre there are also complicated chemical changes to be taken into account.

16.27 Light and radiation

A comparison of the equations (16.18) for the electrostatic force, (16.23) for the electromagnetic force, and (16.25) for the force of induction, shows that in the last equation the force decreases with distance as $1/r$, while in the other two it decreases much more rapidly as the inverse square $1/r^2$. Thus at great distances it will still be possible to detect the effect of the *acceleration* of an electric charge, even when the effect of its velocity or the simple electrostatic effect have become negligible. This is the principle of a wireless transmitter, in which the electrons in the aerial are constantly maintained in a state of rapid oscillation. Other examples of electromagnetic radiation of this sort are light, radiant heat, ultra-violet rays, X-rays and gamma rays.

By using a more exact form of the equations than the ones we have given it can be shown that the radiation, or field produced by acceleration, is not transmitted instantaneously (as implied by equation 16.25) but travels outwards with velocity c in empty space, and $c/\sqrt{(K\mu)}$ in a medium of dielectric constant K and permeability μ. It follows that $\sqrt{(K\mu)}$ is the ratio of the velocity of light in free space to the velocity in the medium, and is called the " refractive index " n of the medium.

The principal property of n is the refraction equation $\sin i/\sin r = n$, where i is the angle of incidence and r that of refraction. (For proof, see any text-book on optics.) When i and r are small this becomes $i/r \simeq n$. From this it can be shown that if a thin lens is made of material of refractive index n and has spherical surfaces of radii of curvature R_1 and R_2 respectively (measured positively if curved away from the object), and focal length f, then

$$\frac{1}{v} - \frac{1}{u} = \frac{1}{f} = (n - 1)\left(\frac{1}{R_1} - \frac{1}{R_2}\right)$$

where u metres = the distance of the object, and v metres is the distance of the image. The reciprocal $1/f$ of the focal length is called the "power" of the lens, and is measured in *dioptres*, where 1 dioptre (D) = 1/metre. This unit is of importance in expressing the degree of short- or long-sightedness of an individual; the necessary correcting lens will have its power calculated in dioptres.

The two most important properties of a beam of light or radiation are the wavelength, the physical equivalent of colour, and the amplitude, the physical equivalent of brightness. The wavelength is usually expressed in Angstroms; 1 *Angstrom* (Å) = 10^{-10} metre. (Frequency = velocity/wavelength = c/λ.) The brightness can be expressed in energy units (as watts/m^2): but this is often not very suitable as it does not take into account the different sensitivity of the eye to different wavelengths. It is usual to take an international *candle-power* defined in terms of a special lamp as the unit of intensity of a source of light. The flux or amount of light per second falling on a surface is measured in *lumens*; 1 candle-power lamp emits 4π lumens. The intensity of illumination of a surface is measured in *lux*: 1 lux = 1 lumen per square metre. At a distance d metres from a source of c candle-power the intensity of illumination of a surface on which the light falls normally is c/d^2 lux; if it falls at an angle θ with the normal to the surface, the intensity will be $(c \cos \theta)/d^2$ lux.

16.28 Assay units

Certain substances, such as newly discovered antiseptics, insecticides, and vitamins cannot be estimated by the usual chemical methods, since either their chemical structure is not known, or no appropriate scheme of quantitative analysis has been worked out. In such a case a bio-assay has to be performed and the strength of the unknown substance estimated by its effect on experimental animals or animal tissue. It is then usual to invent an arbitrary unit describing the ratio of the potency of the specimen under investigation to that of a standard preparation, and in time such an arbitrary unit will become an "international unit". Such international units may remain in use even when the bio-assay procedure has been replaced by a more exact chemical assay.

The mathematical and statistical processes used in assay work form an important but rather complicated and specialized part of mathematical biology. It would be impossible, for reasons of space, to do justice to the subject here. The reader who is interested is referred to other books on the subject, such as D. J. Finney, *Statistical Method in Biological Assay* (Charles Griffin & Co., 1952, 1964) and J. H. Burn, D. J. Finney and L. G. Goodwin, *Biological Standardization* (O.U.P., 1950). A possible simplification of some of the methods has been suggested by J. Berkson, "Application of the logistic function to bio-assay", *Journ. Amer. Stat. Assoc.*, **39** (1944), 375.

APPENDIX TABLES

EXPLANATION OF THE TABLES

Table 1—The Greek alphabet

This table is included for reference. It gives a list of the letters of the Greek alphabet, and their names (in the customary English spelling). (Note that the *ch* in *chi*, χ, is always pronounced hard, i.e. like *k*.)

Table 2—Common logarithms

This is a four-figure table of common logarithms, with mean differences.

Most tables of logarithms do not give four-figure accuracy in the early part of the table if *mean* differences are used. In order to remedy this defect the values of $\log x$ are here tabulated for each increase of ·005 in x when x lies between 1·0 and 1·5, instead of the usual tabulation at intervals of ·01. Thus the table gives log 1·000 = ·0000, log 1·005 = ·0022, log 1·010 = ·0043, log 1·015 = ·0065, and so on. Logically these should be written in order in a row; but so as to save space and to present the tabulation in a convenient form the value of log 1·005 (= ·0022) is printed immediately below that of log 1·000 (= ·0000). The value of log 1·015 (= ·0065) is printed immediately below that of log 1·010 (= ·0043), and so on up to log 1·495 (= ·1746) which is below log 1·490 (= ·1732). After this point the tabulation proceeds in the usual way, since the extra entries are no longer needed to ensure sufficient accuracy. Thus in the row opposite "1·5" in the left-hand margin occur the values of log 1·50 = ·1761, log 1·51 = ·1790, log 1·52 = ·1818, and so on. This latter part of the table is to be used in the customary way.

Example. To find log 2·544.

Opposite "2·5" in column "4" find log 2·54	=	·4048
Mean difference for last figure "4"	=	7
log 2·544	=	·4055

The use of the first part of the table is almost as simple, once the principle of tabulation has been understood. Thus to find log 1·035 we look along the row marked "1·0" as far as column headed "3", finding log 1·030 = ·0128. Underneath this will be found log 1·035 = ·0149. In this part of the table mean differences will only be required up to 4 in the last place, since, for example, "1·037" can be looked on as "1·035 + 2 in the last place", and its logarithm can be found thus—

log 1·035	=	·0149
diff. for 2	=	8
log 1·037	=	·0157

Further accuracy is gained by taking advantage of the available space and giving two sets of mean differences opposite "1·0". The upper set is marked "(1·00–4)", meaning that it is to be used when the first three figures of the number lie between 1·00 and 1·04 inclusive; the second set is marked "(1·05–9)", and is to be used when the first three figures lie between 1·05 and 1·09 inclusive.

Example. To find log 1·034.

$$\begin{aligned} \text{Opposite "1·0" in column "3" find } \log 1{\cdot}03 &= {\cdot}0128 \\ \text{Difference for "4", in upper row,} &= \quad 17 \\ \hline \log 1{\cdot}034 &= {\cdot}0145 \end{aligned}$$

This standard of accuracy will be found sufficient for most purposes. But, if so desired, it can be still further improved in some cases by the use of negative differences (Colson fashion) where the number approaches the next tabulated value.

Example. To find log 1·919. We note that 1·919 is nearer 1·920 than 1·910. Thus

$$\begin{aligned} \log 1{\cdot}920 &= {\cdot}2833 \\ \text{diff. for } -1 &= \quad -2 \\ \hline \log 1{\cdot}919 &= {\cdot}2831 \end{aligned}$$

Example. To find log 1·224. We note that 1·224 is nearer 1·225 than 1·220. Thus

$$\begin{aligned} \log 1{\cdot}225 &= {\cdot}0881 \\ \text{diff. for } -1 &= \quad -4 \\ \hline \log 1{\cdot}224 &= {\cdot}0877 \end{aligned}$$

Antilogarithms can be found by the inverse use of the table.

Example. To find antilog ·7813.

By inspection, the nearest entry in the table is log 6·04 = ·7810. This falls short of the required value ·7813 by 3. The entry "3" in the mean difference table occurs in the column headed "4"; therefore log 6·044 = ·7810 + 3 (in the last figure) = ·7813, or antilog ·7813 = 6·044.

Natural logarithms can be found by the formula

$$\ln x = 2{\cdot}3026 \log x$$

and exponentials by the formula

$$e^x = \text{antilog}\,({\cdot}4343x)$$

Tables 3, 4 and 5—Sines and cosines

Table 3 gives the values of sin $x°$, where x is expressed in degrees and decimals of a degree. (This division has advantages over the customary division into minutes and seconds.) The whole number of degrees in x is to be read on the left, and the first decimal place in the headings at the top of the columns: e.g. to find sin 23·7°, look in the row beginning "23" under the heading "·7", obtaining sin 23·7° = ·4019. A correction for the second decimal place in x is given by the mean difference.

Example. To find sin 23·73°.

$$\begin{aligned} \text{From Table 3, } \sin 23{\cdot}7 &= {\cdot}4019 \\ \text{mean difference for } 3 &= +\ \ 5 \\ \hline \sin 23{\cdot}73 &= {\cdot}4024 \end{aligned}$$

The value of cos $y°$ can also be found from the table; in this case y is to be read from the *right*-hand column and row at the *bottom* of the table, and differences must be *subtracted.* (This is equivalent to the identity $\cos y° = \sin (90 - y)° = \sin x°$.)

Example. To find cos 66·27°.

By looking in the row ending "66" above "·2" in the bottom row we find

$$\begin{aligned} \cos 66{\cdot}2° &= {\cdot}4035 \\ \text{negative difference for "7"} &= -\ 11 \\ \hline \cos 66{\cdot}27° &= {\cdot}4024 \end{aligned}$$

Values of $\sin^{-1} x$ and $\cos^{-1} x$ can be found by inverse use of the table.

Example. To find $\sin^{-1} \cdot 4024$, i.e. the angle $x°$ whose sine is ·4024.

By inspection, the nearest entry in the table is $\sin 23{\cdot}7° = {\cdot}4019$; and ·4024 is greater than this by 5 in the last place. To find the second decimal in x, we must therefore find what heading in the mean differences lies above an entry equal to 5 (or as nearly as possible equal to 5). In this case, the entry "5" comes under the heading "3", and hence $\sin 23{\cdot}73° = {\cdot}4024$, or $\sin^{-1} {\cdot}4024 = 23{\cdot}73°$.

If we require answers in radian measure, the rule for conversion is that 1 radian = 57·30°, or 1° = ·01745 radian. In particular,

$$\int_a^b \frac{dx}{\sqrt{1 - x^2}} = \sin^{-1} b - \sin^{-1} a \quad \text{(in radians)}$$
$$= {\cdot}01745\,(\sin^{-1} b - \sin^{-1} a) \quad \text{(in degrees).}$$

Example. To find $\int_0^{0\cdot9} \frac{dx}{\sqrt{1 - x^2}}$.

This is $\cdot01745\,(\sin^{-1} 0{\cdot}9 - \sin^{-1} 0) = {\cdot}01745 \sin^{-1} {\cdot}9000$ (in degrees). But from the table, $\sin 64{\cdot}1° = {\cdot}8996$, and ·9000 exceeds this by 4 in the last place, which is the mean difference under the heading "5". Hence $\sin^{-1} {\cdot}9000 = 64{\cdot}15°$, and the integral is $\cdot01745 \times 64{\cdot}15 = 1{\cdot}119$.

Tables 4 and 5 are intended to give more accurate values for small angles. Table 4 applies to angles measured in degrees. We have

$$\sin x° = (\sin x°)/x \times x$$

and

$$x = \frac{\sin x°}{(\sin x°)/x}$$

Example. Find $\sin 4{\cdot}011°$.

From Table 4, when $x = 4{\cdot}011$ (and therefore lies in the range from 3·047 to 4·535), $(\sin x°)/x = {\cdot}01744$, whence $\sin x° = \sin 4{\cdot}011° = {\cdot}01744 \times 4{\cdot}011 = {\cdot}06995$, to about 4 significant figure accuracy. (In contrast, direct use of Table 3 gives $\sin 4{\cdot}011 = {\cdot}0700$ with only 3 significant figures.)

Example. Find $\sin^{-1} {\cdot}06995$, i.e. the angle $x°$ for which $\sin x = {\cdot}06995$.

From Table 4 we see that when $\sin x° = {\cdot}06995$ we have $(\sin x°)/x = {\cdot}01744$, whence $x = {\cdot}06995/{\cdot}01744 = 4{\cdot}011$.

Table 5 serves the same purpose in radian measure. We then have for small x,

$$\begin{aligned} \sin x \simeq \quad \sinh^{-1} x &\simeq (1 - A)x \\ \sin^{-1} x \simeq \quad \sinh x &\simeq (1 + A)x \\ \operatorname{cosec} x \simeq \operatorname{cosech}^{-1} x &\simeq (1 + A)/x \\ \operatorname{cosec}^{-1} x \simeq \quad \operatorname{cosech} x &\simeq (1 - A)/x \end{aligned}$$

Example. Find $\sin {\cdot}02$ (radians).

From Table 5, when $x = {\cdot}02$, $A = {\cdot}0001$, whence $\sin {\cdot}02 = {\cdot}02\,(1 - {\cdot}0001) = {\cdot}02000$ to 4 significant figures.

Tables 6, 7 and 8—Tangents and cotangents

Table 6 gives the value of $\tan x°$, where the integral part of x is read in the *left*-hand column and the first decimal place in the headings above the columns, the correction for the second decimal place being given by the mean differences. It also gives $\cot y°$, where y is similarly read in the *right*-hand column (for integral part) and at the *foot* of the columns (for first decimal), and differences are *subtracted*. This corresponds to the identity $\cot y° = \tan (90 - y)° = \tan x°$.

Example. To find $\cot 64{\cdot}42°$.

From Table 6, $\cot 64{\cdot}4° = {\cdot}4748$
negative difference for 2 = − 4

$\cot 64{\cdot}42 = {\cdot}4744$

For angles x between 45° and 90°, or y between 0° and 45°, mean differences are not sufficiently accurate. It is then simplest to use the formulas

$$\tan y^\circ = 1/\cot y^\circ; \qquad \cot x^\circ = 1/\tan x^\circ$$

Example. To find cot 20·33°.

We have

$$\begin{aligned}\cot 20{\cdot}33^\circ &= 1/\tan 20{\cdot}33^\circ \\ &= 1/{\cdot}3705 = 2{\cdot}699.\end{aligned}$$

Alternatively, for $x < 87°$ and for $y > 3°$, tan x and cot y can be found by linear interpolation between successive entries in Table 6. Thus since $\cot 20{\cdot}3^\circ = 2{\cdot}703$ and $\cot 20{\cdot}4^\circ = 2{\cdot}689$ we have

$$\begin{aligned}\cot 20{\cdot}33 &= 2{\cdot}703 + {\cdot}3\,(2{\cdot}689 - 2{\cdot}703) \\ &= 2{\cdot}703 - {\cdot}0004 = 2{\cdot}699.\end{aligned}$$

The table can be used inversely to find $\tan^{-1}x$ and $\cot^{-1}x$ in degrees, and

$$\begin{aligned}\int_a^b \frac{dx}{1+x^2} &= \tan^{-1}b - \tan^{-1}a \qquad \text{(in radians)} \\ &= {\cdot}01745\,(\tan^{-1}b - \tan^{-1}a) \text{ (in degrees).}\end{aligned}$$

(If b is greater than 1, it is more convenient to use $\cot^{-1}(b^{-1})$ instead of $\tan^{-1}b$; similarly for a.)

Example. To find $\displaystyle\int_{0\cdot2}^{2} \frac{dx}{1+x^2} = {\cdot}01745\,(\cot^{-1}\tfrac{1}{2} - \tan^{-1}0{\cdot}2).$

From Table 6,

cot 63·4° = ·5008	tan 11·3° = ·1998
negative difference for 3 = − 7	difference for 1 = + 2
cot 63·43° = ·5001	tan 11·31° = 2·000

Hence the integral is $\cdot01745\,(63{\cdot}43 - 11{\cdot}31) = {\cdot}01745 \times 52{\cdot}12 = {\cdot}9095$.

Table 7 is used to find more accurately the tangents and cotangents of small angles $x°$, or of angles $(90 - x)°$ near to a right angle, according to the formulas:

$$\tan x^\circ = \cot(90 - x)^\circ = x \times (\tan x^\circ)/x$$

$$\cot x^\circ = \tan(90 - x)^\circ = \frac{1}{x \times (\tan x^\circ)/x}$$

$$x = \frac{\tan x^\circ}{(\tan x^\circ)/x} = \frac{1}{(\cot x^\circ).(\tan x^\circ)/x}$$

Example. To find tan 2·020°.

From Table 7, $(\tan 2{\cdot}020^\circ)/2{\cdot}020 = {\cdot}01746$, whence

$$\tan 2{\cdot}020^\circ = 2{\cdot}020 \times {\cdot}01746 = {\cdot}03527.$$

Example. To find tan 87·980°.

$\tan 87{\cdot}980^\circ = \cot(90 - 87{\cdot}980)^\circ = \cot 2{\cdot}020^\circ = 1/(2{\cdot}020 \times {\cdot}01746) = 28{\cdot}35.$

Table 8 is used to find tangents and cotangents of small angles in radian measure, according to the formulas

$$\begin{aligned}\tan x &\simeq \tanh^{-1}x \simeq (1 + B)x \\ \tan^{-1}x &\simeq \tanh x \simeq (1 - B)x \\ \cot x &\simeq \coth^{-1}x \simeq (1 - B)x \\ \cot^{-1}x &\simeq \coth x \simeq (1 + B)x.\end{aligned}$$

Example. To find tan ·02020 (radians).

From Table 8, when $x = {\cdot}02020$, $B = {\cdot}0001$, whence $\tan {\cdot}02020 = 1{\cdot}0001 \times {\cdot}02020 = {\cdot}02020$ to four significant figures.

Table 1—Greek alphabet

Α	α	alpha	Ν	ν	nu
Β	β	beta	Ξ	ξ	xi
Γ	γ	gamma	Ο	ο	omicron
Δ	δ	delta	Π	π	pi
Ε	ϵ	epsilon	Ρ	ρ	rho
Ζ	ζ	zeta	Σ	σ	sigma
Η	η	eta	Τ	τ	tau
Θ	θ	theta	Υ	υ	upsilon
Ι	ι	iota	Φ	ϕ	phi
Κ	κ	kappa	Χ	χ	chi
Λ	λ	lambda	Ψ	ψ	psi
Μ	μ	mu	Ω	ω	omega

Table 2—Common logarithms

	0	1	2	3	4	5	6	7	8	9	1	2	3	4	5	6	7	8	9	
1.0	.0000	.0043	.0086	.0128	.0170	.0212	.0253	.0294	.0334	.0374	4	8	13	17						(1.00-4)
	.0022	.0065	.0107	.0149	.0191	.0233	.0273	.0314	.0354	.0394	4	8	12	16						(1.05-9)
1.1	.0414	.0453	.0492	.0531	.0569	.0607	.0645	.0682	.0719	.0755	4	8	12	16						(1.10-2)
	.0434	.0473	.0511	.0550	.0588	.0626	.0663	.0700	.0737	.0774	4	8	11	15						(1.13-9)
1.2	.0792	.0828	.0864	.0899	.0934	.0969	.1004	.1038	.1072	.1106	4	7	11	14						(1.20-3)
	.0810	.0846	.0881	.0917	.0952	.0986	.1021	.1055	.1089	.1123	3	7	10	14						(1.24-9)
1.3	.1139	.1173	.1206	.1239	.1271	.1303	.1335	.1367	.1399	.1430	3	7	10	13						(1.30-3)
	.1156	.1189	.1222	.1255	.1287	.1319	.1351	.1383	.1414	.1446	3	6	10	13						(1.34-9)
1.4	.1461	.1492	.1523	.1553	.1584	.1614	.1644	.1673	.1703	.1732	3	6	9	12						
	.1477	.1508	.1538	.1569	.1599	.1629	.1658	.1688	.1717	.1746										
1.5	.1761	.1790	.1818	.1847	.1875	.1903	.1931	.1959	.1987	.2014	3	6	8	11	14	17	20	22	25	
1.6	.2041	.2068	.2095	.2122	.2148	.2175	.2201	.2227	.2253	.2279	3	5	8	11	13	16	18	21	24	
1.7	.2304	.2330	.2355	.2380	.2405	.2430	.2455	.2480	.2504	.2529	2	5	7	10	12	15	17	20	22	
1.8	.2553	.2577	.2601	.2625	.2648	.2672	.2695	.2718	.2742	.2765	2	5	7	9	12	14	16	19	21	
1.9	.2788	.2810	.2833	.2856	.2878	.2900	.2923	.2945	.2967	.2989	2	4	7	9	11	13	16	18	20	
2.0	.3010	.3032	.3054	.3075	.3096	.3118	.3139	.3160	.3181	.3201	2	4	6	8	11	13	15	17	19	
2.1	.3222	.3243	.3263	.3284	.3304	.3324	.3345	.3365	.3385	.3404	2	4	6	8	10	12	14	16	18	
2.2	.3424	.3444	.3464	.3483	.3502	.3522	.3541	.3560	.3579	.3598	2	4	6	8	10	12	14	15	17	
2.3	.3617	.3636	.3655	.3674	.3692	.3711	.3729	.3747	.3766	.3784	2	4	6	7	9	11	13	15	17	
2.4	.3802	.3820	.3838	.3856	.3874	.3892	.3909	.3927	.3945	.3962	2	4	5	7	9	11	12	14	16	
2.5	.3979	.3997	.4014	.4031	.4048	.4065	.4082	.4099	.4116	.4133	2	3	5	7	9	10	12	14	15	
2.6	.4150	.4166	.4183	.4200	.4216	.4232	.4249	.4265	.4281	.4298	2	3	5	7	8	10	11	13	15	
2.7	.4314	.4330	.4346	.4362	.4378	.4393	.4409	.4425	.4440	.4456	2	3	5	6	8	9	11	13	14	
2.8	.4472	.4487	.4502	.4518	.4533	.4548	.4564	.4579	.4594	.4609	2	3	5	6	8	9	11	12	14	
2.9	.4624	.4639	.4654	.4669	.4683	.4698	.4713	.4728	.4742	.4757	1	3	4	6	7	9	10	12	13	
3.0	.4771	.4786	.4800	.4814	.4829	.4843	.4857	.4871	.4886	.4900	1	3	4	6	7	9	10	11	13	
3.1	.4914	.4928	.4942	.4955	.4969	.4983	.4997	.5011	.5024	.5038	1	3	4	6	7	8	10	11	12	
3.2	.5051	.5065	.5079	.5092	.5105	.5119	.5132	.5145	.5159	.5172	1	3	4	5	7	8	9	11	12	
3.3	.5185	.5198	.5211	.5224	.5237	.5250	.5263	.5276	.5289	.5302	1	3	4	5	6	8	9	10	12	
3.4	.5315	.5328	.5340	.5353	.5366	.5378	.5391	.5403	.5416	.5428	1	3	4	5	6	8	9	10	11	
3.5	.5441	.5453	.5465	.5478	.5490	.5502	.5514	.5527	.5539	.5551	1	2	4	5	6	7	9	10	11	
3.6	.5563	.5575	.5587	.5599	.5611	.5623	.5635	.5647	.5658	.5670	1	2	4	5	6	7	8	10	11	
3.7	.5682	.5694	.5705	.5717	.5729	.5740	.5752	.5763	.5775	.5786	1	2	3	5	6	7	8	9	10	
3.8	.5798	.5809	.5821	.5832	.5843	.5855	.5866	.5877	.5888	.5899	1	2	3	5	6	7	8	9	10	
3.9	.5911	.5922	.5933	.5944	.5955	.5966	.5977	.5988	.5999	.6010	1	2	3	4	5	7	8	9	10	
4.0	.6021	.6031	.6042	.6053	.6064	.6075	.6085	.6096	.6107	.6117	1	2	3	4	5	6	8	9	10	
4.1	.6128	.6138	.6149	.6160	.6170	.6180	.6191	.6201	.6212	.6222	1	2	3	4	5	6	7	8	9	
4.2	.6232	.6243	.6253	.6263	.6374	.6284	.6294	.6304	.6314	.6325	1	2	3	4	5	6	7	8	9	
4.3	.6335	.6345	.6355	.6365	.6375	.6385	.6395	.6405	.6415	.6425	1	2	3	4	5	6	7	8	9	
4.4	.6435	.6444	.6454	.6464	.6474	.6484	.6493	.6503	.6513	.6522	1	2	3	4	5	6	7	8	9	
4.5	.6532	.6542	.6551	.6561	.6571	.6580	.6590	.6599	.6609	.6618	1	2	3	4	5	6	7	8	9	
4.6	.6628	.6637	.6646	.6656	.6665	.6675	.6684	.6693	.6702	.6712	1	2	3	4	5	6	7	7	8	
4.7	.6721	.6730	.6739	.6749	.6758	.6767	.6776	.6785	.6794	.6803	1	2	3	4	5	5	6	7	8	
4.8	.6812	.6821	.6830	.6839	.6848	.6857	.6866	.6875	.6884	.6893	1	2	3	4	4	5	6	7	8	
4.9	.6902	.6911	.6920	.6928	.6937	.6946	.6955	.6964	.6972	.6981	1	2	3	4	4	5	6	7	8	

Table 2—Common logarithms (*concluded*)

	0	1	2	3	4	5	6	7	8	9	1	2	3	4	5	6	7	8	9
5.0	.6990	.6998	.7007	.7016	.7024	.7033	.7042	.7050	.7059	.7067	1	2	3	3	4	5	6	7	8
5.1	.7076	.7084	.7093	.7101	.7110	.7118	.7126	.7135	.7143	.7152	1	2	3	3	4	5	6	7	8
5.2	.7160	.7168	.7177	.7185	.7193	.7202	.7210	.7218	.7226	.7235	1	2	2	3	4	5	6	7	7
5.3	.7243	.7251	.7259	.7267	.7275	.7284	.7292	.7300	.7308	.7316	1	2	2	3	4	5	6	6	7
5.4	.7324	.7332	.7340	.7348	.7356	.7364	.7372	.7380	.7388	.7396	1	2	2	3	4	5	6	6	7
5.5	.7404	.7412	.7419	.7427	.7435	.7443	.7451	.7459	.7466	.7474	1	2	2	3	4	5	5	6	7
5.6	.7482	.7490	.7497	.7505	.7513	.7520	.7528	.7536	.7543	.7551	1	2	2	3	4	5	5	6	7
5.7	.7559	.7566	.7574	.7582	.7589	.7597	.7604	.7612	.7619	.7627	1	2	2	3	4	5	5	6	7
5.8	.7634	.7642	.7649	.7657	.7664	.7672	.7679	.7686	.7694	.7701	1	1	2	3	4	4	5	6	7
5.9	.7709	.7716	.7723	.7731	.7738	.7745	.7752	.7760	.7767	.7774	1	1	2	3	4	4	5	6	7
6.0	.7782	.7789	.7796	.7803	.7810	.7818	.7825	.7832	.7839	.7846	1	1	2	3	4	4	5	6	6
6.1	.7853	.7860	.7868	.7875	.7882	.7889	.7896	.7903	.7910	.7917	1	1	2	3	4	4	5	6	6
6.2	.7924	.7931	.7938	.7945	.7952	.7959	.7966	.7973	.7980	.7987	1	1	2	3	3	4	5	6	6
6.3	.7993	.8000	.8007	.8014	.8021	.8028	.8035	.8041	.8048	.8055	1	1	2	3	3	4	5	5	6
6.4	.8062	.8069	.8075	.8082	.8089	.8096	.8102	.8109	.8116	.8122	1	1	2	3	3	4	5	5	6
6.5	.8129	.8136	.8142	.8149	.8156	.8162	.8169	.8176	.8182	.8189	1	1	2	3	3	4	5	5	6
6.6	.8195	.8202	.8209	.8215	.8222	.8228	.8235	.8241	.8248	.8254	1	1	2	3	3	4	5	5	6
6.7	.8261	.8267	.8274	.8280	.8287	.8293	.8299	.8306	.8312	.8319	1	1	2	3	3	4	5	5	6
6.8	.8325	.8331	.8338	.8344	.8351	.8357	.8363	.8370	.8376	.8382	1	1	2	3	3	4	4	5	6
6.9	.8388	.8395	.8401	.8407	.8414	.8420	.8426	.8432	.8439	.8445	1	1	2	2	3	4	4	5	6
7.0	.8451	.8457	.8463	.8470	.8476	.8482	.8488	.8494	.8500	.8506	1	1	2	2	3	4	4	5	6
7.1	.8513	.8519	.8525	.8531	.8537	.8543	.8549	.8555	.8561	.8567	1	1	2	2	3	4	4	5	5
7.2	.8573	.8579	.8585	.8591	.8597	.8603	.8609	.8615	.8621	.8627	1	1	2	2	3	4	4	5	5
7.3	.8633	.8639	.8645	.8651	.8657	.8663	.8669	.8675	.8681	.8686	1	1	2	2	3	4	4	5	5
7.4	.8692	.8698	.8704	.8710	.8716	.8722	.8727	.8733	.8739	.8745	1	1	2	2	3	4	4	5	5
7.5	.8751	.8756	.8762	.8768	.8774	.8779	.8785	.8791	.8797	.8802	1	1	2	2	3	3	4	5	5
7.6	.8808	.8814	.8820	.8825	.8831	.8837	.8842	.8848	.8854	.8859	1	1	2	2	3	3	4	5	5
7.7	.8865	.8871	.8876	.8882	.8887	.8893	.8899	.8904	.8910	.8915	1	1	2	2	3	3	4	4	5
7.8	.8921	.8927	.8932	.8938	.8943	.8949	.8954	.8960	.8965	.8971	1	1	2	2	3	3	4	4	5
7.9	.8976	.8982	.8987	.8993	.8998	.9004	.9009	.9015	.9020	.9025	1	1	2	2	3	3	4	4	5
8.0	.9031	.9036	.9042	.9047	.9053	.9058	.9063	.9069	.9074	.9079	1	1	2	2	3	3	4	4	5
8.1	.9085	.9090	.9096	.9101	.9106	.9112	.9117	.9122	.9128	.9133	1	1	2	2	3	3	4	4	5
8.2	.9138	.9143	.9149	.9154	.9159	.9165	.9170	.9175	.9180	.9186	1	1	2	2	3	3	4	4	5
8.3	.9191	.9196	.9201	.9206	.9212	.9217	.9222	.9227	.9232	.9238	1	1	2	2	3	3	4	4	5
8.4	.9243	.9248	.9253	.9258	.9263	.9269	.9274	.9279	.9284	.9289	1	1	2	2	3	3	4	4	5
8.5	.9294	.9299	.9304	.9309	.9315	.9320	.9325	.9330	.9335	.9340	1	1	2	2	3	3	4	4	5
8.6	.9345	.9350	.9355	.9360	.9365	.9370	.9375	.9380	.9385	.9390	1	1	2	2	3	3	4	4	5
8.7	.9395	.9400	.9405	.9410	.9415	.9420	.9425	.9430	.9435	.9440	0	1	1	2	2	3	3	4	4
8.8	.9445	.9450	.9455	.9460	.9465	.9469	.9474	.9479	.9484	.9489	0	1	1	2	2	3	3	4	4
8.9	.9494	.9499	.9504	.9509	.9513	.9518	.9523	.9528	.9533	.9538	0	1	1	2	2	3	3	4	4
9.0	.9542	.9547	.9552	.9557	.9562	.9566	.9571	.9576	.9581	.9586	0	1	1	2	2	3	3	4	4
9.1	.9590	.9595	.9600	.9605	.9609	.9614	.9619	.9624	.9628	.9633	0	1	1	2	2	3	3	4	4
9.2	.9638	.9643	.9647	.9652	.9657	.9661	.9666	.9671	.9675	.9680	0	1	1	2	2	3	3	4	4
9.3	.9685	.9689	.9694	.9699	.9703	.9708	.9713	.9717	.9722	.9727	0	1	1	2	2	3	3	4	4
9.4	.9731	.9736	.9741	.9745	.9750	.9754	.9759	.9763	.9768	.9773	0	1	1	2	2	3	3	4	4
9.5	.9777	.9782	.9786	.9791	.9795	.9800	.9805	.9809	.9814	.9818	0	1	1	2	2	3	3	4	4
9.6	.9823	.9827	.9832	.9836	.9841	.9845	.9850	.9854	.9859	.9863	0	1	1	2	2	3	3	4	4
9.7	.9868	.9872	.9877	.9881	.9886	.9890	.9894	.9899	.9903	.9908	0	1	1	2	2	3	3	4	4
9.8	.9912	.9917	.9921	.9926	.9930	.9934	.9939	.9943	.9948	.9952	0	1	1	2	2	3	3	4	4
9.9	.9956	.9961	.9965	.9969	.9974	.9978	.9983	.9987	.9991	.9996	0	1	1	2	2	3	3	3	4

Table 3—Sines and cosines

Values of sin x and cos y

x	.0	.1	.2	.3	.4	.5	.6	.7	.8	.9		1	2	3	4	5	6	7	8	9	
0	.0000	.0017	.0035	.0052	.0070	.0087	.0105	.0122	.0140	.0157	.0175	2	3	5	7	9	10	12	14	16	89
1	.0175	.0192	.0209	.0227	.0244	.0262	.0279	.0297	.0314	.0332	.0349	2	3	5	7	9	10	12	14	16	88
2	.0349	.0366	.0384	.0401	.0419	.0436	.0454	.0471	.0488	.0506	.0523	2	3	5	7	9	10	12	14	16	87
3	.0523	.0541	.0558	.0576	.0593	.0610	.0628	.0645	.0663	.0680	.0698	2	3	5	7	9	10	12	14	16	86
4	.0698	.0715	.0732	.0750	.0767	.0785	.0802	.0819	.0837	.0854	.0872	2	3	5	7	9	10	12	14	16	85
5	.0872	.0889	.0906	.0924	.0941	.0958	.0976	.0993	.1011	.1028	.1045	2	3	5	7	9	10	12	14	16	84
6	.1045	.1063	.1080	.1097	.1115	.1132	.1149	.1167	.1184	.1201	.1219	2	3	5	7	9	10	12	14	16	83
7	.1219	.1236	.1253	.1271	.1288	.1305	.1323	.1340	.1357	.1374	.1392	2	3	5	7	9	10	12	14	16	82
8	.1392	.1409	.1426	.1444	.1461	.1478	.1495	.1513	.1530	.1547	.1564	2	3	5	7	9	10	12	14	15	81
9	.1564	.1582	.1599	.1616	.1633	.1650	.1668	.1685	.1702	.1719	.1736	2	3	5	7	9	10	12	14	15	80
10	.1736	.1754	.1771	.1788	.1805	.1822	.1840	.1857	.1874	.1891	.1908	2	3	5	7	9	10	12	14	15	79
11	.1908	.1925	.1942	.1959	.1977	.1994	.2011	.2028	.2045	.2062	.2079	2	3	5	7	9	10	12	14	15	78
12	.2079	.2096	.2113	.2130	.2147	.2164	.2181	.2198	.2215	.2233	.2250	2	3	5	7	9	10	12	14	15	77
13	.2250	.2267	.2284	.2300	.2317	.2334	.2351	.2368	.2385	.2402	.2419	2	3	5	7	8	10	12	14	15	76
14	.2419	.2436	.2453	.2470	.2487	.2504	.2521	.2538	.2554	.2571	.2588	2	3	5	7	8	10	12	14	15	75
15	.2588	.2605	.2622	.2639	.2656	.2672	.2689	.2706	.2723	.2740	.2756	2	3	5	7	8	10	12	13	15	74
16	.2756	.2773	.2790	.2807	.2823	.2840	.2857	.2874	.2890	.2907	.2924	2	3	5	7	8	10	12	13	15	73
17	.2924	.2940	.2957	.2974	.2990	.3007	.3024	.3040	.3057	.3074	.3090	2	3	5	7	8	10	12	13	15	72
18	.3090	.3107	.3123	.3140	.3156	.3173	.3190	.3206	.3223	.3239	.3256	2	3	5	7	8	10	12	13	15	71
19	.3256	.3272	.3289	.3305	.3322	.3338	.3355	.3371	.3387	.3404	.3420	2	3	5	7	8	10	11	13	15	70
20	.3420	.3437	.3453	.3469	.3486	.3502	.3518	.3535	.3551	.3567	.3584	2	3	5	7	8	10	11	13	15	69
21	.3584	.3600	.3616	.3633	.3649	.3665	.3681	.3697	.3714	.3730	.3746	2	3	5	6	8	10	11	13	15	68
22	.3746	.3762	.3778	.3795	.3811	.3827	.3843	.3859	.3875	.3891	.3907	2	3	5	6	8	10	11	13	15	67
23	.3907	.3923	.3939	.3955	.3971	.3987	.4003	.4019	.4035	.4051	.4067	2	3	5	6	8	10	11	13	14	66
24	.4067	.4083	.4099	.4115	.4131	.4147	.4163	.4179	.4195	.4210	.4226	2	3	5	6	8	10	11	13	14	65
25	.4226	.4242	.4258	.4274	.4289	.4305	.4321	.4337	.4352	.4368	.4384	2	3	5	6	8	9	11	13	14	64
26	.4384	.4399	.4415	.4431	.4446	.4462	.4478	.4493	.4509	.4524	.4540	2	3	5	6	8	9	11	12	14	63
27	.4540	.4555	.4571	.4586	.4602	.4617	.4633	.4648	.4664	.4679	.4695	2	3	5	6	8	9	11	12	14	62
28	.4695	.4710	.4726	.4741	.4756	.4772	.4787	.4802	.4818	.4833	.4848	2	3	5	6	8	9	11	12	14	61
29	.4848	.4863	.4879	.4894	.4909	.4924	.4939	.4955	.4970	.4985	.5000	2	3	5	6	8	9	11	12	14	60
30	.5000	.5015	.5030	.5045	.5060	.5075	.5090	.5105	.5120	.5135	.5150	2	3	5	6	8	9	11	12	14	59
31	.5150	.5165	.5180	.5195	.5210	.5225	.5240	.5255	.5270	.5284	.5299	1	3	4	6	7	9	10	12	13	58
32	.5299	.5314	.5329	.5344	.5358	.5373	.5388	.5402	.5417	.5432	.5446	1	3	4	6	7	9	10	12	13	57
33	.5446	.5461	.5476	.5490	.5505	.5519	.5534	.5548	.5563	.5577	.5592	1	3	4	6	7	9	10	12	13	56
34	.5592	.5606	.5621	.5635	.5650	.5664	.5678	.5693	.5707	.5721	.5736	1	3	4	6	7	9	10	12	13	55
35	.5736	.5750	.5764	.5779	.5793	.5807	.5821	.5835	.5850	.5864	.5878	1	3	4	6	7	9	10	11	13	54
36	.5878	.5892	.5906	.5920	.5934	.5948	.5962	.5976	.5990	.6004	.6018	1	3	4	6	7	8	10	11	13	53
37	.6018	.6032	.6046	.6060	.6074	.6088	.6101	.6115	.6129	.6143	.6157	1	3	4	6	7	8	10	11	13	52
38	.6157	.6170	.6184	.6198	.6211	.6225	.6239	.6252	.6266	.6280	.6293	1	3	4	5	7	8	10	11	12	51
39	.6293	.6307	.6320	.6334	.6347	.6361	.6374	.6388	.6401	.6414	.6428	1	3	4	5	7	8	9	11	12	50
40	.6428	.6441	.6455	.6468	.6481	.6494	.6508	.6521	.6534	.6547	.6561	1	3	4	5	7	8	9	11	12	49
41	.6561	.6574	.6587	.6600	.6613	.6626	.6639	.6652	.6665	.6678	.6691	1	3	4	5	7	8	9	10	12	48
42	.6691	.6704	.6717	.6730	.6743	.6756	.6769	.6782	.6794	.6807	.6820	1	3	4	5	6	8	9	10	12	47
43	.6820	.6833	.6845	.6858	.6871	.6884	.6896	.6909	.6921	.6934	.6947	1	3	4	5	6	8	9	10	11	46
44	.6947	.6959	.6972	.6984	.6997	.7009	.7022	.7034	.7046	.7059	.7071	1	2	4	5	6	7	9	10	11	45
		.9	.8	.7	.6	.5	.4	.3	.2	.1	.0	−1	−2	−3	−4	−5	−6	−7	−8	−9	y

DIFFERENCES NEGATIVE FOR COSINES

Table 4—Values of (sin x°)/x

x	(sin x°)/x	sin x°
0·000 – 3·046	·01745	·0000 – ·0531
3·047 – 4·535	·01744	·0532 – ·0790
4·536 – 5·644	·01743	·0791 – ·0983
5·645 – 6·570	·01742	·0984 – ·1144
6·571 – 7·380	·01741	·1145 – ·1284
7·381 – 8·110	·01740	·1285 – ·1411
8·111 – 8·779	·01739	·1412 – ·1526
8·780 – 9·402	·01738	·1527 – ·1634
9·403 – 9·986	·01737	·1635 – ·1733
9·987 – 10·538	·01736	·1734 – ·1828

Table 3—Sines and cosines (*concluded*)

Values of sin x and cos y

x	.0	.1	.2	.3	.4	.5	.6	.7	.8	.9		1	2	3	4	5	6	7	8	9	
45	.7071	.7083	.7096	.7108	.7120	.7133	.7145	.7157	.7169	.7181	.7193	1	2	4	5	6	7	9	10	11	44
46	.7193	.7206	.7218	.7230	.7242	.7254	.7266	.7278	.7290	.7302	.7314	1	2	4	5	6	7	8	10	11	43
47	.7314	.7325	.7337	.7349	.7361	.7373	.7385	.7396	.7408	.7420	.7431	1	2	4	5	6	7	8	9	11	42
48	.7431	.7443	.7455	.7466	.7478	.7490	.7501	.7513	.7524	.7536	.7547	1	2	3	5	6	7	8	9	10	41
49	.7547	.7559	.7570	.7581	.7593	.7604	.7615	.7627	.7638	.7649	.7660	1	2	3	5	6	7	8	9	10	40
50	.7660	.7672	.7683	.7694	.7705	.7716	.7727	.7738	.7749	.7760	.7771	1	2	3	4	6	7	8	9	10	39
51	.7771	.7782	.7793	.7804	.7815	.7826	.7837	.7848	.7859	.7869	.7880	1	2	3	4	5	7	8	9	10	38
52	.7880	.7891	.7902	.7912	.7923	.7934	.7944	.7955	.7965	.7976	.7986	1	2	3	4	5	6	7	8	10	37
53	.7986	.7997	.8007	.8018	.8028	.8039	.8049	.8059	.8070	.8080	.8090	1	2	3	4	5	6	7	8	9	36
54	.8090	.8100	.8111	.8121	.8131	.8141	.8151	.8161	.8171	.8181	.8192	1	2	3	4	5	6	7	8	9	35
55	.8192	.8202	.8211	.8221	.8231	.8241	.8251	.8261	.8271	.8281	.8290	1	2	3	4	5	6	7	8	9	34
56	.8290	.8300	.8310	.8320	.8329	.8339	.8348	.8358	.8368	.8377	.8387	1	2	3	4	5	6	7	8	9	33
57	.8387	.8396	.8406	.8415	.8425	.8434	.8443	.8453	.8462	.8471	.8480	1	2	3	4	5	6	7	7	8	32
58	.8480	.8490	.8499	.8508	.8517	.8526	.8536	.8545	.8554	.8563	.8572	1	2	3	4	5	6	6	7	8	31
59	.8572	.8581	.8590	.8599	.8607	.8616	.8625	.8634	.8643	.8652	.8660	1	2	3	4	4	5	6	7	8	30
60	.8660	.8669	.8678	.8686	.8695	.8704	.8712	.8721	.8729	.8738	.8746	1	2	3	3	4	5	6	7	8	29
61	.8746	.8755	.8763	.8771	.8780	.8788	.8796	.8805	.8813	.8821	.8829	1	2	2	3	4	5	6	7	7	28
62	.8829	.8838	.8846	.8854	.8862	.8870	.8878	.8886	.8894	.8902	.8910	1	2	2	3	4	5	6	6	7	27
63	.8910	.8918	.8926	.8934	.8942	.8949	.8957	.8965	.8973	.8980	.8988	1	2	2	3	4	5	5	6	7	26
64	.8988	.8996	.9003	.9011	.9018	.9026	.9033	.9041	.9048	.9056	.9063	1	2	2	3	4	5	5	6	7	25
65	.9063	.9070	.9078	.9085	.9092	.9100	.9107	.9114	.9121	.9128	.9135	1	1	2	3	4	4	5	6	6	24
66	.9135	.9143	.9150	.9157	.9164	.9171	.9178	.9184	.9191	.9198	.9205	1	1	2	3	4	4	5	6	6	23
67	.9205	.9212	.9219	.9225	.9232	.9239	.9245	.9252	.9259	.9265	.9272	1	1	2	3	3	4	5	5	6	22
68	.9272	.9278	.9285	.9291	.9298	.9304	.9311	.9317	.9323	.9330	.9336	1	1	2	3	3	4	4	5	6	21
69	.9336	.9342	.9348	.9354	.9361	.9367	.9373	.9379	.9385	.9391	.9397	1	1	2	2	3	4	4	5	5	20
70	.9397	.9403	.9409	.9415	.9421	.9426	.9432	.9438	.9444	.9449	.9455	1	1	2	2	3	3	4	5	5	19
71	.9455	.9461	.9466	.9472	.9478	.9483	.9489	.9494	.9500	.9505	.9511	1	1	2	2	3	3	4	4	5	18
72	.9511	.9516	.9521	.9527	.9532	.9537	.9542	.9548	.9553	.9558	.9563	1	1	2	2	3	3	4	4	5	17
73	.9563	.9568	.9573	.9578	.9583	.9588	.9593	.9598	.9603	.9608	.9613	1	1	2	2	3	3	4	4	5	16
74	.9613	.9617	.9622	.9627	.9632	.9636	.9641	.9646	.9650	.9655	.9659	0	1	1	2	2	3	3	4	4	15
75	.9659	.9664	.9668	.9673	.9677	.9681	.9686	.9690	.9694	.9699	.9703	0	1	1	2	2	3	3	4	4	14
76	.9703	.9707	.9711	.9715	.9720	.9724	.9728	.9732	.9736	.9740	.9744	0	1	1	2	2	2	3	3	4	13
77	.9744	.9748	.9751	.9755	.9759	.9763	.9767	.9770	.9774	.9778	.9781	0	1	1	1	2	2	3	3	3	12
78	.9781	.9785	.9789	.9792	.9796	.9799	.9803	.9806	.9810	.9813	.9816	0	1	1	1	2	2	2	3	3	11
79	.9816	.9820	.9823	.9826	.9829	.9833	.9836	.9839	.9842	.9845	.9848	0	1	1	1	2	2	2	3	3	10
80	.9848	.9851	.9854	.9857	.9860	.9863	.9866	.9869	.9871	.9874	.9877	0	1	1	1	1	2	2	2	3	9
81	.9877	.9880	.9882	.9885	.9888	.9890	.9893	.9895	.9898	.9900	.9903	0	1	1	1	1	2	2	2	2	8
82	.9903	.9905	.9907	.9910	.9912	.9914	.9917	.9919	.9921	.9923	.9925	0	0	1	1	1	1	2	2	2	7
83	.9925	.9928	.9930	.9932	.9934	.9936	.9938	.9940	.9942	.9943	.9945	0	0	1	1	1	1	1	2	2	6
84	.9945	.9947	.9949	.9951	.9952	.9954	.9956	.9957	.9959	.9960	.9962	0	0	1	1	1	1	1	1	2	5
85	.9962	.9963	.9965	.9966	.9968	.9969	.9971	.9972	.9973	.9974	.9976	0	0	0	1	1	1	1	1	1	4
86	.9976	.9977	.9978	.9979	.9980	.9981	.9982	.9983	.9984	.9985	.9986	0	0	0	0	1	1	1	1	1	3
87	.9986	.9987	.9988	.9989	.9990	.9990	.9991	.9992	.9993	.9993	.9994	0	0	0	0	0	0	1	1	1	2
88	.9994	.9995	.9995	.9996	.9996	.9997	.9997	.9997	.9998	.9998	.9998	0	0	0	0	0	0	0	0	1	1
89	.9998	.9999	.9999	.9999	.9999	1.0000		1.0000		1.0000		0	0	0	0	0	0	0	0	0	0
		.9	.8	.7	.6	.5	.4	.3	.2	.1	.0	−1	−2	−3	−4	−5	−6	−7	−8	−9	y

DIFFERENCES NEGATIVE FOR COSINES

Table 5—A = x²/6

x	A
·00000 – ·01732	·0000
·01733 – ·03000	·0001
·03001 – ·03873	·0002
·03874 – ·04582	·0003
·04583 – ·05196	·0004
·05197 – ·05744	·0005
·05745 – ·06245	·0006
·06246 – ·06708	·0007
·06709 – ·07141	·0008
·07142 – ·07550	·0009

Table 6—Tangents and cotangents

Values of tan *x* and cot *y*

x	.0	.1	.2	.3	.4	.5	.6	.7	.8	.9		1	2	3	4	5	6	7	8	9	
0	.0000	.0017	.0035	.0052.	.0070	.0087	.0105	.0122	.0140	.0157	.0175	2	3	5	7	9	10	12	14	16	89
1	.0175	.0192	.0209	.0227	.0244	.0262	.0279	.0297	.0314	.0332	.0349	2	3	5	7	9	10	12	14	16	88
2	.0349	.0367	.0384	.0402	.0419	.0437	.0454	.0472	.0489	.0507	.0524	2	4	5	7	9	11	12	14	16	87
3	.0524	.0542	.0559	.0577	.0594	.0612	.0629	.0647	.0664	.0682	.0699	2	4	5	7	9	11	12	14	16	86
4	.0699	.0717	.0734	.0752	.0769	.0787	.0805	.0822	.0840	.0857	.0875	2	4	5	7	9	11	12	14	16	85
5	.0875	.0892	.0910	.0928	.0945	.0963	.0981	.0998	.1016	.1033	.1051	2	4	5	7	9	11	12	14	16	84
6	.1051	.1069	.1086	.1104	.1122	.1139	.1157	.1175	.1192	.1210	.1228	2	4	5	7	9	11	12	14	16	83
7	.1228	.1246	.1263	.1281	.1299	.1317	.1334	.1352	.1370	.1388	.1405	2	4	5	7	9	11	12	14	16	82
8	.1405	.1423	.1441	.1459	.1477	.1495	.1512	.1530	.1548	.1566	.1584	2	4	5	7	9	11	13	14	16	81
9	.1584	.1602	.1620	.1638	.1655	.1673	.1691	.1709	.1727	.1745	.1763	2	4	5	7	9	11	13	14	16	80
10	.1763	.1781	.1799	.1817	.1835	.1853	.1871	.1890	.1908	.1926	.1944	2	4	5	7	9	11	13	14	16	79
11	.1944	.1962	.1980	.1998	.2016	.2035	.2053	.2071	.2089	.2107	.2126	2	4	5	7	9	11	13	15	16	78
12	.2126	.2144	.2162	.2180	.2199	.2217	.2235	.2254	.2272	.2290	.2309	2	4	5	7	9	11	13	15	16	77
13	.2309	.2327	.2345	.2364	.2382	.2401	.2419	.2438	.2456	.2475	.2493	2	4	6	7	9	11	13	15	17	76
14	.2493	.2512	.2530	.2549	.2568	.2586	.2605	.2623	.2642	.2661	.2679	2	4	6	7	9	11	13	15	17	75
15	.2679	.2698	.2717	.2736	.2754	.2773	.2792	.2811	.2830	.2849	.2867	2	4	6	8	9	11	13	15	17	74
16	.2867	.2886	.2905	.2924	.2943	.2962	.2981	.3000	.3019	.3038	.3057	2	4	6	8	10	11	13	15	17	73
17	.3057	.3076	.3096	.3115	.3134	.3153	.3172	.3191	.3211	.3230	.3249	2	4	6	8	10	12	13	15	17	72
18	.3249	.3269	.3288	.3307	.3327	.3346	.3365	.3385	.3404	.3424	.3443	2	4	6	8	10	12	14	16	17	71
19	.3443	.3463	.3482	.3502	.3522	.3541	.3561	.3581	.3600	.3620	.3640	2	4	6	8	10	12	14	16	18	70
20	.3640	.3659	.3679	.3699	.3719	.3739	.3759	.3779	.3799	.3819	.3839	2	4	6	8	10	12	14	16	18	69
21	.3839	.3859	.3879	.3899	.3919	.3939	.3959	.3979	.4000	.4020	.4040	2	4	6	8	10	12	14	16	18	68
22	.4040	.4061	.4081	.4101	.4122	.4142	.4163	.4183	.4204	.4224	.4245	2	4	6	8	10	12	14	16	18	67
23	.4245	.4265	.4286	.4307	.4327	.4348	.4369	.4390	.4411	.4431	.4452	2	4	6	8	10	12	14	17	19	66
24	.4452	.4473	.4494	.4515	.4536	.4557	.4578	.4599	.4621	.4642	.4663	2	4	6	8	11	13	15	17	19	65
25	.4663	.4684	.4706	.4727	.4748	.4770	.4791	.4813	.4834	.4856	.4877	2	4	6	9	11	13	15	17	19	64
26	.4877	.4899	.4921	.4942	.4964	.4986	.5008	.5029	.5051	.5073	.5095	2	4	7	9	11	13	15	17	20	63
27	.5095	.5117	.5139	.5161	.5184	.5206	.5228	.5250	.5272	.5295	.5317	2	4	7	9	11	13	15	18	20	62
28	.5317	.5340	.5362	.5384	.5407	.5430	.5452	.5475	.5498	.5520	.5543	2	5	7	9	11	14	16	18	20	61
29	.5543	.5566	.5589	.5612	.5635	.5658	.5681	.5704	.5727	.5750	.5774	2	5	7	9	12	14	16	18	20	60
30	.5774	.5797	.5820	.5844	.5867	.5890	.5914	.5938	.5961	.5985	.6009	2	5	7	9	12	14	16	19	21	59
31	.6009	.6032	.6056	.6080	.6104	.6128	.6152	.6176	.6200	.6224	.6249	2	5	7	10	12	14	17	19	22	58
32	.6249	.6273	.6297	.6322	.6346	.6371	.6395	.6420	.6445	.6469	.6494	2	5	7	10	12	14	17	20	22	57
33	.6494	.6519	.6544	.6569	.6594	.6619	.6644	.6669	.6694	.6720	.6745	3	5	8	10	13	15	18	20	23	56
34	.6745	.6771	.6796	.6822	.6847	.6873	.6899	.6924	.6950	.6976	.7002	3	5	8	10	13	15	18	21	23	55
35	.7002	.7028	.7054	.7080	.7107	.7133	.7159	.7186	.7212	.7239	.7265	3	5	8	11	13	16	18	21	24	54
36	.7265	.7292	.7319	.7346	.7373	.7400	.7427	.7454	.7481	.7508	.7536	3	5	8	11	14	16	19	22	24	53
37	.7536	.7563	.7590	.7618	.7646	.7673	.7701	.7729	.7757	.7785	.7813	3	6	8	11	14	17	19	22	25	52
38	.7813	.7841	.7869	.7898	.7926	.7954	.7983	.8012	.8040	.8069	.8098	3	6	9	11	14	17	20	23	26	51
39	.8098	.8127	.8156	.8185	.8214	.8243	.8273	.8302	.8332	.8361	.8391	3	6	9	12	15	18	21	23	26	50
40	.8391	.8421	.8451	.8481	.8511	.8541	.8571	.8601	.8632	.8662	.8693	3	6	9	12	15	18	21	24	27	49
41	.8693	.8724	.8754	.8785	.8816	.8847	.8878	.8910	.8941	.8972	.9004	3	6	9	12	16	19	22	25	28	48
42	.9004	.9036	.9067	.9099	.9131	.9163	.9195	.9228	.9260	.9293	.9325	3	6	10	13	16	19	22	26	29	47
43	.9325	.9358	.9391	.9424	.9457	.9490	.9523	.9556	.9590	.9623	.9657	3	7	10	13	17	20	23	27	30	46
44	.9657	.9691	.9725	.9759	.9793	.9827	.9861	.9896	.9930	.9965	1.0000	3	7	10	14	17	21	24	27	31	45
		.9	.8	.7	.6	.5	.4	.3	.2	.1	.0	−1	−2	−3	−4	−5	−6	−7	−8	−9	y

DIFFERENCES NEGATIVE FOR COTANGENTS

Table 7—Values of (tan x°)/x

x			(tan x°)/x	tan x°		
0·000	–	1·002	·01745	·0000	–	·0174
1·003	–	2·578	·01746	·0175	–	·0450
2·579	–	3·508	·01747	·0451	–	·0612
3·509	–	4·230	·01748	·0613	–	·0739
4·231	–	4·849	·01749	·0740	–	·0848
4·850	–	5·396	·01750	·0849	–	·0944
5·397	–	5·892	·01751	·0945	–	·1031
5·893	–	6·349	·01752	·1032	–	·1112
6·350	–	6·774	·01753	·1113	–	·1187
6·775	–	7·173	·01754	·1188	–	·1258

Table 6—Tangents and cotangents (*concluded*)

Values of tan *x* and cot *y*

x	.0	.1	.2	.3	.4	.5	.6	.7	.8	.9		
45	1.0000	1.0035	1.0070	1.0105	1.0141	1.0176	1.0212	1.0247	1.0283	1.0319	1.0355	44
46	1.0355	1.0392	1.0428	1.0464	1.0501	1.0538	1.0575	1.0612	1.0649	1.0686	1.0724	43
47	1.0724	1.0761	1.0799	1.0837	1.0875	1.0913	1.0951	1.0990	1.1028	1.1067	1.1106	42
48	1.1106	1.1145	1.1184	1.1224	1.1263	1.1303	1.1343	1.1383	1.1423	1.1463	1.1504	41
49	1.1504	1.1544	1.1585	1.1626	1.1667	1.1708	1.1750	1.1792	1.1833	1.1875	1.1918	40
50	1.1918	1.1960	1.2002	1.2045	1.2088	1.2131	1.2174	1.2218	1.2261	1.2305	1.2349	39
51	1.2349	1.2393	1.2437	1.2482	1.2527	1.2572	1.2617	1.2662	1.2708	1.2753	1.2799	38
52	1.2799	1.2846	1.2892	1.2938	1.2985	1.3032	1.3079	1.3127	1.3175	1.3222	1.3270	37
53	1.3270	1.3319	1.3367	1.3416	1.3465	1.3514	1.3564	1.3613	1.3663	1.3713	1.3764	36
54	1.3764	1.3814	1.3865	1.3916	1.3968	1.4019	1.4071	1.4124	1.4176	1.4229	1.4281	35
55	1.4281	1.4335	1.4388	1.4442	1.4496	1.4550	1.4605	1.4659	1.4715	1.4770	1.4826	34
56	1.4826	1.4882	1.4938	1.4994	1.5051	1.5108	1.5166	1.5224	1.5282	1.5340	1.5399	33
57	1.5399	1.5458	1.5517	1.5577	1.5637	1.5697	1.5757	1.5818	1.5880	1.5941	1.6003	32
58	1.6003	1.6066	1.6128	1.6191	1.6255	1.6319	1.6383	1.6447	1.6512	1.6577	1.6643	31
59	1.6643	1.6709	1.6775	1.6842	1.6909	1.6977	1.7045	1.7113	1.7182	1.7251	1.7321	30
60	1.7321	1.7391	1.7461	1.7532	1.7603	1.7675	1.7747	1.7820	1.7893	1.7966	1.8040	29
61	1.8040	1.8115	1.8190	1.8265	1.8341	1.8418	1.8495	1.8572	1.8650	1.8728	1.8807	28
62	1.8807	1.8887	1.8967	1.9047	1.9128	1.9210	1.9292	1.9375	1.9458	1.9542	1.9626	27
63	1.9626	1.9711	1.9797	1.9883	1.9970	2.0057	2.0145	2.0233	2.0323	2.0413	2.0503	26
64	2.0503	2.0594	2.0686	2.0778	2.0872	2.0965	2.1060	2.1155	2.1251	2.1348	2.1445	25
65	2.1445	2.1543	2.1642	2.1742	2.1842	2.1943	2.2045	2.2148	2.2251	2.2355	2.2460	24
66	2.2460	2.2566	2.2673	2.2781	2.2889	2.2998	2.3109	2.3220	2.3332	2.3445	2.3559	23
67	2.3559	2.3673	2.3789	2.3906	2.4023	2.4142	2.4262	2.4383	2.4504	2.4627	2.4751	22
68	2.4751	2.4876	2.5002	2.5129	2.5257	2.5386	2.5517	2.5649	2.5782	2.5916	2.6051	21
69	2.6051	2.6187	2.6325	2.6464	2.6605	2.6746	2.6889	2.7034	2.7179	2.7326	2.7475	20
70	2.7475	2.7625	2.7776	2.7929	2.8083	2.8239	2.8397	2.8556	2.8716	2.8878	2.9042	19
71	2.9042	2.9208	2.9375	2.9544	2.9714	2.9887	3.0061	3.0237	3.0415	3.0595	3.0777	18
72	3.0777	3.0961	3.1146	3.1334	3.1524	3.1716	3.1910	3.2106	3.2305	3.2506	3.2709	17
73	3.2709	3.2914	3.3122	3.3332	3.3544	3.3759	3.3977	3.4197	3.4420	3.4646	3.4874	16
74	3.4874	3.5105	3.5339	3.5576	3.5816	3.6059	3.6305	3.6554	3.6806	3.7062	3.7321	15
75	3.7321	3.7583	3.7848	3.8118	3.8391	3.8667	3.8947	3.9232	3.9520	3.9812	4.0108	14
76	4.0108	4.0408	4.0713	4.1022	4.1335	4.1653	4.1976	4.2303	4.2635	4.2972	4.3315	13
77	4.3315	4.3662	4.4015	4.4373	4.4737	4.5107	4.5483	4.5864	4.6252	4.6646	4.7046	12
78	4.7046	4.7453	4.7867	4.8288	4.8716	4.9152	4.9594	5.0045	5.0504	5.0970	5.1446	11
79	5.1446	5.1929	5.2422	5.2924	5.3435	5.3955	5.4486	5.5026	5.5578	5.6140	5.6713	10
80	5.6713	5.7297	5.7894	5.8502	5.9124	5.9758	6.0405	6.1066	6.1742	6.2432	6.3138	9
81	6.3138	6.3859	6.4596	6.5350	6.6122	6.6912	6.7720	6.8548	6.9395	7.0264	7.1154	8
82	7.1154	7.2066	7.3002	7.3962	7.4947	7.5958	7.6996	7.8062	7.9158	8.0285	8.1443	7
83	8.1443	8.2636	8.3863	8.5126	8.6427	8.7769	8.9152	9.0579	9.2052	9.3572	9.5144	6
84	9.5144	9.6768	9.8448	10.019	10.199	10.385	10.579	10.780	10.988	11.205	11.430	5
85	11.430	11.664	11.909	12.163	12.429	12.706	12.996	13.300	13.617	13.951	14.301	4
86	14.301	14.669	15.056	15.464	15.895	16.350	16.832	17.343	17.886	18.464	19.081	3
87	19.081	19.740	20.446	21.205	22.022	22.904	23.859	24.898	26.031	27.271	28.636	2
88	28.636	30.145	31.821	33.694	35.801	38.188	40.917	44.066	47.740	52.081	57.290	1
89	57.290	63.657	71.615	81.847	95.489	114.59	143.24	190.98	286.48	572.96	∞	0
		.9	.8	.7	.6	.5	.4	.3	.2	.1	.0	y

Table 8—B = $x^2/3$

x	B
·00000 – ·01224	·0000
·01225 – ·02121	·0001
·02122 – ·02738	·0002
·02739 – ·03240	·0003
·03241 – ·03674	·0004
·03675 – ·04062	·0005
·04063 – ·04415	·0006
·04416 – ·04743	·0007
·04744 – ·05049	·0008
·05050 – ·05339	·0009

ANSWERS TO PROBLEMS

(In the text problems are numbered consecutively within each section)

pp. 31–32 (Section 2.10)

(1) (*b*), (*c*), (*e*) and (*f*).
(2) A^2 must have the same reduced value as the square of the reduced value of A, i.e., as of 0^2, 1^2, 2^2, . . . 9^2 or 10^2.
(3) A^2 must end in the same figure as does the square of the last figure of A. For if $A = 10B + b$, where b is the last figure, $A^2 = 10\,(10B^2 + 2Bb) + b^2$. $62{,}500 = 250^2$; $41{,}616 = 204^2$; all the others are excluded by the results of this or the preceding problem, except for 1210, which cannot be a square because it is divisible by 2 but not by 2^2.
(4) $110{,}592 = 48^3$.
(6) G.C.F. = 12; factors 1, 2, 3, 4, 6, 12.
(8) G.C.F. = 3; L.C.M. = 1440.

p. 43 (Section 3.4)

(1) Minimum 3 when $x = 0$.
(2) Minimum 10 when $x = 2$.
(3) Minimum 20 when $x = 5/2$.
(4) Maximum 44·145 when $x = 3$.

p. 44

(5) $x = \pm\sqrt{37} = \pm\ 6{\cdot}083$.
(6) $x = 2 \pm \sqrt{15} = -\ 1{\cdot}873$ or $5{\cdot}873$.
(7) $x = 2{\cdot}5 \pm \sqrt{5} = {\cdot}264$ or $4{\cdot}736$.
(8) $x = {\cdot}145$ or $5{\cdot}855$.

p. 45

(9) $(x - 2)\,(x + 2)$.
(10) $(x + 2)\,(x + 3)$.
(11) None.
(12) None.
(13) None.
(14) $x(3 - 5x)$.

p. 56 (Section 3.7)

Query: at least one fraction would have a zero denominator, and would therefore be illegitimate.

p. 58 (Section 3.8)

(1) Quotient $x + 4$, remainder 1.
(2) Quotient $3x - 4{\cdot}5$, remainder $17{\cdot}5$.
(3) Quotient $2x + 9$, remainder $10x - 10$.
(4) Quotient $(6x^2 + 15x + 43)/4$, remainder $(230x - 39)/4$.
(6) $x^4 + 2x$.
(7) $a = -3$, $b = 0$.
(8) Similarly for division by $(x - a)$.
(9) $9C + 3B + A$.
(10) $k^2 + 5k + 6$.
(11) When $Ck^2 + Bk + A = 0$.
(12) Similarly for division by $x^2 + Bx + A$.

p. 60 (Section 3.9)

(1) $x + x(x - 1)$.
(2) $1 - 3(x + 1) + (x + 1)(x + 2)$.
(3) $-14 + 7(x + 2) + 0x(x + 2) + x(x + 2)(x - 2)$.
(4) $A = 14$.

p. 62 (Section 3.10)

(1) ·4126.
(2) ·9236.
(3) ·9236 (indistinguishable from the preceding to 4 places).

p. 64

(4) 56·25, 86·25 ft.
(5) 9·485, 10·490 (correct values 9·487, 10·488).
(6) 24·264.

p. 66 (Section 3.11)

(1) 2·26517, 2·26516.
(2) 2·0305.

p. 67 (Section 3.12)

(3) $(x-1)(x+7)$.
(4) $(x-1)(x+2)(x+5)$.
(5) $(x-1)(x^2-2x+4)$.
(6) $C=-5$, $A=4$, $(x+1)(x-2)$.

p. 71 (Section 3.13)

(1) $x=3/2$.
(2) $x=2$.
(3) $x=5$.
(4) $x=1/a$.

p. 74 (Section 4.2)

Query: No. For example, $5>3$ and $4>1$ but $(5-4)<(3-1)$.

p. 75

(2) $A+B>a+b$.
(3) $(A^3+A^2+A)-(a^3+a^2+a)=(A-a)(A^2+Aa+a^2+A+a+1)$. But $A^2+Aa+a^2+A+a+1=\frac{1}{2}(A+a+1)^2+\frac{1}{2}A^2+\frac{1}{2}a^2+\frac{1}{2}$, and $(A-a)>0$. Hence the result follows. Also if $A=-1$, $a=-2$, $A>a$ but $A^2<a^2$.
(4) $x^2+4x-(-4)=(x+2)^2\geqslant 0$; equality if $x=-2$.
(5) If x is negative, $x+1/x\leqslant -2$.
(6) The rule breaks down.
(8) Equality if $xY=Xy$.
(9) $1/\sqrt{x}<1/\sqrt{y}$; $1/(x^2+1)<1/(y^2+1)$; $(x^2+1)/(y^3+2)>(y^2+1)\div(x^3+2)$.
(10) $a^3+b^3+c^3-3abc=(a+b+c)(a^2+b^2+c^2-ab-bc-ca)=\frac{1}{2}(a+b+c)[(a-b)^2+(b-c)^2+(c-a)^2]$.

p. 77 (Section 4.3)

(4) No; e.g. $A=2$, $a=1$, $x=-3$.
(6) Yes. (Divide through by $|Aa|$).
(7) When x and y have the same sign or at least one is zero.

p. 89 (Section 5.5)

(5) $|\operatorname{cosec}\theta|\geqslant 1$, $|\sec\theta|\geqslant 1$.
(6) $\tan\theta$, $\sec\theta$, $\cos\theta$, $-\tan\theta$, $\sin\theta$.
(7) $\sin 45°=\cos 45°=1/\sqrt{2}=·707$.
$\tan 45°=\cot 45°=1$.
$\sec 45°=\operatorname{cosec} 45°=\sqrt{2}=1·41$.
(8) $\sin 30°=\cos 60°=1/2$.
$\operatorname{cosec} 30°=\sec 60°=2$.
$\cos 30°=\sin 60°=\sqrt{3}/2=·866$.
$\sec 30°=\operatorname{cosec} 60°=2/\sqrt{3}=1·155$.
$\tan 30°=\cot 60°=1/\sqrt{3}=·577$.
$\cot 30°=\tan 60°=\sqrt{3}=1·732$.

(9) $\sin 15° = \cos 75° = \cdot 259$.
$\operatorname{cosec} 15° = \sec 75° = 3\cdot 86$.
$\cos 15° = \sin 75° = \cdot 966$.
$\sec 15° = \operatorname{cosec} 75° = 1\cdot 035$.
$\tan 15° = \cot 75° = \cdot 268$.
$\cot 15° = \tan 75° = 3\cdot 73$.
(11) 1·22 km., 1·18 km., 1·37 km.
(12) ·0875 m/sec, 60·1 watts.

p. 100 (Section 5.11)

(3) $y = (10 - x)/7$, $y = -2 + 7x$.

p. 101

(4) Underground $x/0\cdot 8 - y/0\cdot 3 = 1$. British Railways $x/0\cdot 2 + y/0\cdot 4 = 1$. Intersection $(28/95, -18/95)$. tan angle $= -9\cdot 5$, angle $= 96°$. Road $9x = 4y$. Distances to Ox Cross from intersection ·350 km, Underground ·281 km, British Railways ·179 km.
(5) tan angle $= (l'm'' - m'l'')/(l'l'' + m'm'')$. Parallel if $l'm'' = l''m'$, perpendicular if $l'l'' + m'm'' = 0$.
(9) The perpendicular to the line from the origin.

pp. 104–105 (Section 5.14)

(1) $x^2 + (y + 15)^2 = 625$; $(\pm 20, 0)$; $(0, -40)$ and $(0, 10)$; $(7, -39)$ and $(7, 9)$; $(-24, -8)$ and $(15, 5)$; $(-7, -39)$ and $(15, 5)$.
(3) Centre $(5, 0)$, radius 5; $(3, 9)$; $(2, 4)$.
(4) Centre $(-3, 1\cdot 1)$, radius $\sqrt{180\cdot 2}$.
(5) $x^2 + y^2 = a^2 + b^2$; $\tan \angle QPR = b/a$.

p. 110 (Section 5.16)

(3) $F = (\sqrt{7}, 0)$; $F' = (-\sqrt{7}, 0)$.
(4) K, K' have x co-ordinates $= \dfrac{a^2 lm \pm ab\sqrt{[b^2 - l^2 + a^2 m]}}{b^2 + a^2 m}$

and S has x-co-ordinate $-a^2 lm/(b^2 + a^2 m)$. The corresponding y co-ordinates are given by $y = l + mx$.

p. 115 (Section 5.17)

(1) $x^2 - 4y^2 = 1$; $(\pm \frac{1}{2}\sqrt{5}, 0)$; $\frac{1}{2}\sqrt{5}$; $x = \pm 2/\sqrt{5}$.

p. 124 (Section 6.4)

(1) 8·241, 3·970, 2·750, 4·544, 1·880.
(2) 2·533, 7·055, 9·942, 3·472, 4·777.
(3) 1·465, 1·042, 1·500.

p. 125

(5) 1·3820, 2·5340, 2·3651, 3·5315.
(6) 8242, 788200, 45·44.
(7) 2630, 323·6, 15390.
(8) 3·4129, 4·7555.

p. 127 (Section 6.5)

(1) 1·732, 2·450, 2·080.
(5) 1·400.

p. 133 (Section 6.7)

(1) 160 calories.
(2) 100 cm.

p. 140 (Section 6.11)

(1) About 7 km.
(2) $3\cdot 3 \times 10^6$ metres; about 500 km.

p. 143 (Section 6.12)
Query: $M = 1/\ln 10$, or in general $1/\ln$ (base).
(1) ·6931; 1·099; 1·609.
(2) 7·389; 20·08; $2{\cdot}202 \times 10^4$; ·1353; ·0000454.

p. 151 (Section 6.14)
(1) 3·762; 3·627; ·9640; 1·317; 1·444.
(2) $\frac{1}{2}\ln \dfrac{1+x}{1-x}$
(3) (i) $\frac{1}{3}(y-2)$, single-valued.
(ii) $\pm\sqrt{(y-1)}$, double-valued.
(iii) $\ln(1+y)$, single-valued, $y > -1$.
(iv) $(1-y)/(1+y)$, single-valued, $y \neq -1$.
(v) $y = \pm\sqrt{(1-y^2)}$, double-valued.

pp. 157–158 (Section 7.2)
(1) $\log n = 2{\cdot}64 - {\cdot}42t$; $n = 439e^{-{\cdot}97t}$.
(2) $\log r = {\cdot}77 + {\cdot}052t$; $r = 5{\cdot}9e^{{\cdot}12t}$.
(3) $\log S = \bar{1}{\cdot}011 + \frac{2}{3}\log W$; $S = {\cdot}103W^{2/3}$; ·41 litre.
(4) $A = 107e^{-{\cdot}051t}$; 91 days.

p. 161 (Section 7.3)
(2) Probably monomolecular.
(3) $y = 53{\cdot}2/(x + 8{\cdot}3)$; $a = 53{\cdot}2$; $b = 8{\cdot}3$;
$y = 6{\cdot}29e^{-{\cdot}084x}$; $K = 6{\cdot}29$; $B = -{\cdot}084$.

p. 174 (Section 7.7)
(7) When $B_1/U_1 + B_2/U_2 = 0$ the third scale would have to be placed infinitely far away.

p. 177 (Section 8.2)
Query: 8/3 m.p.h.

p. 181
(1) 3·5.
(2) $-1 - 1/t_1t_2$.
(2) $-3(t_1+t_2)/t_1^2t_2^2$.
(4) $-2(1+t_1)^{-1}(1+t_2)^{-1}$.
(5) $(-1+t_1+t_2+t_1t_2)(1+t_1)^{-1}(1+t_2)^{-1}$.
(6) $1+t_1+t_2+t_1^2+t_1t_2+t_2^2$.
(7) $(t_1+t_2)(1-t_1^2)^{-1}(1-t_2^2)^{-1}$.

p. 185 (Section 8.4)
(2) $-2t/(1+t^2)^2$; $1/2\sqrt{t}$; $-1/2t\sqrt{t}$.

p. 188 (Section 8.5)
(1) Derivatives 24·96, 39·00, 56·16 from the formula.

p. 193 (Section 8.7)
(1) The first proof relies on the assertion that $\cos \delta x$ is *very* near to 1; this is correct, but needs stricter definition. E.g. the proof would have failed had $\cos \delta x$ been approximately $1 + \delta x$.

p. 197 (Section 8.9)
(1) $12t + 21t^2$.
(2) $3\sqrt{t} + 1/2\sqrt{t}$.
(3) $2t \sin t + t^2H \cos t$.
(4) $2(1 + \ln t)\sin t + 2tH \ln t \cos t$.
(5) $(1 + \frac{1}{2}\ln t)/\sqrt{t}$.

p. 199 (Section 8.10)

(7) $-2t/(1+t^2)^2$; $1/(1+t)^2$; $-1/t(\ln t)^2$; $\ln t$; $-(\ln t)(t\ln t - t)^{-2}$.

pp. 204–205 (Section 8.12)

(1) $4 + 2t$.
(2) $\cdot 3 \sin\theta$.
(3) $33\cdot 15/\sqrt{283} = 1\cdot 970$.
(4) $D_t x = 2/\pi(6x - x^2)$.
(5) $D_t x = \frac{1}{2}k\sqrt{(Fa/t)}$.
(6) $u - 9\cdot 81t$.
(7) $\cdot 050$ cm/sec.
(8) $\cdot 1458$ radians/sec $= 8\cdot 354°$/sec; $4\cdot 177°$/sec.

p. 208 (Section 8.14)

(1) $t + 1/t$.
(2) $D_t x = v(1+4x^2)^{-\frac{1}{2}}$; $D_t y = 2xv(1+4x^2)^{-\frac{1}{2}}$.

p. 210 (Section 8.15)

(1) Take $\phi(\epsilon) = \epsilon/B$.
(2) Take $\phi(\epsilon)$ to be the smaller of the two numbers ϵ/B, k.
(3) When $X - k < x < X + k$, $|x - X| < k = \phi(\epsilon)$, therefore $|y - Y| < \epsilon$, $|y| < |Y| + \epsilon = B$.
(4) In Problem (3) replace y by $\delta y/\delta x$. This $\rightarrow dy/dx$ and is therefore bounded when $X - k < x < X + k$. The result follows from Problem (2).

pp. 212–213 (Section 9.1)

(1) Errors $\cdot 008727 \cos\theta$, $\cdot 008727 (\sec\theta)^2$; proportional errors $\cdot 008727 \cot\theta$, $\cdot 008727 \sec\theta \operatorname{cosec}\theta$.
(2) $\pi R/50$ cm^2.
(3) $-\cdot 0048$ radians $= -\cdot 275°$.
(4) 4 per cent, $2\frac{2}{3}$ per cent.
(5) $45°$.

p. 214 (Section 9.2)

(1) $\ln y$; x/y.
(2) $2x/y$; $-x^2/y^2$.
(3) $1/2\sqrt{(x+y)}$; $1/2\sqrt{(x+y)}$.

p. 218 (Section 9.3)

(1) $(x\,\delta x + y\,\delta y)/\sqrt{(x^2+y^2)}$.
(2) $\delta m/s^3 - 3m\,\delta s/s^4$.

pp. 227–228 (Section 9.5)

(1) $\delta M = -2K(\delta R/R^2 + \delta L/L^2)$.
(2) $w_x = [\sec(x+y)]^2 + [\sec(x-y)]^2$; $w_y = [\sec(x+y)]^2 - [\sec(x-y)]^2$.
(3) $w_x = -4xy^2/(x^4 - y^4)$; $w_y = 4x^2y/(x^4 - y^4)$.

p. 234 (Section 9.7)

(1) $V_t = \pi\beta^2(4\beta + 5at^{2/3})/3$; $A_t = 8\pi\beta(\beta t^{-1/3} + at^{1/3})/3$.
(2) $r_{x|y} = x/\sqrt{(x^2+y^2)}$; $r_{y|x} = y/\sqrt{(x^2+y^2)}$; $r_t = (13 - 45t + 50t^2) \div \sqrt{(13 - 30t + 25t^2)}$.
(3) $I_x = -2Bx/(x^2+y^2+4)^2$; $I_y = -2By/(x^2+y^2+4)^2$; $I_t = -72B(13t - 240)/(13t^2 - 480t + 4644)^2$.

p. 235

(4) $T_{P|n} = a(\beta + 2P)/nR(\beta - P)^4$; $T_{n|P} = -aP/n^2R(\beta - P)^3$.

p. 242 (Section 10.2)

(1) $3t + t^2 + C$; $\frac{1}{3}t^3 - \cos t + C$; $2t + \ln t + \sin t + C$.
(2) $(5t + \frac{1}{4}t^2 + C)$ mm.
(3) $\frac{1}{3}e^t + C$.

p. 246 (Section 10.3)

(2) $t + \frac{1}{2}t^2 + e^t + C$.
(3) $t^2 + \frac{1}{2}e^{2t} - \frac{1}{2}e^{-2t} + C$.
(4) $-\cos t - \frac{1}{2}\cos 2t - \frac{1}{3}\cos 3t + C$.
(5) $-1/t + 3\ln t + 3t + \frac{1}{2}t^2 + C$.
(6) $2\tan^{-1}t - t + C$.
(7) $9\sinh^{-1}(\frac{1}{2}t) - 4\cosh^{-1}(\frac{1}{3}t) + C$.
(8) $-\cos t + \sin t - \ln(\pm\cos t) + C$.
(9) $\frac{1}{3}(\cosh 3t + \sinh 3t) + C = \frac{1}{3}e^{3t} + C$.
(10) $\frac{1}{3}(\cosh 3t - \sinh 3t) + C = \frac{1}{3}e^{-3t} + C$.
(11) $t + \frac{1}{2}t^2 + \frac{1}{3}t^3 + \frac{1}{4}t^4 + C$.
(12) $t - \frac{1}{2}t^2 + \frac{1}{3}t^3 - \frac{1}{4}t^4 + \frac{1}{5}t^5 + C$.

p. 250 (Section 10.4)

(1) $2e^{\sqrt{t}} + C$.
(2) $\sin(t + 3) + C$.
(3) $-\sqrt{(1 - t^2)} + C$.
(4) $-\frac{1}{4}\cos(2t^2 + 3) + C$.
(5) $-\frac{1}{2}(\ln\cos t)^2 + C$.
(6) $-\frac{1}{4}(\cos t)^4 + C$.
(7) $(1 + e^t)\ln(1 + e^t) - 1 - e^t + C$.
(8) $\frac{1}{3}(2t + 3)^{3/2} + C$.
(9) $-1/2(1 + \sin t)^2 + C$.
(10) $(t + \frac{1}{2})\log(4t + 2) - (t + \frac{1}{2})M + C$.
(11) $\tanh^{-1}\sin x + C$.
(12) $\frac{1}{3}(1 + t^2)^{3/2} + C$.
(13) $\ln(1 + e^t) + C$.
(14) $\frac{2}{3}(1 + \ln t)^{3/2} + C$.
(15) $\frac{1}{2}(\sin t)^2 + C$.
(16) $\frac{2}{9}[\ln(1 + t^3)]^{3/2} + C$.

p. 253 (Section 10.5)

(1) $t\sin t + \cos t + C$.
(2) $\frac{1}{2}t^2\ln t - \frac{1}{4}t^2 + C$.
(3) $(t^2 - 2t + 2)e^t + C$.
(4) $(2 - t^2)\cos t + 2t\sin t + C$.
(5) $\frac{1}{2}t\sqrt{(1 + t^2)} + \frac{1}{2}\sinh^{-1}t + C$. [A first integration by parts gives $\int v\,dt = t\sqrt{(1 + t^2)} - \int t^2(1 + t^2)^{-\frac{1}{2}}dt$; the second integral can be written $\int(1 + t^2)(1 + t^2)^{-\frac{1}{2}}dt - \int(1 + t^2)^{-\frac{1}{2}}dt = \int v\,dt - \sinh^{-1}t + C$. This gives an equation for $\int v\,dt$.]
(6) From Problem (1), $\int\theta\cos\theta\,d\theta = \theta\sin\theta + \cos\theta + C$. Put $\sin\theta = t$; this becomes $\int\sin^{-1}t\,dt = t\sin^{-1}t + \cos(\sin^{-1}t) + C$. So the term $\pm(1 - t^2)^{\frac{1}{2}}$ must be taken to be $\cos(\sin^{-1}t)$.

p. 265 (Section 10.9)

(1) (A), $y = 1/\sqrt{(C - 2t)}$.
(2) (A), $y = \sin^{-1}(t + C)$.
(3) (A), $y = \sqrt{[1 + C(t^2 - 1)]}$.
(4) (B), $y = \sqrt{(Ct - t^2)}$.
(5) (A), $y = 4/(C - t^4)$.
(6) (A), $y = \cos^{-1}(C\operatorname{cosec} t)$.
(7) (C), $2(y + 1)/(t + 1) + \tan\ln[C(y + 1)^2 + C(t + 1)^2] = 0$.
(8) (F), $y = t^2 + Cte^t$.
(9) (H), $y = t + e^{t+t^2}/(C - e^t)$.
(10) (I) or (A), $y = \sin^{-1}(C\operatorname{cosec} t)$.

p. 268 (Section 10.10)

(1) $y = \varphi(tu)$.
(4) $y = \varphi(t^2 - u^2)$.

p. 271 (Section 11.1)

(1) 3/2.
(2) 19/4.
(3) $e^\beta - e^\alpha$.
(4) 1.
(5) 0.

p. 276 (Section 11.2)

(1) $\frac{1}{3}\sqrt{32}$.
(2) $e - 1$.

p. 279 (Section 11.4)

(1) $X_r = 2 + 3r$; $X_7 = 23$.
(2) $\delta = 11$, $X_r = -6 + 11r$.

p. 281

(3) $n(2n - 1)$.

p. 288 (Section 11.6)

(2) $n^2(3 + 2n)$.
(3) $3r^2 - r$; $n^2 + n^3$.

p. 296 (Section 11.9)

(3) $R^2 \cos^{-1}(a/R) - a\sqrt{(R^2 - a^2)}$.
(4) $L^2 \cos^{-1}\dfrac{L}{2R} + R^2 \cos^{-1}\left(1 - \dfrac{L^2}{2R^2}\right) - RL\sqrt{\left(1 - \dfrac{L^2}{4R^2}\right)} =$
$R^2[\pi + 2\theta\cos 2\theta - \sin 2\theta]$ where $L = 2R\cos\theta$; this $= \frac{1}{2}\pi R^2$ when $\theta = \cdot 952$ radians, $L = 1{\cdot}160\,R$.
(6) $RH(\theta + 2) + R^2\theta$.
(7) $\pi(R_2^2 - R_1^2)\sqrt{[1 + T^2/(R_2 - R_1)^2]}$.

p. 305 (Section 11.12)

(2) $\frac{1}{6}\pi L^3$.

pp. 307–308 (Section 11.13)

(1) $1500\ \text{cm}^2$.
(2) $3{\cdot}927 \times 10^{-8}$ cc.
(3) $4{\cdot}4 \times 10^9$.
(4) $9{\cdot}5 \times 10^3$.

p. 314 (Section 11.15)

(1) 2.

p. 319 (Section 11.17)

(1) $\frac{1}{2}$.
(2) e^{-2}.
(3) $\frac{1}{2}\pi$.

p. 324 (Section 11.18)

(1) $e^{-\frac{1}{2}x_1^2} - e^{-\frac{1}{2}x_2^2}$.
(2) 0.
(3) 0.
(4) $\frac{1}{2}\sin 1$.

p. 326 (Section 12.1)

(1) 2.
(2) $2/t^3$.
(3) $-1/t^2$.
(4) $(6t^2 - 2)/(1 + t^2)^3$.

p. 328

(5) $y_t = \sinh t$, $y_{tt} = \cosh t$, $y_{ttt} = \sinh t$, and so on alternately.
(6) $y_t = -K \sin Kt$, $y_{tt} = -K^2 \cos Kt$, $y_{ttt} = K^3 \sin Kt$, $y_{tttt} = K^4 \cos Kt$, and so on.
(7) $y_t = B + 2Ct + e^t$, $y_{tt} = 2C + e^t$, all higher derivatives $= e^t$.
(8) $(D_t)^n y = (-1)^{n+1} |\underline{n-1}\, (1+t)^{-n}$.

pp. 334–335 (Section 12.2)

(1) Width = twice depth.
(2) $t = 2 \pm 1/\sqrt{3}$.
(3) $t = e$.
(4) 0; $e^{1/e}$.
(5) $\frac{1}{2}$; $\frac{1}{4}$.
(6) Two metres from strongest lamp.
(7) Radius of circle = height of rectangle $= 6/(4+\pi)$ metres $= {\cdot}840$ m.
(8) 100 cm.
(9) Square $8/(\pi + 4) = 1{\cdot}120$ m; circle $2\pi/(\pi+4) = {\cdot}880$ m.

p. 337 (Section 12.4)

(1) $-\sin t + t$.
(2) $C_2 + C_1 t + \frac{1}{2}t^2 + \frac{1}{6}t^3$.

p. 339 (Section 12.5)

(1) 1·1; 1·105.
(2) 1·05; 1·04875.

p. 344 (Section 12.7)

(1) $T - T^2/2 + T^3/3 - T^4/4 +$ etc.
(2) $M^{-1}(T - T^2/2 + T^3/3 - T^4/4 +$ etc.)
(3) $T - T^3/|\underline{3} + T^5/|\underline{5} - T^7/|\underline{7} +$ etc.
(4) $1 - T^2/|\underline{2} + T^4/|\underline{4} -$ etc.
(5) $1 + \frac{1}{2}T - \frac{1}{8}T^2 + \frac{1}{16}T^3 - \frac{5}{128}T^4$; 1·048809.

p. 345 (Section 12.8)

(1) $t = -\frac{7}{4}$; minimum; $t = -1$, neither maximum nor minimum.
(2) $t = 0$, minimum; $t = \frac{4}{5}$, maximum.
(3) At beginning of range, minimum if first non-zero derivative > 0, maximum if first non-zero derivative < 0. At end of range, vice versa.

p. 349 (Section 12.11)

(1) $t + u - \frac{1}{6}t^3 - \frac{1}{2}t^2u - \frac{1}{2}tu^2 - \frac{1}{6}u^3$.
(2) $1 + \frac{1}{2}t + tu - \frac{1}{8}t^2$.

p. 358 (Section 13.1)

(1) 1; ·1; $1/10^{n-1}$.
(2) 3·6; $\frac{2}{3}$; $5{\cdot}4(\frac{2}{3})^n$.
(3) 32; $-\frac{3}{4}$; $-\frac{128}{3}(-\frac{3}{4})^n$.

pp. 358–359 (Section 13.2)

(1) 10×2^n animals; $(2^n - 1) \times 10^4$ square metres; 10,230,000 square metres.
(2) $\frac{10}{11}[1 - (-\frac{1}{10})^n]$.

p. 360 (Section 13.3)

(1) $[n(n+1)r^{n+3} - 2n(n+2)r^{n+2} + (n+1)(n+2)r^{n-1} - 2r]/(r-1)^3$.
(2) $[n(n+1)(n+2)r^{n+4} - 3n(n+1)(n+3)r^{n+3} + 3n(n+2)(n+3)r^{n+2}$
$- (n+1)(n+2)(n+3)r^{n+1} + 6r]/(r-1)^4$

p. 362 (Section 13.6)

Query: only the series $0 + 0 + 0 + \ldots$ with sum 0.

p. 367 (Section 13.7)
(1) $(-1)^{m-1}t^{m-1}/\left|\underline{m-1}\right.$
(2) $t^{m-1}/M^{m-1}\left|\underline{m-1}\right.$.
(3) $(-1)^{m-1}t^{2m-2}/\left|\underline{2m-2}\right.$.
(4) $t^{2m-1}/\left|\underline{2m-1}\right.$.
(5) $t^{2m-2}/\left|\underline{2m-2}\right.$.
(6) $-t^m/m$.
(7) mt^{m-1}.

p. 378 (Section 13.11)
(1) The Binomial Series.

p. 389 (Section 14.6)
(1) If $\boldsymbol{a}, \boldsymbol{b}, \boldsymbol{c}, \boldsymbol{d}$ are the four vectors from the origin O to the vertices of the tetrahedron, the vector OG is $\frac{1}{4}(\boldsymbol{a}+\boldsymbol{b}+\boldsymbol{c}+\boldsymbol{d})$.

p. 390 (Section 14.7)
(1) Replace $\boldsymbol{w}$ in (14.11) by $(-\boldsymbol{w})$.

p. 394 (Section 14.12)
(1) $\{1, 0\}, \{1, 2\pi/n\}, \{1, 4\pi/n\}, \ldots, \{1, 2n - 2)\pi/n\}$.

p. 395 (Section 14.13)
(7) $-1, -\omega, -\omega^2$.
(8) $\{1, \pi/3\}$ and $\{1, 4\pi/3\}$.
(9) $\{\sqrt{r}, \frac{1}{2}\theta\}$ and $\{\sqrt{r}, \frac{1}{2}\theta + \pi\}$.

p. 400 (Section 14.17)
(1) $10 + 4i$.
(2) -2.
(3) i.
(4) $3 - 2i$.

p. 401 (Section 14.18)
(14) The vector $z\,\boldsymbol{I}$ is reflected in the x-axis.

p. 405 (Section 14.19)
(3) $\frac{1}{2}\ln 2 + (2r + \frac{1}{4})\pi i$.
(4) $1 + (2r + 1)\pi i$.

p. 405 (Section 14.20)
(1) $\cos 1 + i \sin 1 \simeq \cdot 54 + \cdot 84 i$.
(2) -1.
(3) $-e$.

p. 406
Query: $\operatorname{Ln}\exp z = z + 2n\pi i$ where n is an integer.

p. 408 (Section 14.21)
(4) 0.
(5) $\pm\{\sqrt{2}, \frac{1}{4}\pi\} = \pm(1 + i)$.
(6) Three roots, except when $z = 0$; $\{r^{1/3}, \frac{1}{3}\theta\}, \{r^{1/3}, \frac{1}{3}\theta + \frac{2}{3}\pi\}, \{r^{1/3}, \frac{1}{3}\theta + \frac{4}{3}\pi\}$.

p. 411 (Section 14.23)
(1) $\cosh(x + iy) = \cosh x \cos y + i \sinh x \sin y$;
$\sinh(x + iy) = \sinh x \cos y + i \cosh x \sin y$.

p. 413

(3) $\frac{1}{2}(e^{-1} + e) \cos 1 + \frac{1}{2}i(e^{-1} - e) \sin 1 \simeq \cdot 83 - \cdot 99i$;
$\frac{1}{2}(e^{\frac{1}{2}\pi} + e^{-\frac{1}{2}\pi}) \sin \frac{1}{2} - \frac{1}{2}(e^{\frac{1}{2}\pi} - e^{-\frac{1}{2}\pi})\, i \cos \frac{1}{2} \simeq 1{\cdot}20 - 2{\cdot}02\, i$.
(4) $\cot z = (e^{iz} + e^{-iz})i/(e^{iz} - e^{-iz})$; $\sec z = 2/(e^{iz} + e^{-iz})$; $\operatorname{cosec} z = 2i/(e^{iz} - e^{-iz})$.
(5) $\cot z = i \coth iz$; $\sec z = \operatorname{sech} iz$; $\operatorname{cosec} z = i \operatorname{cosech} iz$.

p. 418 (Section 15.1)

(2) $\frac{1}{5}e^x(\cos 2x + 2 \sin 2x) + C$.

p. 420

(3) $(2 - x^2) \cos x + 2x \sin x + C$.
(4) $-\frac{1}{4}x \cos 2x + \frac{1}{8} \sin 2x + C$.
(5) $\frac{1}{2}e^x(x \cos x + x \sin x - \sin x) + C$.

p. 426 (Section 15.2)

(1) $\frac{1}{2} - 2\pi^{-1}(\sin x + \frac{1}{3} \sin 3x + \frac{1}{5} \sin 5x + \ldots)$.
(2) $\cos x + \cdot 3 \cos 2x + \cdot 52 \sin 2x + \cdot 2 \sin 3x$.

p. 431 (Section 15.3)

(1) $1/4(x - 2) - 1/4(x + 2)$.
(2) $1/6(x - 3) - 1/6(x + 3)$.
(3) $5/2(x - 1) - 1/2(x + 1)$.
(4) $5/3(x - 1) - 2/3(x + 2)$.
(5) $1/2(x - \sqrt{2}) + 1/2(x + \sqrt{2})$.
(6) $3/2x - 1/2(x + 2)$.
(7) $5/2(x - 1) - 8/(x - 2) + 11/2(x - 3)$.
(8) $-3/2(x + 1) + 2/(x - 1) + 7/2(x - 3)$.
(9) $5/6(x - 1) + 2/3(x + 2) - 3/2(x + 5)$.
(10) $-1/2(x - 1) + (3\sqrt{3} - 2)/2(\sqrt{3} - 1)(x - \sqrt{3}) + (3\sqrt{3} + 2)/(\sqrt{3} + 1)(x + \sqrt{3})$.

p. 434 (Section 15.4)

(1) $-\frac{1}{4} \ln(\pm x) - 1/2x + \frac{1}{4} \ln[\pm(x + 2)] + C$.
(2) $\ln[\pm(1 - x)] + 2/(1 - x) + C$.
(3) $\frac{1}{9} \ln[\pm(x - 2)] - 5/3(x - 2) - \frac{1}{9} \ln[\pm(x + 1)] + C$.
(4) $-\frac{1}{9} \ln[\pm(x + 1)] - 1/3(x + 1) + \frac{1}{9} \ln[\pm(x + 4)] + C$.
(5) $\ln[\pm(x + 1)] + 3/(x + 1) - 3/(x + 1)^2 - 1/3(x + 1)^3 + C$.

p. 437 (Section 15.5)

(1) $\frac{1}{2}i \operatorname{Ln}(x + i) - \frac{1}{2}i \operatorname{Ln}(x - i) + C$.
(2) $\frac{1}{2} \operatorname{Ln}(x^2 - \omega^2) + C$.
(3) $\frac{1}{2} \operatorname{Ln}(x^2 + 1) + C$.

p. 443 (Section 15.7)

(1) $\frac{1}{7}[\ln(x - 4) - \ln(x + 3)] + C$.
(2) $\frac{1}{26}[-\ln(x^2 + 4) + 3 \tan^{-1}(\frac{1}{2}x) + 2 \ln(x + 3)] + C$.
(3) $\frac{1}{6}[\ln(x^2 + 1) - \ln(x^2 + 4)] - \frac{1}{2} \tan^{-1}(\frac{1}{2}x) + \tan^{-1} x + C$.
(4) $\frac{1}{2}(x^2 + 1) - \frac{1}{2} \ln(x^2 + 1) + \ln x + C$.
(5) $$2 \ln(x^2 + 1) + 2 \tan^{-1} x - 2 \ln(x^2 + x + 1) - \frac{8}{\sqrt{3}} \tan^{-1} \frac{2x + 1}{\sqrt{3}} + C.$$

p. 463 (Section 16.14)

(1) $(\frac{11}{6}, \frac{1}{3})$.
(2) $(1{\cdot}4, 1{\cdot}1, \cdot 4)$.

INDEX

abscissa, 80
absolute magnitude or value of complex number, 396; of real number, 73–8; of vector, 389
absolute temperature, 173, 475
acceleration, 325–37; as derivative, 322; of electric charge, 489
accuracy, 5–8, 13, 359–60
Achilles and the tortoise, 362
acre, 450
acuity of vision, 140
adaptation, 47–8
addition, on punched cards, 27; by slide rule, 162; of complex numbers, 396–8; of derivatives, 193; of series, 375–6; of vectors, 386–8
adult height, 130
air-cells in lungs, 306–7
algebra, 33–70; in geometry, 79–114; laws of, 34–6, 74–8, 386–9, 391–8
algebraic division, 56–8; fractions, 52; integrals, 427–34, 437–43
alignment chart, *see* nomogram
allometric relations, 132–3
alveoli, 307
ampere, 481
amplitude, of complex number, 396; of harmonic motion, 421
analytical geometry, 79–114
angle, antilogarithm of, 120; circular measure of, 137; eccentric, 106; measurement of, 81–2; radian, 137; trigonometric functions of, 82–96, 138–40, 367, 375–6
Angstrom, 490
anthrax, 154–6
anti-derivative, *see* indefinite integral
antilogarithms, 120; derivative of, 116–17, 192; series for, 367; *see also* logarithms; exponential
antiseptic, 154–7
apothecary's units, 452
approximate equality, 138
approximation, 228–30, 314–17, 341–4, 359–60
arbitrary constant, 241, 254
arc, 136–8, 305
are, 450
area, principal formulas, 305; of circle, 276, 292, 305; of cone, 294–5, 305; of cylinder, 293, 296, 305; of ellipse, 276, 305; of loops, 290, 311–12; of parallelogram, 296, 305; of polygon, 313–14; of prism, 293, 305; of sphere, 295–6, 305; of spheroid, 303; of triangle, 296, 305; under parabola, 275, 290; units and conversion factors, 450; numerical evaluation, 314–17; as an integral, 271–6; sign convention for, 310–11; of air-cells, 306–7; of wound, 158
arithmetic, checking of, 23–5; division in, 15–17; multiplication in, 15, 18–19; simplified, 15–19
arithmetic series or progression, 278–81
arithmetico-geometric series, 359
ascending factorial, 369–72
assay, 490
asymptote, 113–15
atmosphere, 466
atom, size of, 153
atomic weight, 474–5
atto- (prefix), 449
average rate, velocity, 176–81
avoirdupois, 451–2
axis, 103, 105–6, 109–13

bacteria, 117, 131, 223, 227, 236, 254, 306–7
Baggally, 54
balloon, 231
bar, 466
base, of logarithms, 116–17, 122–3
bees, 335
Benedict, 133
Berkson, J., 491
Bernoulli, 263
Bessel's function, 379
Bickley, W. G., 316
binomial coefficient (reduced factorial), 369–72; series, 367–72
blood, 213, 333
blood-groups, 72, 380–1
Bornstein, 478
bound, 78
Boyle's law, 53
brackets, 30–1, 86
Briggs, and logarithms, 116
brightness, 490
British units, 449–55, 466
Brodetsky, S., 165
Brodie, 469–70
Bromwich, T. J. I'A., 286

Bruun, A. Fr., 130
Burn, J. H., 491

cable, 334
calculating machine, 15–18
calculation, *see* evaluation
calculus, *see* derivative; partial derivative; differential equation; integral; indefinite integral; definite integral
calendar, 447–9
Calkins, 40
calorie, 167–8, 470, 474
calorific value, 167–8
Cambridge, 178
candle power, 490
capacitance, capacitor, 483, 486–9
capacity, *see* volume, capacitance
carbohydrate, 167–8
cards, punched, 26–8
Carrel, 158
cartesian co-ordinates, 79–83
casein, 161
casting out nines and elevens, 23–5
catenary, 144
cell, blood, 213
Celsius (centigrade) scale, 38, 475
centi- (prefix), 449
centre of mass (of gravity), centroid, 389, 462–5
C.G.S. units, 452–5
change (*see also* small changes; change of variable), bracket symbol for, 269; delta symbol for, 178, 211; rate of change, 176–90
change of variable, in derivative, 199–203; in partial derivative, 221–2; in integral, 246–50, 319–24
characteristic, 124
checking of arithmetic, 23–5
chemical reaction, 160–1, 477–89
Chick, Dame H., 154–6
circle, area of, 276, 292, 305; equation of, 103–5; length of, 136
circuit, 482, 486–9
circular arc, 136–8, 305; function, *see* trigonometric function; measure, 137
circulation of blood, 333
circumference of circle, 136, 305
Clark, 157
coding of data, 26–8
coil, 485–6
Colson, J., 4, 128, 130
common factor, 31–2
common logarithm, 116
common multiple, 31–2
comparison of magnitudes, 73–8
completing the square, 42–3, 391–3, 415
complex number, 391–421; definition of, 391; amplitude or phase of, 396, 421; addition of, 396–8; conjugate of, 400–2; cube root of 1, 394; derivate of function of, 413–16; division of, 395; exponential of, 405–14; hyperbolic functions of, 411–13; imaginary, 394, 398–400; integration of function of, 416–20; limit of function of, 413; logarithm of, 402–9, 414; magnitude or modulus of, 396; multiplication of, 391–5; powers of, 393, 408–9; real and imaginary parts of, 398–400; roots of, 394, 407–8; subtraction of, 395; Taylor series of function of, 415–17; trigonometric function of, 411–13
component of vector, 384
composite number, 10, 30–2
computation, *see* numerical evaluation
condenser, *see* capacitor
conductivity, electric, 482; heat, 476
cone, area of, 294–5, 305; volume of, 297, 306; double, 110–12
conic sections, 102–15
conjugate complex, 400–2
constant, 33; dielectric, 483; of integration, 241, 254; of dissociation, 477
constraint, 231–7
convergence, 13, 362–4, 373–6, 415; tests for, 364
conversion factors, for area, 450; energy, force, power, 454; heat, 471; length, 449–50; mass, 451–2; pressure, 466; temperature, 475–7; velocity, 451–2; volume, 451
cooling, rate of, 210, 256, 476
co-ordinates, cartesian, 79–83, 222; polar, 82–3, 222; origin of, 39, 80–1
Cope-Chat, 26–8
correction for error, 212
cosecant, cosec, cosech, *see* trigonometric functions *or* hyperbolic functions
cosine, cos, cosh, series, 367, 410–13; *see* trigonometric functions *or* hyperbolic functions
coulomb, 480, 484
cube root, by logarithms, 127; of 1, 394
cubic polynomials, 45–8
Cullwick, E. G., 480
current, 481–8
curve, length of, 300; motion along, 206–8; tangent to, 186
cwt (hundredweight), 451–2
cylinder, area of, 293, 296, 305; volume of, 297, 306; section of, 107–10

D = derivative, 186; = dioptre, 490
dachshund, 47–8
damped oscillation, 489
data, coding of, 26–8
day, 447–9
deci (pre-fix), 449
decimal point, placing of, 8, 22; places, 8, 33; recurring, infinite, unending, 13

decomposition, of casein, 161; of tetanolysin, 160; of vibriolysin, 161; into partial fractions, 427–34, 437–43
definite integral, 269–78, 288–324; area as, 271–6, 290–6; change of variable in, 319–24; by parts, 320; definition of, 270; distance as, 269; infinite, 318–19; length of curve as, 300; limits or termini of, 271, 312, 318–24; numerical evaluation of, 315–17; Simpson's rule for, 315–17; substitution in, 319–24; volume as, 296–300; work as, 289
degree, of angle, 81; of temperature, 38
deka- (prefix), 449
delta, 178, 211
density, 455–65
dependent variable, 213
derivative, derivate (*see also* partial derivative), 182–90, 203–4; standard forms, 203–4; of algebraic functions, 184–5; of hyperbolic functions, 199, 203; of inverse functions, 202–3; of logarithms and antilogarithms, 190–1, 203; of powers, 203; of a product, 194–6; of a quotient, 197, 203; of a series, 375–6; of a sum, 193–4; of trigonometric functions, 192–3, 196, 199, 203; acceleration as, 327; change of variables in, 199–203; complex, 413–16; higher order, 327–55
derived function, *see* derivative *and* partial derivative
derived units, 452–4
Descartes, 78, 81
descending factorial, 369–72
dielectric constant, 483
diet, heat value of, 168
differences, in interpolation, 60–2, 92–4, 123–4; mean, 92–4; negative, 92–4; *see also* small changes
differential coefficient, *see* derivative
differential equation, methods of solution of, 259–64, 378–9, 487–9; exact, 263; Bernoulli, 263; homogeneous, 259; integrating factor of, 264; linear, 262, 482–9; partial, 265–8; Riccati, 263; with separable variables, 259; solvable for variable, 261; arbitrary constant in solution of, 254; of population growth, 254–8; second order, 336, 487–9; series solution of, 378–9
differentiation, *see* derivative *and* partial derivative
digestion, 159
dimensions, method of, 471–4
dioptre, 490
directrix, 103, 105, 109–15
discontinuity, 190
distance between points, 95; between point and line, 99; as integral, 269
divergence, 362–7, 373–6
divisibility, 10, 28–30
division, arithmetical, 15–17; by logarithms, 123–9; by nomogram, 164–165; by slide rule, 162–3; of complex numbers, 395; of polynomials, 56–8, 427, 438; of a straight line, 101–2
dog, 47–8
double integral, 456–62, 465–7
double series, 382
drachm, 451–2
dram, 451–2
Dreyer, 132–3
Duarte, A. J., 356
Dubois, 132, 169–73, 476
du Nouy, 158
dyne, 454

e, 143
Eagle, A., 446
earth, size of, 153
eccentric angle, 106
eccentricity, 109–15
elasticity, 38
electric charges and currents, 480–9; capacitance, capacitor, oscillation, resistance, 486–9; units, 480–7
electrocardiogram, 426
electrolysis, 483
electromagnet, 485
electromagnetic induction, 485–9; radiation, 489; units, 480
electromotive force, 482
ellipse, 105–10; area, 276, 305; as section of cylinder, 107–10; as section of cone, 110
elliptic integral, 446
energy, electric, 483; heat and mechanical, 225; in diet, 168; of gas, 224–6; units of, 454
England, 7, 38
equiangular spiral, 117–22
erg, 454
error, experimental, 4–7; small, 211–37, 335–49; proportional or percentage, 7, 212; correction for, 212; bounds for, 345–6
Euclid, 14
evaluation, numerical, of area, 314–17; of definite integral, 315–17; of logarithm, 135; of trigonometric functions, 91–2; *see also* interpolation; table
evolution, 47
exact differential equation, 263
Exactus, 15
exponential decay, 318; function, 143–5, 214–15, 367, 375–6, 405–14; growth, 245–8; series, 365–6, 375, 410–17
extrapolation, danger of, 46
eyes, of cave fish, 4

factorial, ascending and descending, 369–72; of integer, 22–3, 342, 369–72
factorization, arithmetic, 10–11, 28–32, 44–5, 52–8, 60; algebraic, 60–7; tests for, 10–11, 67
Fahrenheit scale, 38, 475
Famulener, 161
farad, 483
Faraday, 483
fat, 167–8
fathom, 450
Feldman, W. M., 157, 307, 330–4
femto- (prefix), 449
figure, significant, 6–7
Finney, D. J., 491
fish, cave, 4; flying, 130
Fisher, R. A., 316
flagpole, height of, 84
Fletcher, A., 254
flow of liquid, 467–70
fluid drachm, ounce, 451–2
flying fish, 130
focus, 103, 105, 109–15
foetus, 40
foot (unit of length), 450
force, unit of, 454; electromagnetic, 479–88; psychological, 54; triangle of, 386
Fourier series, 424–6
fraction, logarithm of, 128; partial, 427–434, 437–43
function, definition of, 49; arbitrary, 265–8; Bessel, 379; complex, 402–13; derived (*see* derivative; exponential), 143–9, 405–12, 414; hyperbolic, 144–9; implicit, 107; inverse, 149–151; trigonometric, 82–96, 367, 375–6
fundamental note, 423
furlong, 450

gallon, 451
gas constant, 477
gas, energy of, 222–4; equation and properties of (pressure, temperature, volume), 220–5, 233, 476–7; specific heat of, 225–6
Gaussian units, 480
G.C.F., 31–2
geometric rate of growth, 360–1; series (progression), 356–64
geometry, 1; analytic (algebraic), 79–114
germane units, 453
gigo- (prefix), 449
gill, 451
Giorgi, G., 480; system of units, 452–6, 466–8
Glaser, E. G., 476
glomerulus, 469
Goodwin, L. G., 491
grade (unit of angle), 81
gradient (slope) of line, 33, 37, 177, 186
graduation, *see* scale
grain, 452
gram, 451–2
graph, of hyperbolic functions, 145; of trigonometric functions, 87; straight line, 96, 154–62
graph paper, logarithmic, 153
graphical aids, *see* nomogram; slide rule; scale
gravity, 40, 44
greatest common factor, 31–2
Greek letters, 282, 497
Gregorian calendar, 447
growth, exponential, 214–15, 250; geometric, 360–1; line of, 106; logistic, 258; of bacteria, 131, 236; of population, 242–6, 250, 254–8; of schoolgirls, 45
Guldberg, 477

haemoglobin, 479–80
Hardy, G. H., 14
hare, tortoise, snail, 345
harmonic (overtone), 423; motion, 421–426
Hartman, N., 158
Haughton, S., 106
H.C.F., 31–2
healing of wound, 158
heat, 470–7; units of, 225, 470; conversion factors, 471; energy, 225; latent, 470; loss of, 170, 476; produced by infant, 133; value of diet, 168
hecto- (prefix), 449
height, and pulse rate, 156; inheritance of, 36; of adults, 36–7, 130, 153; of distant object, 85, 140; of schoolgirls, 45–6; sitting, 132; of flagpole, 84; of hill, 85
henry, 486
heredity, 36, 47–8
heterogonic relation, 132–3
higher order derivative, 327–35, 346–51
highest common factor, 31–2
Highway Code, 64
Hill, A. V., 480
hill, height of, 85
Hollerith, 26–8
homogeneous equations (differential), 259
honeycomb, 335
Hooke's law, 38, 289
horse-power, 455
hundredweight, 451–2
Hunter, J., 332
hyperbola, definition and properties, 111–15; rectangular, 53, 55, 115
hyperbolic functions, 144–9, 199, 203, 411–13; *see also* trigonometric functions
hyperbolic logarithms, *see* logarithms, natural

identity (equation), 34
imaginary number, 394; part, 398–400
immersion in cold water, 476
immigration, 255–8
implicit function, 107
Ince, 255
inch, 450
indefinite integral, definition of, 240; standard forms, 243–4; arbitrary constant in, 241; by parts, 250–2; change of variable in, 246–50; notation for, 241; of product, 250–2; of series, 375–6; of sum, 241; substitution in, 246–50
index, refractive, 489
inductance, coefficient of, 486
induction, electromagnetic, 485–9
inequality, 73–8
inertia, moment of, 465
infant, heat production, 133
infection of wound, 158
infinite integral, 318; decimal, 13
infinity, 53, 189, 318; sum to, 361
inflexion, point of, 331
inheritance of height, 36
insect, path of, 119
instantaneous velocity, 181–5 (*see also* derivative)
integral (*see also* indefinite integral, definite integral), standard forms, 243–4; complex, 416–20; elliptic, 446; infinite, 318; rational function, 427–34, 437–44; repeated, 458–62, 465–7; multiple (double or surface, triple or volume), 456–62, 465–7; trigonometric, 418–25, 443–5
integrating factor, 264
intensity of illumination, 490
interplanetary rocket, 318
interpolation, 60–6, 92–4, 123–4
inverse function, 149–51; derivative of, 202–3
inverse square law, 318
inverse tangent series, 376
inversion of cane sugar, 477
ion, 55
irrational integrals, 444–6
Isaac, 9

Jacob, 9
jaws, 47–8
Joachimsthal's formulas, 102
joule, 225, 471

Keetch, D. V., 380
Kelvin, 475
Kestelman, H., 433
kidneys, 469
Kjeldahl, 161

Lack, D., 357
lapwing, 357–61
latent heat, 470
law, of cooling, 210, 256; of mass action, 477; Ohm's, 482
Lawler, S. D., 380
laws of algebra, 34–6, 74–8, 386–9, 391–8
L.C.M., least common multiple, 31–2
Lee, A., 36–8
Leibnitz, 210, 241
length, summary of formulas, 305; units and conversion factors, 449–50; of arc, 136, 305; of catenary, 301; of curve, as integral, 300; of circle, 305; of head of locust, 356; of parabola, 301; of vector, 389
lens, 490
Levens, A. S., 165
light, 474, 483, 489
limit, 188, 208–10; complex, 413; formal definition, 208–10; of error, 345–6; (terminus) of integration, 271, 312, 318–24
line, of growth, 106; density, 456; straight, 37–40, 60–1, 96–102, 154–62
linear differential equation, 262
linear function, 36–40
litre, 451
living things, size of, 153
locust, 356
logarithmic comparisons, 130; log. graph-paper, 153; log. scale, 152–8, 163, 169–74
logarithms, 106–43; tables of, 493–4, 498; addition law, 116–17, 121; base of, 116–17, 122–3; characteristic of, 124; common or Briggsian, 116; complex, 402–14; derivative of, 191–203; mantissa of, 124; many-valued, 404, 407–10; modulus of, 141; natural, napierian or hyperbolic, 142; of fractions, 128; of negative numbers, 134; series for, 367, 375–6; systems of, 122–3, 141–2; use in finding powers and roots, 125–30; use in multiplication and division, 123–9
logistic function, 146, 258
London schoolgirls (L.C.C. report), 45
loop, area of, 290, 311–12
loss of heat, 170, 476
lumen, 490
lung, 306–7
lux, 490

machine computation, 14–18
Maclaurin series, 366; *see also* Taylor series
Madsen, Th., 160–1
magnetic properties, 485–9
magnitude, absolute (real), 75; (complex), 396; (vector), 389; comparison of, 73–8; directed, *see* vector; complex number

major axis, 103, 105–6, 109–13
mantissa, 124
manual computation, 18–19
many-valued function, 150–1; logarithm and powers, 402–5, 408–10
mass, units and conversion factors, 451–2; action, 477; centre, *see* centre of mass
Massey, H. S. W., 433
matrix (*plural* matrices), 380–2
maxima and minima, 41–4, 47–8, 328–55; absolute, 43, 317; terminal, 317–18; for several variables, 339–43; of rate of growth of child, 320
mean, weighted, 463
mechanical equivalent of heat, 225
Meeh, 133
mega- (prefix), 449
Mendel, G., 2
Menger, K., 205–6
metabolism in bacterium, 223, 306
metre-kilogram-second (MKS) system, 452–5, 466–8
metric system, 449–55, 466–86
micro- (prefix), 449
microfarad, 483
mil, 450
mile, 450
Miller, J. C. P., 254
Milne-Thomson, L. M., 446
minim, 452
minimum, *see* maxima
minor axis, 103, 105–6, 109–13
minute of angle, 81; of time, 447
MKS system, 452–5, 466–8
modulus, of real number, 75–8; of complex number, 396; of elasticity, 38; of logarithm, 141; of scale, 171
mole, 226, 473, 475
molecular weight, 226, 475
mollusc, 106
moment of inertia, 465–6
month, 447–9
Moronami, 239
mortality of lapwing, 357–61
motion, curved, 206–8; *see also* velocity
multinomial series, 382
multiple integral, 456–62, 465–7
multiplication, by logarithms, 123–9; by nomogram, 164–5; by punched cards, 27; by slide rule, 162–3; of complex number, 391–5; of vector, 385; of series, 375–6
Murray, H. A., 246

nano- (prefix), 449
Napier, 142
napierian, *see* logarithms
natural logarithms, *see* logarithms
negative area, 310–11; power, 128
nerve, 334
Neville, E. H., 446, 484
Newton, Isaac, law of cooling, 210, 256
newton (unit of force), 454
Ngboglus, 422
nitrogen determination, 162
nomogram, 163–74; for addition, 163–5; for heat value of diet, 168; for Dubois formula, 167–73; for heat loss, 170; for multiplication, 164–5; for respiratory quotient, 173; for surface area, 167–73; modulus of scale of, 171; scales and graduations, 152–9
normal temperature and pressure, 466; vision, 140
N.T.P., 466
numerical evaluation of area or definite integral, 314–17; logarithms, 135; trigonometric functions, 91–2; *see also* interpolation; table

oblate spheroid, 106
ohm, Ohm's law, 482
order, of approximation, of small quantity, 340; of differentiation, 349–51; of reaction, 479; second, of differential equation, 336, 421–6, 487–9; second, of derivative, 336
ordinary differential equation, *see* differential equation
ordinate, 80
origin of co-ordinates, 39, 80
oscillation, 421–6, 486–9
ounce, 451
overtone, 423

parabola, 40; properties of, 102–3; area under, 275, 290; length of, 301
parallelogram, area of, 296, 305
parallelopipedon, volume of, 299
parametric form, 107
partial derivative, derivate, or differential coefficient, 213–37; (higher order), 346–51; change of variables in, 221–2; formal definition, 228–30; in constrained variation, 231–7; order of differentiation in, 349–51
partial fractions, 427–34, 437–43
parts, integration by, 250–2, 320; real and imaginary, 398–400
Pascal's triangle, 370–3
Pearl, 258
Pearson, Karl, 36–8
pendulum, 422
pennyweight, 452
pepsin, 159, 205
percentage error, 7, 212
perfect gas, 470
permeability, magnetic, 485
permittivity, 481–3
perpendicular lines, 98
phase, 396, 421
phenol, 154–7
π (pi), 33, 137
Piaggio, H. T., 255

pico- (prefix), 449
pint, 451
plane, 390–1
pleasure, 54
point, co-ordinates of, 79–83; of inflexion, 331; representing vector, 384
poise, 468
Poiseuille, 468–70
polar co-ordinates, 82–3, 222
polygon, area of, 313–14
polynomial, 34–52, 58–67, 193
population of England and Wales, 7, 38; of United Kingdom, 7; with immigration, 255–8
potential, 482
poundal, 454
power (algebraic), complex, 393, 408–10; by logarithms, 125–130; derivative of, 203; many-valued, 404, 407–10; negative, 128; sum of, 286; (rate of work), 454; (of lens), 490
Powers-Samas, 26–8
practical calculation, *see* numerical
pre-natal growth, 40
pressure, 466
prime number, 10, 28–30
prism, area of, 293, 305; volume of, 298, 306
product, derivative of, 194; integral of, 250–2, 320
progression, *see* series
projection on line, 99
prolate spheroid, 106
proportion, 68–72
proportional error, 7, 212
proportional parts, 61, 92–4, 123–4
protein, 167–8
psychological forces, 54
pulse rate, 156
punched cards, 26–8
pyramid, volume of, 298, 306
Pythagoras, 12, 75, 86, 95–6, 294

quadratic, 43–4; polynomial, 40–5
quart, 451
quarter, 451–2
quotient, derivative of, 192; respiratory, 173

rabbit, 47
Race, R. R., 380
radian, 137
radiation, 489
radius, 103, 109
range, 130
rate of change, instantaneous, *see* derivative; average, 176–81
ratio, 68–72; sex, 70
rational function, 52, 427–34, 437–43
reaction, chemical, 477–80
real number, 14, 394; real part, 398–400
Réaumur, 475
rectangular hyperbola, 53, 55, 115
recurring decimal, 13, 349–54
reduction formula, 442, 444
refraction of light, 489–90
relativity, 474
repeated integral, 458–62, 465–7
resistance, electric, 482
respiration, 173, 306–7
Riccati equation, 263
Robertson, 332
Robinson, G., 426
rocket, 318
Rogers, C. A., 282
root, by logarithms, 125–30; complex, 394, 407–8; of algebraic equations, 65–7, 434–7; of minus one, 394
Rosenhead, L., 254
rounding off, 8
Russian letter ∂, 213

saddleback, 354
Sanger, R., 380
scalar, 383
scale, logarithmic, 152–8; square root, 158–9; reciprocal, 158; of nomogram, 163–74
Scammon, 40
schoolgirls, 45–6, 131
Schütz-Borisoff law, 159, 205
scruple, 452
sea urchin, 106
sec, secant, sech, *see* trigonometric functions *and* hyperbolic functions
second, of angle, 81
second moment, 465–6
second order, derivative, 327–55; differential equation, 336, 421, 487–9
section, of cone, 110–12; of cylinder, 107–10
sector, area of, 292, 305
separable variables, 259
series, addition of, 375–6; antilogarithm, 367; arithmetic and generalized arithmetic, 278–81, 283–8; arithmetico-geometric, 359–60; binomial, 367–9; complex, 415–16; convergence of, 362–4, 373–6, 415; differentiation and integration of, 375–6; divergence of, 362–4; double, 382; exponential, 367, 375–6; in solution of differential equations, 378; Fourier (trigonometric), 424–6; geometric, 356–64; inverse tangent, 376; logarithmic, 367, 375–6; Maclaurin, *see* Taylor series; manipulations with, 375–6, 382; multinomial, 382; power, *see* Taylor series; sine and cosine, 367, 413; sinh and cosh, 367, 375; sum to infinity, 361
sex-ratio, 70
shape of cells in honeycomb, 335

shear, 466
sight, normal, 140
sigma notation, 281–8, 380–2
sign of area, 310–11
significant figures, 6–8
simple harmonic motion, 421
Simpson's rule, 315, 317
sine, series for, 367, 415; table of, 494–6, 500–1; *see also* trigonometric functions; hyperbolic functions
sinh, series for, 367, 375; pronunciation, 146; *see* hyperbolic functions
sizes of natural objects, 153
Sjöquist, 159
Skellam, J. G., 379
slide rule, 162–3
slope of line, 33, 37, 177, 186
small changes and errors, 211–37, 335–349
Snellen's type, 140
Soames, H., 239
Soper, H. E., 369
sound, velocity of, 204
space vector, 383
specific heat, 475–7
speed, *see* velocity
sphere, area of, 295, 305; volume of, 299, 306
spheroid, 106; area of, 303; volume of, 303, 306
spiral, equiangular, 117–22
square, completing the, 42–3, 391–3, 415
square root, evaluation of, 127; complex, 407; scale, 158–9; integration of functions containing, 444–6; of minus one, 394
standard derivatives, 203–4; integrals, 243–4
stere, 451
stone (unit of mass), 451–2
straight line(s), 37–40, 60–1, 96–102, 154–62
substitution, in integral, 246–50, 319–24
subtraction, by nomogram, 163–5; by slide rule, 162; of complex number, 396–8; of vectors, 388
successive approximation, to small change, 341–4
sugar, inversion of, 477
sulphur in bacterium, 306
sum, of arithmetic series, 279–80; of arithmetico-geometric series, 359; of generalized arithmetic series, 284–5; of geometric series, 358, 360; of powers, 286; Rogers's convention for, 282; sigma notation for, 281–8, 380–2; to infinity, 361
surface, nomogram for area of, 169–73; area and weight of child, 158; density, 456–60, 463–5; integral, 446–7, 456–8, 463–5; of revolution, 301–6
swimming in cold water, 476
table, of common logarithms, 498–9, (expl.) 493–4; of trigonometric functions, 500–3, (expl.) 494–6; Greek alphabet, 497, (expl.) 493; conversion factors, (area), 450; (force, energy, power), 454; (heat), 471; (length), 449–50; (mass), 451–2; (pressure), 466; (temperature), 475–7; (volume), 451; month numbers, 448; reduced factorials, binomial coefficients, 370–373; atomic weights, 475; standard derivatives, 203–4; standard integrals, 243–4; use of, 60–2, 65–6, 92–4, 123–5, 134, 493–6
Talbot, 133
tangent, to a curve, 186; inverse, 376; *see also* trigonometric functions; hyperbolic functions
tanh, *see* hyperbolic functions; pronunciation, 146
Taylor series, 365–80, 415–17; convergence of, 373–6; uniqueness of, 377; *for particular functions, see under* series
temperature, of gas, 38, 222–4, 233, 476–7; scales and units, 475–7; absolute, 173, 475; and pulse rate, 156
tera- (prefix), 449
terminus or limit of integration, 271, 312, 318–24
test, for factors, 10–11, 71
tetanolysin, 160
tetrahedron, 389
thermal conductivity, 476
Thompson, d'Arcy W., 106, 335, 356
thread, tension in, 38
time, 447–9
ton, 451–2
tonne, 451–2
train, 178
transformer, 485
triangle, area of, 296–305; centroid of, 465; of vectors, 386–8; Pascal, 370–3
triangular pyramid, 389
trigonometric functions, 82–96, 367, 375–6; addition formulas, 90–1, 94; relations between, 86–8, 94, 145–9; complex, 411–13; derivative of, 192–3, 197–9, 203; evaluation of, 91–4; graphs of, 87; integral of, 243–5, 418–25, 443–5; tables of, 494–6, 500–3; of 30° and 45°, 85, 89
triple integral, 460, 465–7
Troy, 451–2
trypsin, 161
two-valued function, 150–1

unending decimal, 133, 149–50
units, of angle, 81, 137; of area, 450; of brightness, 490; electric and magnetic, 480–7; of energy, power, 455, 471; power of lens, 490; of force, 455;

of heat, 470; of length, 450; of pressure, 466; of viscosity, 468; of volume, 451; change of, 453–4, 471–4; British and metric, 449–55, 466–86; derived, 452–4; germane, 453
urine, flow of, 469

value, calorific, 167–8; absolute, 73–8, 389, 396
van der Monde, 377
Vanellus vanellus, 357, 361
variable, 33; dependent or independent, 213; change of, in integral, 246–50, 319–24; separable, in differential equation, 259; differential equation solvable for, 261
variation, constrained, 231–7
vector, 383–90; addition of, 386–8; component of, 384; magnitude or modulus of, 389; plane, 390–1; representing complex number, 383; spatial, 383; subtraction, 385; zero, 383
velocity, instantaneous, 136–85 (*see also* derivative); average, 176–84; of sound, 204; units and conversion factors, 451–2
Verhulst, 258
vertex, 103, 105, 113
vibration, 421–6
violin, 422
viscosity, 467–70
vision, 140
volume, summary of formulas, 306; as definite integral, 296–300; of cone, 298, 306; of cylinder, 297, 306; of lungs, 306–7; of parallelopipedon, 299; of prism, 298, 306; of pyramid, 298, 306; of sphere, 299, 306; of spheroid, 303, 306; of surface of revolution, 303, 306; units and conversion factors of, 451

Waage, 477
Walbum, 160–1
Wales, 7, 38
Watson, G. N., 379
watt, 454
weight, atomic, 474–5
weight, and sitting height, 132; and surface area, 158; of schoolgirls, 45–6
weighted mean, 463
Whittaker, E. T., 379, 420
work, as definite integral, 289; units of, 454
wound, healing of, 158

x-axis, 39, 80
x-co-ordinate (abscissa), 80

yard, 450
Yates, F., 316
y-axis, 39, 80
y-co-ordinate (ordinate), 80
year, 447–9

zero vector, 383